MATHEMATICAL METHODS IN SCIENCE AND ENGINEERING

APPLICATIONS IN OPTICS AND PHOTONICS

First Edition

Masud Mansuripur

University of Arizona

cognella®
SAN DIEGO

Bassim Hamadeh, CEO and Publisher
Jennifer McCarthy, Field Acquisitions Editor
Amy Smith, Project Editor
Alia Bales, Production Editor
Jess Estrella, Senior Graphic Designer
Stephanie Kohl, Licensing Coordinator
Natalie Piccotti, Director of Marketing
Kassie Graves, Vice President of Editorial
Jamie Giganti, Director of Academic Publishing

Printed in the United States of America.

cognella® ACADEMIC PUBLISHING
3970 Sorrento Valley Blvd., Ste. 500, San Diego, CA 92121

Table of Contents

Preface

This book started as a set of lecture notes for a course in mathematical methods designed for our first-year graduate students in the College of Optical Sciences at the University of Arizona in Tucson. In addition to our graduate students, a few senior undergraduates from the Department of Physics and the College of Engineering have also attended these lectures over the years.

To keep the book within a reasonable size limit for a one-semester course, each chapter begins with a brief introduction, then dives into the subject with little regard for mathematical formalities and the conventional approach of definitions, lemmas, theorems, proofs, etc. The emphasis throughout the book is on conceptual appreciation for the topics and acquiring a physical feel and intuition for the mathematical ideas under consideration. Examples appear frequently to clarify concepts and methods, and proofs are given in an informal way, without delving into the minutiae or invoking the so-called "delta-epsilon" type of reasoning, which is prevalent in many mathematical textbooks.

Several examples are provided from the field of optics and photonics to clarify some of the concepts. For instance, the optics of thin-films is used to introduce the student to the subject of infinite series; Snell's law of refraction is derived from Fermat's extremum principle; contour integration is employed to arrive at the Kramers-Kronig relations; diffraction of electromagnetic waves provides examples for the theory of Fourier transformation and stationary-phase approximation; the use of the *comb* function is illustrated via Nyquist's sampling theorem; in the context of ordinary differential equations, the mass-and-spring model is studied as the quintessential harmonic oscillator, which, among other things, underlies the Lorentz model of oscillating dipoles in electromagnetic systems; Bessel functions are examined in great detail because of their importance in fiber-optical systems (as well as in other cylindrical waveguides and resonators); parametric oscillators are discussed in conjunction with the theory of second-order ordinary differential equations; the wave equation and the diffusion equation—prominent in electrodynamics as well as in quantum optics—are treated extensively in the chapter on partial differential equations; a nonlinear partial differential equation that describes the propagation of wave-packets in a nonlinear optical medium is examined in detail; the study of caustics requires familiarity with the Airy functions; the statistics of noise in photodetection is used to illustrate concepts in probability and random variables; matrix algebra is emphasized for its crucial role in quantum optics; and vector calculus is given its own chapter because Maxwellian electrodynamics cannot be mastered without a strong background in this area. By and large, however, the concepts and methods discussed in the present book share many common threads with those in other textbooks on mathematical methods in physics and engineering. What makes this book especially suitable for students of optics is not the variety of examples drawn from the field of optics, but rather the focus on specific mathematical topics that have been deemed most useful for our students in the College of Optical Sciences. I have chosen these topics based on my years of experience teaching courses in geometrical optics, diffraction theory, Fourier optics, classical electrodynamics, optical communication, information theory, and optical data storage.

Chapter 1 deals with elementary concepts such as properties of numbers, and certain methods of mathematical reasoning. In chapter 2 we analyze some of the basic issues in Euclidean geometry, such as the Pythagorean theorem and the properties of circles and triangles. Chapter 3, with the title of "Odds and Ends," is a potpourri of various topics that the student is expected to have seen in high school or in his or her undergraduate curriculum, but may have forgotten—or perhaps not quite have appreciated their significance at an earlier time. Various methods of integration are discussed here, as are the fundamentals of complex number theory, definitions and elementary properties of complex functions, Stirling's approximation, and certain elements of discrete mathematics. In chapter 4, we provide a number of important techniques to deal with finite as well as infinite sums and products. The emphasis is on approaches to finding sums and products, leaving out some of the subtler issues, such as rigorous proofs of convergence. Chapter 5 is focused on a few important topics from multivariate calculus, such as finding the maxima, minima, and saddle points of functions of two variables, constrained optimization with the aid of Lagrange multipliers, and a rudimentary treatment of curvilinear coordinates.

In chapter 6 we turn to complex analysis and discuss the properties of analytic functions of a complex variable, while introducing differentiation and integration in the complex plane. After providing a fairly intuitive proof of Cauchy's fundamental theorem, we give numerous examples of integration using complex-plane techniques. Chapter 7 is concerned with Fourier transformation, where we introduce the properties of the Fourier integral and give an elementary proof of the Fourier theorem, followed by many examples of the Fourier transforms of various functions. The complex-plane techniques from chapter 6 find an important application in evaluating the Fourier integrals, and we provide multiple examples of calculating these Fourier integrals based on Cauchy's theorem. Examples from signal processing, Nyquist's sampling theorem, and diffraction theory, are scattered throughout the chapter. Chapter 8 is concerned with ordinary differential equations. Here we analyze simple electrical circuits, the mass-and-spring model of a harmonic oscillator, and an elementary case of parametric oscillation. Of particular importance in chapter 8 is the section devoted to the solution of Bessel's equation using the method of Frobenius. Methods of solving partial differential equations are discussed in chapter 9, with special attention paid to the wave equation and the diffusion equation in Cartesian as well as cylindrical coordinate systems. Chapter 10 begins with a description of the Sturm-Liouville theorem, giving an overview of the many fascinating properties of functions that are solutions of a certain class of second-order ordinary differential equations. The chapter then proceeds to investigate some of the important properties of such well-known special functions as the Airy function, Euler's Gamma function, Riemann's zeta function, Jacobi's theta function, and Dirichlet's eta function.

Linear algebra, which deals primarily with the properties of matrices as linear operators, is an important topic in many branches of physical sciences; this topic is introduced in chapter 11. In chapter 12 we provide an elementary treatment of vector algebra and vector calculus. Chapter 13 is concerned with the theory of probability and serves as a brief introduction to the subjects of discrete as well as continuous random variables, random functions, and stochastic processes. Elementary concepts from Shannon's information theory, such as the entropy of a discrete random variable and the mutual information content of a pair of discrete random variables, are introduced and briefly discussed in this chapter as well.

It is possible to cover the entire book in a one-semester course. In classroom lectures, I have discussed the main topics and the motivation for each topic, leaving some of the lengthy details for the students to study on their own. A few of the more advanced topics, such as the geometry of a space curve in chapter 5, saddle-point approximation in chapter 7, the nonlinear partial differential equation of chapter 9, or properties of Riemann's zeta function and related functions in chapter 10, may be omitted on a first reading. A large number of homework problems are provided at the end of each chapter, so the students can practice and test their level of understanding of the subject after familiarizing themselves with the topics covered in the chapter.

I am grateful to a large number of students who took my course in *Mathematical Methods in Optics and Photonics* during the last decade at the University of Arizona. Their questions and suggestions have undeniably improved the quality of my lectures, as well as the content of this book. I also would like to thank my wife, Annegret, without whose loving care and patience this book would not have become reality. It is to her and to our children, Kaveh and Tobias, that this book is dedicated.

Masud Mansuripur

Tucson, July 2019

Chapter 1

Elementary Concepts and Methods

"The aim of exact science is to reduce the problems of nature to the determination of quantities by operations with numbers."

James Clerk Maxwell (1831–1879)

"Science is built up with facts, as a house is with stones. But a collection of facts is no more a science than a heap of stones is a house."

Jules Henri Poincaré (1854–1912)

1.1. Introduction. In applied mathematics we often deal with numbers. The numbers can be positive or negative, integer or fractional, rational or irrational, prime or composite, real or complex, etc. In this chapter, we learn some of the basic properties of ordinary real numbers. (Complex numbers will be treated in detail in later chapters.) In dealing with mathematical objects, we often need to *prove* certain statements concerning their properties. For instance, we may want to demonstrate that every integer is either *prime* or can be written as a unique product of its prime factors (the fundamental theorem of arithmetic). Or we may want to prove that, in a right-angle triangle, the lengths a and b of the sides adjacent to the right angle are related to the length c of the hypotenuse via $c^2 = a^2 + b^2$ (Pythagorean theorem). There are no standard ways to prove a mathematical theorem or statement. One often uses examples to familiarize oneself with methods of proof that have been devised by ingenious mathematicians over the course of several millennia. Sometimes one resorts to "proof by contradiction," which starts by making an assumption, then proceeds to show that the assumption results in a logical contradiction or inconsistency. Other times, one invokes a method known as "proof by induction." In general, however, one must use logical arguments and move, in a step-by-step fashion, from a set of assumptions, definitions, and previously proven theorems toward the desired goal of proving or disproving a statement. In this chapter, we use several examples to demonstrate a few powerful methods of proof. In the process, we also learn some fundamental properties of real numbers and functions.

1.2. Elementary concepts. The student is expected to be familiar with mathematical concepts such as rules for adding, subtracting, multiplying, and dividing real and complex numbers. We will not discuss the basic notions of algebra, trigonometry, Euclidean geometry, and integral and differential calculus. This is what you learned in your high school and college days, and we assume that you have a basic knowledge and understanding of these subjects. If you need to refresh your memory and practice your skills at solving quadratic equations, differentiating functions of a single variable, expanding a function in a Taylor series, or integrating a function using a change of variable or employing the method of integration by parts, please try solving some of the problems at the end of the chapter. What we will discuss in the following sections is a random selection of topics that you may or may not be already familiar with, yet they contain some fine points and subtleties that may have escaped you in the past. As you wade through the chapter, you should keep thinking about the various ways in which the concepts and methods can be generalized or applied to different problems. You should also appreciate that you are just seeing the metaphorical tips of several icebergs; a lot is hidden beneath the surface, and you may want to consider diving deeper and discovering things on your own.

1.3. Permutation. The number of different ways in which m distinct objects can be arranged in a row is "m factorial," written as $m! = 1 \times 2 \times 3 \times \cdots \times m$. This is because the first object in the row could be any one of the m available objects, whereas the choice for the second is limited to $m-1$ objects, for the third to $m-2$ objects, and so on. Therefore, the total number of permutations available for m distinct objects is $m \times (m-1) \times (m-2) \times \cdots \times 3 \times 2 \times 1 = m!$.

Note: By definition, $0! = 1$. There also exists a double-factorial notation, which is used when sequential even numbers or sequential odd numbers are multiplied together. For instance, $10!! = 2 \times 4 \times 6 \times 8 \times 10$ and $7!! = 1 \times 3 \times 5 \times 7$.

1.4. Choosing m out of n distinct objects. The notation $\binom{n}{m}$, where both n and m are non-negative integers, indicates the number of different ways in which m objects can be chosen out of a collection of n distinct objects. For example, suppose you have 10 friends and you are asked to choose 3 of them to come along to a party. How many different ways can one pick 3 friends out of 10? The answer is that you have 10 choices for your first pick, 9 choices for the second pick, and 8 choices for the third pick. This amounts to $10 \times 9 \times 8 = 10!/7!$ choices. However, the same 3 friends can be chosen in different permutations, that is, the same 3 can be picked in $3 \times 2 \times 1 = 3!$ different ways. The number of distinct ways in which you can pick 3 friends out of a total of 10 is, therefore, $10!/(3! \times 7!)$. In general, selecting m out of a total of n distinct objects (discounting all the possible permutations) can be done in

$$\binom{n}{m} = \frac{n!}{m!\,(n-m)!}. \tag{1}$$

different ways. Note that, by definition, $0! = 1$, so that the number of unique ways in which one can pick n objects out of a total of n is just $n!/(n! \times 0!) = 1$.

1.5. Taylor series expansion. Consider the function $f(x)$, defined over the real axis x, its value at a given point x_0 being $f(x_0)$, and its various derivatives at x_0 being $f'(x_0)$, $f''(x_0)$, $f'''(x_0)$, $f^{(4)}(x_0), \cdots, f^{(n)}(x_0), \cdots$. The Taylor series expansion of $f(x)$ around the point $x = x_0$ is given by

$$f(x) = f(x_0) + f'(x_0)(x - x_0) + \frac{f''(x_0)}{2!}(x - x_0)^2 + \cdots + \frac{f^{(n)}(x_0)}{n!}(x - x_0)^n + \cdots. \tag{2}$$

Depending on the nature of the function $f(x)$, the Taylor series expansion may be valid (i.e., the series would converge to the correct value of $f(x)$) either for all values of x or for a limited range of x in the neighborhood of x_0. For most practical purposes, one can terminate the series at a finite value of n and obtain a good approximation to the actual value of $f(x)$ for a desired x, or a desired range of x, inside the domain of convergence of the series. The Taylor series is essentially a polynomial function of $x - x_0$, with the degree of the polynomial being allowed to go to infinity. You may want to compare the various derivatives of $f(x)$ at $x = x_0$ with those of its Taylor series expansion to understand why the coefficient $1/n!$ of the nth term is necessary in the above formula.

1.6. Mathematical proof. In mathematics, we must prove things to establish their universal validity. A proof is generally constructed based on the rules of logic and logical reasoning or deduction. Mathematical statements are either true or false—or perhaps undecidable. If A leads to B and B leads to C, then one concludes that A leads to C. If Q is a consequence of P, then the absence of Q is tantamount to the absence of P. The magnificent edifice that is modern mathematics is built on a foundation consisting of a few primitive definitions and axioms (or postulates), using no other tool than logical reasoning. Euclidean geometry, to pick but one example, is a branch of mathematics that has emerged from the concepts of geometrical point, straight line, and flat plane, together with axioms asserting the congruence of all right angles, the possibility of constructing a circle given a straight line segment as its radius, and the so-called "parallel postulate." The parallel postulate asserts that a straight line and a point outside that straight line define a plane, within which it is possible to draw one and only one straight line that passes through the point and is parallel to the original straight line. Everything that is known about Euclidean geometry is a logical consequence of these definitions and primitive axioms in conjunction with the parallel postulate.

This section provides a few examples of methods that brilliant mathematicians, from antiquity to the present time, have discovered and passed on to future generations. Proof by induction and proof by contradiction are two well-known methods in mathematical sciences. However, many other methods exist that cannot be easily described or formalized, but must be learned through examples. You will encounter several such methods in the subsections that follow.

1.6.1. Proof by induction. To prove a statement about something that depends on an integer N, we may begin by demonstrating that the statement is true for $N = 1$. We then *assume* the validity of the statement for an arbitrary integer N and proceed to demonstrate its validity for $N + 1$. Given that the statement has been confirmed for $N = 1$, it will then be proven for $N = 2$. But then it must also be true for $N = 3$, and so on.

Example 1. Prove that $\sum_{n=1}^{N} n = N(N+1)/2$.

Proof. The statement is true for $N = 1$, where $1 = 1(1+1)/2$. Suppose it is true for an arbitrary N. Then for $N + 1$ we will have

$$\textstyle\sum_1^{N+1} n = (\sum_1^N n) + (N+1) = N(N+1)/2 + (N+1) = (N+1)(N+2)/2. \quad (3)$$

The proof is now complete.

Example 2. Prove the binomial expansion theorem, namely,

$$(a+b)^N = \sum_{n=0}^{N} \binom{N}{n} a^n b^{N-n}. \quad (4)$$

Proof. The theorem is obviously valid for $N = 1$, where $a + b = \binom{1}{0} a^0 b^1 + \binom{1}{1} a^1 b^0$. Suppose it is true for N. Then for $N + 1$ we will have

$$(a+b)^{N+1} = (a+b)\sum_{n=0}^{N} \binom{N}{n} a^n b^{N-n} = \sum_{n=0}^{N} \binom{N}{n} a^{n+1} b^{N-n} + \sum_{n=0}^{N} \binom{N}{n} a^n b^{N-n+1}$$

$$= \sum_{n=0}^{N} \binom{N}{n} a^{n+1} b^{N+1-(n+1)} + \sum_{n=0}^{N} \binom{N}{n} a^n b^{N+1-n}$$

$$= \sum_{m=1}^{N+1} \binom{N}{m-1} a^m b^{N+1-m} + \sum_{m=0}^{N} \binom{N}{m} a^m b^{N+1-m}$$

$$= \binom{N}{N} a^{N+1} + \binom{N}{0} b^{N+1} + \sum_{m=1}^{N} \binom{N}{m-1} a^m b^{N+1-m} + \sum_{m=1}^{N} \binom{N}{m} a^m b^{N+1-m}$$

$$= a^{N+1} + b^{N+1} + \sum_{m=1}^{N} \left[\binom{N}{m-1} + \binom{N}{m}\right] a^m b^{N+1-m}. \tag{5}$$

The coefficient of $a^m b^{N+1-m}$ may now be simplified, as follows:

$$\binom{N}{m-1} + \binom{N}{m} = \frac{N!}{(m-1)!(N-m+1)!} + \frac{N!}{m!(N-m)!} = \frac{mN!}{m!(N-m+1)!} + \frac{(N-m+1)N!}{m!(N-m+1)!}$$

$$= \frac{(N+1)!}{m!(N-m+1)!} = \binom{N+1}{m}. \tag{6}$$

We thus arrive at

$$(a+b)^{N+1} = a^{N+1} + b^{N+1} + \sum_{m=1}^{N} \binom{N+1}{m} a^m b^{N+1-m} = \sum_{m=0}^{N+1} \binom{N+1}{m} a^m b^{N+1-m}. \tag{7}$$

The proof is now complete.

Proof by induction gone wrong. We use the method of induction to prove that all the horses in the world are of the same color. Obviously, this is a ridiculous and patently false statement. The student is asked to find the flaw in the following line of reasoning.

For $n = 1$, the set contains only one horse, and clearly one horse has the same color. Now suppose the statement is true for an arbitrary n, that is, in any set containing n horses, all the horses are of the same color. We must then prove that all the horses in a set containing $n + 1$ horses are of the same color. Take any subset of n horses from the set of $n + 1$. By assumption, these n horses all have the same color. Since this is true for any combination of n horses out of the set containing $n + 1$, then all $n + 1$ horses must have the same color. The proof is thus complete! What went wrong?

1.6.2. Proof by contradiction. A great example of proof by contradiction is Euclid's proof that prime numbers are infinite in number, or equivalently, that there is no "largest" prime number. Let $p_1, p_2, p_3, \cdots, p_N$ be consecutive prime numbers starting with $p_1 = 2$ and ending with the largest possible prime, p_N. We form a new integer $q = p_1 p_2 p_3 \cdots p_N + 1$. Obviously, $q > p_N$. If q itself is prime, then our theorem is proven, as we have found a prime greater than p_N, which was initially *assumed* to be the largest possible prime. If, however, q is not a prime, since it is not divisible by $p_2, p_3, \cdots, p_N$ (by construction), it must be divisible by another prime located somewhere between p_N and q. Either way, there exists a prime greater than p_N. The proof is thus complete.

1.7. Fundamental theorem of arithmetic. An integer N that is greater than 1 is either prime, or can be decomposed in a unique way into the product of prime numbers smaller than N.

Proof. If N happens to be prime, then, by definition, it is divisible only by 1 and by itself. If N is not prime, then it is divisible by at least one positive integer other than 1 and N. This means that we can write N as the product of two other integers, say, N_1 and N_2. (N_1 and N_2 may be different numbers, or they may be the same.) Now, if both N_1 and N_2 happen to be prime numbers, then the decomposition of N is complete and we have $N = N_1 \times N_2$. Otherwise, we repeat the process until none of the factors can be further decomposed. Since, in each step of this decomposition, the resulting numbers become smaller, the process must end after a finite number of steps. In the end, we will have $N = N_1^{n_1} \times N_2^{n_2} \times N_3^{n_3} \times \cdots N_k^{n_k}$, which shows the number N decomposed into the product of k prime numbers $(N_1, N_2, \cdots, N_k)$, each having its own multiplicity $(n_1, n_2, \cdots, n_k)$. The proof of uniqueness is left as an exercise for the student; see problem 10 at the end of the chapter.

1.8. Geometric series. We would like to prove that $1 + x + x^2 + x^3 + \cdots = 1/(1-x)$ when $|x| < 1$. This will be shown to be true for any x, be it positive or negative, real or complex, so long as the absolute value of x is strictly less than 1. Note that the absolute value of a complex number $x = x' + \mathrm{i}x''$ is $|x| = \sqrt{x'^2 + x''^2}$. (In general, the complex number x can be written in polar notation as $x = |x| \exp(\mathrm{i}\varphi)$ where $\varphi = \tan^{-1}(x''/x')$.)

Proof. We begin by considering the finite sum $S_n = 1 + x + x^2 + x^3 + \cdots + x^n$, where n is an arbitrary positive integer. At this point we are not restricted to values of x that satisfy the constraint $|x| < 1$. Multiplying and dividing S_n by $1 - x$ yields

$$S_n = \frac{(1-x)(1+x+x^2+x^3+\cdots+x^n)}{1-x} = \frac{(1+x+x^2+x^3+\cdots+x^n) - (x+x^2+x^3+\cdots+x^n+x^{n+1})}{1-x} = \frac{1-x^{n+1}}{1-x}. \quad (8)$$

This formula is valid for any positive integer n and for any value of x, positive or negative, real or complex. We now note that, if $|x|$ happens to be strictly less than 1, then the value of x^{n+1} will diminish with increasing n. For instance, if $x = 0.9$, then $x^2 = 0.81$, $x^3 = 0.729$, $x^4 = 0.6561$, and so on. Also, if x happens to be complex, then $x^n = |x|^n \exp(\mathrm{i}n\varphi)$, which would approach zero as $n \to \infty$ irrespective of the phase angle φ so long as $|x| < 1$. We thus see that, in the limit $n \to \infty$, provided that $|x| < 1$, the sum S_n approaches $1/(1-x)$.

Example 1. For $x = ½$, we find $lim_{n\to\infty} S_n = 1 + \frac{1}{2} + \frac{1}{4} + \frac{1}{8} + \cdots = 1/(1 - ½) = 2$.

Example 2. For $x = -½$, we find $lim_{n\to\infty} S_n = 1 - \frac{1}{2} + \frac{1}{4} - \frac{1}{8} + \cdots = 1/(1 + ½) = ⅔$.

Example 3. For $x = \mathrm{i}$, we have $S_n = 1 + \mathrm{i} + \mathrm{i}^2 + \mathrm{i}^3 + \cdots + \mathrm{i}^n$. The diagram below shows that S_n goes round and round a unit-length square within the complex plane but fails to converge to any specific point. In other words, the limit of S_n does *not* appear to be $1/(1-\mathrm{i}) = (1+\mathrm{i})/(1-\mathrm{i}^2) = ½ + ½\mathrm{i}$. This is because in this example $|x| = |\mathrm{i}|$ is 1 and not strictly less than 1. To use the geometric series formula, we must choose for x something whose absolute value (or magnitude) is strictly less than 1. For instance, if we choose $x = 0.999\mathrm{i}$, then the corresponding S_n will go round and round a diminishing square, as shown in the figure, eventually arriving at a point near the center of the square.

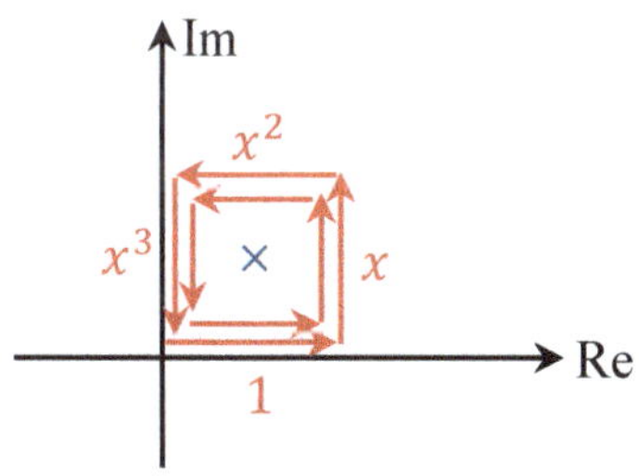

Example 4. It is a well-known fact that a number such as $⅔ = 0.66666\cdots$ may be approximated to any desired accuracy as $⅔ \cong 0.6666\cdots667$. However, the resulting number is only an approximation and cannot be said to be precisely equal to ⅔. The situation is different, however, for a number such as $ab.cdef99999\cdots$, with $f < 9$ followed by an interminable sequence of 9s. In this case, we may write

$$ab.cdef99999\cdots = ab.cdef + 9 \times 10^{-5} + 9 \times 10^{-6} + 9 \times 10^{-7} + \cdots$$

$$= ab.cdef + 9 \times 10^{-5}(1 + 10^{-1} + 10^{-2} + 10^{-3} + \cdots)$$

Geometric series → $$= ab.cdef + 9 \times \frac{10^{-5}}{1-0.1} = ab.cdef + 10^{-4} = ab.cdef'.$$ ← $f' = f + 1$

It is thus legitimate to say, for example, that 0.1257 is exactly equal to $0.125699999\cdots$.

1.9. The Schwarz inequality. It is often useful to have upper and/or lower bounds on entities that are not amenable to direct calculation. The Schwarz inequality, named after the German mathematician Hermann Schwarz (1843–1921), provides an upper bound on a definite integral of the product of two arbitrary functions of the real variable x, namely, $\int_a^b f(x)g(x)\mathrm{d}x$. In this section, we prove the famous inequality at first for a pair of real-valued functions, then extend the proof to complex functions of a real variable using another well-known upper bound on definite integrals. In the context of the present chapter, our emphasis will be on the particular *method of proof* as an instructive example of tools and clever techniques that mathematicians employ to arrive at concrete propositions. Needless to say, the significance of the inequality as a valuable piece of mathematics in its own right should not be overlooked. For an alternative proof of the Schwarz inequality, see problem 8.

Let $f(x)$ and $g(x)$ be two real-valued functions. For an arbitrarily chosen real-valued parameter λ, we may write

$$0 \le \int_a^b [f(x) + \lambda g(x)]^2 \mathrm{d}x = \int_a^b f^2(x)\mathrm{d}x + 2\lambda \int_a^b f(x)g(x)\mathrm{d}x + \lambda^2 \int_a^b g^2(x)\mathrm{d}x. \quad (9)$$

The particular value of λ that minimizes the right-hand side of this expression may be found as follows:

$$\frac{d}{d\lambda}\int_a^b [f(x) + \lambda g(x)]^2 \mathrm{d}x = 2\int_a^b f(x)g(x)\mathrm{d}x + 2\lambda \int_a^b g^2(x)\mathrm{d}x = 0$$

$$\rightarrow \quad \lambda = -\int_a^b f(x)g(x)\mathrm{d}x \Big/ \int_a^b g^2(x)\mathrm{d}x. \quad (10)$$

Substituting this value of λ in the original expression now yields

$$\int_a^b f^2(x)\mathrm{d}x - 2\frac{[\int_a^b f(x)g(x)\mathrm{d}x]^2}{\int_a^b g^2(x)\mathrm{d}x} + \frac{[\int_a^b f(x)g(x)\mathrm{d}x]^2}{\int_a^b g^2(x)\mathrm{d}x} \ge 0.$$

$$\rightarrow \quad \left[\int_a^b f(x)g(x)\mathrm{d}x\right]^2 \le \left[\int_a^b f^2(x)\mathrm{d}x\right]\left[\int_a^b g^2(x)\mathrm{d}x\right]. \quad (11)$$

The last inequality is known as the Schwarz inequality. Equality holds only when $f(x) = \alpha g(x)$ for some (real-valued) constant α.

Next, we prove the inequality for complex-valued functions $f(x)$ and $g(x)$. It is well known that, for any complex function of a real variable, say, $h(x)$, the absolute value of the integral is less than or equal to the integral of the absolute value of the function, that is, $|\int_a^b h(x)\mathrm{d}x| \le \int_a^b |h(x)|\mathrm{d}x$. To prove this inequality, recall that the definite integral of $h(x)$ is the limit of a sum over infinitesimal intervals Δx spanning the domain $[a, b]$ of the integral; in other words, $\int_a^b h(x)\mathrm{d}x = lim_{\Delta x \to 0} \sum_n h(x_n)\Delta x$. Each term in this summation is a small complex number that can be represented by a little arrow in the complex plane. The sum of the little arrows (in the complex plane) is then the big arrow that connects the beginning of the first to the end of the last little arrow, as shown in the figure. While the absolute value of $\int_a^b h(x)\mathrm{d}x$ equals the length of the big arrow, the sum of the lengths of all the little arrows comprise the integral of $|h(x)|$. It is thus clear that $|\int_a^b h(x)\mathrm{d}x|$ must be smaller than $\int_a^b |h(x)|\mathrm{d}x$,

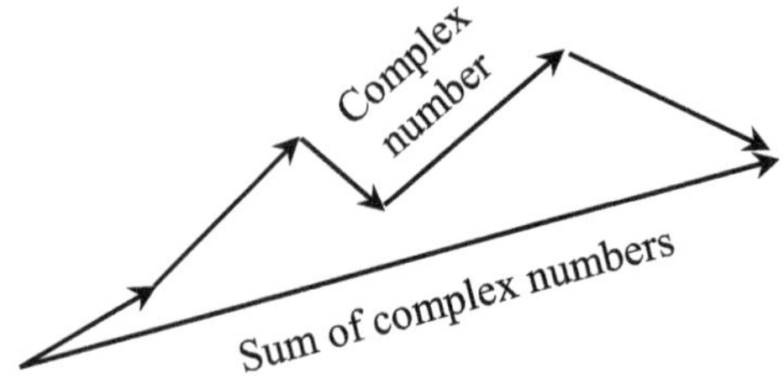

unless all the little arrows line up in the same direction, in which case equality would hold. If we now set $h(x) = f(x)g(x)$ and use the identity $|f(x)g(x)| = |f(x)||g(x)|$, we will have

$$|\int_a^b f(x)g(x)\mathrm{d}x| \leq \int_a^b |f(x)||g(x)|\mathrm{d}x. \tag{12}$$

The functions appearing on the right-hand side of this inequality are now real valued, and we may invoke our earlier result to write

$$\left|\int_a^b f(x)g(x)dx\right|^2 \leq \left[\int_a^b |f(x)||g(x)|dx\right]^2 \leq \left[\int_a^b |f(x)|^2 dx\right]\left[\int_a^b |g(x)|^2 dx\right]. \tag{13}$$

This is the Schwarz inequality for complex functions. The equality holds only when $|f(x)| = |\alpha g(x)|$ for some constant α, and also when $f(x)g(x)$ has a constant phase angle at all values of x within $[a, b]$. This occurs when $f(x) = \alpha g^*(x)$, where α is a complex constant.

1.10. Method of Vieta jumping. Another elegant method of mathematical proof is exemplified by the solution to the following problem:

Let $n_1 > n_2 > 0$ be a pair of integers satisfying $(n_1^2 + n_2^2)/(1 + n_1 n_2) = k$, where $k > 0$ is also an integer. Show that k must be a perfect square.

This is an example from the class of problems involving a pair of positive integers, whose general method of solution has come to be known as *root flipping* or *Vieta jumping*. The Vieta formulas relate combinations of sums and products of the roots of a polynomial equation to its polynomial coefficients. For instance, in the simple case of the second-order (or quadratic) polynomial equation $ax^2 + bx + c = 0$, the roots x_1 and x_2 satisfy the equation $a(x - x_1)(x - x_2) = 0$. Consequently, $x_1 + x_2 = -b/a$ and $x_1 x_2 = c/a$ constitute the complete set of Vieta's formulas for the quadratic equation.

Proof: In the expression that defines k, let us substitute x for n_1, then solve for x, knowing in advance that $x = n_1$ is one of the solutions, and that there exists only one other solution—because the equation is quadratic in x. We will have

$$\frac{x^2 + n_2^2}{1 + n_2 x} = k. \tag{14}$$

Rearranging the terms of the above equation yields

$$x^2 - (n_2 k)x + (n_2^2 - k) = 0. \tag{15}$$

The two roots of this equation, namely, $x = n_1$ and $x = n_3$ (not necessarily a positive integer at this point), must add up to $n_2 k$ and multiply to $(n_2^2 - k)$, that is,

$$n_1 + n_3 = n_2 k, \tag{16a}$$

$$n_1 n_3 = n_2^2 - k. \tag{16b}$$

According to Eq.(16a), the second root, n_3, must be an integer. Now, if $n_3 \leq -1$, substitution into Eq.(14) would make the left-hand side of the equation negative, which is impossible, given that $k > 0$. Alternatively, if $n_3 = 0$, substitution into Eq.(16b) reveals that $k = n_2^2$, proving that k is a perfect square. The only remaining possibility is $n_3 \geq 1$, in which case, according to Eq.(16b), given that $n_1 > n_2$ and $k > 0$, we must have $n_3 < n_2$. We thus have a new pair of integers, n_2 and n_3, which satisfy $n_2 > n_3 > 0$ and $(n_2^2 + n_3^2)/(1 + n_2 n_3) = k$. Repeating this argument will, sooner or later, result in a pair of integers, n_j and n_{j+1}, whose smaller element is zero. Consequently, k must be a perfect square.

The procedure can be traced in reverse to find all pairs of integers (n_1, n_2) corresponding to a given k—provided, of course, that k is a perfect square. The first such pair would be $(\sqrt{k}, 0)$, the next would be $(k^{3/2}, \sqrt{k})$, the next $\left(\sqrt{k}(k^2-1), k^{3/2}\right)$, and so on.

Brain teaser. One hundred light bulbs are arranged in a row and numbered sequentially, from 1 to 100. Each bulb has a toggle-switch that is used to turn it on when it is off, and vice versa. Starting with the first bulb, we turn them all on. We then move to the second bulb and toggle all bulbs identified by an integer-multiple of 2 to the off state. Next we move to the third bulb and toggle the state of all bulbs whose numbers are multiples of 3, and so on. This process continues until we reach the last bulb in the row. Which light bulbs will be on at the end?

1.11. Rationals and irrationals. Figure 1.1 shows the real axis x, on which we have selected an origin to represent the number 0, then used a unit-length ruler to mark off the integers 1, 2, 3, etc. Similarly, we can move to the left of the origin to mark off the negative integers $-1, -2, -3$, and so on. The rational numbers are simply defined as the ratio of two integers. These numbers can also be identified on the real axis, provided that we have rulers of length $1/n$ for various integers n. The decimal representation of a rational may terminate after a finite number of digits, for example, $1/2 = 0.5$ and $1/8 = 0.125$. Alternatively, some rationals may have an infinite number of digits in their decimal representation, such as $1/3 = 0.3333\cdots$ and $3/11 = 0.272727\cdots$ and $461/9900 = 0.0465656565\cdots$. The decimal representation of a rational number will sooner or later settle into a repetitive pattern. A real number whose decimal representation does not terminate, nor becomes periodic (following an initial non-periodic sequence in some cases), is said to be irrational. It can be shown that the numbers $\sqrt{2}$, $-\sqrt{5}$, the base e of the natural logarithm, and π are irrational. Later we will show that there are far more irrationals than there are rationals on the real line.

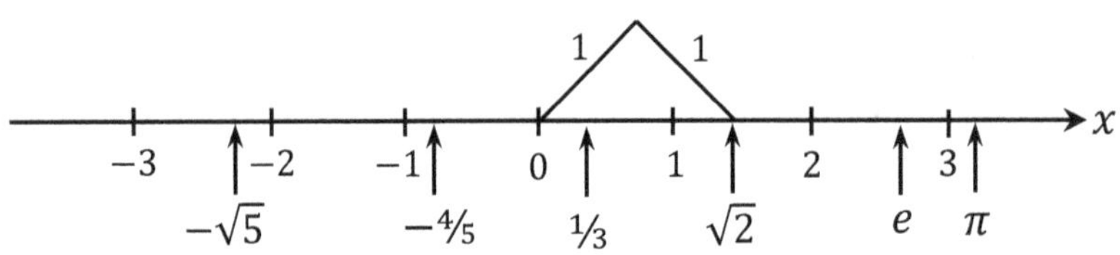

Figure 1.1. The real numbers are located on a straight line extending from $-\infty$ to ∞. Shown are a few positive as well as negative integers, rationals, and irrationals. The unit length is chosen arbitrarily. The irrational number $\sqrt{2}$ is shown as the hypotenuse of a right-angle isosceles triangle.

The set of rational numbers is dense, in the sense that between any pair of rationals one can find an infinite number of rationals. For instance, between m/n and p/q, one can always pick the mid-point $\tfrac{1}{2}[(m/n) + (p/q)] = (mq + np)/(2nq)$, which is also a rational. This process can be repeated ad infinitum, so that an infinite number of rationals are identified between the original pair m/n and p/q. No matter how close two rationals may be to each other on the real line, there always exists an infinite number of rationals between them.

1.11.1. Proof that $\sqrt{2}$ is irrational. This is another example of proof by contradiction, credited to a student of Pythagoras circa 500 BC. Suppose $\sqrt{2}$ is rational. Then the integers n and m must exist such that $\sqrt{2} = n/m$. Assume further that all the common factors between n and m have been eliminated, so that the ratio n/m is irreducible. Considering that $n^2/m^2 = 2$, we can see immediately that $n^2 = 2m^2$, which makes n^2

an even number. However, the square of an odd number must be odd, thus requiring that n itself be even (otherwise n^2 cannot be even). We thus have $n = 2p$, where p is some integer. This yields $n^2 = 4p^2$, and since we already have $n^2 = 2m^2$, we must conclude that $m^2 = 2p^2$. It is now seen that m^2 is even, which demands that m must be even as well. We have reached a contradiction, because both m and n are now required to be even (i.e., they must both be divisible by 2), but this is impossible, because our starting assumption was that n/m is irreducible. We have thus proven that $\sqrt{2}$ is an irrational number.

1.11.2. Can the mid-point between two irrationals be a rational? We already know that the mid-point between two rationals is automatically a rational. The question arises as to whether the mid-point between two irrationals is necessarily an irrational? The answer is no, and a simple example suffices to demonstrate the point. Take any irrational number, say, π, and write down its decimal representation: $\pi = 3.141592653589 \cdots$. By virtue of being an irrational, this sequence of digits does *not* terminate and does *not* fall into a periodic pattern beyond some point. Now pick an arbitrary position along the sequence of digits, say, the fifth digit after the decimal point, and proceed to "flip" every digit that follows. (Upon flipping, a digit is replaced by its complement with respect to 9. For instance, 2 is replaced by 7, 5 is replaced by 4, and 9 is replaced by 0.) In this example, the new number thus created is $\tilde{\pi} = 3.141597346410 \cdots$. The new number is also irrational, because its sequence of digits does not terminate, nor does it become periodic beyond some point. The mid-point between our two irrational numbers thus chosen will be a rational. In this example, the mid-point is at ½$(\pi + \tilde{\pi}) = 6.283189999999/2 = 6.28319/2 = 3.141595$, which is a rational.

Can a^b be rational when both a and b are irrational? The answer is yes. Consider, for example, $a = b = \sqrt{2}$. If $a^b = \sqrt{2}^{\sqrt{2}}$ happens to be rational, then we have a pair of irrationals that satisfy the requirement. If, however, $\sqrt{2}^{\sqrt{2}}$ turns out to be irrational, then we set $a = \sqrt{2}^{\sqrt{2}}$ and $b = \sqrt{2}$, in which case $a^b = \left(\sqrt{2}^{\sqrt{2}}\right)^{\sqrt{2}} = \sqrt{2}^2 = 2$ is rational and, once again, we have found a pair of irrationals that satisfy the requirement.

1.11.3. Interesting irrational number. The number shown below is irrational because its fractional decimal representation does not terminate and is (obviously) non-periodic.

$$0.1234567891011121314151617181920212223242526\ldots$$

A naïve argument might lead one to believe that, since $\sum_{n=1}^{\infty} n/10^n = 10/81$ is the ratio of two integers, the number in question must be rational. The argument is invalid, however, because a little reflection shows that the asserted infinite sum does *not* represent the number in question.

Digression: Upon differentiation with respect to x, the geometric series $\sum_{n=0}^{\infty}(1/x^n) = x/(x-1)$, which is valid for $|x| > 1$, yields

$$\frac{\mathrm{d}}{\mathrm{d}x}\sum_{n=0}^{\infty} x^{-n} = \sum_{n=0}^{\infty} -nx^{-n-1} = -\frac{1}{x}\sum_{n=1}^{\infty}(n/x^n) = \frac{\mathrm{d}}{\mathrm{d}x}\left(\frac{x}{x-1}\right) = -\frac{1}{(x-1)^2}$$

$$\rightarrow \quad \sum_{n=1}^{\infty}(n/x^n) = x/(x-1)^2. \tag{17}$$

For $x = 10$, the above series yields $\sum_{n=1}^{\infty} n/10^n = 10/81$.

1.11.4. Proof that e is irrational. The following proof of irrationality of e, the base of natural logarithm, is due to Jean-Baptiste Joseph Fourier (1768–1830). Here, we have another example of proof by contradiction. Suppose e is rational, that is $e = n/m$ for a pair of positive integers m and n. Given the following Taylor series expansion of e^x,

$$e^x = 1 + x + \frac{x^2}{2!} + \frac{x^3}{3!} + \cdots + \frac{x^m}{m!} + \frac{x^{m+1}}{(m+1)!} + \cdots, \tag{18}$$

we must have (setting $x = 1$):

$$e = \frac{n}{m} = 1 + 1 + \frac{1}{2!} + \frac{1}{3!} + \cdots + \frac{1}{m!} + \frac{1}{(m+1)!} + \frac{1}{(m+2)!} + \cdots. \tag{19}$$

Let us multiply both sides of the preceding equation by $m!$. On the left-hand side we will have $m!\,(n/m) = (m-1)!\,n$, which is an integer. On the right-hand side, all the terms up to and including $1/m!$ will become integers, but the remaining terms will remain fractional, as follows:

$$m!\,e = (m-1)!\,n = m! + m! + (3 \times 4 \times \cdots \times m) + \cdots + 1 + \frac{1}{m+1} + \frac{1}{(m+1)(m+2)} + \cdots. \tag{20}$$

For the above equality to hold, all the fractional numbers appearing at the end of the series must add up to an integer. However, the following line of reasoning shows that the sum of these fractionals cannot be an integer.

$$\frac{1}{m+1} + \frac{1}{(m+1)(m+2)} + \cdots < \frac{1}{m+1} + \frac{1}{(m+1)^2} + \frac{1}{(m+1)^3} + \cdots$$

$$= \frac{1}{m+1}\left[1 + \frac{1}{(m+1)} + \frac{1}{(m+1)^2} + \cdots\right] = \frac{1}{m+1}\left(\frac{1}{1 - \frac{1}{m+1}}\right) = \frac{1}{m}. \tag{21}$$

This strict inequality shows that the sum of the fractional numbers is less than 1 (even if m happens to be 1). We have thus arrived at a contradiction, proving that the base e of the natural logarithm must be irrational.

1.11.5. Both sets of rationals and irrationals are dense. One can always find an irrational number between any pair of rationals, no matter how close those rationals may be to each other. Moreover, it is always possible to find a rational between any pair of irrationals, and an irrational between any pair of irrationals, and so forth. A simple way to see this is by examining a pair of arbitrary numbers whose decimal representation may be written as $\alpha = ab.cdefghi\cdots$ and $\beta = ab.cdef'g'h'i'\cdots$. In this example, the first five digits $ab.cde$ are the same, but, at the sixth digit, β begins to deviate from α. Without loss of generality, we may assume that $f < f'$. Now, it may happen that $f' - f \geq 2$, in which case we choose $f'' = f + 1$, and form a new number $\gamma = ab.cdef''$, which is clearly a rational located between α and β. Any digits that we append to the end of γ will create a new number, such as $\delta = ab.cdef''g''h''\cdots$, which will continue to be greater than α and smaller than β. If we choose the appended digits to be those of an irrational number (say, $g''h''i''\cdots = 31415927\cdots$, which are the digits of π), the resulting number δ will be an irrational located between α and β.

What if $f' = f + 1$? In this case, we form $\gamma = ab.cdef'$, which is greater than α, then add digits after f' that would make the resulting number less than β. (Note that any digits added after f' *cannot* drop γ below α.) The next digit after f' that would make γ

less than β is $g'' = g' - 1$. What if g' is already zero and cannot be reduced any further? Then set $g'' = 0$ and move to the next digit, setting $h'' = h' - 1$. Again, if $h' = 0$, you must set $h'' = 0$ and move to the next digit, and so on. In this way, you will end up with a number such as $\gamma = ab.cdef'00i''$ that is strictly greater than α and smaller than β. The only situation in which one fails to form such a number is when β terminates at f', that is, when all the subsequent digits $g'h'i' \cdots$ of β are zero. Here, one must replace $\beta = ab.cdef'$ with its equivalent $\beta = ab.cdef999999\cdots$, then proceed as above. In any case, once an appropriate γ located between α and β is formed, it will be the sought-after rational. And if one desires to find an irrational between α and β, one must form the number δ by appending the digits of π (or any other irrational number) to the end of γ.

1.12. Countable and uncountable sets. A set is defined as *countable* if its elements can be put into one-to-one correspondence with the set of positive integers $\mathbb{N} = \{1, 2, 3, 4, \cdots\}$. For instance, the set of *all* integers $\mathbb{Z} = \{\cdots, -4, -3, -2, -1, 0, 1, 2, 3, 4, \cdots\}$ is a countable set because one can start by mapping 0 onto 1 $(0 \to 1)$, then continue counting by jumping back and forth between positive and negative integers, as shown in figure 1.2. In this way, all the elements of $\mathbb{Z}$ will be projected (or mapped) onto the elements of $\mathbb{N}$, as follows:

$$0 \to 1, \quad 1 \to 2, \quad -1 \to 3, \quad 2 \to 4, \quad -2 \to 5, \quad 3 \to 6, \quad -3 \to 7, \cdots.$$

Note that all the elements of $\mathbb{Z}$ are counted in this way, and that the counting proceeds smoothly and without interruption as one moves through the set.

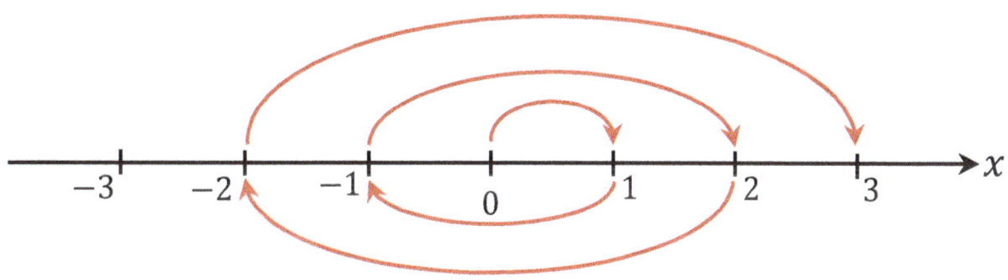

Figure 1.2. The set $\mathbb{Z}$ of all positive and negative integers (including 0) is put into one-to-one correspondence with the set of all positive integers $\mathbb{N}$ by starting to count at 0, then jumping back and forth between positive and negative integers.

Another example of a countable set is the set of all rationals. Here, we consider the points having integer coordinates (m, n) in a two-dimensional plane, as shown in figure 1.3. Clearly, every vertex of this two-dimensional mesh (or grid) is associated with an integer m on the horizontal axis and another integer n on the vertical axis, making the point itself a representation of the rational number m/n. Of course, some rationals are repeated on the grid, but it is possible to systematically eliminate the repetitions as we make our way along a well-defined path through the grid. We start to count at the lower-left corner, where $m = n = 1$, then make our way through the set of all rationals by following along the arrows shown in figure 1.3. We assign sequentially the positive integers from $\mathbb{N}$ to the vertices of the mesh, and whenever we arrive at a rational that we have encountered before, we simply skip it. In this way, the entire set is counted, nothing is left behind, and no rational is counted more than once.

Let us now consider all the rationals residing on the real axis x, and try to cover them with short segments cut from a straight line and overlaid onto the x-axis. Since the rationals are countable, it is possible to assign a unique integer to each and every one of them, then locate them one by one and proceed to cover them by placing a short line segment over each and every one of them. Suppose we cover the first rational on the x-axix by a short line-segment of arbitrary length, say, ε. We then locate the second rational on the x-axis and cover it with a line-segment of length $\frac{1}{2}\varepsilon$, then the third rational and cover it with an even shorter line segment of length $\frac{1}{4}\varepsilon$, and so on. Overlaps among the various line segments are unavoidable, of course, and will be permitted. The total length of the line segments thus placed over portions of the x-axis is less than $L = \varepsilon + \frac{1}{2}\varepsilon + \frac{1}{4}\varepsilon + \frac{1}{8}\varepsilon + \cdots = 2\varepsilon$. Needless to say, ε could be as small as one might wish and, therefore, all the rationals on the real axis are covered under a blanket whose total length is less than 2ε. As a subset of the real numbers represented by points along the x-axis, the rationals are seen to be quite insignificant! We say that the "measure" of the subset containing all the rationals inside the set of all real numbers is zero.

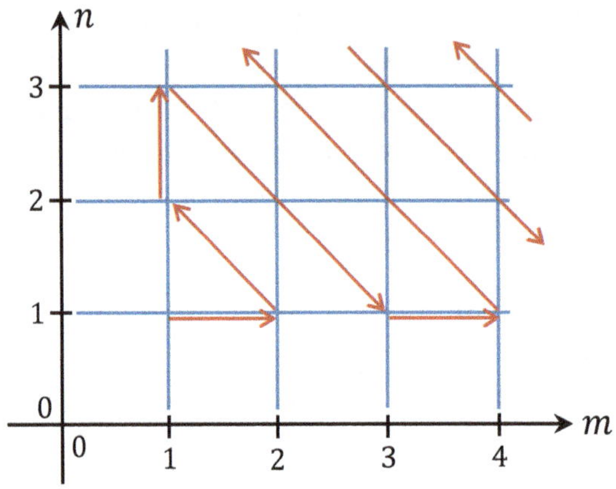

Figure 1.3. A two-dimensional grid in the mn-plane would represent all the rationals if each vertex is assigned its corresponding value of m/n. Starting at $m = n = 1$ and moving along the arrows, one visits every vertex and, therefore, encounters every possible rational number. Some rationals are repeated, of course; for example, $m/n = 2/3$ and $m/n = 4/6$ should not be counted as two different numbers. By moving along the arrows, skipping each vertex whose associated rational has been previously encountered, and assigning successive integers to each unique vertex, one can put all the rationals into one-to-one correspondence with all the positive integers.

A set whose elements cannot be put into a one-to-one correspondence with the set $\mathbb{N}$ of positive integers is said to be *uncountable*. The German mathematician Georg Cantor (1845–1918) showed that, among other things, the set of all real numbers in the interval $(0, 1)$ is uncountable. His ingenious method of proof by contradiction is astonishingly simple. It begins by assuming that the set of all real numbers in the $(0, 1)$ interval is countable. One can then imagine all these numbers written down, say, in their decimal representation, in an ordered, sequential list, as shown in figure 1.4.

Cantor proceeded to construct a new number between 0 and 1 by taking the first digit (after the decimal point) from the first number in the ordered list, modifying it, and using it as the first digit (after the decimal point) of the new number. He took the second digit from the second number in the list and, after changing it, placed it in the second place after the decimal point of the new number. This process can continue indefinitely by moving down the list and picking the nth digit from the nth number, modifying it, and using it as the nth digit of the new number. The new number thus constructed is obviously outside the list, because if, for example, it happened to coincide with the mth element of the list, then its mth digit would be the same as the mth digit of the listed element, which it is not—by construction. Cantor thus concluded that the set of all real numbers between 0 and 1 is uncountable.

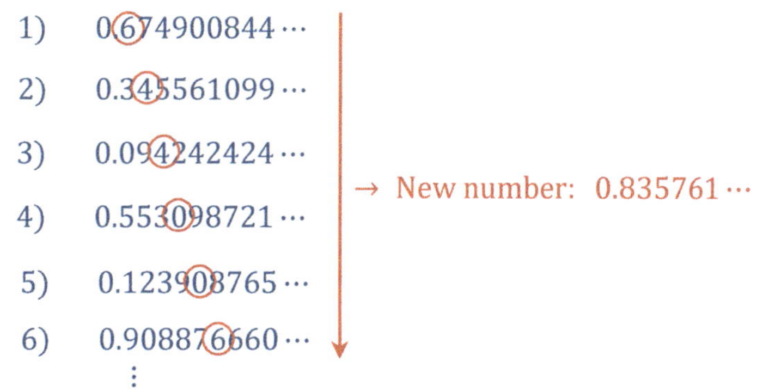

Figure 1.4. A possible listing of *all* the real numbers in the $(0,1)$ interval — assuming that such an ordered list can, in principle, be constructed. Shown to the left of each number are the positive integers 1, 2, 3, 4, 5,⋯ assigned to the various members of the set. To construct a new number between 0 and 1 that would be outside the ordered list, Cantor took the first digit after the decimal point from the first number and modified it to become the first digit (after the decimal point) of his new number. (In the example shown, the first digit of the first number is 6, which has been changed to 8 to qualify as the first digit of the new number.) Cantor then took the second digit of the second number in the list, modified it, and used it for the second digit of his new number. He continued this process by going down the list and adding digits to the new number. Since the new number obviously differs from every single one of the numbers in the list, it cannot possibly be in the list. Nevertheless, the new number is a legitimate real number belonging to the $(0,1)$ interval. In this way, Cantor showed that all the real numbers between 0 and 1 constitute an uncountable set.

A point of caution concerning Cantor's proof. One must be aware that a number whose decimal representation eventually terminates can have two different forms. For example, $0.3680000\cdots$ is the same as $0.36799999\cdots$. In such cases, one must pick only one of these two representations for inclusion in the table. Suppose we decided to consistently pick the representation that ends in $000\cdots$ for all such numbers. Then, when we construct the new number (i.e., the one that will end up not being in the table), we must avoid choosing the digit 9 for the new number at all times. In this way, the newly constructed number will have no 9s in it, and there will never be a danger of it representing a number such as $0.36799999\cdots$.

1.12.1. Does a solid square contain more points than a short line segment? Consider the line segment between 0 and 1 on the x-axis. We already know this line segment as the home to an uncountably infinite number of points. Any point on this segment can be represented by a decimal number such as $x = 0.abcdefgh\cdots$. Next, consider a square in the xy-plane whose adjacent sides coincide with the unit-length line segments between 0 and 1 on the x and y axes. A point within the area of the square may be represented by a pair of decimal numbers such as $(x, y) = (0.a'b'c'd'e'f'\cdots, 0.a''b''c''d''e''f''\cdots)$. The question arises as to which set contains more points: the one representing the unit-length line segment or the one corresponding to the unit-area square?

The answer is that both sets have exactly the same infinite number of points. In fact, there exists a one-to-one correspondence between all the points on the line and all the points within the square. To see this, observe that one can construct an (x, y) pair of numbers for any point x located on the line segment by assigning the even digits of x to one coordinate and the odd digits of x to the other coordinate. For example, the point $x = 0.abcdefgh\cdots$ is mapped onto the point $(x, y) = (0.aceg\cdots, 0.bdfh\cdots)$ of the square. Conversely, one could merge the digits of the x and y coordinates of any point within the square to arrive at the coordinates of a corresponding point on the line segment. For instance, $(x, y) = (0.a'b'c'd'\cdots, 0.a''b''c''d''\cdots)$ is mapped onto the point $x = 0.a'a''b'b''c'c''d'd''\cdots$ of the line segment. In this way, we see that the points on the line and those in the square are in a one-to-one relationship. Consequently, the line segment and the square must possess precisely the same number of points.

Digression. Can the elements of an arbitrary set S be put into one-to-one correspondence with the elements of another set $\tilde{S}$, which consists of all possible subsets of S? Suppose S has a finite number of elements, say, $S = \{s_1, s_2, s_3\}$. Then $\tilde{S} = \{\phi, \{s_1\}, \{s_2\}, \{s_3\}, \{s_1, s_2\}, \{s_1, s_3\}, \{s_2, s_3\}, \{s_1, s_2, s_3\}\}$, where ϕ represents the null (or empty) set. Clearly, for any finite set S, the number of all possible subsets is greater than the number of elements of S and, consequently, a one-to-one correspondence between S and $\tilde{S}$ is not possible. Nevertheless, for the sake of the following argument, let us pick a large set S, say, the set of all the people in the world, and assume that a one-to-one correspondence has (somehow) been established between every individual and a unique subset of S. Any given person s_i is now assigned to a specific subset of human beings, which we may designate as $\tilde{s}_i$. Two possibilities now present themselves: either s_i *is* a member of $\tilde{s}_i$, or s_i is *not* a member of $\tilde{s}_i$. The collection of all the individuals s_i who are *not* a member of their assigned subset $\tilde{s}_i$ is itself a subset of S, which we may call the Groucho set. (The comedian Groucho Marx famously said that he did not want to belong to a club which accepted him as a member!) Can the person assigned to the Groucho set be a member of that set? It is in attempting to answer this question that our original assumption, namely, the existence of a one-to-one correspondence between the elements of S and those of $\tilde{S}$, falls apart. Applied to an uncountable set S (e.g., the set of all real numbers) and its corresponding $\tilde{S}$, this argument proves that the infinite size of $\tilde{S}$ is a larger infinity than that of S.

1.12.2. Cantor's Set and Cantor's Function. The Cantor set is a *measure zero* subset of the $[0,1]$ interval containing an *uncountable* number of points. In fact, the Cantor set can be mapped in a one-to-one fashion onto the entire $[0,1]$ line segment whose measure is equal to 1. The Cantor function $y = C(x)$ is an interesting function that maps the $[0,1]$ interval of the x-axis onto the $[0,1]$ interval of the y-axis. The Cantor function is a continuous, non-decreasing function of x whose derivative $C'(x)$ exists almost everywhere—that is, everywhere within its domain except on a measure zero subset of the $[0,1]$ interval. Despite the fact that $C(x)$ rises continuously from 0 to 1 as x moves from 0 toward 1, the derivative $C'(x)$ is zero wherever it exists (i.e., almost everywhere).

The cantor set is constructed as follows: Starting with the interval $[0,1]$ on the real line, first remove the middle third, $(1/3, 2/3)$. The remaining line segments will be $[0, 1/3]$ and $[2/3, 1]$. Next, remove the middle third of the remaining two segments, that is, eliminate the open interval $(1/9, 2/9)$ from the first segment, and the open interval $(7/9, 8/9)$ from the second segment. Repeat this process ad infinitum. The total length of the segments thus removed will be

$$\ell = \frac{1}{3} + \frac{2}{3^2} + \frac{2^2}{3^3} + \cdots = \frac{1}{3}\left[1 + \frac{2}{3} + \left(\frac{2}{3}\right)^2 + \cdots\right] = \frac{1/3}{1 - 2/3} = 1. \qquad (22)$$

Consequently, the measure of the remaining set (i.e., the Cantor set) is equal to zero. Notice that the points thus removed from the $[0,1]$ interval have the property that their base-3 (or ternary) representation contains the digit 1 somewhere after the decimal point. For instance, all the points in the $(1/3, 2/3)$ interval are written in base-3 as $0.1 \times\times\times \cdots$, where the various $\times$s could be 0, 1, or 2. Similarly, the points in the $(7/9, 8/9)$ interval are written as $0.21 \times\times\times \cdots$, with 1 appearing in second place after the decimal point (and possibly elsewhere as well). It is true that the ternary representation of some of the remaining points (which belong to the Cantor set) may contain the digit 1 as well. For example, $1/3 = 0.100000 \cdots$ and $7/9 = 0.210000 \cdots$. However, these points can be written alternatively as $1/3 = 0.02222 \cdots$ and $7/9 = 0.202222 \cdots$, which contain only 0s and 2s. We conclude that the Cantor set is the set of all points in the $[0,1]$ interval whose ternary representation contains 0s and 2s, but no 1s.

If we now take the ternary representation of the numbers in the Cantor set and replace every 2 with 1, then treat the numbers as base-2 (or binary) representations, we will have a one-to-one mapping from the points in the Cantor set to *all* the numbers in the $[0,1]$ interval (expressed in binary format). The Cantor set is thus seen to be an uncountable set of measure zero.

The Cantor function $y = C(x)$ is a mapping from x in the interval $[0,1]$ onto y, which is also in the interval $[0,1]$. Two equivalent definitions of the Cantor function are given below.

i) The first iteration of the function is the straight line $y_1 = x$. The second iteration is obtained by distorting the first iteration as follows: Replace the central third of the function by $y_2 = ½$, then draw a straight line connecting $(x, y) = (0,0)$ to $(x, y) = (⅓, ½)$ and another straight line connecting $(x, y) = (⅔, ½)$ to $(x, y) = (1,1)$, as shown in figure 1.5a. Next, apply the same procedure to the two slanted lines separately, that is, distort them by flattening the central third of each line and reorienting the slanted segments to arrive at a continuous function. Repeat the procedure at infinitum to arrive at the Cantor function depicted in figure 1.5b.

ii) Take each point on the x-axis and write it out in ternary representation. If the ternary representation contains only 0s and 2s, replace all the 2s with 1s, then read the resulting number as a binary number and call it $y = C(x)$. If the ternary representation of x contains one or more 1s, then move right from the decimal point until you encounter the first 1. Delete all the digits that appear after this first 1, and convert all the 2s that precede it into 1s. Read the resulting number as a binary number and call it $y = C(x)$.

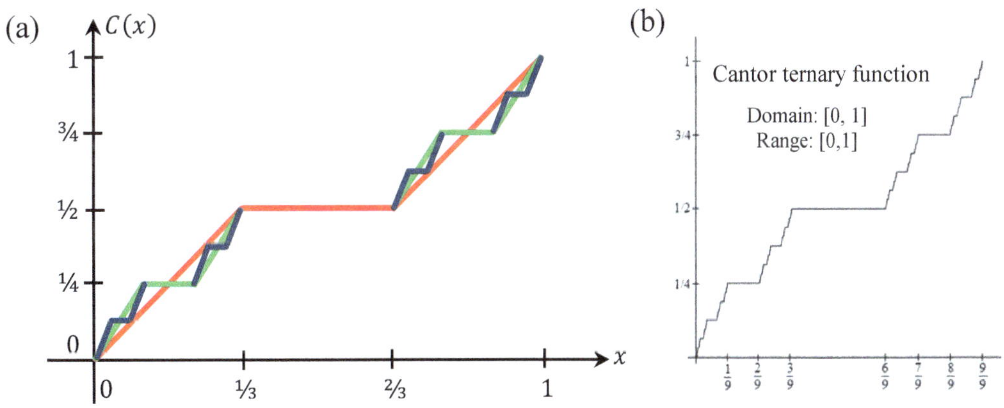

Figure 1.5. The Cantor function $C(x)$ is a mapping of the $[0,1]$ interval of the x-axis onto the $[0,1]$ interval of the y-axis. (a) The first three stages of construction of $C(x)$. (b) The Cantor function is continuous everywhere and differentiable almost everywhere, with $C'(x) = 0$ wherever the derivative exists.

It is not difficult to prove that the Cantor function is non-decreasing and continuous everywhere. It is also differentiable almost everywhere, that is, on the horizontal segments of the graph of $C(x)$, with the derivative being $C'(x) = 0$.

Suggestions for further reading

1. A. Weil, *Number Theory for Beginners*, Springer-Verlag, New York (1979).
2. A. N. Kolmogorov and S. V. Fomin, *Introductory Real Analysis*, Dover, New York (1970).
3. F. B. Hildebrand, *Advanced Calculus for Applications*, 2nd edition, Prentice-Hall, New Jersey (1976).
4. G. H. Hardy and E. M. Wright, *An Introduction to the Theory of Numbers*, Oxford University Press, New York (2008).

Problems

Problem 1) Find the roots of the polynomial equation $ax^2 + bx + c = 0$, where $a \neq 0$, b, and c are arbitrary complex coefficients of the polynomial.

Problem 2) Prove the chain rule of differentiation, namely, $\mathrm{d}f[g(x)]/\mathrm{d}x = g'(x)f'[g(x)]$. Use the chain rule to evaluate the first derivatives of each of the following functions:

$$\sin(\cos x), \qquad \exp(-\kappa x^2), \qquad \tan(\sqrt{ax^2+bx+c}\,),$$

$$\ln(\tan^2 x), \qquad \tanh(\ln x), \qquad x^x, \qquad \sinh^x(x).$$

Problem 3) Find the first few derivatives of the function $f(x) = \exp(-\pi x^2)$. Expand $f(x)$ in a Taylor series around the origin, $x_0 = 0$. Use a qualitative argument to show that, as more terms are used in the Taylor series expansion, it approaches the function $f(x)$ over a wider range of values of x.

Problem 4) Evaluate $\int_{-\infty}^{\infty} \exp(-\pi x^2)\,\mathrm{d}x$. Use your result to determine the value of the following integral, where x_0 is an arbitrary point on the x-axis:

$$\int_{-\infty}^{\infty} \exp(-\pi x^2 + 2\pi x_0 x)\,\mathrm{d}x.$$

Problem 5) Use the product rule of differentiation, namely, $[f(x)g(x)]' = f'(x)g(x) + f(x)g'(x)$, to derive the formula for integration by parts, that is,

$$\int_a^b f'(x)g(x)\mathrm{d}x = f(b)g(b) - f(a)g(a) - \int_a^b f(x)g'(x)\mathrm{d}x.$$

Use the method of integration by parts to evaluate the following integrals ($\kappa > 0$):

$$\int_0^{\infty} x\exp(-\kappa x)\,\mathrm{d}x, \qquad \int_0^{\infty} x^2\exp(-\kappa x)\,\mathrm{d}x, \qquad \int_0^{\infty} \sin(\omega x)\exp(-\kappa x)\,\mathrm{d}x,$$

$$\int_0^{\infty} \cos(\omega x)\exp(-\kappa x)\,\mathrm{d}x, \qquad \int_0^{\infty} x^2\exp(-\pi x^2)\,\mathrm{d}x, \qquad \int_0^{\infty} x^3\exp(-\pi x^2)\,\mathrm{d}x,$$

$$\int_0^{a} x\ln(x)\,\mathrm{d}x, \qquad \int_0^{a} x^{\kappa}\ln^2(x)\,\mathrm{d}x, \qquad \int_0^{\pi} \frac{\sin^3\theta}{(1-\kappa\cos\theta)^3}\,\mathrm{d}\theta.$$

Problem 6) Assuming θ is an arbitrary angle and n an integer, prove the following identity:

$$\frac{\sin(\theta)+\sin(2\theta)+\cdots+\sin(n\theta)}{\cos(\theta)+\cos(2\theta)+\cdots+\cos(n\theta)} = \tan[\tfrac{1}{2}(n+1)\theta].$$

Problem 7) Find closed-form expressions for the following series:

$$S_1(N) = 1+2+3+\cdots+N, \qquad S_2(N) = 1^2+2^2+3^2+\cdots+N^2.$$

Hint: One way to derive the closed-form of $S_2(N)$ is to assume that it is a third-order polynomial in N, that is, $S_2(N) = aN^3 + bN^2 + cN + d$, then determine the a, b, c, d coefficients using the fact that $S_2(1) = 1$ and $S_2(N) - S_2(N-1) = N^2$.

Problem 8) For complex-valued functions $f(x)$ and $g(x)$, and the arbitrary complex-valued parameter λ, the following identity is generally valid:

$$0 \le \int_a^b |f(x) + \lambda g(x)|^2 \mathrm{d}x = \int_a^b |f(x)|^2 \mathrm{d}x + \lambda \int_a^b f^*(x) g(x) \mathrm{d}x$$

$$+\lambda^* \int_a^b f(x) g^*(x) \mathrm{d}x + \lambda\lambda^* \int_a^b |g(x)|^2 \mathrm{d}x.$$

Find the value of λ that minimizes the expression on the right-hand-side of this identity. Substitute the value of λ thus obtained into the expression and derive the Schwarz inequality, namely,

$$\left|\int_a^b f(x) g^*(x) \mathrm{d}x\right|^2 \le \int_a^b |f(x)|^2 \mathrm{d}x \int_a^b |g(x)|^2 \mathrm{d}x.$$

Problem 9) Using the decimal representation of two *rational* numbers, a and b, where $0 < a < b < 1$, show that it is always possible to find at least one rational and one irrational number between a and b. Repeat the argument for the case when both a and b are *irrational*.

Problem 10) Show that the decomposition of an integer into its prime factors is unique.

Hint: You may want to begin by assuming that the smallest integer that can be decomposed in two different ways is $N = p_1^{m_1} p_2^{m_2} \cdots p_j^{m_j} = q_1^{n_1} q_2^{n_2} \cdots q_k^{n_k}$, then arrive at a contradiction.

Problem 11) a) Assuming that the prime number p divides the product $n_1 n_2$ of two integers, prove that p must divide either n_1 or n_2 or both. (This property of prime numbers is commonly referred to as Euclid's lemma.)

b) Extend the proof of part (a) to the product $n_1 n_2 \cdots n_k$ of k integers, assuming the product is known to be divisible by the prime number p.

Problem 12) A straight stick of length L is to be cut into n pieces of lengths $x_1, x_2, \cdots, x_n$. The total length of the various pieces must obviously add up to L, that is, $x_1 + x_2 + \cdots + x_n = L$. We would like to devise a strategy for cutting the stick in such a way as to yield the maximum value for the product $x_1 x_2 \cdots x_n$. Use proof by induction to show that $x_1 x_2 \cdots x_n$ is maximized when all the pieces are of equal length, that is, $x_1 = x_2 = \cdots = x_n = L/n$.

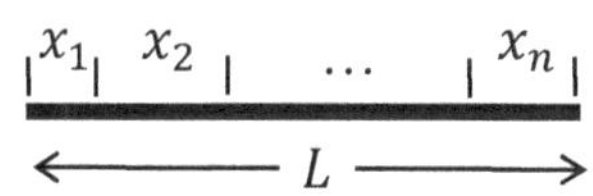

Problem 13) Show that the Taylor series expansion of $f(x) = (1 + x)e^x$ around the point $x = 0$ is given by $f(x) = \sum_{n=0}^{\infty} [(n + 1)/n!] x^n$.

Problem 14) a) Let $S_N^{(m)}$ represent the sum of the mth power of integers from 1 to N, that is, $S_N^{(m)} = \sum_{n=1}^{N} n^m$. Using the fact that $\int_{x=1}^{N+1} x^{k-1} \mathrm{d}x = \sum_{n=1}^{N} \int_{x=n}^{n+1} x^{k-1} \mathrm{d}x$ for any arbitrary integer $k \ge 1$, show that the various $S_N^{(m)}$ satisfy the following recursion relation:

$$\sum_{m=0}^{k-1} \binom{k}{m} S_N^{(m)} = (N + 1)^k - 1.$$

b) Starting with $k = 1$, then repeating the same procedure for $k = 2, 3, 4, \cdots$, find closed form expressions for $S_N^{(0)}, S_N^{(1)}, S_N^{(2)}, S_N^{(3)}, \cdots$.

Problem 15) Which of the following statements are true and which are false? You must provide a brief proof or a counterexample in each case.

a) If x is irrational, then its inverse, $1/x$, is also irrational.

b) If x is irrational, then its square root, $\sqrt{x}$, is also irrational.

c) The sum of a rational number and an irrational number is irrational.

d) The sum of two irrational numbers is irrational.

e) The product of two irrational numbers is irrational.

Problem 16) According to the *fundamental theorem of arithmetic*, any positive integer greater than 1 is either a prime number or can be written as the product of prime numbers in a unique way. For example, $360 = 2 \times 2 \times 2 \times 3 \times 3 \times 5 = 2^3 \times 3^2 \times 5$. Using this fundamental theorem, prove that, given any positive integer N, its square root $\sqrt{N}$ is either an integer or an irrational number. In other words, unless N happens to be a perfect square, its square root is always going to be irrational.

Problem 17) Use proof by induction to demonstrate that $\sum_{n=1}^{N}(2n-1) = N^2$.

Problem 18) Use proof by induction to demonstrate that $n! > 2^n$ for $n \geq 4$.

Problem 19) Each and every one of a countable number of sets $S_1, S_2, S_3, \cdots$ has a countable number of elements. Show that the union of all these sets is a countable set.

Problem 20) Prove that the infinity of points on the entire real line is equal to the infinity of points within the $(0, 1)$ interval, by establishing a one-to-one correspondence between all real numbers between $-\infty$ and ∞, on the one hand, and all the real numbers between 0 and 1, on the other.

Problem 21) Show that $\int_0^{\pi/2} \exp(p \sin x) \sin(2x)\, dx = 2[1 + (p-1)e^p]/p^2$.

Problem 22) Prove that, for a positive integer n and an arbitrary constant p, we will have

$$\int_0^{\pi} \exp(p \cos x) \cos(p \sin x) \cos(nx)\, dx = \int_0^{\pi} \exp(p \cos x) \sin(p \sin x) \sin(nx)\, dx = \frac{\pi p^n}{2n!}.$$

Chapter 2

Elements of Euclidean Geometry

"Philosophy is written in this grand book (I mean the Universe), which stands continually open before our eyes but cannot be understood without first learning to comprehend the language and know the characters in which it is written. It is written in mathematical language, and its characters are triangles, circles and other geometric figures, without which it is impossible to humanly understand a word; without these one is wandering in a dark labyrinth."

Galileo Galilei (1564–1642)

2.1. Introduction. Plane geometry is the oldest branch of mathematics. As far back as 500 BC, Pythagoras of Samos (c. 570–495 BC) had already discovered the precise relation among the lengths of the three sides of right-angle triangles. Around 300 BC, Euclid of Alexandria, a Greek teacher and mathematician, collected all the mathematical knowledge of his day and presented it in a thirteen-volume masterpiece called *Elements*. Additional facts of plane geometry were discovered over the centuries that followed, but also other branches of mathematics such as algebra, trigonometry, analytic geometry, non-Euclidean geometry, real and complex analysis, topology, etc., were discovered and further developed in the intervening years.

Today, Euclidean geometry is no longer a required subject of study in many high school and undergraduate curricula, which is a pity, because, along with the rudiments of geometry, one would learn so much about structured thinking and logical reasoning while studying this fascinating branch of mathematics. In physical sciences, one often encounters situations where a knowledge of Euclidean geometry is essential for analyzing and understanding complex physical systems. In optics and photonics, for example, ray-tracing and aberration theory, geometrical optics, analysis of the behavior of image-forming systems (such as microscopes, telescopes, and photographic cameras), optics of the eye and ophthalmic optics, astronomical and adaptive optics, fiber optical cables and networks, polarization optics, etc., would be impossible without a substantial knowledge of Euclidean geometry.

The goal of the present chapter is to visit a few remarkable concepts and theorems of plane geometry to remind the student of the methods of analysis and logical deduction in this firmly established branch of mathematics. The student is expected to be familiar with elementary concepts and methods, so we will skip over some of the well-known definitions and details, hoping that the student will be able to fill the gaps.

2.2. Complementary and supplementary angles. Two angles that add up to 90° are said to be complementary, and two angles that add up to 180° are said to be supplementary. Figure 2.1a shows a slanted straight line crossing a pair of parallel lines, and also a vertical line that is perpendicular to the parallel pair. The angles formed between the slanted line and the parallel pair are denoted by α and are readily shown to be identical. Also shown is the angle β, which is complementary to α, and the angles γ, which are supplementary to α.

To prove that the three angles of any triangle add up to 180°, we draw a straight line from one of the vertices parallel to the side opposite to that vertex, as in figure 2.1b, where the straight line is drawn from the vertex A parallel to BC. The side AB now crosses the two parallel lines, forming two equal angles β. Similarly, the side AC crosses the two parallel lines, forming two equal angles γ. It is trivial to verify that $\alpha + \beta + \gamma$ now equals 180°.

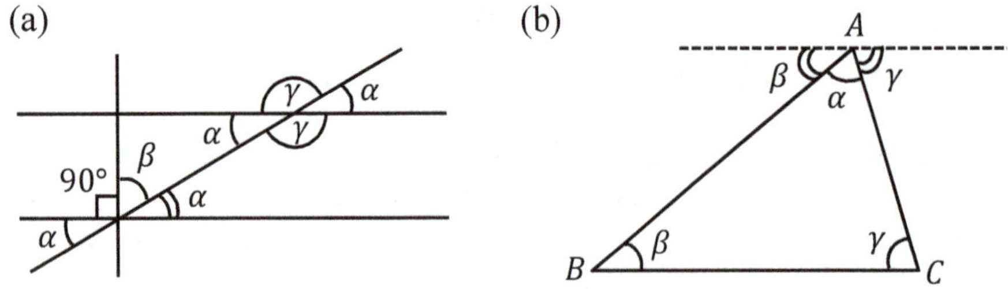

Figure 2.1. (a) A slanted straight line crossing a pair of parallel lines creates equal angles α with each of the parallel lines. Also shown is a straight line perpendicular to the parallel lines, and an angle β that is complementary to α. The angle γ is said to be supplementary to α. (b) To prove that the three angles of any triangle add up to 180°, one must draw a straight line from one of the vertices parallel to the side that is opposite that vertex.

2.3. Similar triangles. The triangles ABC and $A'B'C'$ are said to be *equal* if they can be made to precisely overlap each other. In this case, the corresponding sides and angles of the two triangles will be identical.

By definition, two triangles are *similar* if one is a magnified version of the other, such as those shown in figure 2.2. Here, the corresponding angles will be equal, that is, $\alpha = \alpha'$, $\beta = \beta'$, $\gamma = \gamma'$, and the corresponding sides will be in the same proportion, namely, $\overline{AB}/\overline{A'B'} = \overline{AC}/\overline{A'C'} = \overline{BC}/\overline{B'C'}$.

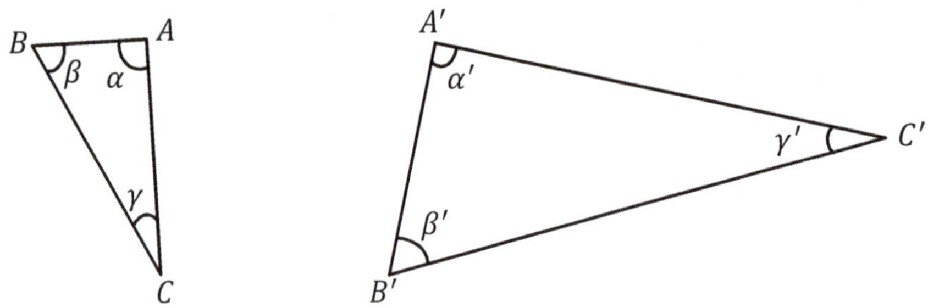

Figure 2.2. The triangles ABC and $A'B'C'$ are similar if one is a magnified version of the other.

2.4. Theorem of Pythagoras. Perhaps the best known fact in all of Euclidean geometry is the Pythagorean theorem, which relates the length c of the hypotenuse of a right-angle triangle to the lengths a and b of the remaining two sides via $c^2 = a^2 + b^2$. For a simple proof of this theorem, consider, with reference to the figure, four identical right-angle triangles arranged in such a way as to form a large square of side-length $a + b$, and a smaller square whose sides have length c. The area of the large square is $(a + b)^2$, that of the small square is c^2, and each of the four triangles has an area equal to $ab/2$. It is easy to see that $(a + b)^2 = c^2 + 4(ab/2)$, which leads to $c^2 = a^2 + b^2$. Although many proofs of the Pythagorean theorem exist, the one shown here is probably the simplest. (Another simple proof is the subject of problem 1 at the end of the chapter.)

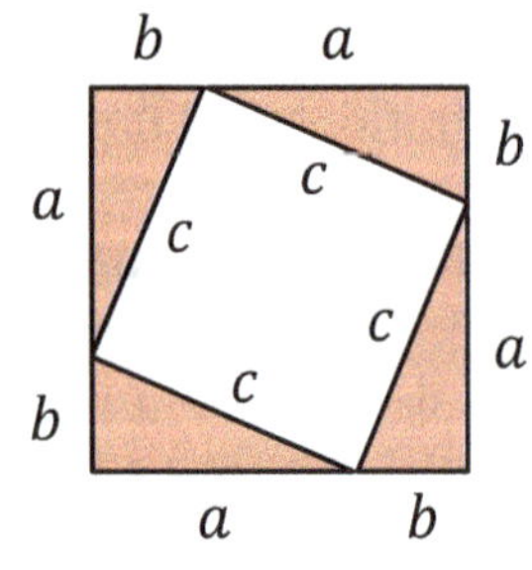

2.5. Three angle bisectors of a triangle cross at a common point inside the triangle. An angle bisector is a straight line that divides an angle into two equal halves. Figure 2.3 shows the bisectors of the angles A and C, crossing at the point Q inside the ABC triangle. From Q, we draw perpendiculars QD, QE, and QF to the three sides of the triangle. The right triangles AQD and AQF are equal to each other; therefore $\overline{QD} = \overline{QF}$. Similarly, the right triangles CQE and CQF are equal to each other; therefore $\overline{QE} = \overline{QF}$. We conclude that $\overline{QD} = \overline{QE}$. A straight line connecting the vertex B to the point Q must now create two equal triangles BQD and BQE. Therefore, BQ is the bisector of the angle at B.

A circle drawn with center at Q and radius QD (which, as we have just shown, equals QE and QF) will be fully contained (or enclosed) within the triangle. The sides AB, AC, and BC of the triangle are tangents of the circle.

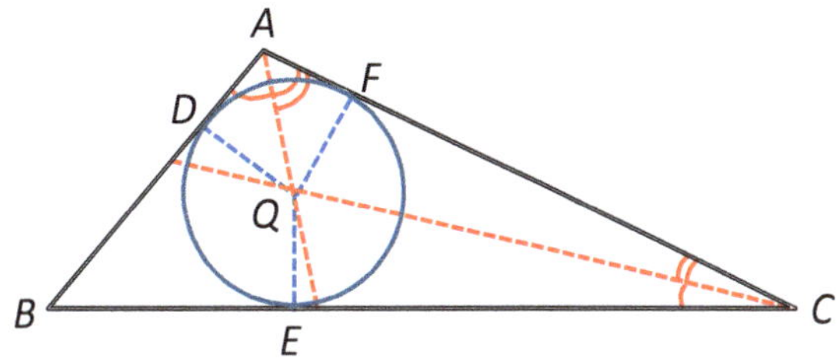

Figure 2.3. The angle bisectors of the triangle ABC cross at the common point Q inside the triangle. QD, QE, and QF are perpendiculars from Q to AB, BC, and AC, respectively. A circle drawn with center at Q and radius equal to $\overline{QD}$ (which is also equal to $\overline{QE}$ and $\overline{QF}$) is fully contained within the triangle. The sides AB, AC, and BC of the triangle are tangents of the circle.

2.6. Three perpendicular bisectors of a triangle always cross at a common point. With reference to figure 2.4, note that all points on the perpendicular bisector QD of AB are equidistant from A and B. In particular, $\overline{QA} = \overline{QB}$. Similarly, all points on the perpendicular bisector QE of BC are equidistant from B and C, and, in particular, $\overline{QB} = \overline{QC}$. The equality of $\overline{QA}$ and $\overline{QC}$ thus ensures that the triangle QAC is isosceles, and that the perpendicular bisector of AC passes through the point Q as well.

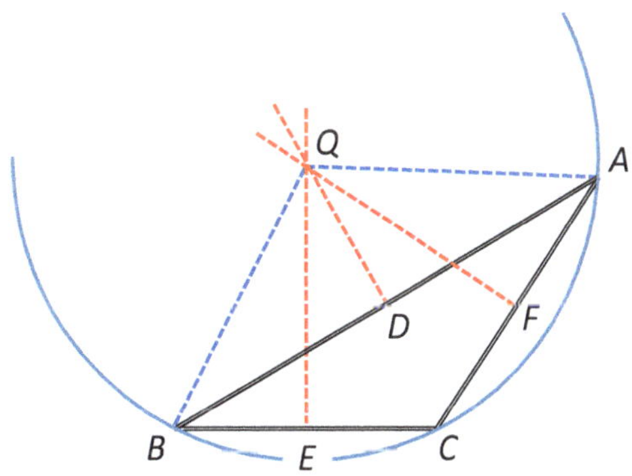

Figure 2.4. The perpendicular bisectors of the triangle ABC are QD, QE, and QF. The crossing-point Q is equidistant from the three vertices of the triangle. A circle drawn with center at Q and radius QA (or QB or QC) will pass through the three vertices and encompass the entire triangle.

2.7. Three altitudes of a triangle always cross at a common point. An altitude is a straight line drawn from one vertex perpendicular to the opposite side of a triangle. The three altitudes of any arbitrary triangle always cross at a common point. One way to demonstrate this is to draw a line parallel to BC through the vertex A, another line parallel to AC through B, and a third line parallel to AB through C. We thus form a new triangle DEF, as shown in figure 2.4. Since $ABCF$ is a parallelogram, its opposite sides are equal to each other; that is, $\overline{AF} = \overline{BC}$ and $\overline{AB} = \overline{FC}$. Similarly, $BEAC$ is also a parallelogram and, therefore, $\overline{EA} = \overline{BC}$ and $\overline{EB} = \overline{AC}$. Moreover, $ABDC$ is a parallelogram whose opposite sides are equal; that is, $\overline{AB} = \overline{CD}$ and $\overline{AC} = \overline{BD}$. We conclude that A is the mid-point of EF. By the same token, B is the mid-point of ED, and C is the mid-point of DF. In addition, the altitude AA' being perpendicular to BC is also perpendicular to its parallel EF. Similarly, BB' and CC' are perpendicular to ED and DF, respectively. The three altitudes of the triangle ABC are seen to be the perpendicular bisectors of the triangle DEF. Considering that the perpendicular bisectors always cross at a common point, we have proven that the three altitudes of any triangle also cross at a common point.

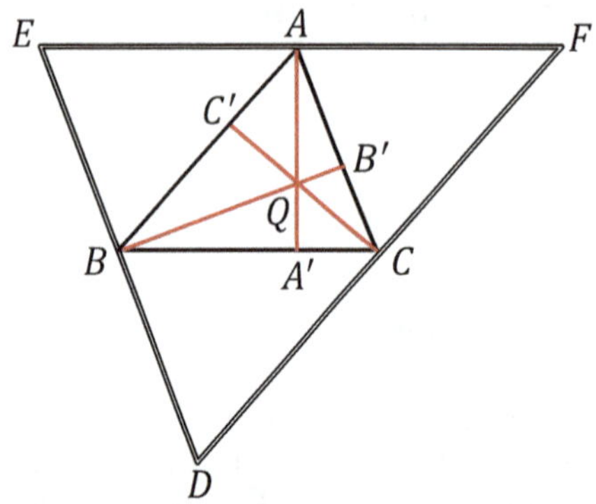

Figure 2.5. The three altitudes AA', BB', and CC' of the triangle ABC cross at the common point Q. To demonstrate this fact, the triangle DEF is constructed such that EF passes through A and is parallel to BC, ED passes through B and is parallel to AC, and DF passes through C and is parallel to AB. The altitudes of ABC are readily seen to be the perpendicular bisectors of DEF.

2.8. Three medians cross at a common point inside the triangle. A median is a straight line connecting a vertex to the midpoint of the opposite side of a triangle. Figure 2.6a shows the median BB' that connects the vertex B to the mid-point of AC, and also the median CC' that connects the vertex C to the midpoint of AB. Since the triangles ABC and $AC'B'$ are similar, $C'B'$ is parallel to BC, and also the length of $C'B'$ is half the length of BC. Next, we notice that the triangles QBC and $QB'C'$ are similar. Consequently, $\overline{QB}/\overline{QB'} = \overline{QC}/\overline{QC'} = \overline{BC}/\overline{B'C'} = 2$. The crossing point Q is thus seen to divide each median (BB' and CC') in a 2:1 ratio. We now argue that the remaining median (i.e., the one connecting A to the midpoint of BC) must pass through Q, because otherwise it will not be able to divide BB' and/or CC' in the required 2:1 ratio.

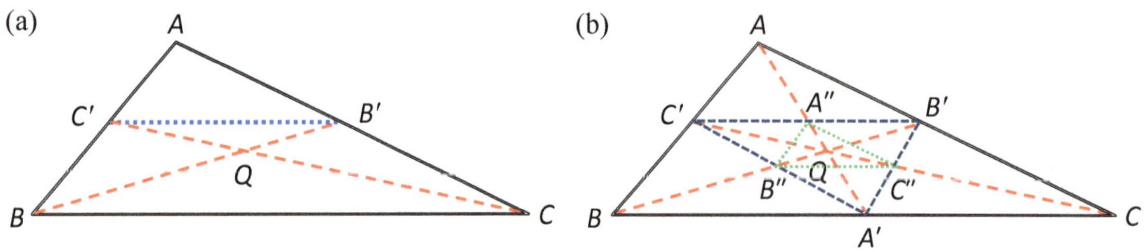

Figure 2.6. (a) The medians BB' and CC' of the triangle ABC cross at Q. The straight line $B'C'$ connecting the midpoints of AB and AC can be shown to be parallel to BC. (b) The triangles ABC, $A'B'C'$, $A''B''C''$ share the same medians. The crossing point Q is inside all such triangles.

Figure 2.6b shows that the triangle $A'B'C'$ has the same medians as the triangle ABC. Connecting the midpoints of $A'B'$, $B'C'$, and $A'C'$ produces yet another triangle, $A''B''C''$, which has the same medians as the parent triangles. In this way, one can create an infinite number of nested triangles, all having the same medians and the same common crossing point Q. The nested triangles approach the point Q as their size continually diminishes.

2.9. Using a straight-edge and a compass to find the square root of a number. An obsession of the ancient Greek geometers was to solve every conceivable mathematical problem using only a compass and a straight-edge (i.e., an unmarked ruler). In this section, we show how they managed to find the square root of a number using these instruments.

One must begin by specifying the unit of length. With reference to figure 2.7, a line segment AB is drawn, whose length is taken to be 1. One then extends AB until the line segment BC acquires the desired length λ (in units of $\overline{AB}$). The next step is to find the midpoint of the line AC and draw a circle of diameter $\overline{AC}$ centered at this midpoint. Finally, a straight line drawn perpendicular to AC from the point B until it crosses the circle at D will have the desired length, that is, $\overline{BD} = \sqrt{\lambda}$.

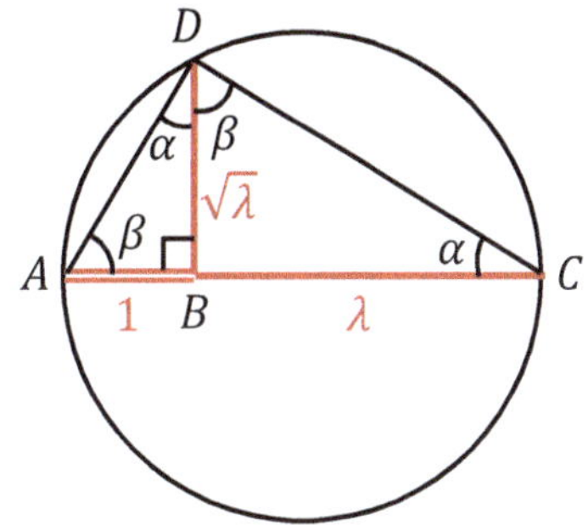

Figure 2.7. Finding the square root of λ using only a straight-edge and a compass. The line segment AB has unit length, while it extension BC has length λ. The circle is centered at the midpoint of the line AC and has a radius equal to $\overline{AC}/2$. A perpendicular to AC, drawn from the point B until it crosses the circle at D, has the desired length, that is, $\overline{BD} = \sqrt{\lambda}$.

To see the logic behind this operation, note that the length of the lower arc AC of the circle (normalized by its radius) is π radians. Therefore, the angle $A\hat{D}C$ facing this arc must be $\pi/2$ radians (90°). This means that the sum of the angles $A\hat{D}B = \alpha$ and $B\hat{D}C = \beta$ equals 90°. Given that ABD is a right-angle triangle, its angle $D\hat{A}B$ must be equal to β. Similarly, since BCD is a right-angle triangle, its angle $B\hat{C}D$ must be equal to α. Consequently, the triangles ABD and BCD are similar. The similarity of the triangles indicates that their corresponding sides are in the same proportion, that is, $\overline{BC}/\overline{BD} = \overline{BD}/\overline{AB}$. Since $\overline{AB} = 1$, we conclude that $\overline{BC} = \overline{BD}^2$, which implies that $\overline{BD} = \sqrt{\lambda}$.

Brain teaser. Build four identical equilateral triangles using only six matches. (You cannot break the matches, and the number of triangles should not exceed four.)

2.10. Theorem. The straight line CD splits the angle C of the ABC triangle, into c_1 and c_2, while dividing AB into AD and BD. Prove that $\overline{AD}/\overline{BD} = (\overline{AC}\sin c_1)/(\overline{BC}\sin c_2)$.

Proof. From the vertices A and B, one can drop the perpendiculars AA' and BB' onto CD. The right-angled triangles $AA'D$ and $BB'D$ are seen to be similar. Therefore,

$$\overline{AD}/\overline{BD} = \overline{AA'}/\overline{BB'} = (\overline{AC}\sin c_1)/(\overline{BC}\sin c_2).$$

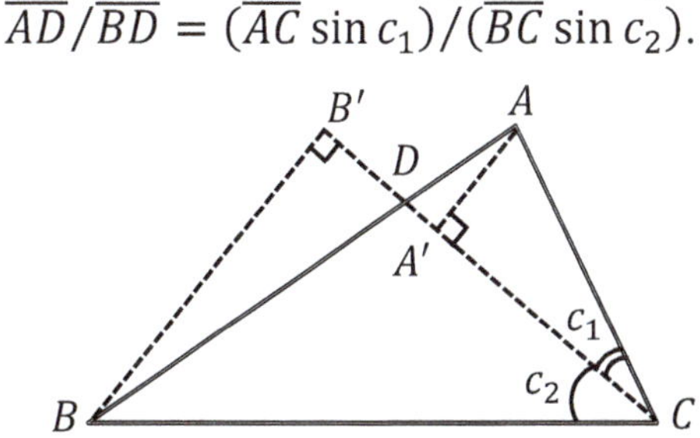

2.11. Theorem. Given an arbitrary triangle, an angle bisector splits the opposite side in proportion to the ratio of the adjacent sides of the bisected angle; that is, for the ABC triangle shown, where CD is the bisector of the angle C, the ratio $\overline{AD}/\overline{DB}$ equals $\overline{AC}/\overline{BC}$.

Proof. Extend BC by an amount equal to AC to arrive at E, as shown. Since $\overline{AC} = \overline{CE}$, the triangle ACE is isosceles and, therefore, $\alpha = \beta$. Considering that $\alpha + \beta$ supplements the angle C of the ABC triangle, and since $c_1 = c_2$, one concludes that $\alpha = c_1 = c_2$. Consequently, $CD \parallel EA$.

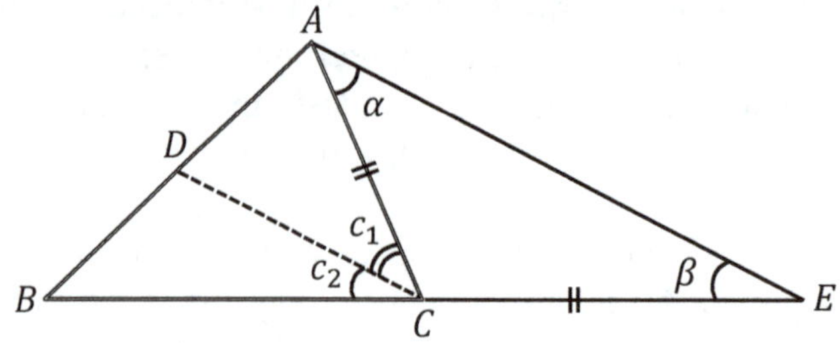

The triangles BCD and BEA are now seen to be similar. Therefore, $\overline{BC}/\overline{CE} = \overline{BD}/\overline{DA}$, or $\overline{BC}/\overline{AC} = \overline{BD}/\overline{DA}$. The proof is thus complete.

2.12. Relating the length of each side of a triangle to the lengths of the other two sides and the opposing angle. In the triangle ABC, whose sides have lengths a, b, and c, the angle θ opposing AC is the angle between AB and BC.

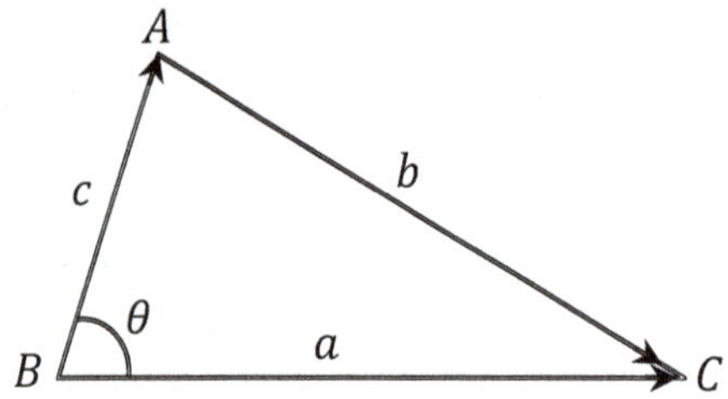

Using vector algebra, we may write $\overrightarrow{AC} = \overrightarrow{BC} - \overrightarrow{BA}$. Consequently,

$$|\overrightarrow{AC}|^2 = \overrightarrow{AC} \cdot \overrightarrow{AC} = (\overrightarrow{BC} - \overrightarrow{BA}) \cdot (\overrightarrow{BC} - \overrightarrow{BA}) = |\overrightarrow{BC}|^2 + |\overrightarrow{BA}|^2 - 2\overrightarrow{BC} \cdot \overrightarrow{BA}$$

$$\rightarrow \quad b^2 = a^2 + c^2 - 2ac\cos\theta \qquad \rightarrow \quad b = \sqrt{a^2 + c^2 - 2ac\cos\theta}.$$

In the special case of a right-angled triangle, where $\theta = 90°$, we will have $\cos\theta = 0$, and the relation among a, b, and c reduces to that of Pythagoras, namely, $b = \sqrt{a^2 + c^2}$.

Brain teaser. Without lifting the pen from the paper, connect these nine dots using four straight lines.

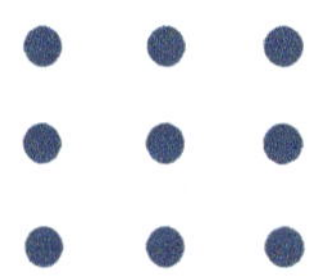

2.13. Ceva's theorem. The following theorem is named after the seventeenth-century Italian mathematician Ceva: Three straight lines drawn from the vertices of the ABC triangle that meet the opposite sides at A', B', C' intersect at a common point S provided that the following identity holds:

$$(\overline{CA'}/\overline{A'B}) \cdot (\overline{AB'}/\overline{B'C}) \cdot (\overline{BC'}/\overline{C'A}) = 1.$$

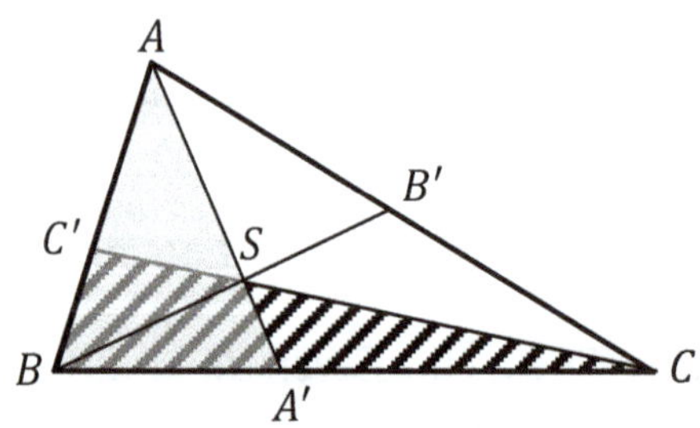

Figure 2.8. An arbitrarily chosen point S is connected by straight lines to the vertices A, B, and C. The lines are subsequently extended to meet the opposing sides of the triangle at A', B', and C'.

Proof. Take a point S and draw the straight-line segments AA', BB', CC' through S. Denote by h_B the altitude dropped from B to AC, and by h_S the altitude dropped from S to AC. We will have

$$\frac{\text{Area of } ABS}{\text{Area of } BCS} = \frac{(\text{Area of } ABB') - (\text{Area of } ASB')}{(\text{Area of } BCB') - (\text{Area of } CSB')} = \frac{(\overline{AB'} \cdot h_B) - (\overline{AB'} \cdot h_S)}{(\overline{B'C} \cdot h_B) - (\overline{B'c} \cdot h_S)} = \overline{AB'}/\overline{B'C}.$$

Similarly,

$$\frac{\text{Area of } ACS}{\text{Area of } ABS} = \overline{CA'}/\overline{A'B} \quad \text{and} \quad \frac{\text{Area of } BSC}{\text{Area of } ASC} = \overline{BC'}/\overline{C'A}.$$

The product of the above ratios now yields

$$\left(\frac{\text{Area of } ABS}{\text{Area of } BCS}\right) \cdot \left(\frac{\text{Area of } ACS}{\text{Area of } ABS}\right) \cdot \left(\frac{\text{Area of } BSC}{\text{Area of } ASC}\right) = 1.$$

One application of Ceva's theorem is in proving that the three medians of a triangle cross at a common point. In this case, $\overline{AB'}/\overline{B'C} = 1$ and $\overline{CA'}/\overline{A'B} = 1$. Therefore, the line connecting C to S must bisect AB at C'.

2.14. Heron's (or Hero's) theorem. Let the sides of a triangle have lengths a, b, and c. According to Heron's theorem, the area of the triangle is $A = \sqrt{s(s-a)(s-b)(s-c)}$, where $s = (a+b+c)/2$ is half the perimeter of the triangle. (For proof see problem 7.)

2.15. Another interesting property of triangles. Not all problems in Euclidean geometry are easy to prove; some problems require ingenuity, and a considerable effort is often required to solve them. As an example, consider the arbitrary triangle ABC, whose sides are divided into three equal parts at (A', A''), (B', B''), and (C', C''), as shown. Find the area of the triangle DEF as a fraction of the area of the original triangle ABC.

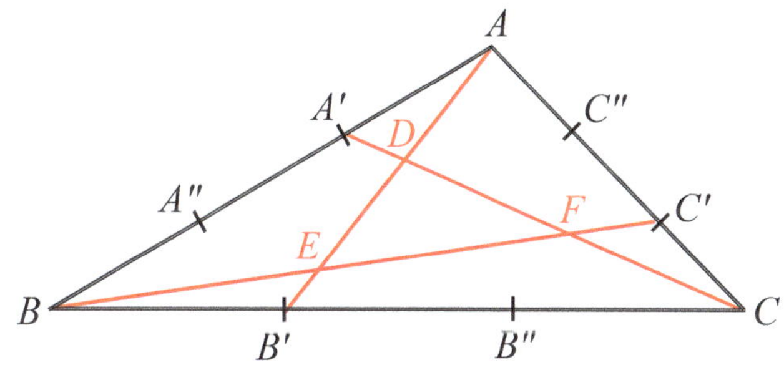

Solution: Consider first the triangle ABB'. Its base BB' is ⅓ of the base BC of ABC, and the two triangles have the same height. Consequently, $\Delta\, ABB' = ⅓(\Delta\, ABC)$, where we have used the symbol Δ to indicate a triangle's area. A similar argument shows that $\Delta\, BCC' = \Delta\, CAA' = ⅓(\Delta\, ABC)$. The sum of the areas of ABB', BCC', and CAA' is thus equal to the area of ABC. These three small triangles do not contain the triangle DEF, but they share the tiny triangles DAA', EBB', and FCC', which were counted twice when we added the areas of ABB', BCC', and CAA' to obtain the area of ABC. Therefore,

$$\Delta\, DEF = \Delta\, DAA' + \Delta\, EBB' + \Delta\, FCC'. \tag{1}$$

Next, we focus on the triangle ABB', whose area is given by

$$\Delta\, ABB' = ½\, \overline{BB'} \times \overline{AB'} \times \sin\varphi.$$

The area of the tiny triangle EBB' enclosed within ABB' is readily seen to be

$$\Delta\, EBB' = ½\, \overline{BB'} \times \overline{EB'} \times \sin\varphi.$$

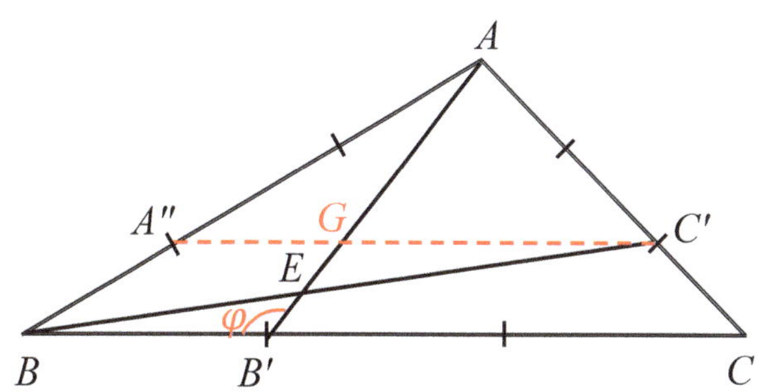

Consequently,

$$\triangle\, EBB' = (\overline{EB'}/\overline{AB'}) \times (\triangle\, ABB') = ⅓(\overline{EB'}/\overline{AB'}) \times (\triangle\, ABC). \qquad (2)$$

To determine the ratio of $\overline{EB'}$ to $\overline{AB'}$, we draw the straight line $A''C'$, which is parallel to BC. The similarity of the triangles EBB' and $EC'G$ then yields

$$\overline{EB'}/\overline{EG} = \overline{BB'}/\overline{GC'} = \overline{BB'}/(⅔ \times \overline{A''C'}) = \overline{BB'}/(⅔ \times ⅔ \times \overline{BC}) = ¾. \qquad (3)$$

We also know, from the similarity of the triangles ABB' and $AA''G$, that the following relation holds

$$\overline{AG}/\overline{AB'} = \overline{AA''}/\overline{AB} = ⅔ \quad \rightarrow \quad (\overline{AB'} - \overline{EG} - \overline{EB'})/\overline{AB'} = ⅔$$

Using Eq.(3)

$$\left(\overline{AB'} - \tfrac{4}{3}\overline{EB'} - \overline{EB'}\right)/\overline{AB'} = 1 - \tfrac{7}{3}(\overline{EB'}/\overline{AB'}) = \tfrac{2}{3} \;\rightarrow\; \overline{EB'}/\overline{AB'} = \tfrac{1}{7}. \quad (4)$$

Combining the results of Eq.(2) and Eq.(4), we will have

$$\triangle\, EBB' = \frac{1}{21}(\triangle\, ABC). \qquad (5)$$

Similar relations hold for $\triangle\, DAA'$ and $\triangle\, FCC'$. Using Eq.(1), we now find

$$\triangle\, DEF = \frac{1}{7}(\triangle\, ABC). \qquad (6)$$

Brain teaser. A rectangular piece is cut out of a round cake, leaving a rectangular hole behind. The hole and the cake are *not* concentric. Describe a single straight cut that would split the remaining cake into two equal halves. (The trivial solution of cutting parallel to the surface is not allowed.) This problem can be generalized to shapes other than circle and rectangle for the cake as well as the hole. Allowed shapes (for both the cake and the hole) are circle, ellipse, hexagon, rectangle, and parallelogram. Shapes that are *not* allowed include triangle, trapezoid, and pentagon.

2.16. Conic section. With reference to figure 2.9, the apex of a right circular cone is at the origin, its half-angle is denoted by ψ, and its axis is tilted away from the z-axis by an angle θ. A plane parallel to the xy-plane at $z = h$ crosses the cone. The cross-sectional ellipse in the $z = h$ plane is described by the equation $(x/a)^2 + [(y - y_c)/b]^2 = 1$. (Considering that $(\theta + \psi) < 90°$, we must have $\cos\psi > \cos(90° - \theta) = \sin\theta$.) In terms of the parameters θ and ψ of the cone and the height h of the plane, we would like to determine the values of a, b, y_c, and the focal length $f = \sqrt{b^2 - a^2}$ of the ellipse.

Note: At a vertex where the ellipse makes contact with its major axis, the sum of the distances to the foci is $(b - f) + (b + f) = 2b$, and where the ellipse contacts its minor axis, the sum of the distances to the foci is $2\sqrt{a^2 + f^2}$. These two entities being equal, we find $f = \sqrt{b^2 - a^2}$.

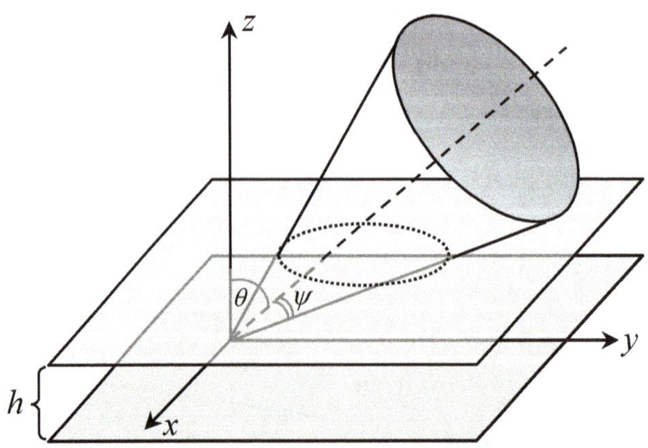

Figure 2.9. The ellipse is the cross-section of a right-circular cone and a plane.

Solution: We solve this problem using the methods of analytic geometry. The vector along the cone axis connecting the apex to the point where the axis crosses the plane is $\boldsymbol{A} = h(\tan\theta\,\hat{\boldsymbol{y}} + \hat{\boldsymbol{z}})$. An arbitrary point on the cross-sectional ellipse is denoted by $\boldsymbol{P} = x\hat{\boldsymbol{x}} + y\hat{\boldsymbol{y}} + h\hat{\boldsymbol{z}}$. Using the fact that $\boldsymbol{A}\cdot\boldsymbol{P} = AP\cos\psi$, we write

$$\boldsymbol{A}\cdot\boldsymbol{P} = h(y\tan\theta + h) = h\sqrt{1+\tan^2\theta}\sqrt{x^2+y^2+h^2}\cos\psi$$

$$\rightarrow\ (y\sin\theta + h\cos\theta)^2 = (x^2+y^2+h^2)\cos^2\psi$$

$$\rightarrow\ x^2\cos^2\psi + y^2(\cos^2\psi - \sin^2\theta) - 2hy\sin\theta\cos\theta = h^2(\cos^2\theta - \cos^2\psi)$$

$$\rightarrow\ x^2\cos^2\psi + (\cos^2\psi - \sin^2\theta)\left(y - \frac{h\sin\theta\cos\theta}{\cos^2\psi - \sin^2\theta}\right)^2 = \frac{(h\sin\psi\cos\psi)^2}{\cos^2\psi - \sin^2\theta}$$

$$\rightarrow\ \left(\frac{\cos^2\psi - \sin^2\theta}{h^2\sin^2\psi}\right)x^2 + \left(\frac{\cos^2\psi - \sin^2\theta}{h\sin\psi\cos\psi}\right)^2\left(y - \frac{h\sin\theta\cos\theta}{\cos^2\psi - \sin^2\theta}\right)^2 = 1.$$

Therefore, the center of the ellipse is at

$$(x_c,\ y_c) = \left(0,\ \frac{h\sin\theta\cos\theta}{\cos^2\psi - \sin^2\theta}\right),$$

its minor and major semi-axes are

$$a = \frac{h\sin\psi}{\sqrt{\cos^2\psi - \sin^2\theta}},$$

$$b = \frac{h\sin\psi\cos\psi}{\cos^2\psi - \sin^2\theta},$$

and its focal length is given by

$$f = \frac{h\sin\theta\sin\psi}{\cos^2\psi - \sin^2\theta}.$$

The extremal points along the major axis of the ellipse are at $y_\pm$, where

$$y_\pm = y_c \pm b = h\left(\frac{\sin\theta\cos\theta \pm \sin\psi\cos\psi}{\cos^2\psi - \sin^2\theta}\right) = h\left(\frac{\sin 2\theta \pm \sin 2\psi}{\cos 2\psi + \cos 2\theta}\right)$$

$$= h\left[\frac{\sin(\theta\pm\psi)\cos(\theta\mp\psi)}{\cos(0+\psi)\cos(0-\psi)}\right] = h\tan(\theta\pm\psi).$$

This result may be readily understood by a quick glance at the diagram in figure 2.9. The distance from the center of the ellipse to the crossing point of the cone axis is

$$y_c - y_A = \frac{h\sin\theta\cos\theta}{\cos^2\psi - \sin^2\theta} - h\tan\theta = \frac{h\tan\theta\sin^2\psi}{\cos^2\psi - \sin^2\theta}.$$

Finally, considering that $y_{f_\pm} = y_c \pm f = h\sin\theta/(\cos\theta \mp \sin\psi)$, it is seen that the cone axis does *not* cross the ellipse at one of its foci. The crossing point of the cone axis at $(x_A, y_A) = (0, h\tan\theta)$ has the following relation to the foci of the ellipse:

$$y_A^{-1} = ½(y_{f_+}^{-1} + y_{f_-}^{-1}).$$

2.16.1. Parabola as the limit of ellipse when $\theta + \psi \rightarrow 90°$. In this limit, one of the foci moves to infinity while the other approaches $y_f = ½y_A$. This may also be deduced from the fact that, when $\cos\theta = \sin\psi$, the remaining focus of the ellipse resides at $y_f = h\sin\theta/(\cos\theta + \sin\psi) = ½h\tan\theta$, which is equal to $½y_A$. The distance from the vertex to the focal point of the parabola is thus given by

$$F = ½h\tan\theta - h\tan(\theta - \psi) = ½h\tan\theta + h\cot 2\theta = ½h\cot\theta.$$

Now, in accordance with our earlier discussion, when $\theta + \psi \to 90°$, the equation of the conic section in the xy-plane at $z = h$ must be

$$2hy \sin\theta \cos\theta = (\cos^2\psi)x^2 - h^2(\cos^2\theta - \cos^2\psi) \;\to\; y = (½h^{-1}\tan\theta)x^2 - h\cot 2\theta.$$

The vertex of the parabola is located at $y = h\tan(\theta - \psi) = h\tan(2\theta - 90°) = -h\cot 2\theta$, which the preceding equation confirms. To determine the focal-length directly, recall that every point on the $y = cx^2$ parabola is equidistant from the focus at $(0, F)$ and the directrix $y = -F$, that is,

$$x^2 + (cx^2 - F)^2 = (cx^2 + F)^2 \quad \to \quad F = 1/(4c).$$

In the present case, since $c = ½h^{-1}\tan\theta$, we arrive at $F = ½h\cot\theta$, as predicted. The conclusion is that the focal point of the parabola coincides with one of the foci of the ellipse in the limit when $\theta + \psi \to 90°$.

Brain teaser. A narrow trail leads from a monastery at the base of a mountain to a temple at the mountaintop. A monk sets out from the monastery one day at dawn, walks up the trail at an uneven pace, reaching the temple at dusk. He spends the night at the temple, then leaves early morning the next day and returns to the monastery by sunset. Show that there exists a point along the trail where the monk has been at precisely the same time on both days.

Suggestions for further reading

1. C. G. Gibson, *Elementary Euclidean Geometry*, Cambridge University Press, United Kingdom (2003).
2. R. A. Johnson, *Advanced Euclidean Geometry*, Dover, New York (1960).
3. A. S. Posamentier and C. T. Salkind, *Challenging Problems in Geometry*, Dover, New York (1996).
4. B. Bold, *Famous Problems of Geometry and How to Solve Them*, Dover, New York (1969).
5. H. Rademacher and O. Toeplitz, *The Enjoyment of Mathematics*, Dover, New York (1990).

Problems

Problem 1) Prove the Pythagorean theorem $(a^2 + b^2 = c^2)$ using the following arrangement of four identical right-angled triangles:

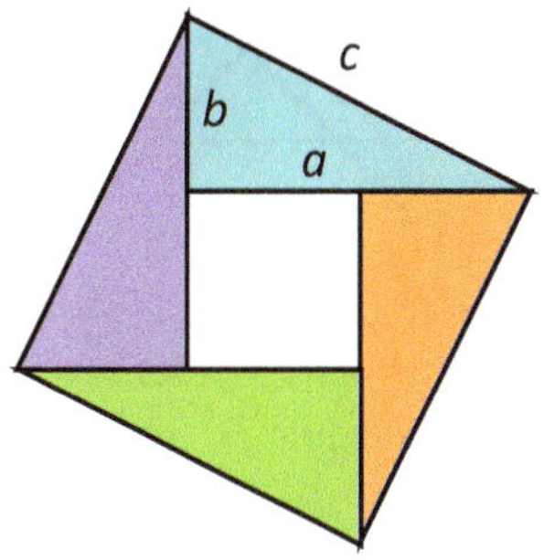

Problem 2) In a right triangle ABC, the median BB' splits the hypotenuse AC into equal halves, that is, $AB' = CB'$. Show that the length of this median is half the length of the hypotenuse, that is $BB' = \frac{1}{2}AC$.

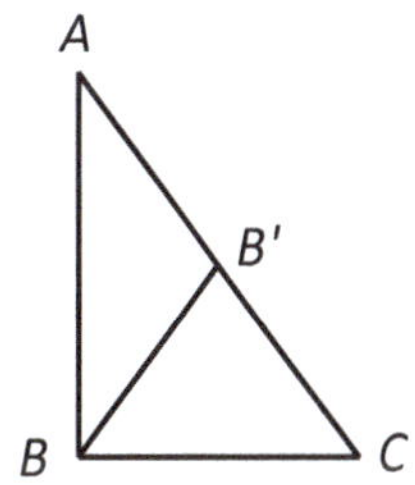

Problem 3) Consider the arbitrary triangle ABC. The median BB' splits the side AC into equal halves, AB' and $B'C$. Similarly, the median CC' splits the side AB into equal halves. Using the properties of similar triangles, show that the point Q at which the two medians intersect divides each median in a 2:1 ratio, that is, $BQ = 2QB'$ and $CQ = 2QC'$. Using this fact, argue that the third median, AA', must necessarily cross the other two medians at Q as well.

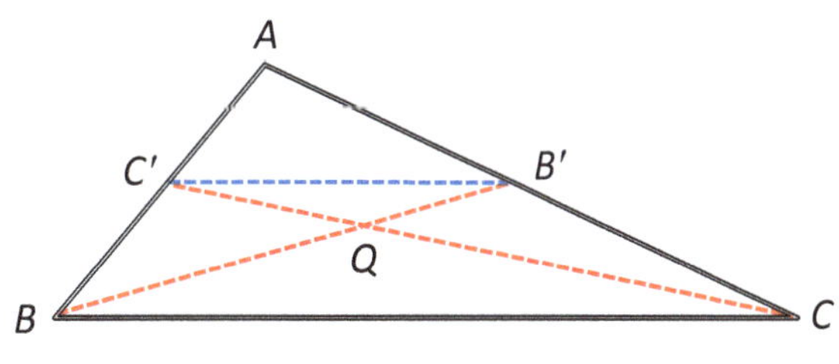

Problem 4) Let AD be the bisector of the angle A of the ABC triangle. Show that the ratio $\overline{BD} : \overline{DC}$ is the same as the ratio $\overline{AB} : \overline{AC}$.

Hint: You may want to draw from D a line parallel to AB and another line parallel to AC.

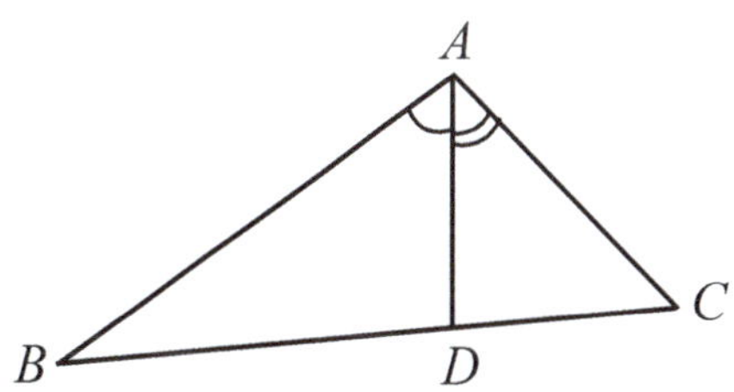

Problem 5) Using only a straight-edge and a compass, find the midpoint of a straight line-segment.

Problem 6) Shown below is a drunkard's zigzag path as he walks from P_1 to P_2—two points that are separated by a distance L along a straight line. The path is a periodic sawtooth function with amplitude A and period $p = L/N$, where N is an integer.

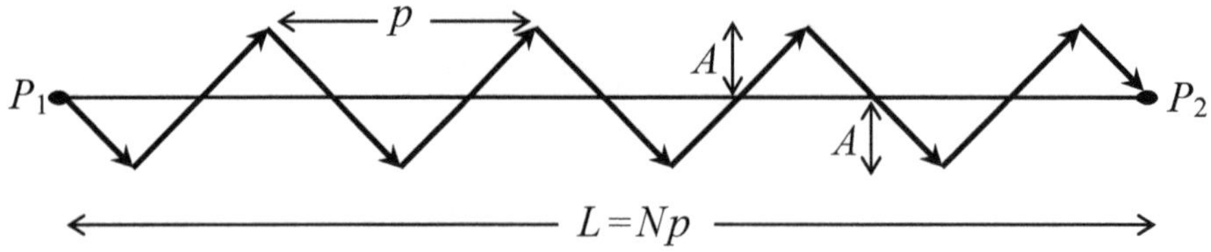

a) Write a formula for the total distance D traveled between P_1 and P_2 along the zigzag path.

b) What is the limit of D when $A \to 0$ provided that N is kept constant?

c) What is the limit of D when $A \to 0$ while $N \to \infty$ in such a way as to keep the product NA constant?

Problem 7) Consider an arbitrary triangle with sides a, b, and c.

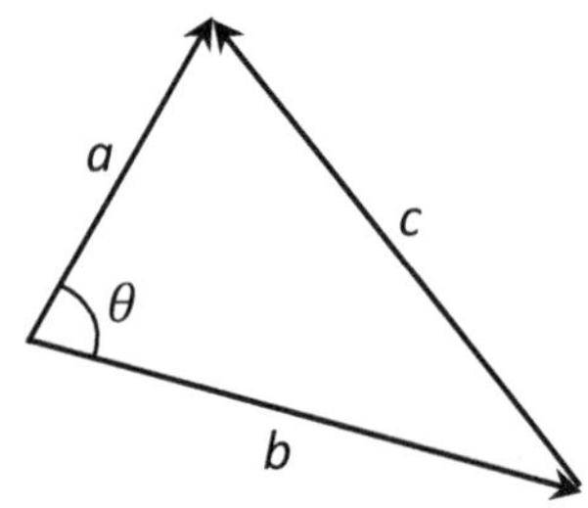

a) Treating each side of the triangle as an ordinary vector in three-dimensional Euclidean space, one may write $\vec{\boldsymbol{c}} = \vec{\boldsymbol{a}} - \vec{\boldsymbol{b}}$. Using elementary vector algebra, derive an expression for the length c in terms of a, b, and the angle θ between a and b.

b) According to the so-called Heron's theorem, the area A and the half-perimeter s of the triangle are related as follows:

$$A = \sqrt{s(s-a)(s-b)(s-c)}.$$

Here, $s = ½(a+b+c)$. Prove Heron's theorem using the result obtained in part (a) and the fact that $A = ½ab\sin\theta$.

Problem 8) a) Show that the area of the ABC triangle is one-half the product of the lengths of any two sides times the sine of the angle between those sides; that is,

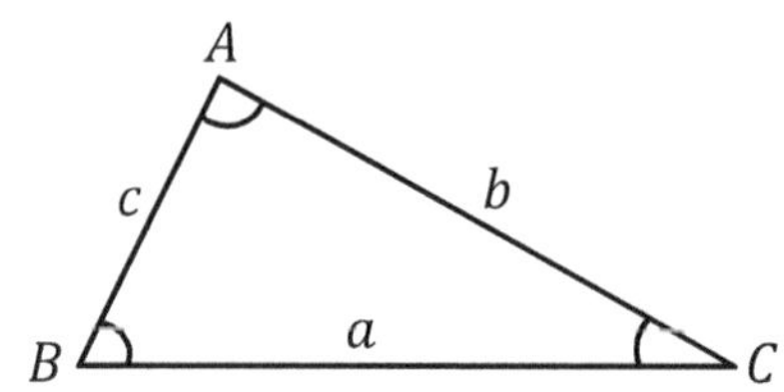

$$\text{Area of } ABC = ½ab\sin C = ½bc\sin A = ½ac\sin B.$$

b) Use the result of part (a) to prove the following identity for any triangle:

$$\frac{\sin A}{a} = \frac{\sin B}{b} = \frac{\sin C}{c}.$$

Problem 9) A circle of radius R is centered at the point O. The points A, B, and C are chosen arbitrarily on the perimeter of the circle, and the angles $\widehat{ABC}$ and $\widehat{AOC}$ are formed, as shown in the figure. Show that, in general, $\widehat{ABC} = ½\widehat{AOC}$.

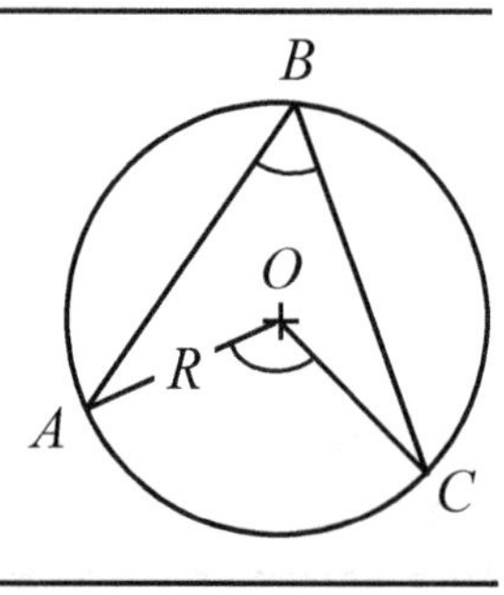

Hint: Draw the diameter of the circle that goes through the point B.

Problem 10) From an arbitrary point O outside a circle of radius R centered at C, a straight-line OAB is drawn such that it crosses the circle at points A and B, as shown.

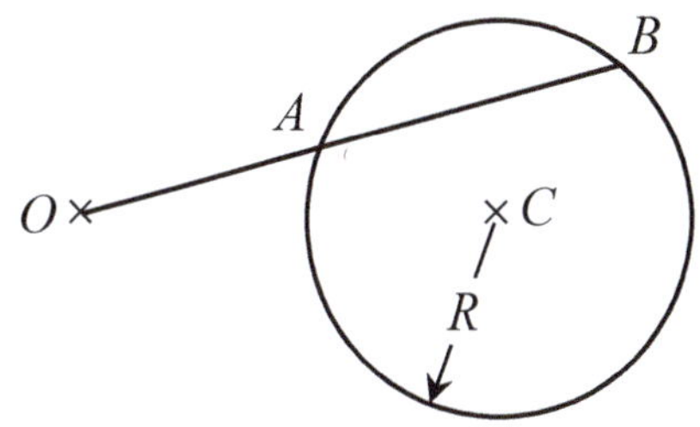

a) Show that the product $\overline{OA} \cdot \overline{OB}$ of the line-segments is the same for all straight lines drawn from O that cross the circle.

b) Show that $\overline{OA} \cdot \overline{OB} = \overline{OC}^2 - R^2$, where $\overline{OC}$ is the length of the straight-line connecting O to C.

c) Draw a tangent OD from the external point O to the circle, and verify that $\overline{OD} \cdot \overline{OD} = \overline{OC}^2 - R^2$.

Hint: Use the properties of similar triangles.

Problem 11) Figure (a) shows a person of height h standing on the beach on a clear day, facing the ocean and trying to see as far as the eye can see. Assuming that the radius of the earth R is much greater than the person's height h, find the distance d to the horizon that is visible to the human eye. (An approximate value for the radius of the earth is $R \cong 6{,}378$ km.)

b) Two small boats, B_1 and B_2, are within the line of sight of the observer, as shown in figure (b). B_1 is on the horizon, while B_2 is closer to the observer. Assuming the angular acuity of the human eye is α (a typical value is $\alpha \cong 0.0003$ radians), find the minimum angular separation θ between B_1 and B_2 that would make the two boats distinguishable to the observer.

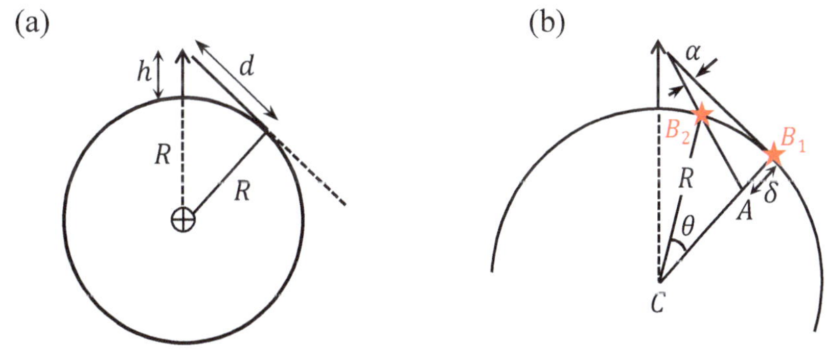

Problem 12) A regular n-sided polygon is constructed from n identical isosceles triangles, each having apex angle $\theta = 2\pi/n$. The sides of the polygon have length s, and the distance from the center of the polygon to each of its vertices is R.

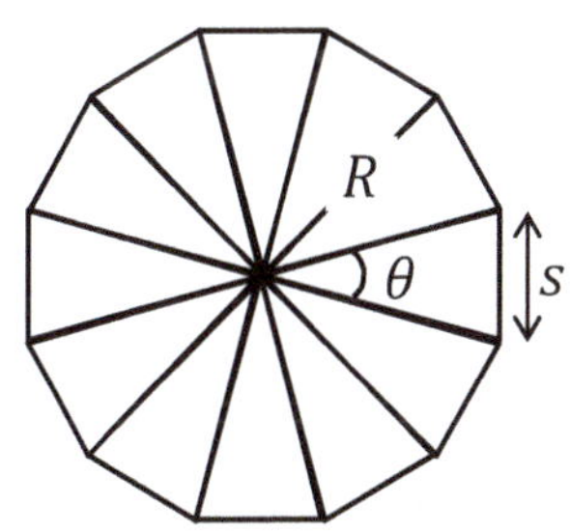

a) Express the area A and the perimeter P of the polygon as functions of R and n.

b) Keeping R as a fixed parameter, for what value of n does the polygon acquire the largest A/P ratio?

Problem 13) A solid equilateral triangle has side-length L. Let an area corresponding to a smaller equilateral triangle, this one of side-length $L/2$, be cut out from the middle of the larger triangle. There will be three smaller triangles left, from which we now remove the midsections in the form of three equilateral triangles, each having side-length $L/4$. This process continues indefinitely, so that, in the end, the original triangle will have an infinite number of holes in the form of a series of smaller and smaller equilateral triangles.

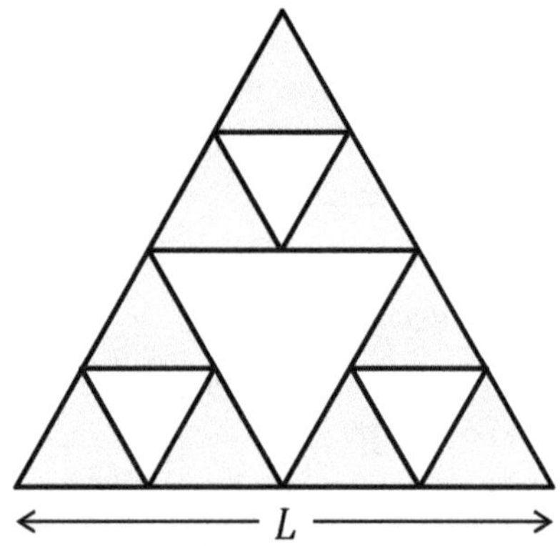

a) What fraction of the original triangle's area will have been removed in the end?

b) The perimeter of the original triangle is $3L$. Once the first triangle has been cut out, an inner boundary of length $3L/2$ will be created. Removing the next three triangles will create additional boundaries of length $9L/4$, and so on. Find the total length of the boundary after n systematic steps of removing the smaller triangles.

Problem 14) a) Show that the center-of-mass of a thin, solid, flat, and homogeneous triangle is at the point inside the triangle where the three medians of the triangle cross each other.

b) Explain why the solid triangle can be balanced (while remaining horizontal) on the sharp tip of a pencil placed immediately beneath the center-of-mass of the triangle.

Problem 15) a) Find the volume of a sphere of radius R by slicing it into N thin disks, each having a thickness of $2R/N$, then summing up the volumes of the individual disks and allowing N to approach infinity. (**Hint**: Use the sum $S_2(N)$ obtained in chapter 1, problem 7.)

b) Using the fact that the volume of a sphere of radius R is $V = 4\pi R^3/3$, show that the surface area of the sphere is given by $S = 4\pi R^2$.

Problem 16) The figure shows a polygon in the xy-plane, whose sides and corners are connected by straight lines to the point $z = h$ on the z-axis. The solid object thus formed is a pyramid whose base area is A and whose height is h.

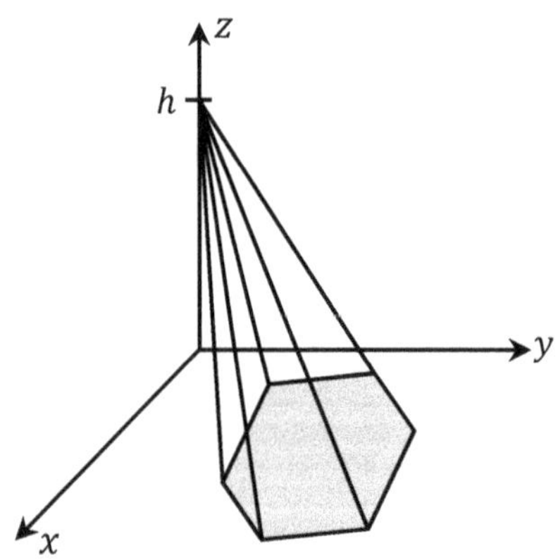

a) Find a formula for the volume V of the pyramid in terms of A and h.

b) How does the above formula for the volume V change if the pyramid is truncated at the top, by a plane that is parallel to the xy-plane and cuts through the pyramid at a height $z = \alpha h$, where α is an arbitrary number between 0 and 1.

Problem 17) Let the surface of a sphere of radius R be covered with a large number of small polygons of arbitrary shape, which fill the entire surface area without overlapping each other. Each polygon may be imagined to be the base of a pyramid whose vertex is at the center of the sphere. Given that the surface area of the sphere is $S = 4\pi R^2$, and that the volume of a pyramid is one-third of the product of its base area and height, show that the volume of the sphere is $V = 4\pi R^3/3$.

Problem 18) The figure shows the cross-sectional profile of a paraboloidal mirror in the xy-plane, whose shape is described by the function $y = ax^2$, with a being a positive real constant. A ray of light propagating along the unit-vector $\boldsymbol{R}_1 = -\hat{\boldsymbol{y}}$ arrives at the point (x_o, y_o) on the surface of the mirror. Also shown in the figure is the unit-vector $\boldsymbol{S}$ that is tangent to the parabola at the point of incidence. The reflected ray propagates along the unit-vector $\boldsymbol{R}_2$, crossing the axis of the paraboloid at a height y_1 above the mirror's vertex.

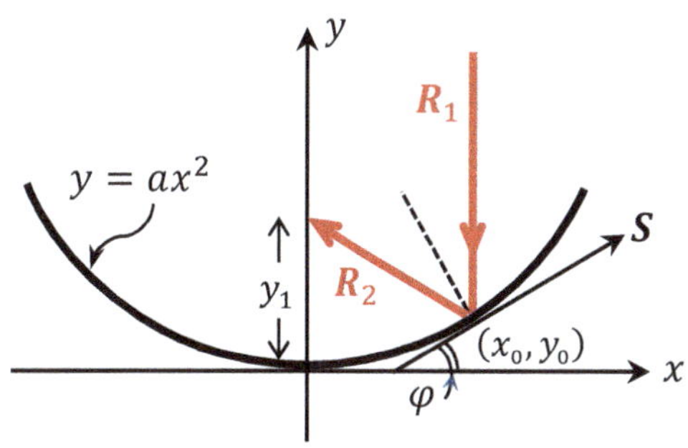

a) Write expressions for unit-vectors $\boldsymbol{S}$ and $\boldsymbol{R}_2$ in terms of the parameters a, x_o, y_o, y_1.

b) Considering that $\boldsymbol{R}_1$ and $\boldsymbol{R}_2$ make equal angles with the tangent vector $\boldsymbol{S}$, use the vector identity $\boldsymbol{v}_1 \cdot \boldsymbol{v}_2 = |\boldsymbol{v}_1||\boldsymbol{v}_2| \cos\theta$ to write an equation relating the various system parameters.

c) Solve the equation found in part (b) to determine the height y_1 of the reflected ray $\boldsymbol{R}_2$ on the axis of the paraboloidal mirror. (You will find that y_1 does *not* depend on the coordinates (x_o, y_o) of incidence. Thus obtained is the location y_1 of the focal point of the mirror as a function of the parameter a, which is related to the curvature of the mirror's surface.)

Problem 19) Let the angle α and the length $\overline{AA'}$ of one of its legs be specified as shown. From the point P outside the angle draw a straight line across the angle such that the perimeter of the ABC triangle thus formed equals twice the length $\overline{AA'}$.

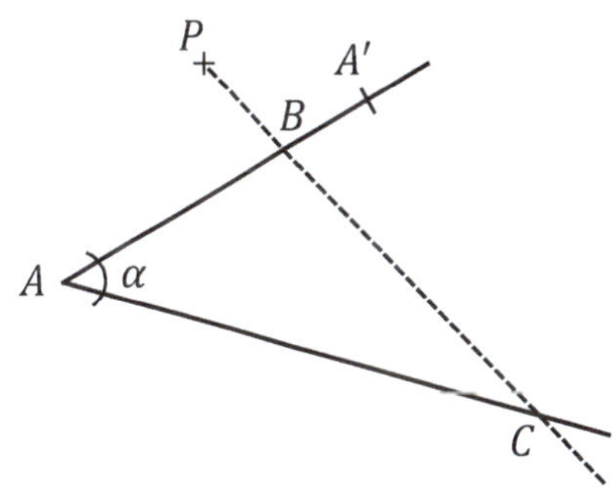

Problem 20) The isosceles triangle ABC has a 20° angle at its vertex A. Assuming that $\overline{AD} = \overline{BC}$, find the value of the angle θ. (**Hint**: The trigonometric solution of the problem is quite general and does not depend on the specific value chosen for the apex angle at A. However, a nice geometric solution exists for the 20° apex angle.)

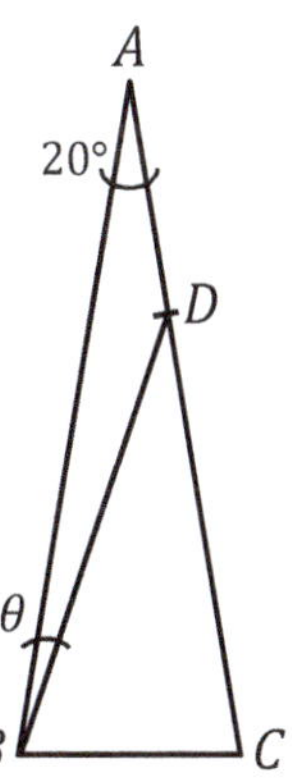

Chapter 3

Odds and Ends

"It's like asking why is Beethoven's Ninth Symphony beautiful. If you don't see why, someone can't tell you. I know numbers are beautiful. If they aren't beautiful, nothing is."

Paul Erdős (1913–1996)

3.1. Introduction. This chapter covers a number of important topics that are not directly related to one another but collectively play an important role in applied mathematics as they provide basic background knowledge and useful tools that should belong to the arsenal of every practitioner of physical and mathematical sciences.

Section 2 provides several examples in which methods of integration such as "change of variable" and "integration by parts" are used to evaluate certain integrals of functions of a real variable. In section 3 we introduce Euler's Gamma function $\Gamma(x)$, which is a continuous and differentiable function of the real variable x, whose value at integer points $x = n + 1$ equals the factorial of n; that is, $\Gamma(n + 1) = n!$. Section 4 is devoted to deriving approximate expressions for $n!$. Specifically, we derive tight upper and lower bounds on $n!$, commonly referred to as Stirling's formulas. Section 5 is an introduction to complex numbers and the arithmetic rules for addition, subtraction, multiplication, and division in the complex plane. This is followed in section 6 by a discussion of elementary functions defined over the complex plane (e.g., polynomials, exponentials, trigonometric functions, the natural logarithm, and so on). A few pathological functions exhibiting extreme and/or unusual characteristics are described in section 7. Finally, section 8 provides a brief introduction to modular arithmetic, which is of value in discrete mathematics, in general, and in the theories of digital signal processing and coding, in particular.

3.2. Methods of integration. "Changing the variable" and "integration by parts" are two standard methods that are frequently used when integrating a function over some specified interval on the real axis. For instance, to find $\int_{-\infty}^{\infty}(1 + x^2)^{-1}\mathrm{d}x$, one could resort to the change of variable $x = \tan\theta$, which yields $\mathrm{d}x = (1 + \tan^2\theta)\mathrm{d}\theta$, and, consequently,

$$\int_{x=-\infty}^{\infty} \frac{\mathrm{d}x}{1 + x^2} = \int_{\theta=-\pi/2}^{\pi/2} \left(\frac{1 + \tan^2\theta}{1 + \tan^2\theta}\right) \mathrm{d}\theta = \pi. \tag{1}$$

As for the method of integration by parts, the basic idea is to use the identity $[f(x)g(x)]' = f'(x)g(x) + f(x)g'(x)$ to convert an integral of $f'(x)g(x)$ to an integral of $f(x)g'(x)$, which might be easier to evaluate than the initial integral. One then writes

$$\begin{aligned}\int_{x_1}^{x_2} f'(x)g(x)\mathrm{d}x &= f(x)g(x)|_{x_1}^{x_2} - \int_{x_1}^{x_2} f(x)g'(x)\mathrm{d}x \\ &= [f(x_2)g(x_2) - f(x_1)g(x_1)] - \int_{x_1}^{x_2} f(x)g'(x)\mathrm{d}x. \end{aligned}\tag{2}$$

As an example, consider $\int_{-\infty}^{\infty} x^2 \exp(-x^2)\,\mathrm{d}x$. Let $f(x) = -½\exp(-x^2)$ and $g(x) = x$. Given that $f'(x) = x\exp(-x^2)$, we will have

$$\int_{-\infty}^{\infty} x^2 \exp(-x^2)\,\mathrm{d}x = -½ \overbrace{x\exp(-x^2)}^{\to 0}|_{-\infty}^{+\infty} + ½\int_{-\infty}^{\infty} \exp(-x^2)\,\mathrm{d}x = \sqrt{\pi}/2. \tag{3}$$

The same procedure, repeated n times, may be used to evaluate $\int_{-\infty}^{\infty} x^{2n} \exp(-x^2)\,\mathrm{d}x$.

Digression: Evaluation of $\int_{-\infty}^{\infty} \exp(-x^2)\,dx$, which was needed to arrive at the final result in Eq.(3), involves a well-known trick in which the integral is first raised to the power two, then converted to a two-dimensional integral in the xy-plane, which is subsequently evaluated by switching from Cartesian to polar coordinates, as follows:

$$\left[\int_{-\infty}^{\infty} \exp(-x^2)\,dx\right]^2 = \int_{-\infty}^{\infty} \exp(-x^2)\,dx \times \int_{-\infty}^{\infty} \exp(-y^2)\,dy$$

$$= \iint_{-\infty}^{\infty} \exp[-(x^2+y^2)]\,dxdy = \int_{r=0}^{\infty} \int_{\varphi=0}^{2\pi} \exp(-r^2) r d\varphi dr$$

$$= 2\pi \int_0^{\infty} \exp(-r^2)\, r dr = -\pi \exp(-r^2)|_0^{\infty} = \pi.$$

We conclude that $\int_{-\infty}^{\infty} \exp(-x^2)\,dx = \sqrt{\pi}$.

We now present a few exmaples in which the methods of "integration by parts" and "change of variable" are used to evaluate several non-trivial integrals.

Example 1) We will show that $\int_0^1 [x/(1-x)] \ln x\,dx = 1 - (\pi^2/6)$. In what follows, we use the geometric series $\sum_{n=0}^{\infty} x^n = 1/(1-x)$, valid for $|x| < 1$, which was derived in chapter 1, section 8. We also need the Basel series $\sum_{n=1}^{\infty}(1/n^2) = \pi^2/6$, which will be discussed in detail in chapter 4, section 8·

Solution: $$\int_0^1 \frac{x \ln x}{1-x} dx = \int_0^1 x \ln x \sum_{n=0}^{\infty} x^n\, dx = \sum_{n=0}^{\infty} \int_0^1 x^{n+1} \ln x\, dx \leftarrow \boxed{\text{Integration by parts}}$$

$$= -\sum_{n=0}^{\infty} 1/(n+2)^2 = 1 - \sum_{n=1}^{\infty} 1/n^2 = 1 - (\pi^2/6).$$

Example 2) The goal here is to show that $\int_0^1 [x/(1+x)] \ln x\,dx = (\pi^2/12) - 1$. Once again, we will need the geometric series $\sum_{n=0}^{\infty}(-x)^n = 1/(1+x)$ and also a variant of the Basel series $\sum_{n=1}^{\infty} (-1)^{n+1}/n^2 = \pi^2/12$ (see chapter 4, section 8).

Solution: $$\int_0^1 \frac{x \ln x}{1+x} dx = \int_0^1 x \ln x \sum_{n=0}^{\infty}(-x)^n\, dx = \sum_{n=0}^{\infty}(-1)^n \int_0^1 x^{n+1} \ln x\, dx$$

$$\boxed{\text{Integration by parts}} \rightarrow = \sum_{n=0}^{\infty} \frac{(-1)^{n+1}}{(n+2)^2} = \left[\sum_{n=1}^{\infty} \frac{(-1)^{n+1}}{n^2}\right] - 1 = \frac{\pi^2}{12} - 1.$$

Example 3) To prove that $\int_0^1 (1-x^2)^{-1} \ln x\,dx = -¼ \int_{-\infty}^{\infty}(x/\sinh x)dx = -\pi^2/8$, we will use the geometric series $\sum_{n=0}^{\infty} x^{2n} = 1/(1-x^2)$, valid for $|x| < 1$, and a variant of the Basel series $\sum_{n=1}^{\infty} 1/(2n-1)^2 = \pi^2/8$ (see chapter 4, section 8).

Solution: $$\int_0^1 \frac{\ln x}{1-x^2} dx = \int_0^1 \ln x \sum_{n=0}^{\infty} x^{2n}\, dx = \sum_{n=0}^{\infty} \int_0^1 x^{2n} \ln x\, dx \leftarrow \boxed{\text{Integration by parts}}$$

$$= -\sum_{n=0}^{\infty} 1/(2n+1)^2 = -\pi^2/8.$$

Relating the above integral to $\int_{-\infty}^{\infty}(x/\sinh x)dx$ requires the change of variable $y = \ln x$, which yields $dy = (1/x)dx$, or $dx = \exp(y)\,dy$. We will have

$$\int_0^1 \frac{\ln x}{1-x^2} dx = \int_{-\infty}^0 \frac{y \exp(y)}{1-\exp(2y)} dy = \int_{-\infty}^0 \frac{y}{\exp(-y)-\exp(y)} dy = -½ \int_{-\infty}^0 (y/\sinh y) dy$$

$$= -¼ \int_{-\infty}^{\infty}(y/\sinh y)dy \quad \rightarrow \quad \int_{-\infty}^{\infty}(x/\sinh x)dx = \pi^2/2.$$

Example 4) We will show that $\int_0^\infty \{x\exp(-x)/[\exp(x)-1]\}\mathrm{d}x = (\pi^2/6)-1$. Once again, we invoke the geometric series and also the Basel series $\sum_{n=1}^\infty (1/n^2) = \pi^2/6$.

Solution: $\displaystyle\int_0^\infty \frac{x\exp(-x)}{\exp(x)-1}\mathrm{d}x = \int_0^\infty \frac{x\exp(-2x)}{1-\exp(-x)}\mathrm{d}x = \int_0^\infty x\exp(-2x)\sum_{n=0}^\infty \exp(-nx)\,\mathrm{d}x$

$$= \sum_{n=2}^\infty \int_0^\infty x\exp(-nx)\mathrm{d}x = \sum_{n=2}^\infty\left[-\cancelto{0}{(x/n)\exp(-nx)}|_0^\infty + (1/n)\int_0^\infty \exp(-nx)\mathrm{d}x\right]$$

$$= -\sum_{n=2}^\infty \exp(-nx)/n^2|_0^\infty = \sum_{n=2}^\infty 1/n^2 = (\pi^2/6)-1.$$

A powerful method of integration by change of the variable involves substituting the new variable $x=\tan(\theta/2)$ for the old variable θ. The following examples demonstrate the application of this method, which is also known as the *Weierstrass substitution*.

Example 5) To evaluate the indefinite integral $\int \frac{\mathrm{d}\theta}{\sin\theta}$, we write $\sin\theta = \frac{2\tan(\theta/2)}{1+\tan^2(\theta/2)}$, then make the substitution $x=\tan(\theta/2)$, with $\mathrm{d}x = ½[1+\tan^2(\theta/2)]\mathrm{d}\theta$. Consequently,

$$\int\frac{\mathrm{d}\theta}{\sin\theta} = \int\frac{2/(1+x^2)}{2x/(1+x^2)}\mathrm{d}x = \int\frac{\mathrm{d}x}{x} = \ln x + C = \ln[\tan(\theta/2)] + C = \ln\left(\frac{\sin\theta}{1+\cos\theta}\right) + C.$$

Example 6) Consider the indefinite integral $\int \mathrm{d}\theta/(a+b\cos\theta)$, where a and b are arbitrary (real-valued) constants. Given that $\cos\theta = [1-\tan^2(\theta/2)]/[1+\tan^2(\theta/2)]$, the Weierstrass substitution $x=\tan(\theta/2)$, which leads to $\mathrm{d}x = ½[1+\tan^2(\theta/2)]\mathrm{d}\theta$, yields

$$\int\frac{\mathrm{d}\theta}{a+b\cos\theta} = \int\frac{2/(1+x^2)}{a+b(1-x^2)/(1+x^2)}\mathrm{d}x = \int\frac{2\mathrm{d}x}{(a+b)+(a-b)x^2} = \frac{2}{a+b}\int\frac{\mathrm{d}x}{1+[(a-b)/(a+b)]x^2}.$$

Case 1: $a=-b\neq 0$. $\quad\displaystyle\int\frac{\mathrm{d}\theta}{a+b\cos\theta} = \frac{1}{a}\int\frac{\mathrm{d}x}{x^2} = -\frac{1}{ax} + C = -(1/a)\cot(\theta/2) + C.$

Case 2: $a=b\neq 0$. $\quad\displaystyle\int\frac{\mathrm{d}\theta}{a+b\cos\theta} = \frac{1}{a}\int\mathrm{d}x = \frac{x}{a} = (1/a)\tan(\theta/2) + C.$

Case 3: $\left(\frac{a-b}{a+b}\right) > 0$. Changing the variable to φ, where $\tan\varphi = \sqrt{\frac{a-b}{a+b}}\,x$, we find

$$\int\frac{\mathrm{d}\theta}{a+b\cos\theta} = \frac{2}{a+b}\int\frac{\sqrt{(a+b)/(a-b)}(1+\tan^2\varphi)}{1+\tan^2\varphi}\mathrm{d}\varphi = \frac{2\varphi}{\mathrm{sign}(a+b)\sqrt{|a^2-b^2|}} + C$$

$$= \frac{2\,\mathrm{sign}(a+b)}{\sqrt{|a^2-b^2|}}\tan^{-1}\left[\sqrt{(a-b)/(a+b)}\tan(\theta/2)\right] + C.$$

Case 4: $\left(\frac{a-b}{a+b}\right) < 0$.

$$\int\frac{\mathrm{d}\theta}{a+b\cos\theta} = \frac{2}{a+b}\int\frac{\mathrm{d}x}{1-(|a-b|/|a+b|)x^2} = \frac{1}{a+b}\int\left(\frac{1}{1+\sqrt{|a-b|/|a+b|}\,x} + \frac{1}{1-\sqrt{|a-b|/|a+b|}\,x}\right)\mathrm{d}x$$

$$= \frac{\mathrm{sign}(a+b)}{\sqrt{|a^2-b^2|}}\ln\left(\frac{1+\sqrt{|a-b|/|a+b|}\,x}{1-\sqrt{|a-b|/|a+b|}\,x}\right) + C$$

$$= \frac{\text{sign}(a+b)}{\sqrt{|a^2-b^2|}} \ln\left[\frac{1+\sqrt{|a-b|/|a+b|}\ \tan(\theta/2)}{1-\sqrt{|a-b|/|a+b|}\ \tan(\theta/2)}\right] + C.$$

$$= \frac{2\ \text{sign}(a+b)}{\sqrt{|a^2-b^2|}} \tanh^{-1}\left[\sqrt{|a-b|/|a+b|}\tan(\theta/2)\right] + C.$$

In all cases, one must pay close attention to the range of integration and the singularity of $\tan(\theta/2)$ at $\theta = \pm\pi$.

Brain teaser. A small boat sits in the middle of a pool. A large rock is in the boat. If the rock is thrown into the pool, does the water level rise, fall, or remain the same?

3.3. Euler's Gamma function. The Gamma function $\Gamma(x)$ is defined for positive values of x as follows:

$$\Gamma(x) = \int_0^\infty y^{x-1} e^{-y} \mathrm{d}y; \qquad x > 0. \tag{4}$$

Using the method of integration by parts, where $f(y) = y^x/x$ and $g(y) = \exp(-y)$, we find

$$\Gamma(x) = (y^x e^{-y}/x)|_{y=0}^{\infty} + (1/x)\int_0^\infty y^x e^{-y} \mathrm{d}y = (1/x)\Gamma(x+1). \tag{5}$$

Consequently, $\Gamma(x+1) = x\Gamma(x)$ for all positive values of x. It is easy to see from Eq.(4) that $\Gamma(1) = \int_0^\infty e^{-y}\mathrm{d}y = 1$. Therefore, $\Gamma(2) = \Gamma(1) = 1$, $\Gamma(3) = 2\Gamma(2) = 2$, $\Gamma(4) = 3\Gamma(3) = 3 \cdot 2 = 3!$, $\cdots$, $\Gamma(n) = (n-1)\Gamma(n-1) = (n-1)!$. Recalling that, by definition, $0! = 1$, we now see in $\Gamma(x)$ a continuous differentiable function, defined for all values of $x > 0$, with the important property that, for $n = 0, 1, 2, 3, \cdots$, its value at $x = n + 1$ precisely equals $n!$. From Eq.(4) we find $lim_{x\to 0}\,\Gamma(x) = \int_0^\infty y^{-1}e^{-y}\mathrm{d}y = \infty$.

It is possible to extend the definition of $\Gamma(x)$ to negative values of x as well, simply by allowing Eq.(5) to *define* $\Gamma(x)$ over the interval $-1 < x < 0$. Since the values of $\Gamma(x+1)$ for this range of x are already given by Eq.(4), one readily obtains $\Gamma(x)$ over the interval $(-1, 0)$. But now the same procedure can be used to determine $\Gamma(x)$ for the values of x belonging to the interval $(-2, -1)$. The process thus continues indefinitely, until $\Gamma(x)$ is extended to the entire negative x-axis. A plot of $\Gamma(x)$ for positive as well as negative values of x is shown in figure 3.1.

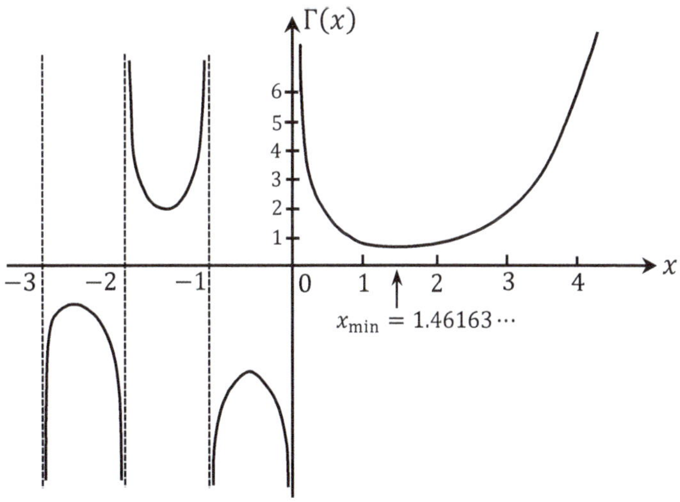

Figure 3.1. A plot of Euler's Gamma function $\Gamma(x)$.

Not only is $\Gamma(x)$ a continuous and differentiable function of the real variable x which, for all positive integers n, smoothly interpolates between adjacent values of $n!$, and not only does it have important applications in the theory of numbers, but also it shows up frequently in many applied problems of science and engineering. Chapter 10 describes certain interesting as well as useful properties of the Gamma function.

3.4. Stirling's approximation. In many applications, one finds it necessary to have a good estimate for the number $N!$ corresponding to some large integer N. Stirling's formula provides the following tight upper and lower bounds on the value of $N!$:

$$(N/e)^N\sqrt{N}e^{7/8} < N! < (N/e)^N\sqrt{N}e. \qquad (6)$$

It is not difficult to obtain a lower bound on $N!$ by approximating the area under the function $f(x) = \ln x$ over the interval $1 \le x \le N$ using the rectangular areas shown in figure 3.2(a). We will find

$$\ln 2 + \ln 3 + \ln 4 + \cdots + \ln N > \int_1^N \ln x\, dx = (x\ln x - x)|_1^N = N\ln N - N + 1$$

$$\rightarrow \quad \ln(N!) > \ln(N^N) - N + 1 \quad \rightarrow \quad N! > N^N e^{1-N} = (N/e)^N e. \qquad (7)$$

Similarly, an upper bound on $N!$ can be found by approximating the same area under the function $f(x) = \ln x$ using trapezoids that fit beneath the function, as shown in figure 3.2(b). We will find

$$\tfrac{1}{2}\ln 2 + \tfrac{1}{2}(\ln 2 + \ln 3) + \tfrac{1}{2}(\ln 3 + \ln 4) + \cdots + \tfrac{1}{2}[\ln(N-1) + \ln N] < \int_1^N \ln x\, dx$$

$$\rightarrow \quad \ln(N!) - \tfrac{1}{2}\ln N < (x\ln x - x)|_1^N = N\ln N - N + 1$$

$$\rightarrow \quad N! < N^{(N+\frac{1}{2})}e^{1-N} = (N/e)^N\sqrt{N}e. \qquad (8)$$

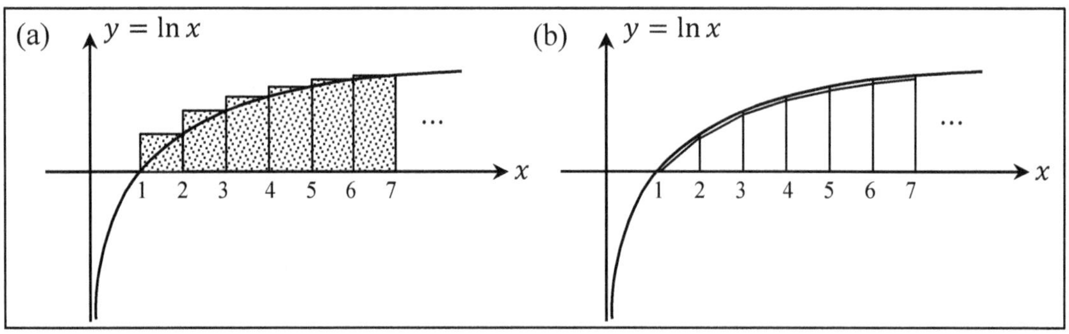

Figure 3.2. (a) The area under the function $f(x) = \ln x$ over the interval $1 \le x \le N$ is less than the areas of the shaded rectangles. (b) The area under the function $f(x) = \ln x$ over the same interval is greater than the areas of the trapezoids beneath the function.

The lower bound obtained in Eq.(7) can be tightened substantially if we shift the rectangles depicted in figure 3.2(a) by ½ unit to the right, as shown in figure 3.3. We will then need to add a small triangle to cover the region $1 \le x \le 1.5$. Considering that the slope of $\ln x$ at $x = 1$ equals 1, it is seen immediately that the height of the small triangle is equal to its base and that, therefore, the triangle's area equals ⅛. Also, since the slope of $\ln x$ continually declines with an increasing x, it is not difficult to see in figure 3.3 that each rectangular area is greater than the corresponding area under $\ln x$. We thus have

$$⅛ + \ln 2 + \ln 3 + \ln 4 + \cdots + \ln(N-1) + ½\ln N > \int_1^N \ln x\, dx = N\ln N - N + 1$$

$$\rightarrow \ln(N!) - ½\ln N + ⅛ > \ln(N^N) - N + 1$$

$$\rightarrow N! > N^{(N+½)}e^{⅞-N} = (N/e)^N\sqrt{N}e^{⅞}. \quad (9)$$

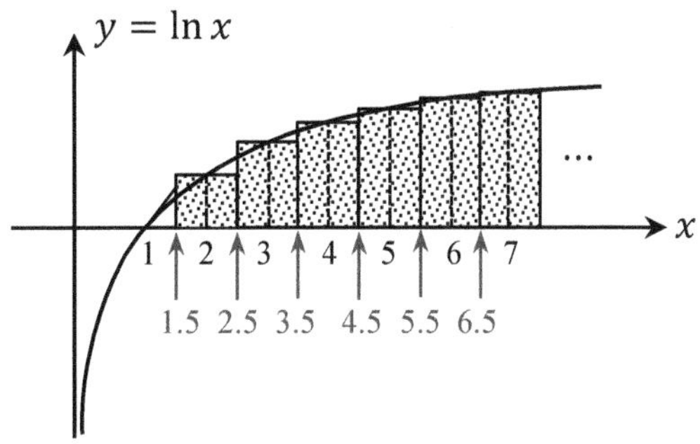

Figure 3.3. The area under the function $f(x) = \ln x$ over the interval $[1, N]$ is less than the corresponding rectangular areas plus the area of the small triangle shown over the region $1 \le x \le 1.5$. The rectangular area between $x = n - ½$ and $x = n$ is greater than the area under $\ln x$, whereas the rectangular area between $x = n$ and $x = n + ½$ is less than the corresponding area under $\ln x$. However, since the slope of $\ln x$ equals $1/x$, which is a decreasing function of x, the total area of the rectangle between $x = n - ½$ and $x = n + ½$ remains greater than $\int_{n-½}^{n+½} \ln x\, dx$.

Equations (8) and (9) thus yield the tight upper and lower bounds given in Eq.(6).

Brain teaser. A short piece of rope is known to take exactly one hour to burn from one end to the other, but the burn rate is uneven. Given two such pieces of rope and a cigarette lighter, how does one go about measuring a 45-minute interval?

3.5. Numbers in the complex plane. A first-order polynomial equation such as $ax + b = 0$, where a and b are real-valued constants, can always be solved for x, yielding $x = -b/a$, which is also a real number (i.e., a point identified on the real axis x). In contrast, a second-order polynomial equation such as $ax^2 + bx + c = 0$, where a, b, and c are real-valued constants, may or may not have solutions that can be identified as points on the real axis x. A simple example is the equation $x^2 + 1 = 0$, in which the quadratic expression on the left-hand side is always greater than the zero on the right-hand side, irrespective of the chosen value of x. In general, one can always *construct* an nth order polynomial equation with real coefficients, say, $a_n(x - x_1)(x - x_2)\cdots(x - x_n) = 0$, which has n real-valued solutions, namely, $x_1, x_2, \cdots, x_n$. However, an arbitrary n^{th} order polynomial equation with real coefficient such as $a_n x^n + a_{n-1}x^{n-1} + \cdots + a_2 x^2 + a_1 x + a_0 = 0$ may or may not have n real-valued solutions.

The main motivation behind the invention of complex numbers was to generalize the concept of "number" to allow polynomial equations of any (finite) order n to admit precisely n solutions. For this to happen, the "number" could no longer be a point confined to the real axis x, but had to *move out* of the x-axis and become a point identified by its Cartesian (x, y) coordinates in a two-dimensional xy-plane. Once the number became a point in an xy-plane, it had to obey certain rules for addition, subtraction, multiplication, and division, which not only had to be powerful enough to allow the solution of polynomial equations, but also had to reduce to the ordinary rules of conventional arithmetic if the numbers being added, subtracted, multiplied, or divided happened to lie on the x-axis of the (newly invented) xy-plane. The generalized rules of arithmetic for the numbers $z_1 = (x_1, y_1)$ and $z_2 = (x_2, y_2)$ that the mathematicians finally settled on can be summarized as follows:

Addition: $z_1 + z_2 = (x_1 + x_2,\ y_1 + y_2)$. (10)

Subtraction: $z_1 - z_2 = (x_1 - x_2,\ y_1 - y_2)$. (11)

Multiplication: $z_1 z_2 = (x_1 x_2 - y_1 y_2,\ x_1 y_2 + y_1 x_2)$. (12)

Division: $z_1/z_2 = z_3 \rightarrow z_1 = z_2 z_3 \rightarrow (x_1, y_1) = (x_2 x_3 - y_2 y_3,\ x_2 y_3 + y_2 x_3)$

$$\rightarrow \begin{pmatrix} x_1 \\ y_1 \end{pmatrix} = \begin{pmatrix} x_2 & -y_2 \\ y_2 & x_2 \end{pmatrix} \begin{pmatrix} x_3 \\ y_3 \end{pmatrix} \rightarrow \begin{pmatrix} x_3 \\ y_3 \end{pmatrix} = \begin{pmatrix} x_2 & -y_2 \\ y_2 & x_2 \end{pmatrix}^{-1} \begin{pmatrix} x_1 \\ y_1 \end{pmatrix}$$

$$\rightarrow \begin{pmatrix} x_3 \\ y_3 \end{pmatrix} = \frac{1}{x_2^2 + y_2^2} \begin{pmatrix} x_2 & y_2 \\ -y_2 & x_2 \end{pmatrix} \begin{pmatrix} x_1 \\ y_1 \end{pmatrix} = \frac{1}{x_2^2 + y_2^2} \begin{pmatrix} x_1 x_2 + y_1 y_2 \\ x_2 y_1 - x_1 y_2 \end{pmatrix}$$

$$\rightarrow \quad z_3 = (x_3, y_3) = \left(\frac{x_1 x_2 + y_1 y_2}{x_2^2 + y_2^2}, \frac{x_2 y_1 - x_1 y_2}{x_2^2 + y_2^2} \right). \tag{13}$$

It is not difficult now to verify that, in the special case when both z_1 and z_2 reside on the x-axis, that is, when $z_1 = (x_1, 0)$ and $z_2 = (x_2, 0)$, these rules reduce to those of ordinary arithmetic for the real numbers x_1 and x_2. It is also trivial to see that the new rules become easier to work with (and also easier to remember) if one *invented* a symbol, say, i, and *assigned* it the property $\mathrm{i}^2 = -1$, then proceeded to write $z = (x, y) = x + \mathrm{i}y$ and extended the conventional distributive, associative, and commutative properties of algebraic operations to symbolic operations on z_1 and z_2, as follows:

$$z_1 \pm z_2 = (x_1 + \mathrm{i}y_1) \pm (x_2 + \mathrm{i}y_2) = (x_1 + x_2) \pm \mathrm{i}(y_1 + y_2). \tag{14}$$

$$\begin{aligned} z_1 z_2 = (x_1 + \mathrm{i}y_1)(x_2 + \mathrm{i}y_2) &= x_1 x_2 + \mathrm{i}^2 y_1 y_2 + \mathrm{i}x_1 y_2 + \mathrm{i}y_1 x_2 \\ &= (x_1 x_2 - y_1 y_2) + \mathrm{i}(x_1 y_2 + y_1 x_2). \end{aligned} \tag{15}$$

Division, being the inverse of multiplication, is readily seen from Eq.(13) to be achievable by multiplying both the numerator (x_1, y_1) and the denominator (x_2, y_2) with a modified version $(x_2, -y_2)$ of the denominator. Thus, in our new symbolic representation, division of z_1 by z_2 may proceed as follows:

$$\begin{aligned} z_1/z_2 = \frac{x_1 + \mathrm{i}y_1}{x_2 + \mathrm{i}y_2} = \frac{(x_1+\mathrm{i}y_1)(x_2-\mathrm{i}y_2)}{(x_2+\mathrm{i}y_2)(x_2-\mathrm{i}y_2)} &= \frac{x_1x_2 - \mathrm{i}^2 y_1 y_2 + \mathrm{i}(x_2 y_1 - x_1 y_2)}{x_2^2 - \mathrm{i}^2 y_2^2 + \mathrm{i}(x_2 y_2 - x_2 y_2)} \\ &= \frac{(x_1 x_2 + y_1 y_2) + \mathrm{i}(x_2 y_1 - x_1 y_2)}{x_2^2 + y_2^2}. \end{aligned} \tag{16}$$

The operation in which the number $z = (x, y) = x + \mathrm{i}y$ is converted to a new number $z = (x, -y) = x - \mathrm{i}y$ is called "conjugation," which is formally represented by placing an asterisk (*) over the number being conjugated; that is,

$$z^* = (x, y)^* = (x + \mathrm{i}y)^* = x - \mathrm{i}y. \tag{17}$$

With this notation, one might now write $z_1/z_2 = z_1 z_2^*/(z_2 z_2^*)$. For historical reasons, the number $z = (x, y) = x + \mathrm{i}y$ is called a "complex" number, with its first component x being referred to as the "real" part of z, and its second component y being called the "imaginary" part of z. Of course, there is nothing "real" or "imaginary" about either of these numbers; both x and y are ordinary numbers (i.e., integers, rationals, or irrationals), which are bound together in the pair (x, y) and obligated to follow the specific rules of addition, subtraction, multiplication, and division given by Eqs.(10)–(13). The symbol i, which stands for imaginary, is simply *assigned* the property $\mathrm{i}^2 = -1$ to facilitate the implementation of arithmetic operations on complex numbers.

Once a point z is identified in the complex xy-plane, it can be represented by its Cartesian coordinates (x, y), or by its polar coordinates (r, φ), where $r = \sqrt{x^2 + y^2}$ and $\varphi = \tan^{-1}(y/x)$. The complex number z thus becomes associated with an arrow in the xy-plane that connects the origin $(x, y) = (0, 0)$ to the point z, as shown in figure 3.4. The length r of the arrow is now said to represent the "magnitude" of the complex number z, which, by convention, is written as $|z| = r$. Also by convention, the angle φ of the arrow, which is measured from the positive x-axis in the counterclockwise direction, is said to represent the "phase" of z.

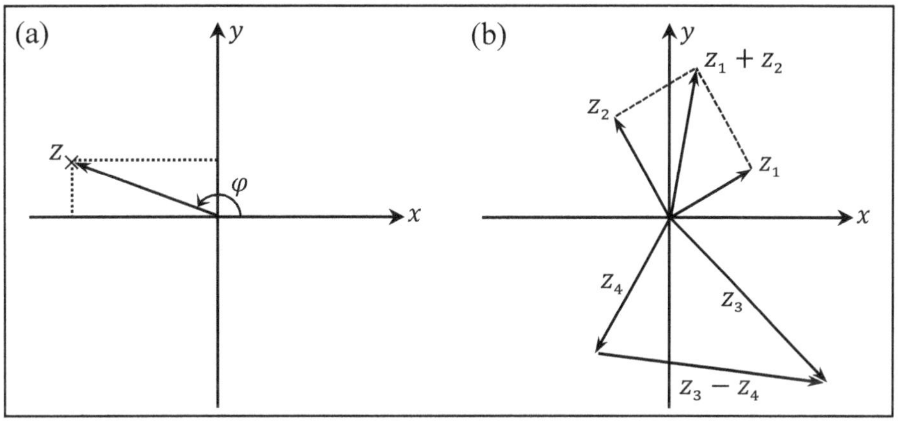

Figure 3.4. (a) The number $z = x + \mathrm{i}y$ in the complex z-plane is represented by an arrow pointing from the origin $(0, 0)$ of coordinates to the point z. (b) The sum of the complex numbers z_1 and z_2 is the arrow drawn from the origin to the corner of a parallelogram formed by the arrows of z_1 and z_2. The difference between the complex numbers z_3 and z_4 is an arrow drawn from the end of z_4 to the end of z_3. To locate the point $z_3 - z_4$ in the complex plane, one translates the corresponding arrow until its starting point coincides with the origin $(0, 0)$ of the xy-plane.

Invoking the polar representation, the complex number $z = x + \mathrm{i}y$ can be written equivalently as $z = |z|(\cos\varphi + \mathrm{i}\sin\varphi)$. Using the Taylor series expansions of $\cos\varphi$ and $\sin\varphi$ around the point $\varphi = 0$, and recalling that $\mathrm{i}^2 = -1$, $\mathrm{i}^3 = -\mathrm{i}$, etc., we may write

$$\cos\varphi + \mathrm{i}\sin\varphi = \left(1 - \frac{\varphi^2}{2!} + \frac{\varphi^4}{4!} - \cdots\right) + \mathrm{i}\left(\varphi - \frac{\varphi^3}{3!} + \frac{\varphi^5}{5!} - \cdots\right)$$

$$= 1 + (\mathrm{i}\varphi) + \frac{(\mathrm{i}\varphi)^2}{2!} + \frac{(\mathrm{i}\varphi)^3}{3!} + \frac{(\mathrm{i}\varphi)^4}{4!} + \cdots = \exp(\mathrm{i}\varphi). \qquad (18)$$

Equation (18) thus establishes the famous Euler identity, $\exp(\mathrm{i}\varphi) = \cos\varphi + \mathrm{i}\sin\varphi$, which allows one to write $z = |z|\exp(\mathrm{i}\varphi)$. Considering that the conjugation operation results in a sign change of φ while leaving the magnitude $|z|$ of the complex number intact, we can readily see that

$$z^* = (x + \mathrm{i}y)^* = x - \mathrm{i}y = |z|(\cos\varphi - \mathrm{i}\sin\varphi) = |z|\exp(-\mathrm{i}\varphi). \qquad (19)$$

One can also demonstrate that the multiplication and division rules in Eqs.(12) and (13) lead to $z_1 z_2 = |z_1||z_2|\exp[\mathrm{i}(\varphi_1 + \varphi_2)]$ and $z_1/z_2 = (|z_1|/|z_2|)\exp[\mathrm{i}(\varphi_1 - \varphi_2)]$. In practice, it is sometimes convenient to work in the Cartesian representation $z = x + \mathrm{i}y$, and at other times it is advantageous to invoke the polar representation $z = |z|\exp(\mathrm{i}\varphi)$. The results of calculations, of course, always turn out the same, irrespective of how one may choose to represent complex numbers during the intermediate steps of the process.

It is imperative to remember that the arrows associated with various numbers $z = x + \mathrm{i}y$ in the complex xy-plane are *not* vectors in a two-dimensional vector space. Vectors have their own rules for addition, subtraction, and multiplication, and no rule for division. In particular, vector multiplication—be it inner multiplication (dot-product) or outer multiplication (cross-product)—has nothing in common with complex number multiplication. One must always be aware of the fact that the arrows used to represent numbers in a complex z-plane only superficially resemble 2D vectors in a 2D Euclidean xy-space. Complex numbers must never be confused with real-valued vectors in a two-dimensional space.

Having defined complex numbers and described arithmetic operations on these numbers, we can now broaden the notion of polynomial equations of an arbitrary order n to include equations such as $a_n z^n + a_{n-1} z^{n-1} + \cdots + a_2 z^2 + a_1 z + a_0 = 0$, where all the coefficients $a_0, a_1, a_2, \cdots, a_{n-1}, a_n$ are complex constants. We will prove in chapter 6 that such polynomial equations have exactly n solutions (or roots), all of which reside in the complex z-plane.

3.6. Elementary functions of a complex variable. The function $f(z)$ of the complex variable $z = x + \mathrm{i}y$ is a mapping from the entire z-plane, or from a subset of the z-plane, onto another complex plane, such as the w-plane, where $w = u + \mathrm{i}v$. Each point z in the "domain" of the function is then associated with a point w in the "range" of the function, and the mapping is expressed as $f(z) = w$, as shown in figure 3.5. By definition, the only constraint on a function is that it cannot associate *more than* one point in the w-plane with any given point in the z-plane. Listed below are a few simple mappings, each constituting a function.

i) $f(z) = 2.0 + \mathrm{i}7.0$ maps every point $z = x + \mathrm{i}y$ onto the single point $w = 2.0 + \mathrm{i}7.0$.

ii) $g(z) = z^*$ associates each point in the z-plane with its own conjugate, that is, $w = z^*$.

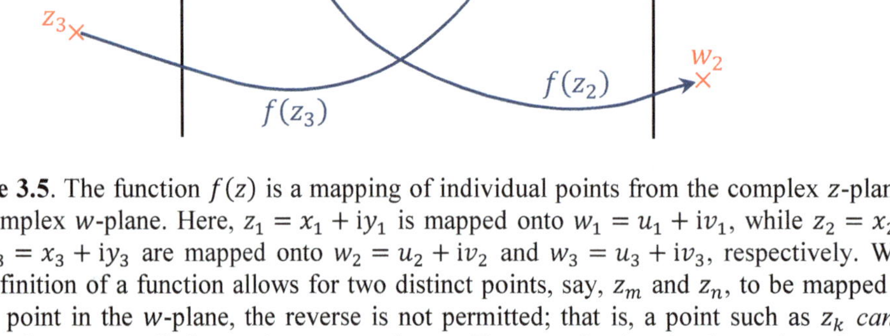

Figure 3.5. The function $f(z)$ is a mapping of individual points from the complex z-plane onto the complex w-plane. Here, $z_1 = x_1 + \mathrm{i}y_1$ is mapped onto $w_1 = u_1 + \mathrm{i}v_1$, while $z_2 = x_2 + \mathrm{i}y_2$ and $z_3 = x_3 + \mathrm{i}y_3$ are mapped onto $w_2 = u_2 + \mathrm{i}v_2$ and $w_3 = u_3 + \mathrm{i}v_3$, respectively. Whereas the definition of a function allows for two distinct points, say, z_m and z_n, to be mapped onto a single point in the w-plane, the reverse is not permitted; that is, a point such as z_k *cannot* be projected onto two distinct points w_{k1} and w_{k2}. The set of points in the z-plane for which a mapping is defined is called the "domain" of the function $f(\cdot)$, while the set of points in the w-plane, representing the projections of all the points within the domain of the function, is referred to as the "range" of $f(\cdot)$.

iii) $h(z) = |z| = \sqrt{zz^*}$ maps each point in the z-plane onto a real number, which is the magnitude $|z|$ of z.

iii) $p(z) = az^2 + bz + c$, where $|z| < 1$, and where a, b, c are specified complex constants, is a mapping that takes each point z within a circle of radius 1.0, centered at the origin of the z-plane, to a point $w = az^2 + bz + c$ in the w-plane. Here, the function $p(\cdot)$ is *not* defined for the points outside the unit-circle, where $|z| \geq 1$.

iv) $q(z) = (z - z_0)^{-3}$ is defined everywhere in the complex z-plane except at the fixed point $z = z_0$. The value assigned to each z by the function $q(z)$ is $w = 1/(z - z_0)^3$.

It is possible, of course, to form new functions by combining two or more existing functions using addition, subtraction, multiplication, division, complex conjugation, etc. For example, $p(z) = af(z) + bg(z)h^*(z)$, where a and b are complex constants, is a function over a subset of the z-plane where $f(\cdot)$, $g(\cdot)$, and $h(\cdot)$ are defined. One might also form composite functions such as $f[g(z)]$ by first projecting points from the z-plane onto the w-plane using the function $g(\cdot)$, followed by a projection of points from the w-plane onto the t-plane using the function $f(\cdot)$. Figure 3.6 shows a typical action of such a composite function on two arbitrary points z_1 and z_2 of the complex z-plane.

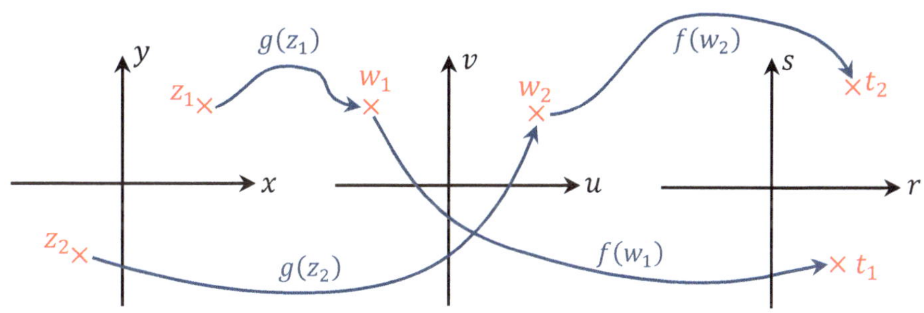

Figure 3.6. The composite function $f[g(z)]$ projects the points z_1 and z_2 of the z-plane onto w_1 and w_2 in the w-plane by means of the function $g(\cdot)$. Subsequently, the function $f(\cdot)$ takes over and maps w_1 and w_2 onto the points t_1 and t_2 in the complex t-plane. The net result is a projection of points in the z-plane onto points in the t-plane via the composite function $f[g(z)]$.

3.6.1. Polynomial functions. The function $f(z) = \sum_{n=0}^{N} a_n z^n$, where N is a positive integer and a_n are arbitrary complex constants, is a polynomial function of order N.

3.6.2. Complex exponential. In analogy with the real-valued exponential function $\exp(x)$, the complex exponential $\exp(z)$ is defined as a polynomial of infinite order with coefficients $a_n = 1/n!$; that is,

$$\exp(z) = 1 + z + (z^2/2!) + (z^3/3!) + \cdots + (z^n/n!) + \cdots. \qquad (20)$$

It turns out that the infinite series in Eq.(20) converges for all values of $z = x + iy$. Consequently, $\exp(z)$ is a well-defined function over the entire z-plane.

An important property of the real-valued function $\exp(x)$ is that $\exp(x_1 + x_2)$ equals the product of $\exp(x_1)$ and $\exp(x_2)$ for all values of x_1 and x_2. We now prove that $\exp(z)$ has a similar property, namely that $\exp(z_1 + z_2) = \exp(z_1)\exp(z_2)$ for all values of the complex numbers z_1 and z_2. To this end, we write

$$\exp(z_1)\exp(z_2) = [\textstyle\sum_{m=0}^{\infty}(z_1^m/m!)][\sum_{n=0}^{\infty}(z_2^n/n!)] = \sum_{m=0}^{\infty}\sum_{n=0}^{\infty}[z_1^m z_2^n/(m!\,n!)]$$

$$\boxed{k = m + n} \rightarrow = \textstyle\sum_{k=0}^{\infty}\sum_{n=0}^{k}[z_1^{k-n} z_2^n/(k-n)!\,n!] = \sum_{k=0}^{\infty}(1/k!)\sum_{n=0}^{k}\binom{k}{n} z_1^{k-n} z_2^n$$

$$= \sum_{k=0}^{\infty}[(z_1 + z_2)^k/k!] = \exp(z_1 + z_2). \quad (21)$$

Thus, in the special case where $z_1 = x$ and $z_2 = \mathrm{i}y$, we may invoke Eq.(21) in conjunction with the Euler identity to write

$$\exp(x + \mathrm{i}y) = \exp(x)\exp(\mathrm{i}y) = \exp(x)\,(\cos y + \mathrm{i}\sin y). \quad (22)$$

3.6.3. Hyperbolic functions. Having defined the function $\exp(z)$, we now proceed to define the hyperbolic sine and cosine in analogy with the corresponding functions of a real variable, namely,

$$\sinh(z) = ½[\exp(z) - \exp(-z)] \quad \text{and} \quad \cosh(z) = ½[\exp(z) + \exp(-z)]. \quad (23)$$

Similarly, the hyperbolic tangent and cotangent functions are defined as $\tanh(z) = \sinh(z)/\cosh(z)$ and $\coth(z) = \cosh(z)/\sinh(z)$.

3.6.4. Trigonometric functions. The sine and cosine of a real variable x are related to the complex exponential $\exp(\mathrm{i}x)$ through the well-known Euler identity as follows: $\sin x = [\exp(\mathrm{i}x) - \exp(-\mathrm{i}x)]/(2\mathrm{i})$ and $\cos x = [\exp(\mathrm{i}x) + \exp(-\mathrm{i}x)]/2$. By analogy, the sine and cosine of a complex variable z are defined as

$$\sin(z) = [\exp(\mathrm{i}z) - \exp(-\mathrm{i}z)]/(2\mathrm{i}), \quad (24)$$

$$\cos(z) = [\exp(\mathrm{i}z) + \exp(-\mathrm{i}z)]/2. \quad (25)$$

Similarly, the tangent and cotangent of the complex variable z are defined as $\tan(z) = \sin(z)/\cos(z)$ and $\cot(z) = \cos(z)/\sin(z)$. It is now easy to arrive at various trigonometric identities using these definitions in conjunction with the fundamental property of the complex exponential given by Eq.(22). For instance, one can show that

$$\sin(z) = \sin(x + \mathrm{i}y) = \sin(x)\cosh(y) + \mathrm{i}\cos(x)\sinh(y), \quad (26)$$

$$\cos(z) = \cos(x + \mathrm{i}y) = \cos(x)\cosh(y) - \mathrm{i}\sin(x)\sinh(y). \quad (27)$$

3.6.5. Natural logarithm. The natural logarithm of a real variable is the inverse function of the real-valued exponential $\exp(x)$ —which is occasionally written as e^x as well. This means that, if $y = \exp(x)$, then $x = \ln y$. The fundamental property of the natural logarithm, namely $\ln(y_1 y_2) = \ln(y_1) + \ln(y_2)$, is thus a direct consequence of the fundamental property of the exponential function, namely $e^{x_1+x_2} = e^{x_1}e^{x_2}$. It is also trivial to see why, for any real variable x, we must have $\ln(e^x) = x$. Given that $\exp(x)$ is always positive, the natural logarithm is defined only for positive real numbers—be they constants or variables.

Digression: An alternative (but equivalent) definition of the natural logarithm is based on an integral of the function $f(y) = 1/y$ over the interval $[1, x]$, that is, $\ln(x) = \int_1^x y^{-1}\,dy$, where $0 < x < \infty$. Considering that the identity $\int_1^x y^s\,dy = (x^{s+1} - 1)/(s + 1)$ is valid for all values of s except when $s = -1$, we set $s = -1 + \varepsilon$, where $|\varepsilon| \ll 1$, to arrive at $\ln x = lim_{\varepsilon\to 0}\,(x^\varepsilon - 1)/\varepsilon$.

By definition, the base e of the natural logarithm has the property that $\ln(e) = 1$. Consequently, $lim_{\varepsilon\to 0}[(e^\varepsilon - 1)/\varepsilon] = 1$, which yields $e = lim_{\varepsilon\to 0}(1 + \varepsilon)^{1/\varepsilon}$. A by-product of this result is the well-known identity $lim_{n\to\infty}(1 + n^{-1})^n = e$. Also note, in accordance with these definitions, that $\int_1^e dx/x = 1$, that the derivative with respect to x of the function $\ln(x)$ is $1/x$, and that $\ln(x_1 x_2) = \ln(x_1) + \ln(x_2)$. For a proof of the latter identity, see problem 6 at the end of the chapter.

To extend the concept of natural logarithm to the complex plane, where, in polar representation, a number is written as $z = |z| \exp(\mathrm{i}\varphi)$, we use *analogy* with the natural logarithm of real-valued entities to write

$$\ln(z) = \ln(|z|e^{\mathrm{i}\varphi}) = \ln(|z|) + \ln(e^{\mathrm{i}\varphi}) = \ln(|z|) + \mathrm{i}\varphi. \tag{28}$$

This equation now serves as the standard definition of the natural logarithm of a complex number z. Note that Eq.(28) is consistent with the conventional definition of the natural logarithm for (positive) real numbers x, whose complex plane representation is $z = x + \mathrm{i}0 = |x|e^{\mathrm{i}0}$.

A problem with the defining identity for $\ln(z)$ given in Eq.(28) is that the phase φ of a complex number z is *not* uniquely specified. In fact, any number z may be written as $|z| \exp[\mathrm{i}(2n\pi + \varphi)]$, where n could be any positive, zero, or negative integer. Needless to say, z is always associated with the *same* point in the complex plane, irrespective of the chosen value of n. However, depending on the value of n, the imaginary part $(2n\pi + \varphi)$ of $\ln(z)$ will land it at different locations in a complex plane. This will create an infinite number of values for $\ln(z)$ associated with any given z, which is not admissible if $\ln(z)$ is to be a legitimate function. One must, therefore, limit the range of values of φ to within a 2π interval so that a unique value can be assigned to $\ln(z)$. In complex analysis, it is typical to choose $-\pi < \varphi \le \pi$, in which case the negative real-axis of the z-plane becomes a "branch-cut" of the function $\ln(z)$; see figure 3.7a. There is, of course, nothing wrong with choosing $0 \le \varphi < 2\pi$ as the allowed range, in which case the positive real axis would become the branch-cut. In fact, any meandering line starting at the origin of coordinates and heading to infinity along an arbitrary path, say, the path shown in figure 3.7b, could be designated as the branch-cut. A point z is then assigned a unique value for its natural logarithm, $\ln(z)$, depending on the crossing point of the branch-cut with a circle of radius $|z|$ centered at the origin. The bottom line is that, each point in the complex z-plane for which it is desired to define $\ln(z)$, must have a uniquely assigned phase φ within some designated 2π interval.

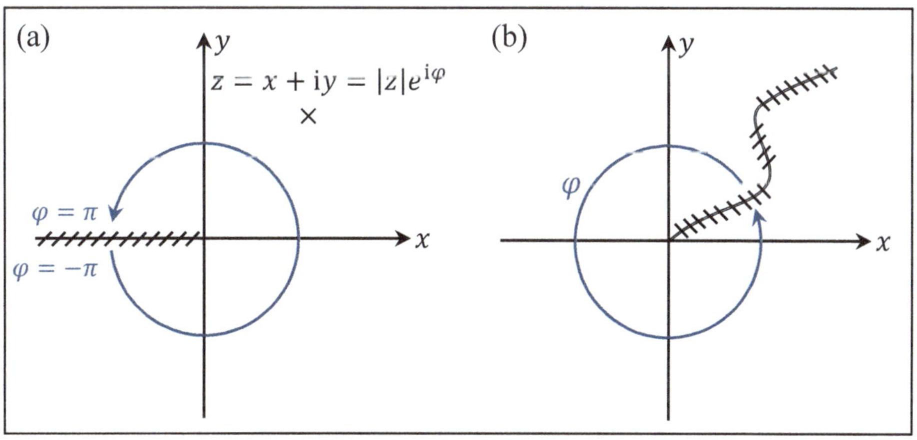

Figure 3.7. (a) The negative x-axis serves as a branch-cut that confines the phase φ of the complex number z to the interval $-\pi < \varphi \le \pi$. (b) A meandering line, starting at the origin of coordinates and heading to infinity along an arbitrary path, could also serve as a branch-cut by cutting across the z-plane in such a way as to allow a unique phase φ to be assigned to each point $z = |z|e^{\mathrm{i}\varphi}$. The allowed range of φ, which now depends on $|z|$, is $\varphi_0(|z|) \le \varphi < 2\pi + \varphi_0(|z|)$.

3.6.6. Square root and higher-order roots. The square root $z^{1/2}$ of the complex number z is, by definition, a number whose square equals z. Writing $z = |z| \exp[\mathrm{i}(2n\pi + \varphi)]$, where n is any arbitrary positive, zero, or negative integer, it is not difficult to see that

$$z^{½} = \sqrt{|z|}\exp[\mathrm{i}(n\pi + ½\varphi)] = \pm\sqrt{|z|}\exp(½\mathrm{i}\varphi). \tag{29}$$

The plus and minus signs on the right-hand side of this equation correspond to even and odd values of n, respectively. Considering that the function $f(z) = z^{½}$ is not allowed to assign more than one value to each z, it becomes necessary to define a branch-cut, such as the negative x-axis shown in figure 3.7a, or the meandering line depicted in figure 3.7b, in order to confine the phase φ of z to within a specific 2π interval. Depending on the chosen branch-cut, the square root of a given point z in the complex plane would assume either the plus sign or the minus sign (but not both), in accordance with Eq.(29).

It must be clear at this point that higher-order roots of a complex number can be obtained similarly to the square root described in the preceding paragraph. For example, the cube root of z will be

$$z^{⅓} = \sqrt[3]{|z|}\exp[\mathrm{i}(2n\pi + \varphi)/3] = \sqrt[3]{|z|}\exp(\mathrm{i}\varphi/3) \times \begin{cases} 1.0; \\ \exp(+\mathrm{i}2\pi/3); \\ \exp(-\mathrm{i}2\pi/3). \end{cases} \tag{30}$$

The cube root of z thus has three possible values, all having the same magnitude $\sqrt[3]{|z|}$ but different phase angles: $\varphi/3$, or $(\varphi/3) + 120°$, or $(\varphi/3) - 120°$. A branch-cut similar to those shown in figure 3.7 must, therefore, be adopted to confine the phase φ of z to within a 2π interval, thereby eliminating two of the three roots given in Eq.(30).

3.6.7. Non-integer powers of polynomials. Consider the function $f(z) = P_n^\alpha(z)$, where $P_n(z) = \sum_{m=0}^{n} a_m z^m$ is an nth-order polynomial with arbitrary complex coefficients a_m, while α is a real-valued, non-integer constant. According to the fundamental theorem of algebra, any n^{th}-order polynomial has n zeros in the complex plane. One may thus expand the polynomial into a product of first-order terms, i.e., $P_n(z) = a_n \prod_{m=1}^{n}(z - z_m)$. We limit the following discussion to a few simple polynomials raised to non-integer powers such as z^α, $(z - z_0)^\alpha$, and $[(z - z_1)(z - z_2)]^\alpha$.

Recall that a branch-cut in the complex z-plane, where $z = |z|\exp(\mathrm{i}\varphi)$, limits the range of the phase-angle φ to a 2π interval, so that certain functions defined over the entire complex plane (or a subset thereof) will end up having unique values. In general, a function defined with the aid of a branch-cut cannot be differentiable on the branch-cut because the immediate vicinity of any point z located on the branch-cut contains points on both sides of that cut.

The function z^α, where α is a real non-integer constant, needs a branch-cut if it is to be uniquely defined for each z. One way to see this is to recognize that $z^\alpha = \exp(\alpha \ln z)$, and that $\ln z$ requires a branch-cut. (Of course, if α happens to be an integer n, then $z^n = |z|^n \exp(\mathrm{i}n\varphi)$ is well-defined and there will be no need to resort to the logarithmic function.) Alternatively, we may write $z^\alpha = |z|^\alpha \exp[\mathrm{i}\alpha(2m\pi + \varphi)]$ and note that different values of m could give rise to different phase-angles $2\alpha m\pi$ when α is *not* an integer. Any straight or meandering line drawn in the complex plane, starting at the origin and heading to infinity along an arbitrary path, would provide an appropriate branch-cut for $f(z) = z^\alpha$.

The function $f(z) = \sqrt{z - z_0}$, where z_0 is a fixed point in the complex plane, may be specified as follows:

i) Draw a straight line connecting z_0 to z in the complex plane.

ii) Determine the magnitude $|z - z_0|$ and phase-angle φ of $z - z_0$, where $0 \le \varphi < 2\pi$.

iii) Write $\sqrt{z - z_0} = |z - z_0|^{½} \exp[\mathrm{i}(m\pi + ½\varphi)]$.

It is clear that, depending on the integer m being odd or even, the square root of $z - z_0$ will have different values. A branch-cut in the form of a straight or meandering line, starting from z_0 and heading to infinity along an arbitrary path, is thus needed to uniquely specify the square root of $z - z_0$. The phase of $z - z_0$ is no longer taken to be in the interval $[0, 2\pi)$; instead, for each z, it must be chosen within a different 2π-interval referenced to the crossing-point of the (meandering) branch-cut with a circle of radius $|z - z_0|$ centered at z_0.

The function $f(z) = \sqrt{(z - z_1)(z - z_2)}$, where z_1 and z_2 are two fixed points in the complex plane, can be uniquely specified with the aid of two branch-cuts, one starting at z_1, the other at z_2, both moving along different (arbitrary) paths to infinity. The two paths, of course, may join at some point in the complex plane, then overlap for the rest of their journey to infinity. While this overlapping section of the two branch-cuts is still needed for the unique identification of the phase angles of $\sqrt{z - z_1}$ and $\sqrt{z - z_2}$, it no longer serves as a boundary between two regions of analyticity. This is because the π phase jumps of $\sqrt{z - z_1}$ and $\sqrt{z - z_2}$ cancel out across the overlapping segment of the two branch-cuts. The boundary separating different regions of analyticity is thus confined to the (straight or meandering) line connecting z_1 to z_2. Note that the overlapping segments of the two branch-cuts would remain a boundary between two regions of analyticity if the function, instead of being the square-root, happens to be the cube-root (or some other root) of $(z - z_1)(z - z_2)$.

Similar considerations apply to functions of the form $f(z) = P_n^{\alpha}(z)/P_m^{\beta}(z)$, where α and β are real-valued non-integers. The function now has n zeros and m poles, each of which could serve as the starting point of a straight or meandering branch-cut. As before, two or more branch-cuts are allowed to have overlapping segments. In general, these overlapping segments may or may not act as boundaries separating adjacent regions of analyticity, depending on whether the overall phase of $f(z)$ remains discontinuous across the (overlapped) branch-cut.

3.6.8. Inverse trigonometric functions. The inverse function of $\sin(z)$ is $\arcsin(z)$, which is sometimes written as $\sin^{-1}(z)$. By definition, $w = \arcsin(z) = \sin^{-1}(z)$ implies that $z = \sin(w)$. Inverse functions exist for other trigonometric functions as well. Thus, $f(z) = \arccos(z) = \cos^{-1}(z)$ assigns a complex number $w = f(z)$ to any given complex number z such that $\cos(w) = z$. Similarly, $f(z) = \arctan(z) = \tan^{-1}(z)$ assigns $w = f(z)$ to a given z in such a way that $\tan(w) = \sin(w)/\cos(w)$ equals z.

Considering the definition of the complex function $\sin(z)$ as $(e^{\mathrm{i}z} - e^{-\mathrm{i}z})/(2\mathrm{i})$, the complex arcsine function $f(z) = \sin^{-1}(z)$ may be evaluated as follows:

$$z = \sin[f(z)] = \{\exp[\mathrm{i}f(z)] - \exp[-\mathrm{i}f(z)]\}/(2\mathrm{i})$$

$$\rightarrow \quad \exp[2\mathrm{i}f(z)] - 2\mathrm{i}z\exp[\mathrm{i}f(z)] - 1 = 0$$

$$\rightarrow \quad \exp[\mathrm{i}f(z)] = \mathrm{i}z \pm \sqrt{(\mathrm{i}z)^2 + 1} \quad \leftarrow \boxed{\text{Solution of quadratic equation}}$$

$$\rightarrow \quad f(z) = -\mathrm{i}\ \ln\left(\pm\sqrt{1 - z^2} + \mathrm{i}z\right). \tag{31}$$

We choose at the outset to confine all relevant phase angles φ to the interval $[0, 2\pi)$. This, of course, is an arbitrary choice, because φ could be restricted to *any* 2π interval, say, $(-\pi, \pi)$, $(\pi, 3\pi)$, etc.

Digression. The ordinary arcsine function associated with real angles θ is confined to the interval $[-1,1]$ of the real axis x. If one defines $\theta = \sin^{-1}(x)$ as the angle in the interval $[-\frac{1}{2}\pi, \frac{1}{2}\pi]$ that satisfies $\sin\theta = x$, one will have $\cos\theta = \sqrt{1-x^2}$ and $\cos(\pi-\theta) = -\sqrt{1-x^2}$. It is also clear that $\sin(\pi-\theta) = \sin\theta = x$. Consequently, Eq.(31) yields

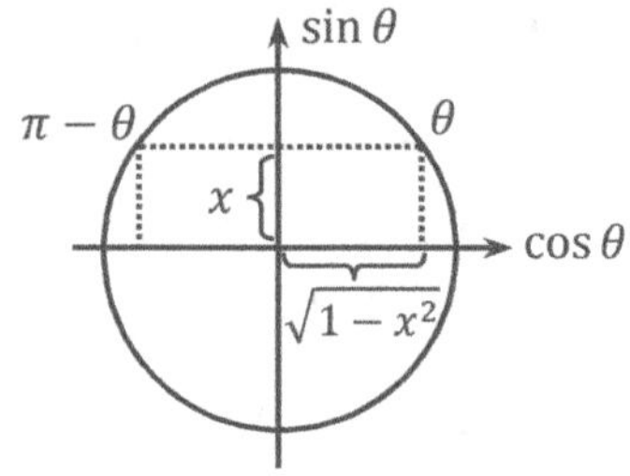

$$+\text{ sign: } f(x) = \sin^{-1}(x) = -\mathrm{i}\ln(\cos\theta + \mathrm{i}\sin\theta) = -\mathrm{i}\ln(e^{\mathrm{i}\theta}) = \theta;$$

$$-\text{ sign: } f(x) = \sin^{-1}(x) = -\mathrm{i}\ln[e^{\mathrm{i}(\pi-\theta)}] = \pi - \theta.$$

Returning to Eq.(31), let us first consider the function $\pm\sqrt{1-z^2}$, with the understanding that the $\pm$ sign should be ignored throughout the analysis. This is because selecting the allowed interval for φ followed by the choice of the branch-cut(s) automatically fixes the sign of the square root at either $+$ or $-$ within various regions of the complex z-plane. When $1-z^2$ happens to be a positive real number, say, α, we will be on a branch-cut defined by the boundary between $\varphi = 0$ and $\varphi = 2\pi$. We thus examine the equation $1-z^2 = \alpha$, which yields $z = \pm\sqrt{1-\alpha}$. If $0 \le \alpha \le 1$, then z will be in the interval $[-1,1]$, and when $\alpha \ge 1$, z will be on the imaginary axis. Consequently, the branch-cuts in this example are the real-line-segment $[-1,1]$ and the entire imaginary axis, as depicted in figure 3.8a.

Alternatively, if we choose to treat $\sqrt{1-z^2}$ as the product of $\sqrt{1-z}$ and $\sqrt{1+z}$, then the real-line-segment $(-\infty,+1)$ becomes the branch-cut for $\sqrt{1-z}$, while the real-line-segment $(-1,+\infty)$ becomes the branch-cut for $\sqrt{1+z}$. Subsequently, the overlap region, namely, the real-line-segment $(-1,1)$, should be ignored, because the addition of two π-jumps across this segment is tantamount to no jump at all. Under this treatment, therefore, the branch-cut for $\sqrt{1-z^2}$ would consist of the real-line segments $(-\infty,-1)$ and $(1,\infty)$, as depicted in figure 3.8b.

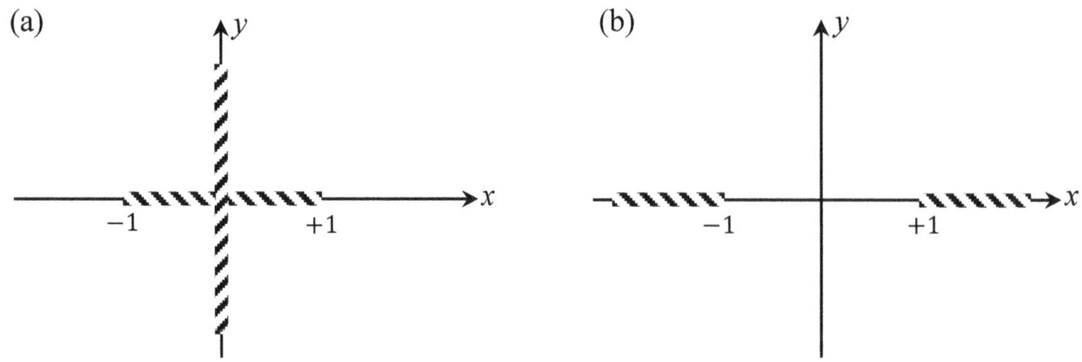

Figure 3.8. Two possible branch-cuts for the function $f(z) = \arcsin(z)$.

Next, we examine the argument of the log function, and determine the values of z for which $\sqrt{1-z^2} + \mathrm{i}z = \alpha$, where, once again, α is an arbitrary positive real number (associated with the boundary between $\varphi = 0$ and $\varphi = 2\pi$). Solving for z, we find $z = \mathrm{i}(\alpha^{-1} - \alpha)/2$, which ranges over the entire imaginary axis when α assumes all positive real numbers in the range $(0,\infty)$. Note, however, that the same thing happens for negative values of α as well, so one has to be careful in designating the imaginary axis as a branch-cut for the log function.

Suppose that $z = \mathrm{i}y$ is a point on the imaginary axis. Then, according to our first approach to computing $\sqrt{1-z^2}$, we will have $\sqrt{1-z^2} + \mathrm{i}z = \sqrt{1+y^2} - y > 0$, indicating that the imaginary axis is indeed a branch-cut for the log function. In other words, crossing from one side of the imaginary axis to the other side will cause the phase of $\sqrt{1-z^2}+\mathrm{i}z$ to jump from $\varphi=0$ to $\varphi=2\pi$, or vice-versa. Consequently, figure 3.8a shows the complete set of branch-cuts for the function $f(z)=\sin^{-1}(z)$.

In contrast, if we were to compute $\sqrt{1-z^2}$ in accordance with our second approach, then $\sqrt{1-z^2}+\mathrm{i}z=\sqrt{1-\mathrm{i}y}\times\sqrt{1+\mathrm{i}y}-y=-\sqrt{1+y^2}-y<0$, which indicates that the imaginary axis is *not* a branch-cut for the log function. This is because the phase of $\sqrt{1-z^2}+\mathrm{i}z$ in the vicinity of the imaginary axis hovers around $\varphi=\pi$, which does *not* produce a phase discontinuity on crossing the imaginary axis. Therefore, in accordance with our second approach, figure 3.8b shows the complete set of branch-cuts of the function $f(z)=\sin^{-1}(z)$.

In conclusion, depending on how we choose to evaluate $\sqrt{1-z^2}$, the branch-cuts for $f(z)=\sin^{-1}(z)$ would be those depicted either in figure 3.8a or in figure 3.8b.

3.7. Pathological functions. Pathological functions exhibit abnormalities that prevent their use in certain mathematical operations such as differentiation, integration, Fourier transformation, and so on. A simple example is

$$f(x)=\begin{cases}0, & \text{when } x \text{ is rational,}\\ 1, & \text{when } x \text{ is irrational.}\end{cases} \tag{32}$$

This function is discontinuous everywhere and does not have a derivative or an integral in the sense defined in standard calculus—although it does have a Lebesgue integral.

Another pathological function is $f(x)=\sin(1/x)$, which oscillates wildly in the vicinity of $x=0$, thus preventing it from having a limit when $x\to 0$, or having a Fourier transform. Many variants of this function with unusual properties can be constructed, for instance, $f(x)=x\sin(1/x)$ or $f(x)=1/\sin(1/x)$, or $f(x)=\ln[\sin^2(1/x)]$.

Let us examine a specific variant, namely, $f(x)=x^3|\sin(1/x)|$, in some detail. This function is zero at $x=0$ and also at $x=1/(n\pi)$, where $n=\pm1,\pm2,\pm3,\cdots$. Clearly, the derivative $f'(x)$ of the function does not exist at $x=1/(n\pi)$, but, with the possible exception of $x=0$, the function is obviously differentiable elsewhere. We would like to show that $f(x)$ is also differentiable at $x=0$. To this end we write $f(x)=x^3\sqrt{\sin^2(1/x)}$ and proceed as follows:

$$f'(x)=3x^2|\sin(1/x)|-\frac{x\cos(1/x)\sin(1/x)}{|\sin(1/x)|}. \tag{33}$$

Considering that $\sin(1/x)/|\sin(1/x)|$ is a discrete function that switches between -1 and $+1$ at $x=1/(n\pi)$, and that both $x^2|\sin(1/x)|$ and $x\cos(1/x)$ approach zero when $x\to 0$, we conclude that $f'(0)=0$. Here, we have a function that is differentiable at $x=0$, but in every neighborhood of $x=0$ there exists an infinite number of points $x_n=1/(n\pi)$, with $|n|$ being greater than some integer N, at which the function does *not* have a derivative.

A good example of a pathological function of a complex variable is $f(z)=1/\cosh(1/z)$. Considering that

$$\cosh(z)=\cosh(x+\mathrm{i}y)=\cosh(x)\cos(y)+\mathrm{i}\sinh(x)\sin(y),$$

we note that all the zeros of $\cosh(z)$ are located on the imaginary axis at $z_n=\mathrm{i}(2n+1)\pi/2$, where $n=0,\pm1,\pm2,\cdots$.

Let us examine the rays $z = \alpha \exp(i\varphi)$ along which z approaches the origin of the complex plane (i.e., $z = 0$) when α, a positive real number, approaches zero while $\varphi = \varphi_0$ remains constant. If the ray coincides with either half of the imaginary axis, that is, if $\varphi_0 = \pm\pi/2$, then, as $\alpha \to 0$, $f(z) = f(\pm i\alpha) = 1/\cosh(\mp i/\alpha) = 1/\cos(\alpha^{-1})$ will pass through an infinite number of singularities at $\alpha = 2/[(2n+1)\pi]$. However, for any other ray, we will have

$$f(z) = \frac{1}{\cosh(\alpha^{-1}\cos\varphi_0 - i\alpha^{-1}\sin\varphi_0)}$$

$$= \frac{2}{\exp(\alpha^{-1}\cos\varphi_0)\exp(-i\alpha^{-1}\sin\varphi_0) + \exp(-\alpha^{-1}\cos\varphi_0)\exp(i\alpha^{-1}\sin\varphi_0)}. \tag{34}$$

When $\alpha \to 0$, depending on the sign of $\cos\varphi_0$, one of the real-valued exponentials in Eq.(34) goes to zero, while the other real-valued exponential goes to infinity. Either way, $f(z) \to 0$ when the point z approaches the origin along the chosen ray.

3.8. Fermat's Little Theorem and several related facts in modular arithmetic. Fermat's little theorem is so named to distinguish it from his famous last theorem. According to the little theorem, for any integer a and prime number p, the number $a^p - a$ is an integer-multiple of p. In the language of modular arithmetic, $a^p = a \,(\mathrm{mod}\, p)$. An alternative (although equivalent) version of the theorem states that $a^{p-1} = 1 \,(\mathrm{mod}\, p)$ provided that a is not divisible by p. Several proofs of the theorem are known, perhaps the simplest one (due to Euler) is given below.

Proof by induction: Fermat's little theorem is obviously valid for $a = 1$. Suppose the theorem is valid for some integer $a \ge 1$, that is, $a^p - a = mp$, where $m \ge 0$ is an integer. Then, for $a + 1$ we will have

$$(a+1)^p = \sum_{n=1}^{p} \binom{p}{n} a^n, \quad \text{where} \quad \binom{p}{n} = \frac{p!}{n!\,(p-n)!} = \frac{(p-n+1)\times(p-n+2)\times\cdots\times p}{1\times2\times3\times\cdots\times n}. \tag{35}$$

Considering that $\binom{p}{n}$ is an integer, and that p appearing in its expression cannot be further reduced (because p is prime), we conclude that $\binom{p}{n}$ is an integer-multiple of p except for $n = 0$ and $n = p$. The part of $(a+1)^p$ that remains to be examined is thus $a^p + 1$. Now, by assumption, $a^p = a + mp$; therefore, the remaining term is $a + mp + 1$. We have thus shown that $(a+1)^p - (a+1) = kp$ for some integer k, which completes the proof by induction.

Proof by combinatorics. An alternative proof starts by looking into an alphabet of size a, from which all words of length p are constructed. The total number of words, of course, is a^p. For example, for $a = 3$ and $p = 5$, the alphabet consists of the letters A, B, C, and there exist $3^5 = 243$ words, namely, $AAAAA$, $AAAAB$,⋯, $CABCB$, ⋯, $CCCCC$. There are a words that contain only one letter, such as $AAAAA$, $BBBBB$, and $CCCCC$ in this example. Removing these words from the set of all possible words leaves us with $(a^p - a)$ p-letter words, which contain at least two different letters. We now form rings out of these p-letter words by joining the first and last letter of each word, then placing the word on a circle, as shown. Each ring will have p friends,

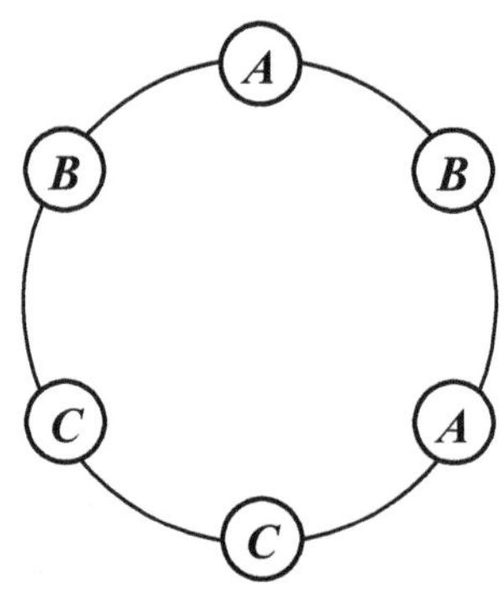

formed by shifting the ring (say, clockwise) by one letter at a time. These friends all differ from each other. There cannot be friends formed by cyclic shifts of letter pairs, or letter triplets, etc., because p is a prime number and pairs, triplets, quadruplets, etc., cannot divide it without some left-over. Consequently, all the remaining rings can be grouped into groups consisting of p friends. In other words, $a^p - a$ is an integer-multiple of p, and the theorem is, therefore, proved.

3.8.1. Theorem 1 (Diophantus). The product of two integers, each expressed as the sum of two perfect squares, is itself the sum of two perfect squares.

$$\begin{aligned}(a^2 + b^2)(c^2 + d^2) &= (a + \mathrm{i}b)(a - \mathrm{i}b)(c + \mathrm{i}d)(c - \mathrm{i}d)\\ &= (a + \mathrm{i}b)(c + \mathrm{i}d)(a - \mathrm{i}b)(c - \mathrm{i}d)\\ &= [(ac - bd) + \mathrm{i}(ad + bc)][(ac - bd) - \mathrm{i}(ad + bc)]\\ &= (ac - bd)^2 + (ad + bc)^2. \qquad (36)\end{aligned}$$

3.8.2. Theorem 2 (Wilson). If p is prime, then $(p - 1)! + 1$ is divisible by p. In other words, $(p - 1)! = -1 \pmod p$.

Proof: The theorem is obviously valid for $p = 2$. For $p \geq 3$, the integers $0, 1, 2, 3, \cdots, (p - 1)$ form a *Galois field* under modulo p arithmetic. In such a field, every element $a \neq 0$ has a unique inverse a^{-1}, such that $aa^{-1} = 1 \pmod p$. If the inverse of an element happens to coincide with itself, that is, if $a^2 = 1 \pmod p$, we will have $a^2 - 1 = (a - 1)(a + 1) = 0 \pmod p$, which implies that $+1$ and -1 (i.e., $p - 1$) are the only elements whose inverses are themselves. Therefore, aside from the first and last terms in the expression $(p - 1)! = 1 \cdot 2 \cdot 3 \cdots (p - 2) \cdot (p - 1)$, each term of the product can be paired with its own inverse within the same product. The products of individual pairs are thus equal to 1 $\pmod p$. Therefore, $(p - 1)! = -1 \pmod p$. The proof is now complete.

As an example, consider the case of $p = 17$, where $16! = 1 \cdot (2 \cdot 9) \cdot (3 \cdot 6) \cdot (4 \cdot 13) \cdot (5 \cdot 7) \cdot (8 \cdot 15) \cdot (10 \cdot 12) \cdot (11 \cdot 14) \cdot 16$. Clearly, the pairs in parentheses are inverses of each other (mod 17), leaving the unpaired terms 1 and 16, whose product is $1 \cdot 16 = -1 \pmod{17}$.

3.8.3. Theorem 3. If $n > 4$ is a composite (i.e., non-prime) number, then $(n - 1)! = 0 \pmod n$. The reason for exempting $n = 4$ will become clear once the following method of proof is understood.

Proof: If $n = a \times b$, then both a and b are smaller than n, which means that both a and b appear somewhere along the sequence of terms comprising $(n - 1)!$. Consequently, $(n - 1)!$ contains $a \times b = n$, which makes the product equal to zero $\pmod n$. An exception to this argument occurs when $a = b$, in which case $2a < a^2 = n$. Now, both a and $2a$ appear somewhere along the sequence, which indicates that $(n - 1)!$ contains $a \times 2a = 2a^2 = 2n$. Once again $(n - 1)!$ turns out to be zero $\pmod n$.

3.8.4. Theorem 4 (Euler). If p is a prime number in the form of $4k + 1$, where k is an integer, then an integer m exists such that $m^2 + 1$ is divisible by p. In the language of modular arithmetic, $m^2 = -1 \pmod p$.

Proof: Let $m = (2k)!$. We will have

$$\begin{aligned}
m^2 &= (1\cdot 2\cdot 3\cdots 2k)\times(1\cdot 2\cdot 3\cdots 2k)\\
&= (1\cdot 2\cdot 3\cdots 2k)\times(-1)\cdot(-2)\cdot(-3)\cdots(-2k)\\
&= (1\cdot 2\cdot 3\cdots 2k)\times(p-1)\cdot(p-2)\cdot(p-3)\cdots(p-2k) \pmod p\\
&= (1\cdot 2\cdot 3\cdots 2k)\times(4k)\cdot(4k-1)\cdot(4k-2)\cdots(2k+1) \pmod p\\
&= 1\cdot 2\cdot 3\cdots 2k\cdot(2k+1)\cdot(2k+2)\cdots(4k-1)\cdot(4k) \pmod p\\
&= (p-1)! \pmod p\\
&= -1 \pmod p. \qquad (37)
\end{aligned}$$

The last step is a consequence of Wilson's theorem. The present theorem may therefore be considered a corollary of Wilson's theorem.

3.8.5. Definitions. A Gaussian integer is a complex number $c = a + ib$, whose real and imaginary parts a and b are integers (positive, negative, or zero). The norm of c is defined as $N(c) = cc^* = a^2 + b^2$. Addition, subtraction, and multiplication operations are defined for Gaussian integers in exactly the same way as they are for ordinary complex numbers, that is $c_1 \pm c_2 = (a_1 \pm a_2) + \mathrm{i}(b_1 \pm b_2)$. Similarly, $c_1c_2 = (a_1a_2 - b_1b_2) + \mathrm{i}(a_1b_2 + a_2b_1)$. It is easy to show that $N(c_1c_2) = N(c_1)N(c_2)$.

The zero of Gaussian integers is $\mathbf{0} = 0 + \mathrm{i}0$, which is uniquely identified in the complex plane. A "unity" Gaussian integer is, by definition, any Gaussian integer whose norm is equal to one, that is, $N(\mathbf{1}) = a^2 + b^2 = 1$. Consequently, there exist exactly four distinct unities in the domain of Gaussian integers, namely, $\mathbf{1} = \pm 1 + \mathrm{i}0$ and $\mathbf{1} = 0 \pm \mathrm{i}$.

A Gaussian prime p is a Gaussian integer with the following properties: (i) $N(p) > 1$; (ii) if p can be decomposed into the product of any two Gaussian integers, say, if $p = \alpha\beta$, then either $\alpha = \mathbf{1}$ or $\beta = \mathbf{1}$. Under this definition, $2 = (1 + \mathrm{i})(1 - \mathrm{i})$ is *not* a Gaussian prime, nor is $5 = (1 + 2\mathrm{i})(1 - 2\mathrm{i})$. However, it is fairly easy to show that 3 is a Gaussian prime and so is 7.

Every Gaussian integer c has three associates obtained by multiplying c with -1, i, and $-\mathrm{i}$. In particular, every Gaussian prime p is associated with four points in the complex plane, obtained by successive 90° rotations of p around the origin $(0, 0)$. Aside from this four-fold multiplicity, each Gaussian prime is unique in the sense that no other Gaussian prime exists that has the same norm as p. The preceding statement is a direct consequence of theorem 9 discussed in section 3.8.10.

3.8.6. Theorem 5. If the norm of a Gaussian integer happens to be an *ordinary* prime number $p \geq 2$, that is, if $N(c) = c'^2 + c''^2 = p$, then c is a Gaussian prime.

Proof: Suppose that $c = ab$. Then $N(c) = N(a)N(b) = p$. Considering that p is an ordinary prime, it can be decomposed only as $p \times 1$. Consequently, either a or b must be unity. We have thus proven that c is a Gaussian prime. As an example, consider $c = 1 + 2\mathrm{i}$, whose norm, 5, is an ordinary prime. Therefore c is a Gaussian prime. Note, however, that 5 is not a Gaussian prime because $5 = (1 + 2\mathrm{i})(1 - 2\mathrm{i})$.

3.8.7. Theorem 6. If $p = 4k - 1$ is an ordinary prime, then p is also a Gaussian prime.

Proof: Suppose $p = ab$. Then $N(a)N(b) = p^2$. If we decompose p^2 as $p^2 \times 1$, then either a or b will be $\mathbf{1}$, in which case p would not be properly decomposed. The only other way to decompose p^2 is by writing it as $p \times p$. Therefore, $N(a) = a'^2 + a''^2 = p$ (a similar relation would exist for b as well.) Now, if a' and a'' are both odd or both even, the sum of their squares will be even, whereas p is odd. If one of a' and a'' is odd and the other one even, then $a'^2 + a''^2$ will be in the form of $4k + 1$, which contradicts our assumption about the form of p. We conclude that $p = 4k - 1$ cannot be properly decomposed into the product of a and b, and that, therefore, p is a Gaussian prime.

3.8.8. Theorem 7. If a Gaussian prime c divides the product ab of two Gaussian integers, then c must divide either a or b (or both).

Proof: Since ab is divisible by c, we can write $ab = \gamma c$, where γ is a Gaussian integer. Now, γ may be decomposed into $\gamma_1\gamma_2 \cdots \gamma_n$, but c, being a Gaussian prime, cannot be decomposed. It is thus seen that any groupings of $\gamma_1\gamma_2 \cdots \gamma_n c$ (to associate one group with a and another group with b) will have c appearing in one group or the other. (In case one of the factors of γ happens to be c, then both a and b could become divisible by c.)

3.8.9. Theorem 8 (Fermat). If the prime number p has the form $4k + 1$, then it can be written as the sum of two perfect squares in a unique way; that is, $p = m^2 + n^2$, where m and n have a unique value determined by p alone. For example, $29 = 4 \times 7 + 1 = 2^2 + 5^2$ and $37 = 4 \times 9 + 1 = 1^2 + 6^2$ and $53 = 4 \times 13 + 1 = 2^2 + 7^2$.

Proof: Considering that the integer ℓ exists such that $\ell^2 + 1$ is divisible by p, we can decompose $\ell^2 + 1$ into the product of Gaussian integers $(\ell + \mathrm{i})(\ell - \mathrm{i})$. This product, however, is not divisible by p, because the imaginary part of each factor, namely $+1$ and -1, is not divisible by p. Consequently, p is decomposable into the product of two Gaussian numbers, say, $p = (a' + \mathrm{i}a'')(b' + \mathrm{i}b'')$. The norm of p is thus given by $p^2 = (a'^2 + a''^2)(b'^2 + b''^2)$. But p is a prime, therefore, p^2 is decomposable in only one way, which requires that $a'^2 + a''^2 = b'^2 + b''^2 = p$. In other words, $p = (m + \mathrm{i}n)(m - \mathrm{i}n)$, which yields $p = m^2 + n^2$.

3.8.10. Theorem 9. All Gaussian primes belong to a set that contains the following Gaussian integers: (i) $1 + \mathrm{i}$ and its four-fold associates; (ii) all ordinary primes in the form of $p = 4k - 1$ and their corresponding associates; (iii) the pair of conjugate primes, $\wp$ and $\wp^*$, obtained by decomposing ordinary primes of the form $p = 4k + 1$, that is, $p = \wp\wp^*$, and their corresponding associates.

Proof: Let c be an arbitrary Gaussian prime. Then $N(c) = cc^* = p_1 p_2 \cdots p_n$, where the p_i are ordinary prime numbers, some of which might appear more than once in the decomposition of $N(c)$. If p_i happens to be 2, then it should be decomposed as $(1 + \mathrm{i})(1 - \mathrm{i})$. If $p_i = 4k - 1$, then it is already a Gaussian prime which cannot be decomposed further. If $p_i = 4k + 1$, then it can be decomposed as $\wp\wp^*$, where both $\wp$ and $\wp^*$ are Gaussian primes. Considering that each of the resulting Gaussian primes appearing in this expansion of $N(c)$ divides cc^*, we conclude that either c or c^* (or both) must be divisible by each and every one of the resulting Gaussian primes. Consequently, c and c^* cannot be Gaussian primes unless $N(c) = 2$, or $N(c) = p$, where the ordinary prime p has the form of $4k + 1$, or $N(c) = p^2$, where the ordinary prime p has the form of $4k - 1$.

A corollary of this theorem is that $p = 4k + 1$ can be decomposed as $p = \wp\wp^*$ in only one way, otherwise $p = \wp_1\wp_1^* = \wp_2\wp_2^*$ would imply that $\wp_1$ divides either $\wp_2$ or $\wp_2^*$, contrary to the assumption that $\wp_2$ and $\wp_2^*$ are Gaussian primes.

Problems

Problem 1) While $(a^b)^c = a^{bc}$, the standard interpretation of a^{b^c} is $a^{(b^c)}$. Given this interpretation, use the chain rule of differentiation to find the derivative $f'(x)$ of the function $f(x) = e^{e^{e^x}} = \exp\{\exp[\exp(x)]\}$.

Problem 2) Consider a straight line-segment of length L. Take out the middle third of the line, then replace it with two equal-length segments at a 60° angle, as shown in the figure. The total length of the line after this first step will be $4L/3$. Repeat the process with each of the four line-segments, so that the overall length of the (broken) line after the second step becomes $16L/9$. Find a formula for the overall length of the broken line after n repetitions of the same procedure.

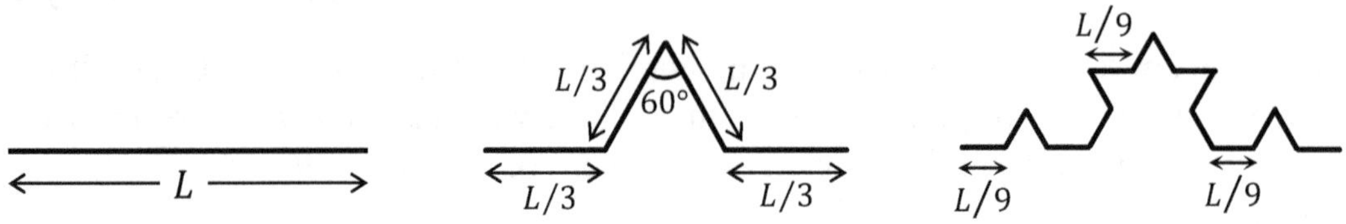

Problem 3) Use the method of integration by parts to verify the following identities:

a) $$\int_0^1 x^n \ln(x)\, dx = -\frac{1}{(n+1)^2}; \qquad n = 0, 1, 2, 3, \cdots.$$

b) $$\int_0^1 \frac{\ln x}{(1+x)^2} dx = -\ln 2.$$

c) Make a change of variable to convert the result in part (b) to the following identity:

$$\int_0^\infty \frac{x}{\cosh^2(x)} dx = \ln 2.$$

Problem 4) a) Plot the function $\ln(1+x)$, which is defined over the semi-infinite interval $x > -1$. Find the Taylor series expansion of the function around the point $x_0 = 0$.

b) Plot the function $\ln(1-x)$, which is defined over the semi-infinite interval $x < +1$. Find the Taylor series expansion of the function around $x_0 = 0$.

c) Plot the function $\ln\left(\frac{1+x}{1-x}\right)$, which is defined over the interval $-1 < x < +1$. Write the Taylor series expansion of the function around $x_0 = 0$.

d) Using the result of part (c) and setting $x = ½$, find an infinite series that converges to ln 3.

e) Using the result of part (c) and setting $x = ⅓$, find an infinite series that converges to ln 2.

Problem 5) The Gamma function $\Gamma(x)$ is defined for positive values of x as follows:

$$\Gamma(x) = \int_0^\infty y^{x-1} e^{-y} dy; \qquad x > 0.$$

a) Show that $\Gamma(1) = 1$.

b) Show that $\Gamma(2) = 1$.

c) Use the method of integration by parts to show that $\Gamma(x+1) = x\Gamma(x)$.

d) Using the results obtained in parts (a)–(c), argue that $\Gamma(n) = (n-1)!$.

e) What is the value of $\Gamma(½)$?

f) What is the value of $\Gamma(n + ½)$, where $n \geq 1$ is an integer?

Problem 6) Use the definition of the natural logarithm as $\ln(x) = \int_1^x y^{-1}dy$ to show that $\ln(x_1x_2) = \ln(x_1) + \ln(x_2)$ for any pair of positive real numbers x_1 and x_2.

Problem 7) Prove that (a) $|\sin z| \geq |\sin x|$, and (b) $|\cos z| \geq |\cos x|$, where $z = x + \mathrm{i}y$.

Problem 8) Find all the zeros of $\sin(z)$, $\cos(z)$, $\sinh(z)$, and $\cosh(z)$.

Problem 9) Show that (a) $\tan^{-1}(z) = \frac{\mathrm{i}}{2}\ln\left(\frac{\mathrm{i}+z}{\mathrm{i}-z}\right)$ and (b) $\tanh^{-1}(z) = \frac{1}{2}\ln\left(\frac{1+z}{1-z}\right)$.

Problem 10) Assuming z_1 and z_2 are arbitrary complex numbers, explain how $z_1^{z_2}$ is calculated.

Problem 11) A complex function $w = f(z)$ can be said to project certain contours in the complex z-plane to the corresponding contours in the complex w-plane. For example, as shown in the figure, $f(z) = \exp(z) = \exp(x + \mathrm{i}y) = \exp(x)\exp(\mathrm{i}y)$ projects straight vertical lines of the z-plane onto circles of radius $r = \exp(x)$ in the w-plane. Similarly, straight horizontal lines of the z-plane are projected by $\exp(z)$ onto straight, semi-infinite lines emanating from the origin of the w-plane, which make an angle $\varphi = y$ with the u-axis.

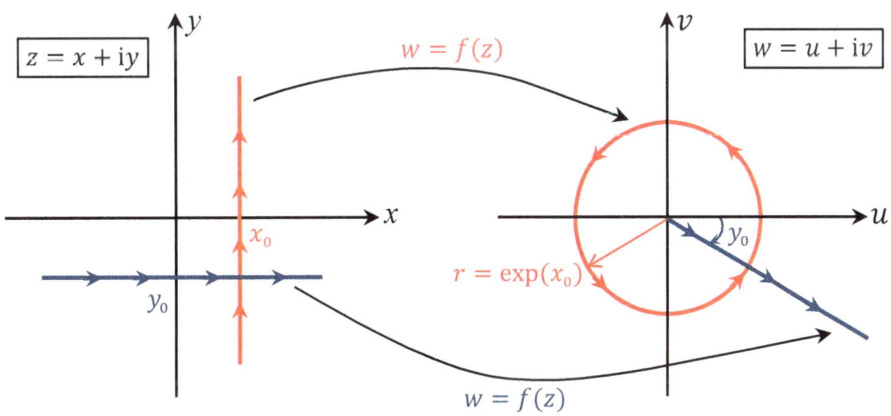

Draw similar contour projections for the following complex functions:

a) $f_1(z) = \ln z$.

b) $f_2(z) = 1/(z-1)$.

c) $f_3(z) = 1/\sqrt{z - \mathrm{i}}$.

Problem 12) Plot the general behavior of the function $f(x) = x^x$ for all real values of x ranging from $-\infty$ to ∞. Identify the point(s) where $f(x)$ has a local maximum or minimum, paying particular attention to the behavior of $f(x)$ in the vicinity of $x = 0$.

Hint: For $x < 0$, you may use the identity $\ln x = \ln|x| + i\pi$.

Problem 13) Find an expression for $I_n = \int_0^{\pi/2} \sin^n(x)\,dx$ when n is an even integer, and also a corresponding expression when n is odd. Show that the sequence $I_0, I_1, I_2, \cdots$ is monotonically decreasing and that I_{n+1}/I_n approaches 1.0 in the limit when $n \to \infty$. Use the latter result to prove Wallis's product formula, namely,

$$\frac{\pi}{2} = \prod_{n=1}^{\infty} \left(\frac{2n}{2n-1} \cdot \frac{2n}{2n+1}\right) = \frac{2\times2}{1\times3} \cdot \frac{4\times4}{3\times5} \cdot \frac{6\times6}{5\times7} \cdots.$$

Problem 14) We showed in section 3.4 that Stirling's approximation to $N!$ yields

$$(N/e)^N\sqrt{N}e^{7/8} < N! < (N/e)^N\sqrt{N}e.$$

It is also known that, in the limit when $N \to \infty$, the factorial of N asymptotically approaches $c\sqrt{N}(N/e)^N$, where the constant coefficient c is somewhere between $e^{7/8}$ and e. Use Wallis's product formula (described in problem 13) to prove that $c = \sqrt{2\pi}$.

Problem 15) Stirling's asymptotic formula, $n! \sim \sqrt{2\pi n}(n/e)^n$, which was the subject of problem 14, states that, in the limit when $n \to \infty$, the factorial of n asymptotically approaches $\sqrt{2\pi n}(n/e)^n$. Here, you are being asked to confirm Stirling's asymptotic formula using an alternative approach based on Euler's expression for the Gamma function, namely, $n! = \Gamma(n+1) = \int_0^\infty e^{-t}t^n dt$.

a) Show that the integrand, $f(t) = e^{-t}t^n$, has a peak at $t_o = n$ and inflection points at $t_{1,2} = n \pm \sqrt{n}$. The function $f(t)$ thus resembles a Gaussian, $e^{-(x-x_0)^2/w_0^2}$, having mean value x_o and a width parameter w_0.

b) Change the integration variable from t to $x = (t-n)/\sqrt{n}$ and confirm that, in the limit when $n \to \infty$, the integral approaches Stirling's asymptotic expression for $n!$.

Problem 16) Show that $\int_0^\infty \exp(-ax^2 - bx^{-2})\,dx = \sqrt{\pi/(4a)}\exp(-2\sqrt{ab})$, where a and b are real positive constants. Use this result to prove the following identities:

a) $$\int_0^T \exp\left(-\frac{a}{T-\tau} - \frac{b}{\tau}\right)\frac{d\tau}{\sqrt{(T-\tau)\tau^3}} = \sqrt{\pi/(bT)}\exp[-(\sqrt{a}+\sqrt{b})^2/T].$$

b) $$\int_0^T \exp\left(-\frac{a}{T-\tau} - \frac{b}{\tau}\right)\frac{d\tau}{[(T-\tau)\tau]^{3/2}} = \sqrt{\frac{\pi}{T^3}}\left(\frac{1}{\sqrt{a}} + \frac{1}{\sqrt{b}}\right)\exp[-(\sqrt{a}+\sqrt{b})^2/T].$$

Problem 17) Show that $\int_0^1 (x^n/\sqrt{1-x})dx = 2[(2n)!!/(2n+1)!!]$.

Hint: You may use the identity $\int_0^{\pi/2} \sin^{2n+1}(\theta)\,d\theta = (2n)!!/(2n+1)!!$, which is readily verifiable by a repeated application of the method of integration by parts; see also problem 13.

Problem 18) Show that $\int_0^1 x^{-x}dx = \int_0^1 e^{-x\ln x}\,dx = \sum_{n=1}^{\infty}(n^{-n})$.

Hint: $\int_0^\infty x^n e^{-\alpha x}dx = n!/\alpha^{n+1}$, $[\text{Re}(\alpha) > 0]$, is obtained using integration-by-parts repeatedly.

Chapter 4

Finite and Infinite Sums and Products

"The measure of greatness in a scientific idea is the extent to which it stimulates thought and opens up new lines of research."

Paul A.M. Dirac (1902–1984)

4.1. Introduction. Many situations arising in physical sciences and engineering require the addition (or multiplication) of a sequence of numbers that represent numerous, even countless, occurrences of a given event, where each occurrence contributes a different amount of some entity to an evolving process. Consider, for instance, an elastic ball bouncing off a solid floor, then rising to a fraction of its previous height after each bounce. There will be an infinite number of bounces, with each bounce taking less time than the previous one. Upon adding the durations of all the bounces, we find the total length of time that the ball will take to come to a standstill—even though the number of its bounces is beyond our ability to count. Another example is the behavior of a ray of light which, upon entering a Fabry-Perot cavity, undergoes numerous back and forth reflections between the parallel pair of mirrors that constitute the cavity. The light trapped within the cavity eventually reaches a steady state, where its amplitude and phase are determined by the sum of an infinite number of terms, each term corresponding to one specific round-trip of the light inside the cavity.

There are no general rules or recipes for finding the value of a sum (or product) that one might encounter in practice. Even deciding whether a particular series converges to a finite value is not always easy. The goal of the present chapter is to provide a few examples of useful as well as interesting series and to describe the ingenious tricks and sophisticated arguments that mathematicians over the years have discovered for dealing with such objects. We have already encountered the archetypal geometric series in chapter 1, where both finite and infinite sequences and their corresponding sums (or series) were examined in some detail. In what follows, we shall begin with a problem in geometrical optics whose solution requires the evaluation of an infinite series. We will then provide other examples, each belonging to a different category of mathematical series, and proceed to demonstrate in each case one or more of the powerful methods that have been invented to tackle such problems. The treatment will not be exhaustive, but it is hoped that the students, after going through the chapter, will have gained insights and acquired a number of tools for their arsenal that will enable them to confront mathematical problems of the sort that typically arise in practice.

4.2. Infinite series in optics. A good way to treat monochromatic geometric-optical rays is to assign a complex amplitude $A = |A|\exp(\mathrm{i}\varphi)$ to each ray, where $|A|$ and φ are the magnitude and phase of the ray, respectively. (Monochromatic means "single-color," which refers to the fact that the ray, when propagating in vacuum, is associated with a single wavelength λ_0.) The refractive indices of optical materials may also be represented by complex numbers, say, $n = n' + \mathrm{i}n''$, where n' is commonly referred to as the refractive index and n'' as the absorption coefficient. (In general, n' and n'' are lumped together as a complex number n and referred to as the "complex" refractive index.)

When a ray propagates a distance d within a homogeneous medium of (complex) refractive index n, its (complex) amplitude A gets multiplied by $\eta = \exp(\mathrm{i}2\pi n d/\lambda_0)$. At the interface between two media, say, medium 1 and medium 2, the ray is partially reflected and partially transmitted. Denoting the (complex) reflection coefficient by ρ_{12}

and the (complex) transmission coefficient by τ_{12}, the incident ray amplitude A (arriving from the incidence medium 1 at the interface with medium 2), splits into a reflected ray $\rho_{12}A$ and a transmitted ray $\tau_{12}A$. The Fresnel reflection and transmission coefficients at normal incidence, namely, $\rho_{ij} = (n_i - n_j)/(n_i + n_j)$ and $\tau_{ij} = 2n_i/(n_i + n_j)$ may be used to simplify the final results of the following calculations, although, strictly speaking, the Fresnel coefficients are not necessary and the end results may be expressed in terms of $\rho_{12}, \tau_{12}, \rho_{21}, \tau_{21}, \rho_{23}, \tau_{23}$, and η.

Figure 4.1 shows that a ray of amplitude A and vacuum wavelength λ_o arrives from the incidence medium of refractive index n_1 at a thin layer of thickness d and refractive index n_2. The layer is coated atop a homogeneous, semi-infinite medium of refractive index n_3. In terms of the parameters $\rho_{12}, \tau_{12}, \rho_{21}, \tau_{21}, \rho_{23}, \tau_{23}$, and the propagation factor $\eta = \exp(\mathrm{i}2\pi n_2 d/\lambda_o)$, we now write an infinite series for the total reflected amplitude at the interface between media 1 and 2, and also another infinite series for the transmitted amplitude at the interface between media 2 and 3.

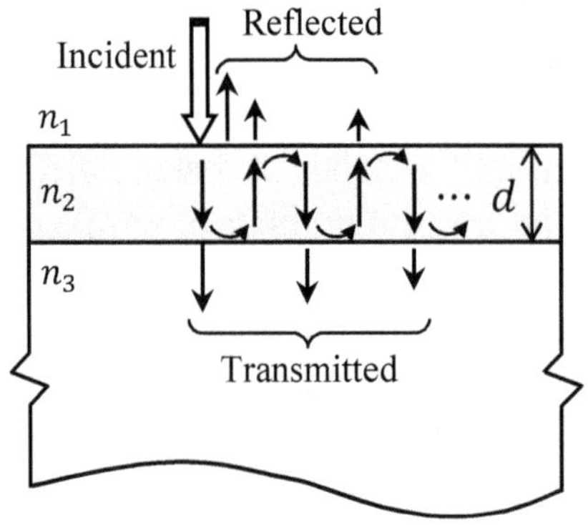

Figure 4.1. A geometric-optical ray of complex amplitude A arrives from the incidence medium of refractive index n_1 at the surface of a thin layer of thickness d and refractive index n_2. The light exiting from the thin layer enters a semi-infinite medium of refractive index n_3. Inside the thin layer, multiple reflections take place at both $2 - 1$ and $2 - 3$ interfaces.

For the reflected light, the first reflection coefficient at the interface between media 1 and 2 is ρ_{12}. The ray then enters the layer (which entails multiplication by τ_{12}), propagates to the bottom of the layer (multiplication by η), gets reflected at the $2 - 3$ interface (multiplication by ρ_{23}), returns to the top of the thin layer (another multiplication by η), and gets transmitted to the incidence medium (multiplication by τ_{21}). The process then continues, with each additional round trip through the thin layer involving another multiplication by $\rho_{21}\eta\rho_{23}\eta$. The overall reflection coefficient r at the top surface of the thin layer is thus given by the following infinite series:

$$\begin{aligned} r &= \rho_{12} + \tau_{12}(\eta\rho_{23}\eta)\tau_{21} + \tau_{12}(\eta\rho_{23}\eta)(\rho_{21}\eta\rho_{23}\eta)\tau_{21} + \tau_{12}(\eta\rho_{23}\eta)(\rho_{21}\eta\rho_{23}\eta)^2\tau_{21} + \cdots \\ &= \rho_{12} + \tau_{12}\rho_{23}\eta^2\tau_{21}[1 + (\rho_{21}\rho_{23}\eta^2) + (\rho_{21}\rho_{23}\eta^2)^2 + (\rho_{21}\rho_{23}\eta^2)^3 + \cdots]. \qquad (1) \end{aligned}$$

As for the transmitted light, the incident ray must enter the top layer (multiplication by τ_{12}), propagate to the bottom of the layer (multiplication by η), then exit from medium 2 to medium 3 (multiplication by τ_{23}). The ray arriving at the $2 - 3$ interface also gets reflected (multiplication by ρ_{23}), propagates upward (multiplication by η), bounces back from the $2 - 1$ interface (multiplication by ρ_{21}), propagates to the bottom of the thin layer (multiplication by η), then exits at the $2 - 3$ interface (multiplication by τ_{23}). The process continues, with each round trip involving a multiplication by $\rho_{23}\eta\rho_{21}\eta$. The overall transmission coefficient t at the bottom of the thin layer is thus given by the following infinite series:

$$t = \tau_{12}\eta\tau_{23} + \tau_{12}\eta(\rho_{23}\eta\rho_{21}\eta)\tau_{23} + \tau_{12}\eta(\rho_{23}\eta\rho_{21}\eta)^2\tau_{23} + \tau_{12}\eta(\rho_{23}\eta\rho_{21}\eta)^3\tau_{23} + \cdots$$
$$= \tau_{12}\eta\tau_{23}[1 + (\rho_{23}\rho_{21}\eta^2) + (\rho_{23}\rho_{21}\eta^2)^2 + (\rho_{23}\rho_{21}\eta^2)^3 + \cdots]. \quad (2)$$

We now use the geometric series formula, $\sum_{k=0}^{\infty} x^k = 1/(1-x)$, where $|x| < 1$, to express in closed form the total reflection and transmission coefficients of the thin layer depicted in figure 4.1; that is,

$$r = \rho_{12} + \frac{\tau_{12}\rho_{23}\eta^2\tau_{21}}{1 - \rho_{21}\rho_{23}\eta^2} = \frac{\rho_{12} - (\rho_{12}\rho_{21} - \tau_{12}\tau_{21})\rho_{23}\eta^2}{1 - \rho_{21}\rho_{23}\eta^2}. \quad (3)$$

$$t = \frac{\tau_{12}\tau_{23}\eta}{1 - \rho_{21}\rho_{23}\eta^2}. \quad (4)$$

These results can be further simplified with the aid of the explicit formulas for the Fresnel reflection and transmission coefficients. For instance, it is not difficult to verify that, in Eq.(3), $\rho_{12}\rho_{21} - \tau_{12}\tau_{21} = -1$. All in all, we have obtained closed-form expressions for the reflection and transmission coefficients of a thin film residing atop a homogeneous substrate under the conditions of normal incidence. As a matter of fact, it can be shown that Eqs.(3) and (4) represent exact mathematical solutions of Maxwell's electrodynamic equations for the material system depicted in figure 4.1.

4.3. Convergent and divergent series. The infinite series $\sum_{n=1}^{\infty} a_n$ is said to converge to a finite value A if, for any arbitrarily chosen small $\varepsilon > 0$, an integer N_ε can be found such that, when $N \geq N_\varepsilon$, the sum $a_1 + a_2 + \cdots + a_N$ lands in the ε-neighborhood of A; that is, $|A - \sum_{n=1}^{N} a_n| < \varepsilon$. For example, $\sum_{n=1}^{\infty}(1/n)^2$ can be shown to converge to $\pi^2/6$ (the famous Basel problem), whereas the series $\sum_{n=1}^{\infty}(1/n)$ is known to be divergent. The series $\sum_{n=1}^{\infty}(-1)^n$ does not converge because its value oscillates between -1 and 0 as the number of terms included in the sum goes through successive odd and even integers.

A series such as $\sum_{n=1}^{\infty}(1/n)$ is easily shown to be divergent, because the finite sum $\sum_{n=1}^{N}(1/n)$ is greater than the area under the function $f(x) = 1/x$ within the interval $(1, N)$; see figure 4.2a. Considering that $\int_1^N (1/x)\mathrm{d}x = \ln x|_1^N = \ln(N) - \ln(1) = \ln(N)$, the partial sum $\sum_{n=1}^{N}(1/n)$ must increase indefinitely as $N \to \infty$.

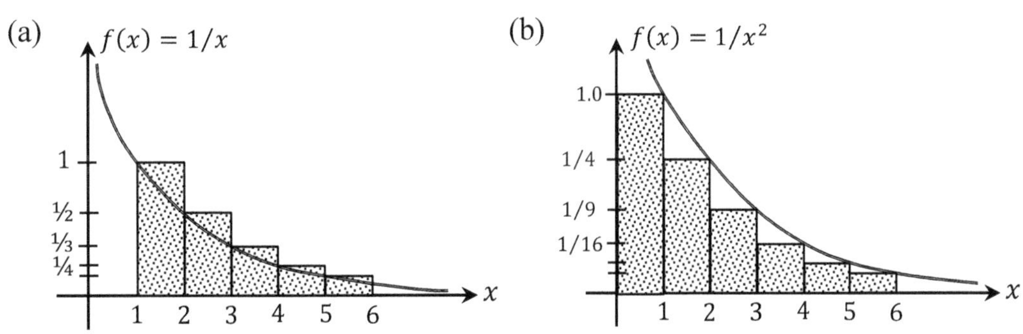

Figure 4.2. (a) The series $\sum_{n=1}^{\infty}(1/n)$ is greater than the area under the graph of $f(x) = 1/x$. (b) The series $\sum_{n=2}^{\infty}(1/n^2)$ is less than the area under the graph of $f(x) = 1/x^2$. (Vertical axes are distorted.)

In contrast, $\sum_{n=2}^{N}(1/n^2)$ is less than the area under the function $g(x) = 1/x^2$ within $1 \leq x \leq N$; see figure 4.2b. Considering that $\int_1^{\infty}(1/x^2)\mathrm{d}x = -(1/x)|_1^{\infty} = 1$, the partial sum $\sum_{n=1}^{N}(1/n^2) = 1 + \sum_{n=2}^{N}(1/n^2)$ must converge to a number below 2 as $N \to \infty$.

Divergent series can lead to counterintuitive results. Consider the series $S = 1 + 2 + 3 + \cdots = \sum_1^{\infty} n$. In what follows we shall try to convince the reader that $S = -1/12$, a surprising result, to say the least. To do so, we note that $S_1 = 1 - 2 + 3 - 4 + \cdots$ can be evaluated as follows:

$$\begin{array}{rl} 2S_1 = & 1 - 2 + 3 - 4 + 5 - 6 + \cdots \\ & + \quad 1 - 2 + 3 - 4 + 5 - \cdots \\ \hline 2S_1 = & 1 - 1 + 1 - 1 + 1 - 1 + \cdots = ½. \end{array}$$

The latter result can be seen from the fact that the partial sums alternate between 0 and 1; therefore, their average value must approach ½. Alternatively, the geometric series $1 + x + x^2 + \cdots = 1/(1 - x)$, in the "extreme" case of $x = -1$, yields $1 - 1 + 1 - 1 + \cdots = 1/(1 + 1) = ½$. In any case, with these sleights of hand, we hope to have convinced the reader that $S_1 = ¼$. We now write

$$\begin{array}{rl} S - S_1 = & \cancel{1} + 2 + \cancel{3} + 4 + \cancel{5} + 6 + \cdots \\ & -\cancel{1} + 2 - \cancel{3} + 4 - \cancel{5} + 6 - \cdots \\ \hline S - S_1 = & \quad 4 \quad + \quad 8 \quad + \quad 12 \quad + \cdots = 4(1 + 2 + 3 + \cdots) = 4S. \end{array}$$

Consequently, $S - S_1 = 4S \rightarrow 3S = -S_1 \rightarrow S = -S_1/3 \rightarrow S = -1/12$. We have thus "shown" that the sum of an infinite number of positive integers is a negative fractional number! Be careful with your logical reasoning and also with how you rearrange the terms of a series!

Brain teaser. A wine bottle has just been uncorked. You are handed the bottle, the cork, a large wineglass, and a marker with which to write on the bottle. You are asked to pour *exactly* half of the wine in the glass. How would you go about it?

4.4. Alternative methods of handling an infinite series. In general, there are no simple, universal rules for evaluating a finite or an infinite series. We learn by studying examples of different methods that have been invented to solve various problems. Occasionally, we find that multiple approaches can be applied to a given class of problems. Consider, for instance, the infinite series $\sum_{k=1}^{\infty} kx^k$, where $|x| < 1$. One way to handle this series is to begin with the geometric series $f(x) = \sum_{k=1}^{\infty} x^k = x/(1 - x)$. Differentiating both sides of the geometric series yields $f'(x) = \sum_{k=1}^{\infty} kx^{k-1} = 1/(1 - x)^2$, which, upon multiplication with x gives the desired solution as $xf'(x) = \sum_{k=1}^{\infty} kx^k = x/(1 - x)^2$.

An alternative approach involves arranging the various terms of the series $\sum_{k=1}^{\infty} kx^k$ in the following way:

$$\Bigg\Downarrow \quad \begin{array}{l} x = x \\ 2x^2 = x^2 + x^2 \\ 3x^3 = x^3 + x^3 + x^3 \\ 4x^4 = x^4 + x^4 + x^4 + x^4 \\ \vdots \end{array}$$

Adding these equations one column at a time (as indicated by the arrow), we find

$$\sum_{k=1}^{\infty} kx^k = \sum_{k=1}^{\infty} \sum_{n=k}^{\infty} x^n = \sum_{k=1}^{\infty}[x^k/(1 - x)] = (\sum_{k=1}^{\infty} x^k)/(1 - x) = x/(1 - x)^2.$$

Both methods thus yield the same answer, which bolsters confidence in the validity of rearranging the terms of the series in accordance with our second approach. Such rearrangements, however, are *not* always legitimate, as will be seen in another example further below. One must always remain vigilant that a particular rearrangement will not change the nature of the series, nor will it violate the fundamental notion as to what the infinite series is meant to represent.

4.5. Taylor series expansion of $\ln x$. Any function that can be expanded in a Taylor series will produce many convergent series. The function $\ln x$ provides a good example. This function may be expanded around the point $x_0 = 1$ as follows:

$$\ln x_0 = 0,$$

$$\mathrm{d}\ln x/\mathrm{d}x|_{x=x_0} = x^{-1}|_{x=x_0} = 1,$$

$$\mathrm{d}^2\ln x/\mathrm{d}x^2|_{x=x_0} = -x^{-2}|_{x=x_0} = -1,$$

$$\mathrm{d}^3\ln x/\mathrm{d}x^3|_{x=x_0} = 2x^{-3}|_{x=x_0} = 2,$$

$$\vdots$$

$$\mathrm{d}^n\ln x/\mathrm{d}x^n|_{x=x_0} = (-1)^{n-1}(n-1)!\,x^{-n}|_{x=x_0} = (-1)^{n-1}(n-1)!.$$

Therefore,

$$\ln x = (x-1) - \tfrac{1}{2}(x-1)^2 + \tfrac{1}{3}(x-1)^3 - \cdots + (-1)^{n-1}\tfrac{1}{n}(x-1)^n + \cdots. \qquad (5)$$

This series is known to be convergent when $|x-1| < 1$. Let us examine the behavior of the series for a few different values of x. At $x = 2$, Eq.(5) yields

$$\ln 2 = 1 - \tfrac{1}{2} + \tfrac{1}{3} - \tfrac{1}{4} + \cdots = \sum_{n=1}^{\infty}(-1)^{n-1}/n. \qquad (6)$$

It is thus seen that the alternating sign version of the divergent series $\sum_{n=1}^{\infty}(1/n)$ is convergent. Another infinite series that adds up to the same value ($\ln 2$) is obtained from Eq.(5) by setting $x = ½$. We find

$$\ln ½ = -\tfrac{1}{2} - \tfrac{1}{2}\left(\tfrac{1}{2}\right)^2 - \tfrac{1}{3}\left(\tfrac{1}{2}\right)^3 - \cdots - \tfrac{1}{n}\left(\tfrac{1}{2}\right)^n - \cdots = -\sum_{n=1}^{\infty} 1/(2^n n). \qquad (7)$$

Considering that $\ln ½ = -\ln 2$, we now see that the series $\sum_{n=1}^{\infty} 1/(2^n n)$ also converges to $\ln 2$. It is not unusual for substantially different series to converge to the same value.

As a final example, let us set $x = 1 + \mathrm{i}$, a complex number for which the logarithm is well defined (see chapter 3, section 3.6.5). Equation (5) yields

$$\ln(1+\mathrm{i}) = \mathrm{i} - \tfrac{1}{2}\mathrm{i}^2 + \tfrac{1}{3}\mathrm{i}^3 - \tfrac{1}{4}\mathrm{i}^4 + \cdots + (-1)^{n-1}\tfrac{1}{n}\mathrm{i}^n + \cdots$$

$$= \left[\tfrac{1}{2} - \tfrac{1}{4} + \tfrac{1}{6} - \cdots + (-1)^{n-1}\tfrac{1}{2n} + \cdots\right] + \mathrm{i}\left[1 - \tfrac{1}{3} + \tfrac{1}{5} - \cdots + (-1)^{n-1}\tfrac{1}{2n-1} + \cdots\right]$$

$$= \left[½\sum_{n=1}^{\infty}(-1)^{n-1}/n\right] + \mathrm{i}\left[\sum_{n=1}^{\infty}(-1)^{n-1}/(2n-1)\right]. \qquad (8)$$

With reference to Eq.(6), the real part of the above series is now seen to be just $½\ln 2$. The imaginary part is an alternating series consisting of the inverses of odd integers, 1, ⅓, ⅕, and so on. To see what this series converges to, consider that, on the left-hand-side of Eq.(8), we have $\ln(1+\mathrm{i})$, which is $\ln[\sqrt{2}\exp(\mathrm{i}\pi/4)]$, which can be simplified as $\tfrac{1}{2}\ln 2 + \mathrm{i}(\pi/4)$. A comparison with Eq.(8) reveals that the second sum on the right-hand-side of Eq.(8) must converge to $\pi/4$. We have thus found the following interesting series for π:

$$4\left(1 - \tfrac{1}{3} + \tfrac{1}{5} - \cdots + \frac{(-1)^{n-1}}{2n-1} + \cdots\right) = \pi. \qquad (9)$$

It is seen that the Taylor series expansion of a well-known function such as $\ln x$ can lead to some interesting (and occasionally unexpected) series. More examples of this process are given in the next section—and also in the problems at the end of the chapter.

Digression: Consider the geometric series $\sum_{n=0}^{\infty} y^n = 1/(1-y)$, which is valid for $|y| < 1$. Integrating both sides of the equation from $y = 0$ to $y = x$ yields

$$\sum_{n=0}^{\infty}\left(\int_0^x y^n \mathrm{d}y\right) = \sum_{n=1}^{\infty}(x^n/n) = \int_0^x \mathrm{d}y/(1-y) = -\ln(1-x); \qquad |x| < 1. \quad (10)$$

This equation shows that $\ln(1-x)$ can be expanded as $-\sum_{n=1}^{\infty}(x^n/n)$ within a convergence radius given by $|x| < 1$. This is essentially the same series as obtained by a Taylor series expansion of $\ln(x)$ around the point $x_0 = 1$, which is given in Eq.(5); their equivalence can be seen simply by replacing x with $1 - x$.

4.6. Taylor series expansions of $\sin(x)$ and $\cos(x)$. It is well known that the derivative with respect to x of $\sin x$ is $\cos x$, whereas that of $\cos x$ is $-\sin x$. These facts, together with the fact that $\sin(0) = 0$ and $\cos(0) = 1$, suffice to enable one to write the Taylor series expansions of $\sin x$ and $\cos x$ around the point $x_0 = 0$ as

$$\sin x = x - (x^3/3!) + (x^5/5!) - (x^7/7!) + \cdots, \qquad (11)$$

$$\cos x = 1 - (x^2/2!) + (x^4/4!) - (x^6/6!) + \cdots. \qquad (12)$$

Note that the arguments of the trigonometric sine and cosine functions *must* be in units of *radians*. This is because

$$\frac{\mathrm{d}}{\mathrm{d}\theta}\sin\theta = lim_{\Delta\theta\to 0}\frac{\sin(\theta+\Delta\theta) - \sin(\theta)}{\Delta\theta}$$

$$= lim_{\Delta\theta\to 0}\frac{\sin(\theta)\underbrace{\cos(\Delta\theta)}_{\cong 1} + \cos(\theta)\underbrace{\sin(\Delta\theta)}_{\cong \Delta\theta} - \sin(\theta)}{\Delta\theta} = \cos\theta. \qquad (13)$$

The approximation $\sin(\Delta\theta) \cong \Delta\theta$ for small values of $\Delta\theta$ used in the above derivation is valid only when $\Delta\theta$ is expressed in units of radians. A similar argument applies to $\mathrm{d}\cos\theta/\mathrm{d}\theta$ as well.

The series given in Eqs.(11) and (12) are valid for all values of x, not just over the interval $[0, 2\pi]$ or $[-\pi, \pi]$. For instance,

$$\sin(1.0 \text{ radian}) = 1 - \frac{1}{3!} + \frac{1}{5!} - \frac{1}{7!} + \cdots \quad \text{and} \quad \cos(1.0 \text{ radian}) = 1 - \frac{1}{2!} + \frac{1}{4!} - \frac{1}{6!} + \cdots,$$

$$\sin(2\pi) = 2\pi - \frac{(2\pi)^3}{3!} + \frac{(2\pi)^5}{5!} - \frac{(2\pi)^7}{7!} + \cdots = 0,$$

$$1 - \cos(2\pi) = \frac{(2\pi)^2}{2!} - \frac{(2\pi)^4}{4!} + \frac{(2\pi)^6}{6!} - \frac{(2\pi)^8}{8!} + \cdots = 0.$$

The periodicity of $\sin x$ and $\cos x$ means that each of the series in Eqs.(11) and (12) will have precisely the same value for x as it will have for $x + 2\pi$, $x - 2\pi$, $x + 4\pi$, etc.

4.7. The method of partial fractions. Certain mathematical expressions can be decomposed into the sum or difference of two (or more) simpler expressions. For instance, one can always write $1/[x(x+1)]$ as $(1/x) - [1/(x+1)]$. A series whose individual terms can be thus broken up into simpler elements can often be evaluated by rearranging and combining the simpler elements. As an example, consider the infinite series $\sum_{n=1}^{\infty} 1/[n(n+1)]$. It is instructive to write out the first few terms of this series to see how the various neighboring terms cancel each other out.

$$\sum_{n=1}^{\infty} \frac{1}{n(n+1)} = \sum_{n=1}^{\infty} \left(\frac{1}{n} - \frac{1}{n+1}\right) = \left(1 - \frac{1}{2}\right) + \left(\frac{1}{2} - \frac{1}{3}\right) + \left(\frac{1}{3} - \frac{1}{4}\right) + \cdots = 1. \qquad (14)$$

Another example is provided by the infinite series $\sum_{n=2}^{\infty} 1/(n^2 - 1)$, as follows:

$$\begin{aligned}\sum_{n=2}^{\infty} \frac{1}{n^2-1} &= \sum_{n=2}^{\infty} \frac{1}{2}\left(\frac{1}{n-1} - \frac{1}{n+1}\right)\\ &= \frac{1}{2}\left[\left(1 - \frac{1}{3}\right) + \left(\frac{1}{2} - \frac{1}{4}\right) + \left(\frac{1}{3} - \frac{1}{5}\right) + \left(\frac{1}{4} - \frac{1}{6}\right) + \left(\frac{1}{5} - \frac{1}{7}\right) + \cdots\right]\\ &= \frac{1}{2}\sum_{n=2}^{\infty}\left(\frac{1}{n-1}\right) - \frac{1}{2}\sum_{n=2}^{\infty}\left(\frac{1}{n+1}\right)\\ &= \frac{1}{2}\sum_{n=1}^{\infty}\left(\frac{1}{n}\right) - \frac{1}{2}\sum_{n=3}^{\infty}\left(\frac{1}{n}\right) = \frac{1}{2}\left(1 + \frac{1}{2}\right) = \frac{3}{4}. \qquad (15)\end{aligned}$$

A slightly more complicated example involves higher-order terms of the form $1/[x(x+1)(x+2)]$, which must be broken up as $(A/x) + [B/(x+1)] + [C/(x+2)]$. To find the A, B, C coefficients, we write

$$\frac{1}{x(x+1)(x+2)} = \frac{A}{x} + \frac{B}{x+1} + \frac{C}{x+2} = \frac{A(x^2+3x+2)+B(x^2+2x)+C(x^2+x)}{x(x+1)(x+2)} = \frac{(A+B+C)x^2+(3A+2B+C)x+2A}{x(x+1)(x+2)}.$$

Equality of the two sides of this equation requires that we set $A + B + C = 0$, $3A + 2B + C = 0$, and $2A = 1$. These three linear equations in thrcc unknowns may now be solved to yield $A = ½$, $B = -1$, and $C = ½$. Applying these results to the infinite series $\sum_{n=1}^{\infty} 1/[n(n+1)(n+2)]$, we find

$$\begin{aligned}\sum_{n=1}^{\infty} \frac{1}{n(n+1)(n+2)} &= ½\sum_{n=1}^{\infty}\left(\frac{1}{n}\right) - \sum_{n=1}^{\infty}\left(\frac{1}{n+1}\right) + ½\sum_{n=1}^{\infty}\left(\frac{1}{n+2}\right)\\ &= ½\sum_{n=1}^{\infty}\left(\frac{1}{n}\right) - \sum_{n=2}^{\infty}\left(\frac{1}{n}\right) + ½\sum_{n=3}^{\infty}\left(\frac{1}{n}\right)\\ &= ½(1 + ½) - ½ + (½ - 1 + ½)\sum_{n=3}^{\infty}\left(\frac{1}{n}\right) = ¼. \qquad (16)\end{aligned}$$

When applying the method of partial fractions, one must be careful to allow only adjacent terms (or terms that are near each other) to cancel out. A cautionary example is provided by the infinite series $\sum_{n=1}^{\infty} 1/[n(2n+1)]$, which yields an incorrect sum if the terms are broken up and carelessly rearranged. To see how things could go wrong, let us begin by noting that $1/[x(2x+1)] = (1/x) - [2/(2x+1)]$. We will have

$$\sum_{n=1}^{\infty} \frac{1}{n(2n+1)} = \sum_{n=1}^{\infty} \left(\frac{1}{n}\right) - \sum_{n=1}^{\infty} \left(\frac{2}{2n+1}\right)$$

Separating odd and even terms → $$= \left[\sum_{m=0}^{\infty} \left(\frac{1}{2m+1}\right) + \sum_{m=1}^{\infty} \left(\frac{1}{2m}\right)\right] - 2\sum_{n=1}^{\infty} \left(\frac{1}{2n+1}\right)$$

$$= 1 - \sum_{m=1}^{\infty} \left(\frac{1}{2m+1}\right) + \sum_{m=1}^{\infty} \left(\frac{1}{2m}\right)$$

See expansion of ln 2 in Eq.(6) → $$= 2 - \ln 2 = 1.30685 \cdots. \tag{17}$$ ← Wrong answer

The above answer is *not* correct, as can be readily verified by adding up the first few terms of the series using a calculator. The reason is that, while rearranging and then cancelling some of the terms of the series in Eq.(17), we allowed some of the terms to run ahead of other terms—while their distance kept growing. This procedure is incompatible with the fundamental *definition* of the sum of an infinite series, which requires that, at first, the partial sum up to $n = N$ be calculated, then the number of terms, N, be allowed to approach infinity. Jumping over some of the terms and skipping ahead, as was done in arriving at the final result of Eq.(17), violates this fundamental requirement.

The correct way to treat the series must ensure that the terms that cancel out always remain close to each other. This requires a different splitting of the terms of the series at the outset, as follows:

$$\sum_{n=1}^{\infty} \frac{1}{n(2n+1)} = \sum_{n=1}^{\infty} \frac{2}{2n(2n+1)} = 2\left[\sum_{n=1}^{\infty} \left(\frac{1}{2n}\right) - \sum_{n=1}^{\infty} \left(\frac{1}{2n+1}\right)\right]$$

Correct answer → $$= 2(1 - \ln 2) = 0.6137 \cdots. \tag{18}$$ ← See expansion of ln 2 in Eq.(6)

Here, the partial fractions $1/(2n)$ and $1/(2n+1)$ remain adjacent to each other as $n \to \infty$, in stark contrast to the incorrect approach followed in Eq.(17), where the odd terms of the first series, $\sum_{m=0}^{\infty} 1/(2m+1)$, did *not* keep pace with the corresponding terms of the second series, $\sum_{n=1}^{\infty} 1/(2n+1)$, while the index n kept rising.

Brain teaser. In this infinite network of identical resistors, what is the electric resistance between A and B?

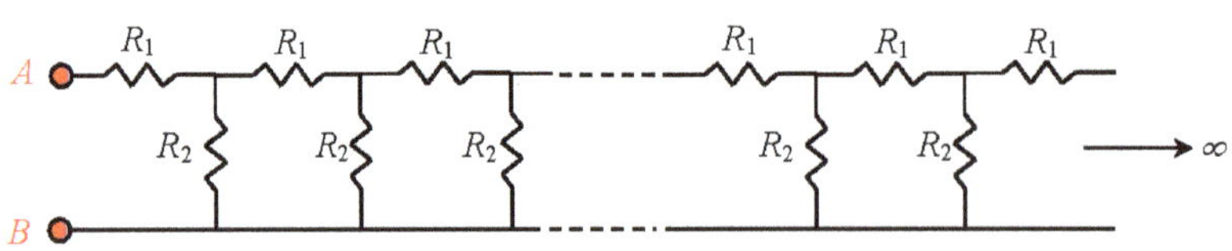

4.8. Riemann's zeta function and the Basel problem. The Riemann zeta function is defined as the infinite sum

$$\zeta(s) = \sum_{n=1}^{\infty}(1/n^s); \quad \mathrm{Re}(s) > 1. \tag{19}$$

It is easy to show that the series converges for any complex number s whose real part is greater than one. This is because $|n^s| = |n^{\mathrm{Re}(s)+\mathrm{i}\,\mathrm{Im}(s)}| = n^{\mathrm{Re}(s)}$ and, consequently,

$$\left|\sum_{n=n_0}^{\infty}(1/n^s)\right| \le \sum_{n_0}^{\infty}|1/n^s| = \sum_{n_0}^{\infty} n^{-\mathrm{Re}(s)} < \int_{n_0-1}^{\infty} x^{-\mathrm{Re}(s)}\mathrm{d}x$$

$$= \left.\frac{x^{1-\mathrm{Re}(s)}}{1-\mathrm{Re}(s)}\right|_{n_0-1}^{\infty} = \frac{(n_0-1)^{1-\mathrm{Re}(s)}}{\mathrm{Re}(s)-1}; \qquad \mathrm{Re}(s) > 1. \tag{20}$$

Thus, considering that for sufficiently large n_0, the sum of the terms corresponding to $n \geq n_0$ can be made as small as desired, we conclude that the infinite sum that defines the zeta function forms a convergent series.

The celebrated Basel problem, whose solution in 1735 brought acclaim to the twenty-seven-year-old Leonhard Euler (1707–1783), concerned the special case of the Riemann zeta function for $s = 2$. Euler observed that an Nth order polynomial $\sum_{n=0}^{N} a_n x^n$ whose zeros are known to be at $x = x_1, x_2, \cdots, x_N$ can be expressed as $a_0 \prod_{n=1}^{N}[1 - (x/x_n)]$. He then conjectured that the function $\sin(x)/x$, which equals 1.0 at $x = 0$ and has an infinite number of zeros at $x = \pm\pi, \pm 2\pi, \cdots, \pm n\pi, \cdots$, could be written as follows:

$$\sin(x)/x = \prod_{n=1}^{\infty}\left(1 - \tfrac{x}{n\pi}\right)\left(1 + \tfrac{x}{n\pi}\right) = \prod_{n=1}^{\infty}\left(1 - \tfrac{x^2}{n^2\pi^2}\right). \tag{21}$$

The lowest-order term of the infinite product on the right-hand-side of Eq.(21) is 1, and the next term is $[-\sum_{n=1}^{\infty}(1/n\pi)^2]x^2$. The remaining terms contain higher powers of x^2, which are not of interest here. Given that $\sin(x) = x - (x^3/3!) + (x^5/5!) - \cdots$, we may write the left-hand-side of Eq.(21) as $\sin(x)/x = 1 - (x^2/3!) + \cdots$. Comparing the two sides of the equation, we conclude that $\sum_{n=1}^{\infty}(1/n^2) = \pi^2/6$. This is Euler's famous solution to the Basel problem.

It is worthwhile to consider the graphical representation, depicted in figure 4.3, of the various quadratic functions (or parabola) appearing on the right-hand-side of Eq.(21), to observe how their product approaches the function $\sin(x)/x$ when more and more terms of the infinite product are included in the graphical representation.

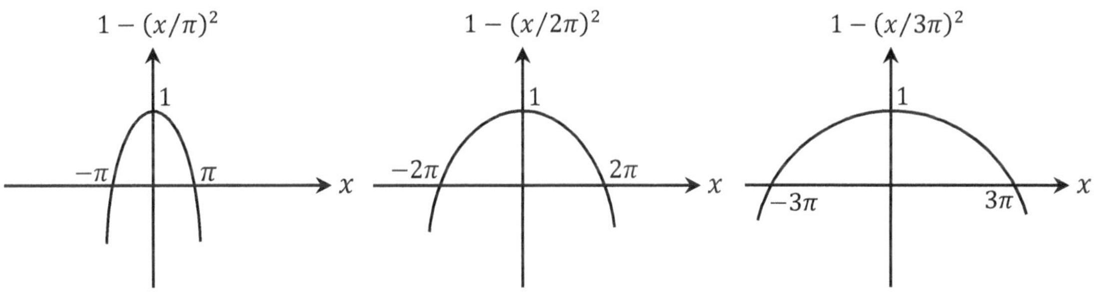

Figure 4.3. Plots of the first few quadratic factors $1 - (x/n\pi)^2$ appearing in Euler's expansion of the function $\sin(x)/x$ given in Eq.(21). When multiplied together, these parabolic functions reproduce more and more of the zeros of $\sin(x)/x$, while retaining their value of 1.0 at $x = 0$.

One could apply Euler's method to the function $\cos x$ as well. Here, the zeros of the function are at $x = \pm\pi/2, \pm 3\pi/2, \pm 5\pi/2, \cdots$, and, as before, the value of the function at $x = 0$ is 1. We will have

$$\cos x = \prod_{n=1}^{\infty}\left[1 - \tfrac{x}{(n-½)\pi}\right]\left[1 + \tfrac{x}{(n-½)\pi}\right] = \prod_{n=1}^{\infty}\left[1 - \tfrac{4x^2}{(2n-1)^2\pi^2}\right]. \tag{22}$$

While the lowest-order terms of the infinite product on the right-hand side of Eq.(22) are $1 - (4/\pi^2)\sum_{n=1}^{\infty} 1/(2n-1)^2$, the corresponding terms in the Taylor series expansion of $\cos x$ are $1 - ½x^2$. We conclude that $\sum_{n=1}^{\infty} 1/(2n-1)^2 = \pi^2/8$.

The results just obtained are not independent and can be related to each other in a simple way. Upon separating the odd and even terms of the Basel series, one finds

$$\pi^2/6 = \sum_{n=1}^{\infty}(1/n^2) = \left[\sum_{m=1}^{\infty} 1/(2m-1)^2\right] + \left[\sum_{m=1}^{\infty} 1/(2m)^2\right]$$

$$= \left[\sum_{m=1}^{\infty} 1/(2m-1)^2\right] + \tfrac{1}{4}\left[\sum_{m=1}^{\infty}(1/m^2)\right]$$

$$= \left[\sum_{m=1}^{\infty} 1/(2m-1)^2\right] + (\pi^2/24) \quad \rightarrow \quad \textstyle\sum_{m=1}^{\infty} 1/(2m-1)^2 = \pi^2/8. \qquad (23)$$

Another method of evaluating the Basel series is discussed in chapter 7, problems 25 and 26. Chapter 10 contains more advanced topics related to Riemann's zeta function.

An interesting application of the Basel series involves the evaluation of the following integral:

$$\int_0^1 \ln(x)\ln(1-x)\,dx = 2 - (\pi^2/6).$$ ← Gradshteyn & Ryzhik 4.221-1

Recalling the Taylor series expansion $\ln(1-x) = -\sum_{n=1}^{\infty}(x^n/n)$, we write

$$\int_0^1 x^n \ln x\,dx = \frac{x^{n+1}}{n+1}\left(\ln x - \frac{1}{n+1}\right)\Big|_0^1 = -1/(n+1)^2\,.$$

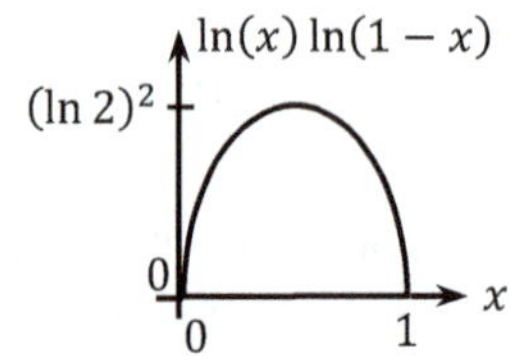

Consequently, ($x^{n+1}\ln x \to 0$ when $x \to 0$)

$$\int_0^1 \ln(x)\ln(1-x)\,dx = \sum_{n=1}^{\infty} 1/[n(n+1)^2]$$

$$= \sum_{n=1}^{\infty}\{(1/n) - [(n+2)/(n+1)^2]\}$$

$$= \textstyle\sum_{n=1}^{\infty}(1/n) - \sum_{n=1}^{\infty}[1/(n+1)] - \sum_{n=1}^{\infty}[1/(n+1)^2]$$

$$= 1 - \textstyle\sum_{n=2}^{\infty}(1/n^2) = 2 - (\pi^2/6).$$

Digression: Euler's prime number product and its relation to Riemann's zeta function. The zeta function is defined as $\zeta(s) = \sum_{n=1}^{\infty}(1/n^s)$, which, for real values of s, converges when $s > 1$. Multiplying $\zeta(s)$ by $(1 - 2^{-s})$ yields

$$\zeta(s)\left(1 - \frac{1}{2^s}\right) = 1 + \frac{1}{2^s} + \frac{1}{3^s} + \cdots - \left(\frac{1}{2^s} + \frac{1}{4^s} + \frac{1}{6^s} + \cdots\right).$$

Thus, multiplication by $(1 - 2^{-s})$ eliminates all the terms $1/n^s$ where n is a multiple of 2. If we now multiply the above product by $(1 - 3^{-s})$, we will find

$$\zeta(s)\left(1 - \frac{1}{2^s}\right)\left(1 - \frac{1}{3^s}\right) = 1 + \frac{1}{3^s} + \frac{1}{5^s} + \frac{1}{7^s} + \cdots - \left(\frac{1}{3^s} + \frac{1}{9^s} + \frac{1}{15^s} + \cdots\right).$$

We have thus eliminated all the remaining terms where n is a multiple of 3. Continuing in this way and multiplying by $(1 - 5^{-s})(1 - 7^{-s})\cdots(1 - p^{-s})$, where p is a prime number, all the terms $1/n^s$ in which n is a multiple of any integer up through p will be eliminated. As $p \to \infty$, we will have

$$\zeta(s)\prod_{\substack{p=\text{prime}\\(2,3,5,7,11,\cdots)}}^{\infty}\left(1 - \frac{1}{p^s}\right) = 1 \quad \rightarrow \quad \zeta(s) = \textstyle\sum_{n=1}^{\infty}(1/n^s) = \prod_{\text{all primes}}\left(\frac{p^s}{p^s - 1}\right).$$

The properties of the zeta function are intimately tied to those of the prime numbers.

4.9. Euler's proof of divergence of the sum of the inverses of primes. The fundamental theorem of arithmetic states that every integer n can be written as a unique product $p_1^{k_1}p_2^{k_2}\cdots p_m^{k_m}$ of its prime factors $p_\ell \geq 2$, where each factor is raised to some integer power $k_\ell \geq 1$. Euler thus wrote

$$\textstyle\sum_{n=1}^{\infty}(1/n) = \prod_{\text{primes}}\left(1 + \frac{1}{p} + \frac{1}{p^2} + \frac{1}{p^3} + \cdots\right) = \prod_{\text{primes}}\frac{1}{1-(1/p)}. \qquad (24)$$

Taking the natural logarithm of both sides of Eq.(24), then expanding $\ln(1-x)$ into its Taylor series $-[x + (x^2/2) + (x^3/3) + \cdots]$, which is valid for $|x| < 1$, we will have

$$\ln[\textstyle\sum_{n=1}^{\infty}(1/n)] = -\sum_{\text{primes}} \ln[1-(1/p)] = \sum_{\text{primes}} \left(\frac{1}{p}+\frac{1}{2p^2}+\frac{1}{3p^3}+\frac{1}{4p^4}+\cdots\right). \quad (25)$$

Now, the sum on the left-hand side of the equation is *divergent*, and it is fairly easy to show that the sum on the right-hand side would be *convergent* if the first term $(1/p)$ were left out.† We conclude that $\sum_{\text{primes}}(1/p)$ must diverge. Note that this is a stronger statement than Euclid's assertion that the sequence of primes is interminable. The above proof of Euler's shows that the primes *cannot* be rare; in other words, they must be fairly densely distributed amongst positive integers because, otherwise, the sum of their inverses would have been convergent.

4.10. The binomial expansion. The function $f(x) = (1+x)^n$, where n is a positive integer, can be expanded in powers of x and written as an nth order polynomial $f(x) = a_0 + a_1 x + a_2 x^2 + \cdots + a_n x^n$. To find the coefficients of this polynomial, differentiate $f(x)$ repeatedly (up to n times), then set $x=0$ for each of the derivatives thus obtained. You will find

$$a_0 = (1+x)^n|_{x=0} = 1,$$

$$a_1 = \frac{\mathrm{d}}{\mathrm{d}x}(1+x)^n|_{x=0} = n(1+x)^{n-1}|_{x=0} = n,$$

$$2!\,a_2 = \frac{\mathrm{d}^2}{\mathrm{d}x^2}(1+x)^n|_{x=0} = n(n-1)(1+x)^{n-2}|_{x=0} = n(n-1) \quad \rightarrow \quad a_2 = \binom{n}{2},$$

$$\vdots$$

$$a_m = \binom{n}{m},$$

$$\vdots$$

$$a_n = \binom{n}{n} = 1.$$

A similar procedure may be applied to the function $f(x) = (1+x)^s$, where s is a *non*-integer, which may be positive or negative. We will have

$$(1+x)^s = \textstyle\sum_{m=0}^{\infty} a_m x^m, \quad (26a)$$

where

$$a_m = (1/m!)\frac{\mathrm{d}^m}{\mathrm{d}x^m}(1+x)^s\Big|_{x=0} = s(s-1)(s-2)\cdots(s-m+1)/m!. \quad (26b)$$

This expansion of $(1+x)^s$ can be shown to be convergent when $|x|<1$. (The condition for convergence is a sufficient but not necessary condition.)

Example 1. Let us examine the case of $s = -½$, for which Eq.(26) yields

$$(1+x)^{-1/2} = 1 - \frac{1}{2}x + \frac{(-1/2)(-3/2)}{2!}x^2 + \frac{(-1/2)(-3/2)(-5/2)}{3!}x^3 + \cdots$$

$$= 1 - \frac{1}{2}x + \frac{3!!}{4!!}x^2 - \frac{5!!}{6!!}x^3 + \cdots, \qquad |x|<1. \quad (27)$$

In formal notation, one writes the preceding expansion as

---

† $\sum_{\text{primes}}\left(\frac{1}{2p^2}+\frac{1}{3p^3}+\frac{1}{4p^4}+\cdots\right) < \frac{1}{2}\int_1^{\infty}\frac{\mathrm{d}x}{x^2} + \frac{1}{3}\int_1^{\infty}\frac{\mathrm{d}x}{x^3} + \frac{1}{4}\int_1^{\infty}\frac{\mathrm{d}x}{x^4} + \cdots = \frac{1}{1\times 2}+\frac{1}{2\times 3}+\frac{1}{3\times 4}+\cdots < \infty.$

$$(1+x)^{-1/2} = 1 + \sum_{m=1}^{\infty}(-1)^m \frac{(2m-1)!!}{(2m)!!} x^m. \tag{28}$$

A simple change of sign from x to $-x$ now gives the Taylor series expansion of the function $(1-x)^{-1/2}$, as follows:

$$(1-x)^{-1/2} = 1 + \frac{1}{2}x + \frac{3!!}{4!!}x^2 + \frac{5!!}{6!!}x^3 + \cdots = \sum_{m=0}^{\infty} \binom{2m}{m} \left(\frac{x}{4}\right)^m, \qquad |x| < 1. \tag{29}$$

Example 2. We derive the Taylor series expansion of the function $f(x) = \sin^{-1} x$, which is sometimes written in the alternative form $f(x) = \arcsin(x)$, where $|x| \le 1$. We write

$$\sin[f(x)] = x \quad \rightarrow \quad f'(x)\cos[f(x)] = 1 \quad \rightarrow \quad \frac{\mathrm{d}f(x)}{\mathrm{d}x} = \frac{1}{\cos[f(x)]} = 1/\sqrt{1-x^2}.$$

It is now possible to express $f(x)$ as the integral of a function with a known Taylor series expansion, namely,

$$f(x) = \int_0^x (1-t^2)^{-½}\mathrm{d}t = \int_0^x \{1 + \sum_{m=1}^{\infty}[(2m-1)!!/(2m)!!]t^{2m}\}\mathrm{d}t$$

$$\rightarrow \quad \sin^{-1}(x) = x + \sum_{m=1}^{\infty} \frac{(2m-1)!!}{(2m+1)(2m)!!} x^{2m+1}, \quad |x| < 1. \tag{30}$$

Figure 4.4 shows a plot of $\arcsin(x)$ versus x, and also that of the function $1/\sqrt{1-t^2}$ versus t, which was used in the above Taylor series expansion of $\arcsin(x)$.

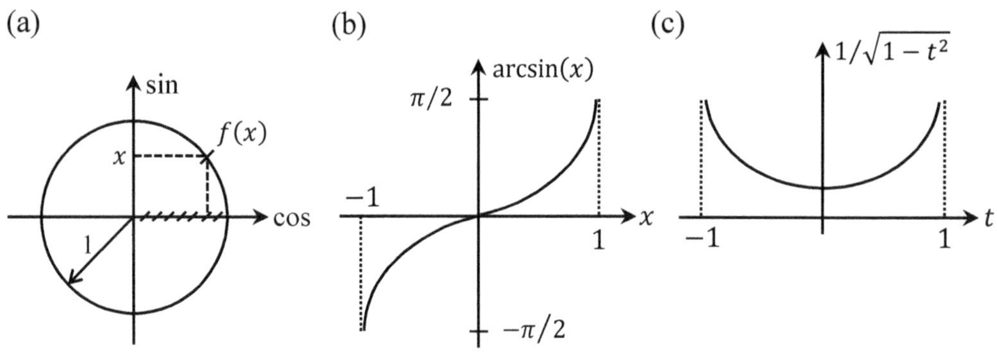

Figure 4.4. (a) The function $f(x) = \arcsin(x)$ is restricted to lie on the right half of the unit circle, when x on the vertical axis ranges from -1 to $+1$. The cosine of $f(x)$ thus lands on the shaded segment of the horizontal axis, which is always positive. (b) Plot of $f(x) = \arcsin(x)$ versus x for $-1 \le x \le 1$. Viewed with a tilted head from the left-hand side, the same graph represents a plot of a half period of the sine function. (c) A plot of the function $(1-t^2)^{-1/2}$, whose integral from 0 to x yields the function $f(x) = \arcsin(x)$.

4.11. The Bernoulli numbers. The Bernoulli numbers B_n are the coefficients appearing in the Taylor series expansion of the function $f(x) = x/(e^x - 1)$, namely,

$$x/[\exp(x) - 1] = \sum_{n=0}^{\infty} B_n x^n/n!. \tag{31}$$

Considering that $x/(e^x - 1)$ can be written as $x/\sum_{n=1}^{\infty}(x^n/n!)$, which reduces to $1/\sum_{n=0}^{\infty}[x^n/(n+1)!]$, our task here is essentially to invert the infinite series in the denominator in order to determine the Bernoulli numbers B_n. Although the Bernoulli numbers have important applications in number theory, the goal of the present discussion is *not* to explore these properties, but rather to use the Bernoulli numbers as an excuse to demonstrate some useful techniques for the manipulation of infinite series.

Returning to Eq.(31), we rename the dummy of one of the infinite series from n to m, then write the product of the two series in the following streamlined form:

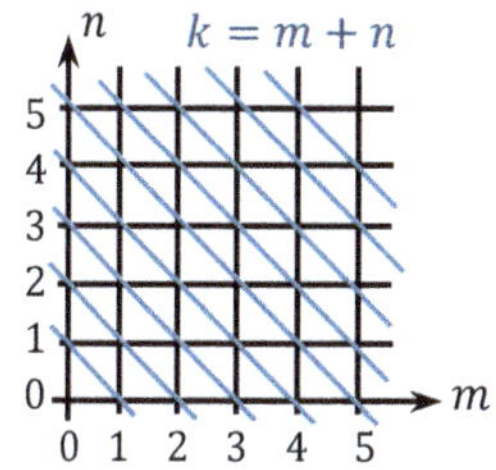

$$1 = \sum_{m=0}^{\infty}[x^m/(m+1)!] \times \sum_{n=0}^{\infty}(B_n x^n/n!)$$

$$= \sum_{m=0}^{\infty}\sum_{n=0}^{\infty} B_n x^{m+n}/[(m+1)!\,n!] \quad \leftarrow \boxed{m+n=k}$$

$$= B_0 + (B_1 + ½B_0)x + \sum_{k=2}^{\infty}\sum_{n=0}^{k} B_n x^k/[(k-n+1)!\,n!]. \tag{32}$$

Setting the coefficients of the various powers of x on both sides of Eq.(32) equal to each other, we find $B_0 = 1$, $B_1 = -½B_0 = -½$, and, for the coefficient of x^k,

$$\frac{B_0}{(k+1)!} + \frac{B_1}{k!} + \frac{1}{(k+1)!}\sum_{n=2}^{k}\binom{k+1}{n}B_n = 0; \qquad k \geq 2. \tag{33}$$

In Eq.(33), we finally have a recursion relation for computing B_k for $k \geq 2$ from $B_0, B_1, \cdots, B_{k-1}$. This recursion relation yields $B_2 = 1/6$, $B_3 = 0$, $B_4 = -1/30$, $B_5 = 0$, $B_6 = 1/42$, $B_7 = 0$, and so on. Aside from $B_1 = -½$, all B_n associated with odd indices of n turn out to be zero. (It is not difficult to show that $[x/(e^x - 1)] - B_1 x$ is equal to $(x/2)\coth(x/2)$, an even function of x, which cannot have any odd powers of x in its Taylor series expansion.) Upon replacing k with $2k$, and n with $2n$, Eq.(33) simplifies to

$$\sum_{n=1}^{k}\binom{2k+1}{2n}B_{2n} = k - ½; \qquad k \geq 1. \tag{34}$$

Equation (34) is a recursion relation for B_{2k}, which yields the value of B_2 when k is set to 1, then the value of B_4 when $k = 2$, then the value of B_6 when $k = 3$, and so on.

Brain teaser. Thirteen highly intelligent men are incarcerated in the same jailhouse. The prisoners may observe each other but are forbidden to communicate by direct or indirect means. Shortly after noon one day, the jail keeper places either a red mark or a green mark on each man's forehead, without informing him of the color of the mark. Each prisoner can see the faces of all the others, but cannot see his own. The jail keeper then informs the prisoners that those with a green mark are free to leave provided that they determine (using pure logic) that they have received a green mark. There will be severe punishment if a red-marked prisoner attempts to leave. Following this announcement, the guards open the door once every hour at the hour, for a few seconds each time, so the green-marked prisoners may exit. Nothing happens until 7:00 p.m. that night, at which point all seven men who have received the green marks get up and leave. How did they recognize the color of their mark, and why did it take them so long?

4.12. Relation between Riemann's zeta function and the Bernoulli numbers. For s, an even integer, the value of $\zeta(s) = \zeta(2n)$ can be obtained in terms of the Bernoulli number B_{2n}. To derive the relation between $\zeta(2n)$ and B_{2n}, we begin by rewriting Eq.(31) as

$$x/[\exp(x) - 1] = 1 - (x/2) + \sum_{n=1}^{\infty} B_{2n}x^{2n}/(2n)!. \tag{35}$$

Replacing x with $2ix$ in the above equation yields

$$\frac{2ix}{\exp(2ix) - 1} - 1 + ix = \sum_{n=1}^{\infty}(-1)^n B_{2n}(2x)^{2n}/(2n)!. \tag{36}$$

The left-hand-side of Eq.(36) is easily shown to reduce to $x\cot x - 1$. Next, we use Euler's expansion of $\sin x$ in Eq.(21), namely, $\sin x = x\prod_{m=1}^{\infty}[1 - (x/m\pi)^2]$, to write

$$\ln(\sin x) = \ln x + \sum_{m=1}^{\infty} \ln[1 - (x/m\pi)^2]. \tag{37}$$

Differentiating both sides of the equation, we find

$$\frac{\cos x}{\sin x} = \frac{1}{x} + \sum_{m=1}^{\infty} \frac{-2x/(m\pi)^2}{1 - (x/m\pi)^2}. \tag{38}$$

Consequently,

$$x \cot x - 1 = -2 \sum_{m=1}^{\infty} \frac{(x/m\pi)^2}{1 - (x/m\pi)^2} = -2 \sum_{m=1}^{\infty} \left(\frac{x}{m\pi}\right)^2 \sum_{n=0}^{\infty} \left(\frac{x}{m\pi}\right)^{2n} \leftarrow \boxed{\sum_{n=0}^{\infty} \alpha^n = \frac{1}{1-\alpha}}$$

$$= -2 \sum_{m=1}^{\infty} \sum_{n=1}^{\infty} (x/m\pi)^{2n} = -2 \sum_{n=1}^{\infty} \zeta(2n)(x/\pi)^{2n}$$

$$= -2 \sum_{n=1}^{\infty} \zeta(2n)(2x)^{2n}/(2\pi)^{2n}. \tag{39}$$

Comparing Eq.(39) with Eq.(36), and recalling that the left-hand side of Eq.(36) equals $x \cot x - 1$, we finally arrive at

$$\zeta(2n) = (-1)^{n+1} \frac{(2\pi)^{2n}}{2(2n)!} B_{2n}. \tag{40}$$

In the special case of $n = 1$, given that $B_2 = ⅙$, this equation yields $\zeta(2) = \pi^2/6$.

4.13. The Bernoulli polynomials. These polynomials, which are denoted by $B_n(x)$, are defined over the interval $x \in [0,1]$ by the recursion relation $B_n'(x) = nB_{n-1}(x)$, where $B_0(x) = 1$; the successive constants of integration are chosen to ensure that, for all the indices $n \geq 1$, $\int_0^1 B_n(x)\mathrm{d}x = 0$. The first few Bernoulli polynomials are listed below:

$$B_0(x) = 1; \quad B_1(x) = x - \frac{1}{2}; \quad B_2(x) = x^2 - x + \frac{1}{6}; \quad B_3(x) = x^3 - \frac{3}{2}x^2 + \frac{1}{2}x;$$

$$B_4(x) = x^4 - 2x^3 + x^2 - \frac{1}{30}; \quad B_5(x) = x^5 - \frac{5}{2}x^4 + \frac{5}{3}x^3 - \frac{1}{6}x. \tag{41}$$

The Bernoulli numbers, which we encountered in the preceding sections, are defined as $B_n = B_n(0)$. With the exception of $B_1(x)$, all the Bernoulli polynomials can be shown to satisfy the identity $B_n(1) = B_n(0) = B_n$; $(n \neq 1)$. Also, with the exception of $B_1 = -½$, all odd-index Bernoulli numbers are zero, that is, $B_3 = B_5 = B_7 = \cdots = 0$. Using the defining property of the Bernoulli polynomials, the kth derivative of $B_N(x)$ is readily found to be

$$B_N^{(k)}(x) = N(N-1)(N-2)\cdots(N-k+1)B_{N-k}(x) = \frac{N!}{(N-k)!} B_{N-k}(x). \tag{42}$$

Upon replicating the polynomial function $B_n(x)$ along the x-axis, one creates the periodic Bernoulli function $\tilde{B}_n(x) = B_n(x - \lfloor x \rfloor)$; here, $\lfloor x \rfloor$ represents the largest integer less than or equal to x. Bernoulli functions are continuous, except for $\tilde{B}_1(x)$, which is a saw-tooth function.

Several properties of the Bernoulli polynomials, including those mentioned, can be proven using the symmetry of $B_n(x)$ with respect to the center point $x = ½$ of their domain. Listed below are equivalent expressions for the polynomials appearing in Eq.(41):

$$B_0(x) = 1; \qquad B_1(x) = (x - ½); \qquad B_2(x) = (x - ½)^2 - \frac{1}{12};$$

$$B_3(x) = (x - ½)^3 - ¼(x - ½); \qquad B_4(x) = (x - ½)^4 - ½(x - ½)^2 + \frac{7}{240};$$

$$B_5(x) = (x - ½)^5 - \frac{5}{6}(x - ½)^3 + \frac{7}{48}(x - ½). \tag{43}$$

Note that the polynomials having an even index are even functions of x with respect to $x = ½$. Consequently, for even values of n, $B_n(0) = B_n(1)$. The added constant at the end of each such polynomial is needed to ensure that the total area under the function vanishes. In contrast, odd-index polynomials are odd functions of x relative to $x = ½$. This feature ensures that the area under odd-index functions is zero, thereby obviating the need for an additive constant term. Since an even-index polynomial is even (relative to $x = ½$), and since it has zero area, each half of the function must have zero area as well.

Now, to obtain an odd-index polynomial, one integrates the immediately preceding (even-index) polynomial with respect to x, starting at $x = 0$ and moving continuously toward $x = 1$. The resulting integral is zero at the starting point, becomes zero at $x = ½$, and zero again at $x = 1$ (because the areas under both halves of the even-index polynomial integrand are zero). Since the result of integration is an odd function (relative to $x = ½$), its area is necessarily zero, thus obviating the need for adding a constant to the integral of an even-index polynomial. Given that the value of the integral is zero at $x = 0$, and is zero again at $x = 1$, we have $B_n(0) = B_n(1) = 0$ when n is odd—except of course in the case of $n = 1$, for which the integrand $B_0(x)$ has a non-zero area.

4.14. Darboux's formula and the Euler-MacLaurin Expansion. In 1876, Jean-Gaston Darboux (1842–1917) introduced a formula for evaluating finite or infinite sums using integrals and, conversely, for evaluating an integral by computing a series. Let $f(z)$ be an analytic function in a region of the complex plane containing the points z_0 and z, and assume that $\phi(t) = \sum_{n=0}^{N} a_n t^n$ is an arbitrary polynomial of the Nth order. (*Note to the reader*: *The various properties of complex functions defined over the complex plane, including differentiability and analyticity, will be treated in detail in chapter 6. For the present discussion, however, an elementary knowledge of the complex plane and complex functions should suffice.*) We arrive at one of Darboux's formulas by a repeated application of the method of integration by parts, as follows:

$$\int_0^1 \phi(t) f^{(N+1)}[z_0 + (z - z_0)t]dt$$

$$= \frac{\phi(1)f^{(N)}(z) - \phi(0)f^{(N)}(z_0)}{z-z_0} - \frac{1}{z-z_0}\int_0^1 \phi'(t) f^{(N)}[z_0 + (z - z_0)t]dt$$

$$= \frac{\phi(1)f^{(N)}(z) - \phi(0)f^{(N)}(z_0)}{z-z_0} - \frac{\phi'(1)f^{(N-1)}(z) - \phi'(0)f^{(N-1)}(z_0)}{(z-z_0)^2}$$

$$+ \frac{1}{(z-z_0)^2}\int_0^1 \phi''(t) f^{(N-1)}[z_0 + (z - z_0)t]dt$$

$$= \sum_{k=0}^{2} (-1)^k \frac{[\phi^{(k)}(1)f^{(N-k)}(z) - \phi^{(k)}(0)f^{(N-k)}(z_0)]}{(z-z_0)^{k+1}}$$

$$- \frac{1}{(z-z_0)^3}\int_0^1 \phi'''(t) f^{(N-2)}[z_0 + (z - z_0)t]dt$$

$$= \sum_{k=0}^{N-1} (-1)^k \frac{[\phi^{(k)}(1)f^{(N-k)}(z) - \phi^{(k)}(0)f^{(N-k)}(z_0)]}{(z-z_0)^{k+1}} + (-1)^N \frac{N!a_N}{(z-z_0)^{N+1}}[f(z) - f(z_0)]. \tag{44}$$

In the last line of the above equation, we have replaced $\phi^{(N)}(t)$ with $N!\,a_N$, which is a constant (i.e., independent of t). Rearranging Eq.(44) now yields

$$N!\,a_N[f(z)-f(z_0)] = \sum_{k=0}^{N-1}[\phi^{(k)}(0)f^{(N-k)}(z_0)-\phi^{(k)}(1)f^{(N-k)}(z)](z_0-z)^{N-k}$$
$$-(z_0-z)^{N+1}\int_0^1 \phi(t)f^{(N+1)}[z_0+(z-z_0)t]dt, \qquad (45)$$

or, equivalently,

$$N!\,a_N[f(z)-f(z_0)] = \sum_{k=1}^{N}[\phi^{(N-k)}(0)f^{(k)}(z_0)-\phi^{(N-k)}(1)f^{(k)}(z)](z_0-z)^{k}$$
$$-(z_0-z)^{N+1}\int_0^1 \phi(t)f^{(N+1)}[z_0+(z-z_0)t]dt. \qquad (46)$$

It is thus clear that $f(z)-f(z_0)$ may be approximated with an Nth order polynomial in $(z-z_0)$, whose coefficients depend on the values of the first N derivatives of $f(z)$ at the terminal points z_0 and z. The last term in Eq.(46) gives the remainder (i.e., the difference between the actual value of $f(z)-f(z_0)$ and the approximate value provided by the polynomial).

Taylor series expansion. As an example, consider the special case of $\phi(t)=(t-1)^N$. Here, $a_N=1$ and

$$\phi^{(k)}(t)=\frac{N!}{(N-k)!}(t-1)^{N-k} \quad \rightarrow \quad \phi^{(N-k)}(t)=\frac{N!}{k!}(t-1)^{k}. \qquad (47)$$

Substitution into Eq.(3) yields

$$f(z)=f(z_0)+\sum_{k=1}^{N}\frac{f^{(k)}(z_0)}{k!}(z-z_0)^k$$
$$+\frac{(z-z_0)^{N+1}}{N!}\int_0^1(1-t)^N f^{(N+1)}[z_0+(z-z_0)t]dt. \qquad (48)$$

This, of course, is a Taylor series expansion of $f(z)$ around the point z_0, with the deviation of $f(z)$ from the Nth order polynomial given by the last term on the right-hand side of Eq.(48).

Returning to Darboux's method, if we repeat the derivation of Eq.(44), but use a polynomial of order $N+1$ (instead of N), we arrive at

$$\int_0^1 \phi(t)f^{(N+1)}[z_0+(z-z_0)t]dt = \sum_{k=0}^{N}(-1)^k\frac{[\phi^{(k)}(1)f^{(N-k)}(z)-\phi^{(k)}(0)f^{(N-k)}(z_0)]}{(z-z_0)^{k+1}}$$
$$+(-1)^{N+1}\frac{(N+1)!a_{N+1}}{(z-z_0)^{N+1}}\int_0^1 f[z_0+(z-z_0)t]dt. \quad (49)$$

Consequently,

$$(N+1)!\,a_{N+1}\int_0^1 f[z_0+(z-z_0)t]dt = [\phi^{(N)}(1)f(z)-\phi^{(N)}(0)f(z_0)]$$
$$+\sum_{k=1}^{N}[\phi^{(N-k)}(1)f^{(k)}(z) - \phi^{(N-k)}(0)f^{(k)}(z_0)](z_0-z)^k$$
$$+(z_0-z)^{N+1}\int_0^1 \phi(t)f^{(N+1)}[z_0+(z-z_0)t]dt. \qquad (50)$$

Here, we have a formula for approximating the integral of $f(z)$ along the straight line connecting z_0 to z, in terms of the values of the function $f(z)$ and its first N derivatives at the terminal points z_0 and z. The last term on the right-hand side of Eq.(50) gives the remainder, which is the difference between the actual value of the integral and its approximate value.

A uniquely suitable class of polynomials that can be deployed in conjunction with Eq.(50) is the class of Bernoulli polynomials, which were suggested (independently) by Leonhard Euler and Colin MacLaurin around 1735 in relation to a primitive version of Eq.(50). We have already seen the Bernoulli polynomials in section 4.13, and are familiar with their relevant properties in the context of the present application. Let us then set $\phi(t) = B_N(t)$ in Darboux's first formula, Eq.(46), and arrive, after straightforward manipulations, at the following expansion:

$$f(z) = f(z_0) + B_1[f'(z_0) + f'(z)](z_0 - z) + \sum_{k=2}^{N} \frac{B_k}{k!}\left[f^{(k)}(z_0) - f^{(k)}(z)\right](z_0 - z)^k$$
$$- \frac{(z_0-z)^{N+1}}{N!}\int_0^1 B_N(t) f^{(N+1)}[z_0 + (z - z_0)t]\mathrm{d}t. \tag{51}$$

This could be a useful expansion of $f(z)$ in terms of its first N derivatives at both z and z_0, provided that the correction term in Eq.(51) turns out to be reasonably small. The far more interesting application of the Bernoulli polynomials, however, arises from setting $\phi(t) = B_{N+1}(t)$ in Darboux's second formula, Eq.(50), which yields

$$\int_0^1 f[z_0 + (z - z_0)t]\mathrm{d}t = -B_1[f(z_0) + f(z)]$$
$$- \sum_{k=1}^{N} \frac{B_{k+1}}{(k+1)!}\left[f^{(k)}(z_0) - f^{(k)}(z)\right](z_0 - z)^k$$
$$+ \frac{(z_0-z)^{N+1}}{(N+1)!}\int_0^1 B_{N+1}(t) f^{(N+1)}[z_0 + (z - z_0)t]\mathrm{d}t. \tag{52}$$

Next, suppose that the integration path along the straight line connecting z_0 to z within the complex plane is divided into equal-length segments $\Delta z = \delta \exp(\mathrm{i}\theta)$, where $\delta > 0$ is the length and $0 \le \theta < 2\pi$ is the orientation angle of each (identical) segment. Denoting by $z_0, z_1, z_2, \cdots, z_m$ the equidistant points along the integration path, we may now add up the contributions of the various segments to the overall integral along the path, and arrive at the following Euler-MacLaurin formula:

$$\int_{t=0}^{m} f[z_0 + \delta \exp(\mathrm{i}\theta)\, t]\mathrm{d}t = \sum_{j=0}^{m} f(z_j) - ½[f(z_0) + f(z_m)]$$
$$+ \sum_{k=1}^{N} \frac{(-\delta)^k \exp(\mathrm{i}k\theta) B_{k+1}}{(k+1)!}\left[f^{(k)}(z_m) - f^{(k)}(z_0)\right]$$
$$+ \frac{(-\delta)^{N+1}\exp[\mathrm{i}(N+1)\theta]}{(N+1)!}\int_{t=0}^{m} \tilde{B}_{N+1}(t) f^{(N+1)}[z_0 + \delta \exp(\mathrm{i}\theta)\, t]\mathrm{d}t. \tag{53}$$

Considering that all the odd-index Bernoulli numbers (aside from B_1) are zero, we can simplify Eq.(53) by eliminating every other term in the second line of the equation. Assuming, for the sake of simplicity, that N is even, we will have

$$\int_{t=0}^{m} f[z_0 + \delta \exp(\mathrm{i}\theta)\, t]\mathrm{d}t = \left[\sum_{j=0}^{m} f(z_j)\right] - ½[f(z_0) + f(z_m)]$$ ← Trapezoid approximation

N is even →

$$- \sum_{k=1}^{N/2} \frac{\delta^{2k-1}\exp[\mathrm{i}(2k-1)\theta] B_{2k}}{(2k)!}\left[f^{(2k-1)}(z_m) - f^{(2k-1)}(z_0)\right]$$
$$- \frac{\delta^{N+1}\exp[\mathrm{i}(N+1)\theta]}{(N+1)!}\int_{t=0}^{m} \tilde{B}_{N+1}(t) f^{(N+1)}[z_0 + \delta \exp(\mathrm{i}\theta)\, t]\mathrm{d}t. \tag{54}$$

Note that the integrals in Eq.(54) are taken over the variable t (not z). As such, they must be multiplied by $\Delta z = \delta \exp(i\theta)$ if one desires to integrate straightforwardly in the z-plane.

4.15. Convergence acceleration of series—Euler's method. The Euler transformation was introduced in the year 1755 by Leonhard Euler as a means of accelerating the convergence of alternating series (i.e., series whose consecutive terms alternate between positive and negative numbers). In general, one tries to break up each term of an infinite series by expressing it as an infinite sum in its own right. Subsequently, the doubly infinite sum is rearranged, first by adding up the terms diagonally, then by summing over all the diagonals, as shown in figure 4.5. In the succinct formulation given by Eq.(55), each term a_k of the series under consideration is written as $\sum_{m=0}^{\infty} c_{k,m} a_k$, where $\sum_{m=0}^{\infty} c_{k,m} = 1$. The doubly infinite sum over k and m is then converted to an infinite sum over n, where $n = k + m$ specifies individual diagonals in the (k, m) plane, and where each diagonal contributes a finite sum containing $n + 1$ terms, as follows:

$$\sum_{k=0}^{\infty} a_k = \sum_{k=0}^{\infty} \sum_{m=0}^{\infty} c_{k,m} a_k = \sum_{n=0}^{\infty} \sum_{k=0}^{n} c_{k,n-k} a_k. \tag{55}$$

Warning: As a general rule, one must be aware that rearranging the terms of a convergent series is *not* always permissible and *could* lead to erroneous results.

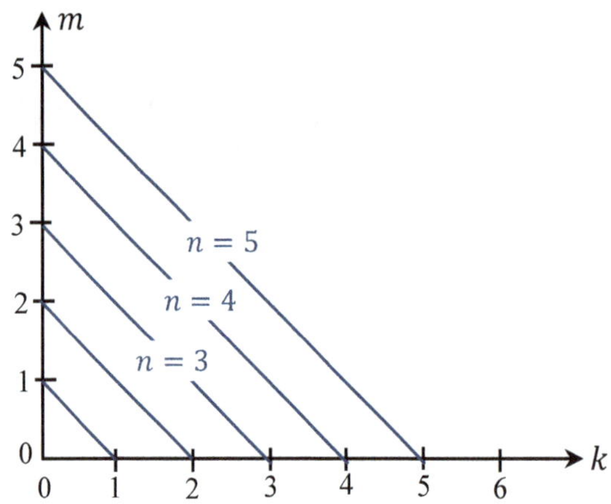

Figure 4.5. The doubly-infinite sum over the (k, m) plane is rearranged by first summing over each diagonal, where $k + m = n$, then adding up the diagonals' contributions.

The coefficients $c_{k,m}$ may be constructed in a variety of ways. The specific set used by Euler is obtained from a Taylor series expansion (around $x = 0$) of the function $f(x) = 1/(1 + x)^{k+1}$, where $k \geq 0$ is an arbitrary integer. We have

$$f(x) = \frac{1}{(1+x)^{k+1}} = 1 - \frac{(k+1)}{1!}x + \frac{(k+1)(k+2)}{2!}x^2 - \frac{(k+1)(k+2)(k+3)}{3!}x^3 + \cdots$$

$$= \sum_{m=0}^{\infty} \binom{k+m}{k} (-x)^m. \tag{56}$$

At $x = -½$, the series is convergent, yielding

$$\sum_{m=0}^{\infty} \frac{1}{2^{k+m+1}} \binom{k+m}{k} = 1. \tag{57}$$

Here, we have numerous ways for breaking up the number 1 into an infinite series; in other words, each integer $k \geq 0$ has its own unique way of producing an infinite sum that converges to 1. (*Note*: The series in Eq.(56) diverges at $x = 1$.) Inserting into Eq.(55) $c_{k,m} = \binom{k+m}{k}/2^{k+m+1}$ gives rise to the following alternative expression for $\sum_{k=0}^{\infty} a_k$:

$$\sum_{k=0}^{\infty} a_k = \sum_{n=0}^{\infty} \sum_{k=0}^{n} c_{k,n-k} a_k = \sum_{n=0}^{\infty} \sum_{k=0}^{n} \frac{1}{2^{n+1}} \binom{n}{k} a_k$$

$$= \sum_{n=0}^{\infty} \frac{1}{2^{n+1}} \sum_{k=0}^{n} \binom{n}{k} a_k = \sum_{n=0}^{\infty} \frac{1}{2^{n+1}} \sum_{k=0}^{n} \binom{n}{k} a_{n-k}. \qquad (58)$$

The last series in this equation is obtained from the preceding one by switching the dummy of the inner sum from k to $n-k$, recognizing that $\binom{n}{k} = \binom{n}{n-k} = \frac{n!}{k!(n-k)!}$.

Whether the transformation described by Eq.(58) accelerates the convergence of the original series depends, of course, on the specific values of a_k. The transformation could be highly effective for some slowly converging *alternating* series. The explicit formula for Euler's transformation of an alternating series is given below.

$$\sum_{k=0}^{\infty}(-1)^k a_k = \sum_{k=0}^{\infty} \sum_{m=0}^{\infty}(-1)^k c_{k,m} a_k$$

$$= \sum_{n=0}^{\infty} \sum_{k=0}^{n}(-1)^k c_{k,n-k} a_k = \sum_{n=0}^{\infty} \sum_{k=0}^{n} \frac{(-1)^k}{2^{n+1}} \binom{n}{k} a_k$$

$$= \sum_{n=0}^{\infty} \frac{1}{2^{n+1}} \sum_{k=0}^{n}(-1)^k \binom{n}{k} a_k = \sum_{n=0}^{\infty} \frac{(-1)^n}{2^{n+1}} \sum_{k=0}^{n}(-1)^k \binom{n}{k} a_{n-k}. \qquad (59)$$

Once again, the last expression in Eq.(59) is obtained from the preceding one by switching the dummy of the inner sum from k to $n-k$, recognizing that $\binom{n}{k} = \binom{n}{n-k}$ and $(-1)^{-k} = (-1)^k$. Note that the terms of the inner sum in Eq.(59) are successive forward differences of the original series, namely, a_0, $(a_0 - a_1)$, $(a_0 - 2a_1 + a_2)$, $(a_0 - 3a_1 + 3a_2 - a_3)$, and so on.

Example 1. Let $S = \sum_{k=0}^{\infty}(-1)^k \rho^k$, where $|\rho| < 1$. Substitution into Eq.(59) yields

$$S = \sum_{n=0}^{\infty} \frac{1}{2^{n+1}} \sum_{k=0}^{n}(-1)^k \binom{n}{k} \rho^k = \sum_{n=0}^{\infty} \frac{(1-\rho)^n}{2^{n+1}} = ½ \sum_{n=0}^{\infty}[(1-\rho)/2]^n. \qquad (60)$$

The rate of convergence of the series improves if $\rho = ½$, remains the same if $\rho = ⅓$, and becomes worse if $\rho = ¼$. Needless to say, the original as well as the modified series in the present example are geometric series, allowing one to verify that, in both cases, $S = 1/(1+\rho)$.

Example 2. The well-known alternating series $\frac{\pi}{4} = 1 - \frac{1}{3} + \frac{1}{5} - \frac{1}{7} + \cdots$ converges rather slowly. To accelerate this series, we calculate the inner sum in Eq.(5) for successive values of n, as follows:

$n = 0$: $\qquad a_0 = 1;$

$n = 1$: $\qquad a_0 - a_1 = 1 - \frac{1}{3} = \frac{2}{3};$

$$n = 2: \qquad a_0 - 2a_1 + a_2 = 1 - \tfrac{2}{3} + \tfrac{1}{5} = \tfrac{2\cdot 4}{3\cdot 5};$$

$$n = 3: \qquad a_0 - 3a_1 + 3a_2 - a_3 = 1 - \tfrac{3}{3} + \tfrac{3}{5} - \tfrac{1}{7} = \tfrac{2\cdot 4\cdot 6}{3\cdot 5\cdot 7}.$$

A pattern thus emerges, which enables one to predict the general expression of the inner sum as $(2^n n!)^2/(2n+1)!$. The accelerated series now becomes

$$\tfrac{1}{4}\pi = \tfrac{1}{2}\sum_{n=0}^{\infty} 2^n (n!)^2/(2n+1)!. \qquad (61)$$

Let us now pick a different set of numbers $c_{k,m}$ by evaluating the Taylor series of $f(x) = 1/(1+x)^{k+1}$ at $x = -1/\beta$, with $\beta > 1$ being an arbitrary parameter. From Eq.(56) we find

$$\sum_{m=0}^{\infty} \binom{k+m}{k} (1-\beta^{-1})^{k+1} (1/\beta)^m = 1. \qquad (62)$$

Using $c_{k,m} = \binom{k+m}{k} (1-\beta^{-1})^{k+1} (1/\beta)^m$ in conjunction with Eq.(55) yields

$$\sum_{k=0}^{\infty} (-1)^k a_k = \sum_{n=0}^{\infty} \frac{\beta-1}{\beta^{n+1}} \sum_{k=0}^{n} \binom{n}{k} (1-\beta)^k a_k. \qquad (63)$$

It must be clear that there are many different choices for $c_{k,m}$ and that each choice affects the convergence rate in its own special way. Listed below are a few examples of functions whose Taylor series expansions yield different sets of values for the $c_{k,m}$ coefficients.

i) $f(x) = \exp(kx) = \sum_{m=0}^{\infty} \left(\frac{k^m}{m!}\right) x^m \xrightarrow[(x=\delta)]{} \sum_{m=0}^{\infty} \frac{k^m \delta^m}{m!\exp(k\delta)} = 1. \qquad (64)$

ii) $f(x) = \sinh(kx) = \sum_{m=0}^{\infty} \frac{k^{2m+1}}{(2m+1)!} x^{2m+1} \xrightarrow[(x=\delta)]{} \sum_{m=0}^{\infty} \frac{k^{2m+1}\delta^{2m+1}}{(2m+1)!\sinh(k\delta)} = 1. \qquad (65)$

iii) $f(x) = \cosh(kx) = \sum_{m=0}^{\infty} \frac{k^{2m}}{(2m)!} x^{2m} \xrightarrow[(x=\delta)]{} \sum_{m=0}^{\infty} \frac{k^{2m}\delta^{2m}}{(2m)!\cosh(k\delta)} = 1. \qquad (66)$

iv) $f(x) = \ln \frac{1}{1-(x/k)} = \sum_{m=0}^{\infty} \frac{x^{m+1}}{(m+1)k^{m+1}} \xrightarrow[(x=\delta;\ |\delta|<1)]{} \sum_{m=0}^{\infty} \frac{(\delta/k)^{m+1}}{(m+1)\ln[1-(\delta/k)]} = -1. \qquad (67)$

Absent any systematic analysis of the convergence rate of the transformed series, it is difficult to decide a priori which choice of the $c_{k,m}$ would yield the best results. The following generalization of Euler's transformation provides a powerful tool for selecting the coefficients with the prior knowledge of an estimate for the convergence rate.

4.16. Generalized method of acceleration for alternating series. Suppose the elements a_k of the convergent alternating series $S = \sum_{k=0}^{\infty}(-1)^k a_k$ can be expressed as the moments of a *positive* generating function $G(x)$ defined over the interval $[0,1]$; that is,

$$a_k = \int_0^1 x^k G(x)dx; \qquad (k = 0, 1, 2, \cdots \text{ and } G(x) \geq 0). \qquad (68)$$

For example, in the case of the series $\ln 2 = 1 - ½ + ⅓ - ¼ + \cdots$, where $a_k = 1/(k+1)$, the generating function is $G(x) = 1$. Similarly, $G(x) = ½x^{-½}$ for $¼\pi = \sum_{k=0}^{\infty}(-1)^k/(2k+1)$, and $G(x) = [\ln(1/x)]^{s-1}/\Gamma(s)$ for Dirichlet's eta function $\eta(s) = \sum_{k=0}^{\infty}(-1)^k/(k+1)^s$.‡

Substituting for a_k from Eq.(68) into the alternating series, and invoking the geometric series $\sum_{k=0}^{\infty}(-x)^k = 1/(1+x)$, we find

$$S = \sum_{k=0}^{\infty}(-1)^k a_k = \int_0^1 \frac{G(x)}{1+x}\mathrm{d}x. \tag{69}$$

Next, define the nth order polynomial $P_n(x) = \sum_{k=0}^{n} p_k(-x)^k$ with the as-yet-unspecified coefficients p_k, and note that

$$S_n = \int_0^1 P_n(x)G(x)dx = \sum_{k=0}^{n}(-1)^k p_k a_k. \tag{70}$$

Since our goal is to have S_n converge to S in the limit when $n \to \infty$, we now try to find an upper bound on the difference between S and S_n, as follows:

$$|S - S_n| = \left|\int_0^1 \frac{1-(1+x)P_n(x)}{1+x} G(x)\mathrm{d}x\right| \le \mathrm{Max}_{x\in[0,1]}|1-(1+x)P_n(x)| \int_0^1 \frac{G(x)}{1+x}\mathrm{d}x. \tag{71}$$

Considering that the integral appearing on the right-hand side of Eq.(71) is equal to S, we conclude that the ratio S_n/S approaches 1.0 at least as rapidly as the peak value of the $n+1^{st}$ order polynomial $Q_{n+1}(x) = 1-(1+x)P_n(x)$ approaches zero (within the interval $0 \le x \le 1$) as $n \to \infty$. Note that $Q_{n+1}(x)$ must satisfy the requirement $Q_{n+1}(-1) = 1$. The coefficients p_k of $P_n(x)$ are derived straightforwardly from the coefficients q_k of $Q_{n+1}(x)$, as shown below:

$$\begin{aligned}
Q_{n+1}(x) &= \sum_{k=0}^{n+1} q_k(-x)^k = 1-(x+1)\sum_{k=0}^{n} p_k(-x)^k \\
&= (1-p_0) + \sum_{k=1}^{n}(p_{k-1}-p_k)(-x)^k + p_n(-x)^{n+1} \\
\to \quad & 1-p_0 = q_0, \quad p_{k-1}-p_k = q_k, \quad p_n = q_{n+1} \\
\to \quad & p_k = \sum_{m=k+1}^{n+1} q_m; \qquad (0 \le k \le n).
\end{aligned} \tag{72}$$

Note that the last expression for p_0 is consistent with the one given in the penultimate line of Eq.(72), namely, that $p_0 = \sum_{m=1}^{n+1} q_m$ is equivalent to $p_0 = 1 - q_0$. This is because the constraint $Q_{n+1}(-1) = 1.0$ ensures that $\sum_{m=0}^{n+1} q_m = 1.0$. We mention in passing that the generalized method does *not* impose our earlier constraint on the coefficients, namely, $\sum_{m=0}^{\infty} c_{k,m} = 1$. Nevertheless, it is easy to see that this property is implied by the constraint $\sum_{m=0}^{n+1} q_m = 1.0$ imposed on the coefficients of $Q_{n+1}(x)$ in cases where all q_m are non-negative.

‡Starting with $\int_0^{\infty} t^{s-1}\exp[-(k+1)t]\,\mathrm{d}t = (k+1)^{-s}\int_0^{\infty} t'^{s-1}\exp(-t')\,\mathrm{d}t' = \Gamma(s)/(k+1)^s$, introduce the new variable $x = \exp(-t)$ on the left-hand side to arrive at $\int_0^1 x^k[\ln(1/x)]^{s-1}\mathrm{d}x = \Gamma(s)/(k+1)^s$. Note that the generating function $G(x) = [\ln(1/x)]^{s-1}/\Gamma(s)$ is real and positive only when s is real and positive. In contrast, Dirichlet's eta-function series is known to be convergent for complex values of s as well, provided that $\mathrm{Re}(s) > 0$.

Example 1. The polynomial $Q_{n+1}(x) = [(1-x)/2]^{n+1}$ satisfies the condition $Q_{n+1}(-1) = 1$. In addition, its maximum value in the $[0,1]$ interval is $2^{-(n+1)}$, which approaches zero rapidly as $n \to \infty$. The corresponding polynomial $P_n(x)$ is readily obtained, as follows:

$$Q_{n+1}(x) = \frac{1}{2^{n+1}} \sum_{k=0}^{n+1} \binom{n+1}{k} (-x)^k$$

$$\to \quad p_k = \frac{1}{2^{n+1}} \sum_{m=k+1}^{n+1} \binom{n+1}{m}; \qquad (0 \le k \le n). \tag{73}$$

Given that $\binom{n+1}{m} = \binom{n+1}{n-m+1}$, the formula for p_k may also be written as

$$p_k = \frac{1}{2^{n+1}} \sum_{m=0}^{n-k} \binom{n+1}{m}; \qquad (0 \le k \le n). \tag{74}$$

The transformation applied to the alternating series $\sum_{k=0}^{\infty}(-1)^k a_k$ in the present example is identical to the Euler transformation described earlier. To demonstrate their equivalence, we rename the dummy variable of the outer sum in Eq.(59) to m, then terminate the outer sum at $m = n$ to arrive at

$$\sum_{m=0}^{n} \frac{1}{2^{m+1}} \sum_{k=0}^{m} (-1)^k \binom{m}{k} a_k = \sum_{k=0}^{n} (-1)^k a_k \sum_{m=k}^{n} \frac{1}{2^{m+1}} \binom{m}{k}. \tag{75}$$

Comparison with Eqs.(70) reveals that the inner sum on the right-hand side of Eq.(75) must be equal to p_k given by Eq.(74); that is,

$$\sum_{m=k}^{n} \frac{1}{2^{m+1}} \binom{m}{k} = \frac{1}{2^{n+1}} \sum_{m=0}^{n-k} \binom{n+1}{m}. \tag{76}$$

This is indeed the case, as can be readily verified by repeated application of Pascal's identity[§] to the right-hand side of Eq.(76); that is,

$$\sum_{m=k}^{n} 2^{n-m} \binom{m}{k} = \binom{n+1}{0} + \sum_{m=1}^{n-k} \left[\binom{n}{m} + \binom{n}{m-1}\right]$$

$$= 2 \sum_{m=0}^{n-k-1} \binom{n}{m} + \binom{n}{n-k}. \tag{77}$$

The last term on the right-hand side is now cancelled by the last term on the left, leaving the equation in the same form as before, except that n is now lowered by 1. Repeated application of the procedure eliminates the remaining terms one by one.

[§]The Pascal identity $\binom{n}{m} = \binom{n-1}{m} + \binom{n-1}{m-1}$, where $1 \le m \le n-1$, is a direct consequence of the fundamental property of binomial coefficients, namely $\binom{\mu}{\nu} = \mu!/[\nu!\,(\mu-\nu)!]$. In fact, it is easy to see how the identity comes about, considering that selecting m out of $n+1$ distinct objects is equivalent to choosing either all m out of n, or selecting at first $m-1$ out of n then adding the remaining one to the selected $m-1$. In this way it is also possible to generalize the Pascal identity, for example, to $\binom{n}{m} = \binom{n-2}{m} + 2\binom{n-2}{m-1} + \binom{n-2}{m-2}$, where $2 \le m \le n-2$.

Example 2. The polynomial $Q_{n+1}(x) = [(1-2x)/3]^{n+1}$ satisfies the condition $Q_{n+1}(-1) = 1$. In addition, the maximum of its absolute value in the [0,1] interval is $3^{-(n+1)}$, which rapidly approaches zero as $n \to \infty$. The corresponding polynomial $P_n(x)$ is obtained, as follows:

$$Q_{n+1}(x) = \frac{1}{3^{n+1}}\sum_{k=0}^{n+1}\binom{n+1}{k}2^k(-x)^k \;\rightarrow\; p_k = \frac{1}{3^{n+1}}\sum_{m=k+1}^{n+1}\binom{n+1}{m}2^m. \tag{78}$$

Given that $\binom{n+1}{m} = \binom{n+1}{n-m+1}$, the preceding formula for p_k may also be written as

$$p_k = \frac{1}{3^{n+1}}\sum_{m=0}^{n-k}\binom{n+1}{m}2^{n-m+1}; \qquad (0 \le k \le n). \tag{79}$$

As in the preceding example, it is possible to demonstrate that this transformation is the same as that described by Eq.(63) in the case of $\beta = 3$.

4.17. Infinite products. This section provides a brief introduction to the topic of infinite products. The examples that follow show a few ways in which such products can be evaluated.

Example 1. The infinite product $\left(1-\frac{2}{2\times3}\right)\left(1-\frac{2}{3\times4}\right)\cdots\left(1-\frac{2}{n(n+1)}\right)\cdots$ is evaluated as follows:

$$\prod_{n=2}^{\infty}\left[1-\frac{2}{n(n+1)}\right] = \prod_{n=2}^{\infty}\left[\frac{n^2+n-2}{n(n+1)}\right] = \prod_{n=2}^{\infty}\left[\frac{(n+2)(n-1)}{n(n+1)}\right]$$

$$= \prod_{n=2}^{\infty}(n+2) \times \prod_{n=2}^{\infty}(n-1)/\left[\prod_{n=2}^{\infty}(n) \times \prod_{n=2}^{\infty}(n+1)\right]$$

$$= \frac{1}{3!}\prod_{n=1}^{\infty}(n) \times \prod_{n=1}^{\infty}(n)\Big/\left[\prod_{n=1}^{\infty}(n) \times \frac{1}{2!}\prod_{n=1}^{\infty}(n)\right] = \frac{1}{3}. \tag{80}$$

Example 2. Computing the value of the infinite product $\left(1-\frac{1}{4}\right)\left(1-\frac{1}{9}\right)\cdots\left(1-\frac{1}{n^2}\right)\cdots$ proceeds as follows:

$$\prod_{n=2}^{\infty}\left(1-\frac{1}{n^2}\right) = \prod_{n=2}^{\infty}\left[\frac{(n-1)(n+1)}{n \times n}\right]$$

$$= \prod_{n=2}^{\infty}(n-1) \times \prod_{n=2}^{\infty}(n+1)/\left[\prod_{n=2}^{\infty}(n) \times \prod_{n=2}^{\infty}(n)\right]$$

$$= \prod_{n=1}^{\infty}(n) \times \frac{1}{2!}\prod_{n=1}^{\infty}(n)/\left[\prod_{n=1}^{\infty}(n) \times \prod_{n=1}^{\infty}(n)\right] = \frac{1}{2}. \tag{81}$$

Example 3. The infinite product $\left(1+\frac{1}{2}\right)\left(1-\frac{1}{3}\right)\left(1+\frac{1}{4}\right)\left(1-\frac{1}{5}\right)\cdots\left[1+\frac{(-1)^n}{n}\right]\cdots$ may be computed by first separating its odd and even terms, then canceling the resulting common factors; that is,

$$\prod_{n=2}^{\infty}\left[1+\frac{(-1)^n}{n}\right] = \prod_{m=1}^{\infty}\left(1+\frac{1}{2m}\right)\left(1-\frac{1}{2m+1}\right) = \prod_{m=1}^{\infty}\left(\frac{2m+1}{2m}\right)\left(\frac{2m}{2m+1}\right) = 1. \tag{82}$$

Conditions for convergence of $\prod_{n=1}^{\infty} a_n$: It is not difficult to see that, if $lim_{n\to\infty}\, a_n > 1$, the infinite product will diverge to ∞. In contrast, if $lim_{n\to\infty}\, a_n < 1$, the product will approach zero. It is thus convenient to write the product as $\prod_{n=1}^{\infty}(1+b_n)$. The condition

$b_n \to 0$ as $n \to \infty$ is thus necessary (but not sufficient) for convergence to a non-zero value. Assuming $b_n \geq 0$, a lower bound on the infinite product may be found by writing

$$\prod_{n=1}^{\infty}(1+b_n) = 1 + \sum_{n=1}^{\infty} b_n + \sum_{i=1}^{\infty}\sum_{j=1}^{\infty} b_i b_j + \cdots. \tag{83}$$

Now, if $\sum_{n=1}^{\infty} b_n$ diverges, then so does $\prod_{n=1}^{\infty}(1+b_n)$. To find an upper bound on the infinite product when $b_n \geq 0$, note that $1 + b_n \leq \exp(b_n)$. Consequently,

$$\prod_{n=1}^{\infty}(1+b_n) \leq \exp\left(\sum_{n=1}^{\infty} b_n\right). \tag{84}$$

It must be clear, in this case, that if $\sum_{n=1}^{\infty} b_n$ converges, then so does $\prod_{n=1}^{\infty}(1+b_n)$.

Suggestions for further reading

1. G.B. Arfken and H.J. Weber, *Mathematical Methods for Physicists*, 6th edition, Academic Press, Cambridge, MA (2005).
2. F.B. Hildebrand, *Advanced Calculus for Applications*, 2nd edition, Prentice-Hall, Upper Saddle River, NJ (1976).
3. J. Mathews and R.L. Walker, *Mathematical Methods of Physics*, 2nd edition, Benjamin/Cummings Publishing, San Francisco, CA (1970).
4. G. Stephenson and P.M. Radmore, *Advanced Mathematical Methods for Engineering and Science Students*, Cambridge University Press, Cambridge, UK (1990).

Problems

Problem 1) Use expansion in partial fractions to verify the sum given for each of the following series:

$$\sum_{n=1}^{\infty} \frac{1}{n(n+1)(n+2)} = \frac{1}{4}; \quad \sum_{n=1}^{\infty} \frac{1}{n(n+1)(n+2)(n+3)} = \frac{1}{18}; \quad \sum_{n=1}^{\infty} \frac{1}{(2n-1)(2n+1)} = \frac{1}{2}.$$

Problem 2) The Riemann zeta function is defined as $\zeta(s) = \sum_{n=1}^{\infty}(1/n^s)$, where $\text{Re}(s) > 1$. Show that

$$\sum_{m=2}^{\infty}[\zeta(m) - 1] = 1; \qquad \sum_{m=2}^{\infty}(-1)^m\,[\zeta(m) - 1] = ½.$$

Problem 3) The Taylor series expansion of the exponential function is given by

$$\exp(x) = \sum_{n=0}^{\infty}(x^n/n!).$$

Multiply the Taylor series expansions of $\exp(x)$ and $\exp(y)$ to obtain a double series (summed over n from 0 to ∞ and also over m from 0 to ∞). Rearrange the double series to obtain a single series that would represent the expansion of $\exp(x+y)$.

Hint: The binomial expansion $(x+y)^k = \sum_{n=0}^{k}\binom{k}{n}x^n y^{k-n}$ will be helpful in this derivation.

Problem 4) Euler's infinite-product representations of $\sin x$ and $\cos x$ are

$$\sin x = x\prod_{n=1}^{\infty}\left(1 - \frac{x^2}{n^2\pi^2}\right); \qquad \cos x = \prod_{n=1}^{\infty}\left[1 - \frac{4x^2}{(2n-1)^2\pi^2}\right].$$

Using these identities, show that $(\sin x \cos x)/x$ is equal to $\sin(2x)/(2x)$.

Problem 5) Show that $\sum_{n=1}^{\infty} \frac{1}{n(2n-1)(2n+1)} = 2\ln 2 - 1$.

Problem 6) Riemann's zeta function is defined as $\zeta(s) = \sum_{n=1}^{\infty}(1/n^s)$, where $s > 1$. With the aid of the Bernoulli numbers, it can be shown that $\zeta(4) = \pi^4/90$. Show that $\sum_{n=1}^{\infty} 1/(2n-1)^4 = \pi^4/96$.

Problem 7) Show that $\sum_{n=1}^{\infty}(-1)^{n+1}\frac{n}{(n+1)^2} = \frac{\pi^2}{12} - \ln(2)$.

Hint: Use the fact that $\sum_{n=1}^{\infty}\frac{(-1)^{n+1}}{n^2} = \frac{\pi^2}{12}$ and also $\ln(2) = \sum_{n=1}^{\infty}\frac{(-1)^{n+1}}{n}$.

Problem 8) Show that $\sum_{n=1}^{\infty} nx^n = x/(1-x)^2$, where $|x| < 1$.

Problem 9) Show that $\sum_{k=0}^{n}\frac{1}{k+1}\binom{n}{k} = \frac{2^{n+1}-1}{n+1}$.

Hint: Integrate $(x+1)^n$ with respect to x from 0 to 1.

Problem 10) a) Write the Taylor series expansion of the function $f(x) = \exp(x)$ around the point $x = 0$.

b) Use the result of part (a) to expand $g(x) = \exp(ix)$ in a Taylor series around $x = 0$.

c) What is the value of the infinite sum $S_1 = 1 - \frac{1}{2!} + \frac{1}{4!} - \frac{1}{6!} + \frac{1}{8!} - \cdots$?

d) What is the value of the infinite sum $S_2 = 1 - \frac{1}{3!} + \frac{1}{5!} - \frac{1}{7!} + \frac{1}{9!} - \cdots$?

Problem 11) Find the value of the following (infinite) series:

$$S = \frac{1}{1\times 5} - \frac{1}{2\times 6} + \frac{1}{3\times 7} - \frac{1}{4\times 8} + \cdots - \frac{(-1)^n}{n\times(n+4)} + \cdots$$

Problem 12) The function $f(x) = x/\ln(x+1)$ is defined for the real variable x where $x \geq -1$.

a) Use the Taylor series expansion of $\ln(x+1)$ around the point $x = 0$ to estimate the values of $f(x)$ at $x = 0$ and $x = -1$.

b) Find a recursion relation for the coefficients a_n of the Taylor series expansion of $f(x)$ around $x = 0$, namely, $f(x) = \sum_{n=0}^{\infty} a_n x^n$. What are the numerical values of a_0, a_1, a_2, a_3, and a_4?

c) Plot the general shape of the function $f(x)$ versus x for $x \geq -1$. Identify some of the important/interesting features of the curve.

Hint: The method needed in part (b) is similar to that used for obtaining the Bernoulli numbers.

Problem 13) The hyperbolic sine and cosine functions are defined as follows:

$$\sinh(x) = \tfrac{1}{2}[\exp(x) - \exp(-x)],$$

$$\cosh(x) = \tfrac{1}{2}[\exp(x) + \exp(-x)].$$

a) Find the Taylor series expansions of $\sinh(x)$ and $\cosh(x)$ around the point $x = 0$.

b) Show that, in general, the Taylor series expansion of an even function of x around $x = 0$ contains only *even* powers of x, whereas that of an odd function contains only *odd* powers.

c) Considering the symmetry of the function $\tanh(x) = \sinh(x)/\cosh(x)$, write the general form of its Taylor series expansion around $x = 0$.

Note: At this point, you are *not* being asked to determine the coefficients of the Taylor series, only the *general form* of the series based on the oddness or evenness of the function.

d) Use the results obtained in parts (a) and (c) to find a recursion relation for the coefficients of the Taylor series of $\tanh(x)$ so that each coefficient a_n can be derived from the preceding coefficients $a_0, a_1, a_2, \ldots, a_{n-1}$.

Problem 14) Show that $\sum_{m=0,2,4,6,\cdots}^{n} \binom{n}{m} = \sum_{m=1,3,5,7,\cdots}^{n} \binom{n}{m} = 2^{n-1}$.

Hint: Use the fact that $(1+1)^n = 2^n$ and $(1-1)^n = 0$.

Problem 15) Prove the following identities:

a) $f(x) = \sum_{n=1}^{\infty}(x^n/n) = -\ln(1-x)$; $|x| < 1$.

b) $g(x) = \sum_{n=1}^{\infty}[x^{n+1}/n(n+1)] = x + (1-x)\ln(1-x)$; $|x| < 1$.

c) $h(x) = \sum_{n=1}^{\infty}\left[\frac{x^{n+2}}{n(n+1)(n+2)}\right] = \frac{3}{4}x^2 - \frac{1}{2}x - \frac{1}{2}(1-x)^2\ln(1-x)$; $|x| < 1$.

Problem 16) Show that $\sum_{k=1}^{n} k\binom{n}{k} = n2^{n-1}$.

Hint: Differentiate $(x+1)^n$ with respect to x.

Problem 17) Using the binomial expansions of $[1+\exp(i2\pi/3)]^n$, $[1+\exp(i4\pi/3)]^n$, and $(1+1)^n$, verify the following identities:

a) $\binom{n}{0} + \binom{n}{3} + \binom{n}{6} + \cdots = \frac{1}{3}[2^n + 2\cos(n\pi/3)]$,

b) $\binom{n}{1} + \binom{n}{4} + \binom{n}{7} + \cdots = \frac{1}{3}\{2^n + 2\cos[(n-2)\pi/3]\}$,

c) $\binom{n}{2} + \binom{n}{5} + \binom{n}{8} + \cdots = \frac{1}{3}\{2^n + 2\cos[(n-4)\pi/3]\}$.

Problem 18) Use the binomial expansion of $(x+y)^n$, square both sides, then show that

$$\sum_{k=0}^{n}\binom{n}{k}^2 = \binom{2n}{n}.$$

Problem 19) Use Euler's expansion of $\sin x/x$, given in section 4.8, Eq.(21), to prove Wallis's product formula, namely,

$$\frac{\pi}{2} = \prod_{n=1}^{\infty}\left(\frac{2n}{2n-1}\cdot\frac{2n}{2n+1}\right) = \frac{2\times 2}{1\times 3}\cdot\frac{4\times 4}{3\times 5}\cdot\frac{6\times 6}{5\times 7}\cdots.$$

Note: An alternative proof of Wallis's product formula was the subject of problem 13, chapter 3.

In problems 20, 21, and 22, the following Taylor series expansions could be helpful:

$\ln(1+x) = x - ½x^2 + ⅓x^3 - ¼x^4 + \cdots = \sum_{n=1}^{\infty}(-1)^{n+1}x^n/n$, $(-1 < x \le 1)$,

$\ln(1-x) = -(x + ½x^2 + ⅓x^3 + ¼x^4 + \cdots) = -\sum_{n=1}^{\infty}x^n/n$; $(-1 \le x < 1)$.

Problem 20) Show that $\int_0^{\infty}\ln(1+e^{-x})\,dx = \pi^2/12$ and $\int_0^{\infty}\ln(1-e^{-x})\,dx = -\pi^2/6$.

Problem 21) Show that $\ln x = 2\left[\left(\frac{x-1}{x+1}\right) + ⅓\left(\frac{x-1}{x+1}\right)^3 + ⅕\left(\frac{x-1}{x+1}\right)^5 + \cdots\right]$; $(x > 0)$.

Problem 22) a) Plot the function $f(\theta) = \ln(\sin\theta)$ over the interval $0 < \theta < \pi$.

b) Show that $\int_0^{\pi/2}\ln(\cos\theta)\,d\theta = \int_0^{\pi/2}\ln(\sin\theta)\,d\theta = -(\pi/2)\ln 2$.

c) Invoking the trigonometric identity $\cos\theta = \sqrt{1-\sin^2\theta}$, use the result obtained in part (b) to arrive at an infinite series for ln 2.

Chapter 5

Multivariate Calculus

"I seem to have been only like a boy playing on the seashore, and diverting myself in now and then finding a smoother pebble or a prettier shell than ordinary, whilst the great ocean of truth lay all undiscovered before me."

Isaac Newton (1643–1727)

5.1. Introduction. Functions of a single, real-valued variable, typically denoted by $f(x)$, have many interesting properties. We study the continuity of such functions, their differentiability, their local maxima and minima, their integral (or the area under the function when plotted against the variable x, which is represented by points along a straight line called the x-axis). Sometimes we change the variable, from x to $\eta = g(x)$, and discover that it is easier to manipulate $f(\cdot)$ if we treat it as a function of the new variable η rather than the original variable x. Occasionally, we encounter so-called "pathological" functions, which behave in unexpected ways—such as when we examine the function $f(x) = \sin(1/x)$ in the vicinity of $x = 0$, and discover that an infinite number of maxima and minima of the function are squeezed into a tiny interval $(-\varepsilon, \varepsilon)$ surrounding the origin, irrespective of how small the value of ε may be. These and many other properties are also exhibited by functions of two or more variables, although the analysis of such functions, which forms the subject of multivariate calculus, is often much more challenging than the corresponding analysis of functions of a single variable.

As a simple example, consider the real-valued function $f(x, y)$ of two independent, real variables x and y. Suppose for a moment that f is the height of a mountain that rises above a flat plane, on which each point is identified by its Cartesian coordinates x and y. The mountain can have peaks, where $f(x, y)$ reaches a local maximum, and also valleys, where $f(x, y)$ reaches a local minimum. But there could also be saddle points on the mountain, where moving away from the saddle point along some directions could cause a rise in the elevation, whereas moving in some other directions would have the opposite effect of reducing the elevation. In section 2 we address the question of how one should go about identifying the maxima, minima, and saddle points of functions of two variables.

Section 3 is devoted to an exposition of the method of Lagrange multipliers, which is used when one is asked to optimize a function of two or more variables subject to one or more constraints. Think of the function $g(x, y)$ as representing the height of a mountain that rises above the xy-plane. Imagine that the mountain is transparent (e.g., it is made of clear glass), and that it is noon and the sun is shining straight down on this g-mountain from above. If you stand somewhere on the mountain, your shadow on the ground will indicate your (x, y) coordinates on the plane below. Now, suppose you move around on the mountain while trying to maintain your elevation at a constant value of c. Your shadow on the ground below will then trace a path, or follow a contour, that represents all the points in the xy-plane where $g(x, y) = c$. Having identified the contour $g(x, y) = c$ on the flat plane at the base of the mountain, you may now erase the g-mountain from your memory and imagine in its place a different mountain, $f(x, y)$, that rises above the same xy-plane. You are now instructed to walk on the f-mountain while trying to keep your shadow on the previously identified contour $g(x, y) = c$ at all times. In other words, the path that you are allowed to take on the f-mountain is *not* arbitrary; rather, your shadow on the ground below is constrained to stay on a trajectory that has been defined in advance by the shape of the g-mountain—specifically by the contour that is the locus of all the points in the xy-plane at which the elevation of the g-mountain is some specified constant c. The goal of constrained optimization is to identify one or more locations on the f-mountain where you find yourself on a locally flat region of the path that is constrained by the requirement that your shadow on the plane at the base of the mountain cannot deviate from the contour $g(x,y)=c$ at any time during your travels. Note that "local flatness" is no longer a property of the f-mountain as a whole, but rather a characteristic of the specific trail on the f-mountain that is defined by the constraint imposed on the trajectory of the traveler's shadow on the xy-plane at the base of the f-mountain. A few examples involving constrained optimization will be given in section 3; several more examples appear in the problems at the end of the chapter.

In section 4 we discuss the general problem of curvilinear coordinates, with emphasis on converting integrals over a domain specified in the Cartesian (x,y,z) coordinate system to integrals over a different coordinate system, say, (u,v,w). There will be functional relations in the form of $u(x,y,z)$, $v(x,y,z)$, $w(x,y,z)$, and also the corresponding inverse relations in the form of $x(u,v,w)$, $y(u,v,w)$, $z(u,v,w)$, which provide a unique, one-to-one mapping between the xyz and uvw coordinate systems. We will derive a general formula for the Jacobian of such transformations, and give examples of curvilinear coordinates that are designed to simplify some of the integrals that appear in certain problems of practical interest.

Finally, in section 5, we examine the motion of a point-particle along a space curve in three dimensions in terms of the characteristic radius of curvature and torsion of the curve.

5.2. Maxima, minima, and saddle points of a function of two variables. Let $f(x,y)$ be a continuous and differentiable function of x and y over some contiguous domain within the xy-plane. In the vicinity of an arbitrary point (x_0,y_0), the function may be expressed as a Taylor series expansion up to and including the second-order terms, as follows:

$$
\begin{aligned}
f(x,y) = f(x_0,y_0) &+ \partial_x f|_{(x_0,y_0)}(x-x_0) + \partial_y f|_{(x_0,y_0)}(y-y_0)\\
&+\tfrac{1}{2!}\partial_{xx} f|_{(x_0,y_0)}(x-x_0)^2 + \tfrac{1}{2!}\partial_{yy} f|_{(x_0,y_0)}(y-y_0)^2\\
&+\partial_{xy} f|_{(x_0,y_0)}(x-x_0)(y-y_0)+\cdots. \qquad (1)
\end{aligned}
$$

Note that all the partial derivatives of $f(x,y)$ are evaluated at (x_0,y_0). Note also that, while the coefficients of $\partial_{xx}f=\partial^2 f/\partial x^2$ and $\partial_{yy}f=\partial^2 f/\partial y^2$ are $1/2!$, the coefficient of $\partial_{xy}f=\partial^2 f/\partial x\partial y$ is equal to 1. This can be readily understood if one differentiates both sides of Eq.(1) with respect to the relevant variables, then compares the resulting expressions after setting (x,y) equal to (x_0,y_0).

The function $f(x,y)$ will be locally flat in the immediate neighborhood of (x_0,y_0) if the first-order partial derivatives $\partial_x f(x,y)$ and $\partial_y f(x,y)$ vanish at that location. To see this, observe that the first-order terms in Eq.(1) can be written as the dot product of the vectors $\boldsymbol{V}_1=[\partial_x f|_{(x_0,y_0)}]\hat{\boldsymbol{x}}+[\partial_y f|_{(x_0,y_0)}]\hat{\boldsymbol{y}}$ and $\boldsymbol{V}_2=(x-x_0)\hat{\boldsymbol{x}}+(y-y_0)\hat{\boldsymbol{y}}$. Now, the dot product $\boldsymbol{V}_1\cdot\boldsymbol{V}_2$ of any two vectors may be written as $V_{1x}V_{2x}+V_{1y}V_{2y}$ or, equivalently, as $|\boldsymbol{V}_1||\boldsymbol{V}_2|\cos\theta$, where θ is the angle between $\boldsymbol{V}_1$ and $\boldsymbol{V}_2$. Since the lengths $|\boldsymbol{V}_1|$ and $|\boldsymbol{V}_2|$ of the vectors are nonnegative, the dot product will assume both positive and negative values when θ covers all possible angles between 0 and 2π. This, however, is not acceptable if $f(x,y)$ is to be locally flat in the vicinity of (x_0,y_0). Consequently, $\boldsymbol{V}_1$ must vanish, which requires that $\partial_x f(x_0,y_0)=\partial_y f(x_0,y_0)=0$.

Having eliminated the first-order derivatives from Eq.(1), we now turn to the second-order derivatives and examine the conditions under which the flat region around (x_0, y_0) would represent either a local minimum, a local maximum, or a saddle point. To simplify the notation, let us define the parameters $A = \partial_{yy} f|_{(x_0,y_0)}$, $B = \partial_{xy} f|_{(x_0,y_0)}$, $C = \partial_{xx} f|_{(x_0,y_0)}$, $\Delta x = x - x_0$, $\Delta y = y - y_0$, and $s = \Delta y/\Delta x$. In the vicinity of (x_0, y_0), where Δx and Δy are sufficiently small, Eq.(1) yields, to a good approximation,

$$f(x, y) - f(x_0, y_0) \cong ½(\Delta x)^2(As^2 + 2Bs + C). \quad (2)$$

Note that, while Δx and Δy are relatively small positive or negative numbers, their ratio $s = \Delta y/\Delta x$ may fall anywhere in the range from $-\infty$ to ∞. In accordance with Eq.(2), the sign of $f(x, y) - f(x_0, y_0)$ is the same as that of the quadratic expression $As^2 + 2Bs + C$. Depending on the value of $B^2 - AC$, this quadratic expression may have zero, one, or two real-valued roots. If $B^2 - AC < 0$, the quadratic will have no real roots, in which case the sign of $As^2 + 2Bs + C$ will be the same as the sign of A for all values of s. This is because a plot of $As^2 + 2Bs + C$ versus s will be a parabola, which is always above the s-axis, if A happens to be positive, or below the s-axis, if A happens to be negative. We conclude that (x_0, y_0) is a minimum of $f(x, y)$ if $AC > B^2$ and $A > 0$, in which case C turns out to be positive as well. The function will be a maximum at (x_0, y_0) if $AC > B^2$ and both A and C turn out to be negative.

When $B^2 - AC > 0$, the quadratic equation $As^2 + 2Bs + C = 0$ will have two real-valued roots, namely, $s_{1,2} = (-B \pm \sqrt{B^2 - AC})/A$, in which case a plot of $As^2 + 2Bs + C$ will cross the s-axis at $s = s_1$ and $s = s_2$. The quadratic will then have one sign if s falls between s_1 and s_2, and the opposite sign if s falls outside the interval (s_1, s_2). Thus, if one draws two straight lines through (x_0, y_0) in the xy-plane, one with a slope of s_1, the other with a slope of s_2, as shown in the figure, then $f(x, y)$ will either rise above or drop below $f(x_0, y_0)$ depending on where, in the neighborhood of (x_0, y_0), the point (x, y) happens to be relative to the two straight lines. Under such circumstances, the point (x_0, y_0) will be a saddle point of $f(x, y)$.

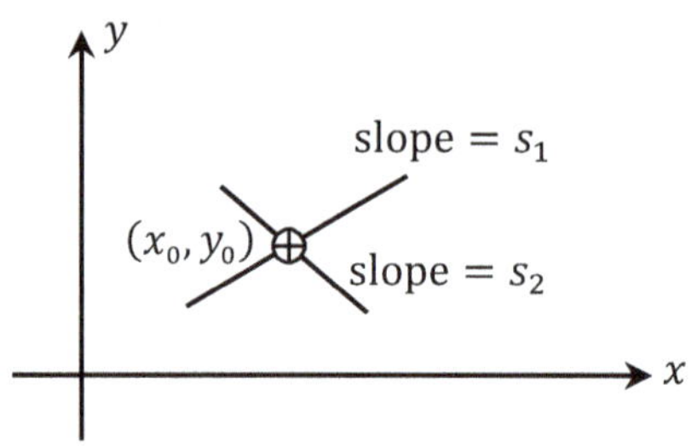

Finally, if $B^2 - AC = 0$, the quadratic expression will have a single root at $s = s_1 = s_2 = -B/A$. If one now draws a short straight line in the xy-plane with slope $s = -B/A$ through (x_0, y_0), then $f(x, y)$ will remain flat along this line but will otherwise rise above $f(x_0, y_0)$ if $A > 0$, or drop below $f(x_0, y_0)$ if $A < 0$. Needless to say, if all the first and second partial derivatives of the function vanish at (x_0, y_0), it will be necessary to examine higher-order derivatives to determine the local character of the function.

Brain teaser. Two hundred ants are placed randomly at various points along a one-meter-long rod. The ants move on a straight line along the length of the rod at a constant speed of 1.0 cm/sec. When two ants collide, they turn around and move in the opposite direction (at the same speed). Ants reaching either end of the rod fall off the edge, never to return. How long does it take for all the ants to fall off?

5.3. Constrained optimization using the method of Lagrange multipliers. Suppose it is desired to locate the point(s) in the xy-plane where the function $f(x, y)$ is locally flat subject to the constraint that $g(x, y) = c$, where c is a given constant. The constraint $g(x, y) = c$ describes a contour in the xy-plane, which has the property that, so long as the point (x, y) stays on the contour, the value of the function $g(x, y)$ remains constant at c. We shall first describe a recipe for finding the solution to this constrained optimization problem following the method of Lagrange multipliers. Afterward, we will explain why the method succeeds in finding a solution. Finally, we generalize the method to functions of more than two variables and to optimization subject to more than one constraint.

The recipe for optimizing a function of two variables, $f(x,y)$, subject to a single constraint, $g(x,y) = c$, is as follows:

i) Form the composite function $f(x,y) + \lambda g(x,y)$, where the parameter λ, the so-called Lagrange multiplier, is presently unspecified.

ii) Set the partial derivatives of the composite function $f + \lambda g$ with respect to both variables x and y equal to zero, that is, $\partial_x(f + \lambda g) = 0$ and $\partial_y(f + \lambda g) = 0$.

iii) Solve the resulting equations for x and y as functions of λ. At this point you will have $x(\lambda)$ and $y(\lambda)$, which satisfy the equations obtained in step (ii).

iv) Choose λ such that $x(\lambda)$ and $y(\lambda)$ obtained in step (iii) will satisfy the constraint $g(x,y) = c$. Denoting the value of λ found in the present step by λ_0, the corresponding values of $x(\lambda)$ and $y(\lambda)$ should be written as $x(\lambda_0)$ and $y(\lambda_0)$.

v) The desired optimal solution to the problem is $[x(\lambda), y(\lambda)]$ evaluated at $\lambda = \lambda_0$, which was found in step (iv).

The reason this recipe yields the correct answer is that, by construction, the composite function $f + \lambda g$ is locally flat at $[x(\lambda), y(\lambda)]$ for any arbitrarily chosen value of λ. However, when λ is chosen in step (iv) such that the constraint $g(x,y) = c$ is satisfied, the point $[x(\lambda), y(\lambda)]$ is brought to lie on the contour $g(x,y) = c$. From here, if one moves slightly away from $[x(\lambda), y(\lambda)]$ while staying on the contour, the value of $g(x,y)$ will not change (because it is forced to remain constant at c). But we know that the composite function $f + \lambda g$ is flat around $[x(\lambda), y(\lambda)]$. Considering that $g(x,y)$ is not changing, the function $f(x,y)$ must remain flat in this neighborhood to ensure the flatness of the composite function.

Example 1. Find a point (x_0, y_0) on the straight line $ax + by = c$ such that the distance between (x_0, y_0) and the origin of coordinates $(x,y) = (0,0)$ is minimized.

Solution: Here, the function to optimize is $f(x,y) = x^2 + y^2$ subject to the constraint $g(x,y) = ax + by = c$. The composite function is $f + \lambda g = x^2 + y^2 + \lambda(ax + by)$, whose partial derivatives with respect to x and y are required to vanish. We thus find

$$\begin{cases} \partial_x(f + \lambda g) = 2x + \lambda a = 0 \\ \partial_y(f + \lambda g) = 2y + \lambda b = 0 \end{cases} \quad \rightarrow \quad \begin{cases} x = -\lambda a/2, \\ y = -\lambda b/2. \end{cases}$$

The next step is to enforce the constraint; that is,

$$g(x,y) = ax + by = -½\lambda a^2 - ½\lambda b^2 = c.$$

This yields $\lambda_0 = -2c/(a^2 + b^2)$ and, consequently, $x_0 = -\lambda_0 a/2 = ac/(a^2 + b^2)$ and $y_0 = -\lambda_0 b/2 = bc/(a^2 + b^2)$. Note that the point (x_0, y_0) thus obtained has a simple geometrical interpretation: It is the foot of the perpendicular dropped from the origin of coordinates in the xy-plane onto the straight line $ax + by = c$.

In the more general case of constrained optimization, where $f(x_1, \cdots, x_n)$ is a function of n variables, and where there may be multiple constraints such as

$g(x_1, \cdots, x_n) = c_1$, $h(x_1, \cdots, x_n) = c_2$, $\cdots$, and $q(x_1, \cdots, x_n) = c_m$, the procedure will be similar to that outlined above. Here, the composite function will be $f + \lambda_1 g + \lambda_2 h + \cdots + \lambda_m q$. Upon setting the partial derivatives of this composite function with respect to $x_1, x_2, \cdots, x_n$ equal to zero, we will have n equations, which must be solved for the n unknowns, namely, $x_1, x_2, \cdots, x_n$. The solutions, of course, turn out to be functions of the Lagrange multipliers, that is, $x_1(\lambda_1, \lambda_2, \cdots, \lambda_m)$, $x_2(\lambda_1, \lambda_2, \cdots, \lambda_m)$, and so on. The optimal values of $\lambda_1, \lambda_2, \cdots, \lambda_m$ will subsequently be determined by enforcing the constraints $g = c_1, h = c_2, \cdots, q = c_m$. At this point, we have found the solutions $x_1, x_2, \cdots, x_n$ that not only satisfy all the constraints, but also ensure that the function $f(x_1, x_2, \cdots, x_n)$ is flat in the immediate vicinity of each solution—so long as the slight variations of $x_1, x_2, \cdots, x_n$ are such that the constraints would remain satisfied at all times.

Example 2. A spherical shell of radius R is centered at (x_c, y_c, z_c) within a Cartesian coordinate system. Find the points on the sphere that are either closest to or farthest away from the origin $(x, y, z) = (0, 0, 0)$ of the coordinate system.

Solution: We are searching for a point (x_o, y_o, z_o) that either minimizes or maximizes the function $f(x, y, z) = x^2 + y^2 + z^2$ subject to the constraint that the desired point be located on the surface of a spherical shell of radius R and center coordinates (x_c, y_c, z_c). In other words, the optimization problem is subject to the constraint $g(x, y) = (x - x_c)^2 + (y - y_c)^2 + (z - z_c)^2 = R^2$. Note that we have chosen to optimize *not* the distance to the origin, but rather the *square* of the distance, which is represented by our function $f(x, y, z)$. This will not affect the optimal solution of the problem, but will make the algebra a little easier to handle. We begin by forming the composite function

$$f(x, y, z) + \lambda g(x, y, z) = (x^2 + y^2 + z^2) + \lambda[(x - x_c)^2 + (y - y_c)^2 + (z - z_c)^2].$$

Setting the partial derivatives of the composite function with respect to x, y, and z equal to zero yields

$$\begin{cases} \partial_x(f + \lambda g) = 2x + 2\lambda(x - x_c) = 0 \\ \partial_y(f + \lambda g) = 2y + 2\lambda(y - y_c) = 0 \\ \partial_z(f + \lambda g) = 2z + 2\lambda(z - z_c) = 0 \end{cases} \rightarrow \quad \begin{cases} x(\lambda) = \lambda x_c/(1 + \lambda), \\ y(\lambda) = \lambda y_c/(1 + \lambda), \\ z(\lambda) = \lambda z_c/(1 + \lambda). \end{cases}$$

To arrive at the optimum value of λ, we now enforce the constraint $g(x, y, z) = R^2$ using the above values of $x(\lambda)$, $y(\lambda)$, and $z(\lambda)$. We will have

$$g(x, y, z) = \left(\tfrac{\lambda}{1+\lambda} - 1\right)^2 (x_c^2 + y_c^2 + z_c^2) = R^2 \ \rightarrow \ \lambda_o = \pm\left(\sqrt{x_c^2 + y_c^2 + z_c^2}/R\right) - 1.$$

Substituting the optimum value of λ in the preceding expressions for $x(\lambda)$, $y(\lambda)$, $z(\lambda)$, and denoting by r_c the distance $\sqrt{x_c^2 + y_c^2 + z_c^2}$ between the center of the sphere and the origin of coordinates, we now find

$$[x(\lambda_o), y(\lambda_o), z(\lambda_o)] = \left(\tfrac{r_c \pm R}{r_c}\right)(x_c, y_c, z_c).$$

Note that the optimal points found on the spherical shell are on the straight line that connects the origin to the center of the sphere. The solution associated with the plus sign is the point that is farthest away from the origin, while that associated with the minus sign is the closest point of the shell to the origin of coordinates. Note also that the method of Lagrange multipliers fails to yield a solution to the present problem in the special case when the shell happens to be centered at the origin, that is, when $r_c = 0$, in which case every point on the spherical surface is equidistant from the origin and no special points exist that can be identified as being closest to or farthest away from the origin.

Brain teaser. A teacher walks into a classroom where 50 students are sitting at their desks. The students know each other well, but the teacher knows nothing about them except that one student is a truth-teller while the remaining 49 are togglers. (A toggler is someone who alternates between lying and telling the truth. If the toggler gave the correct answer to your first question, he or she would lie in response to the second and vice versa. It is not known whether a toggler's first answer is true or false, but if you posed two questions, you are assured of the veracity of exactly one of the two answers.) The teacher is allowed to ask two questions. She may ask both questions from the same student or direct each of her questions to a different student. How does she find the truth-teller?

5.4. Curvilinear coordinates and the Jacobian of coordinate transformation. Let us consider the integral $\int_a^b f(x)\mathrm{d}x$ of the function $f(\cdot)$ over the interval $[a, b]$ of the real variable x. Changing the variable from x to y, where $x = g(y)$, requires that we replace $\mathrm{d}x$ by $g'(y)\mathrm{d}y$ and also that we change the upper and lower limits of the integral accordingly. Consequently,

$$\int_a^b f(x)\mathrm{d}x = \int_{g^{-1}(a)}^{g^{-1}(b)} f[g(y)]g'(y)\mathrm{d}y. \qquad (3)$$

A quick glance at figure 5.1 should reveal the mechanism of this one-dimensional coordinate transformation and, specifically, the necessity of introducing the derivative $g'(y)$ of $g(y)$ into the integrand. One could call $g'(y)$ the *Jacobian* of the transformation from x to y, although the name is usually reserved for higher-dimensional coordinate transforms.

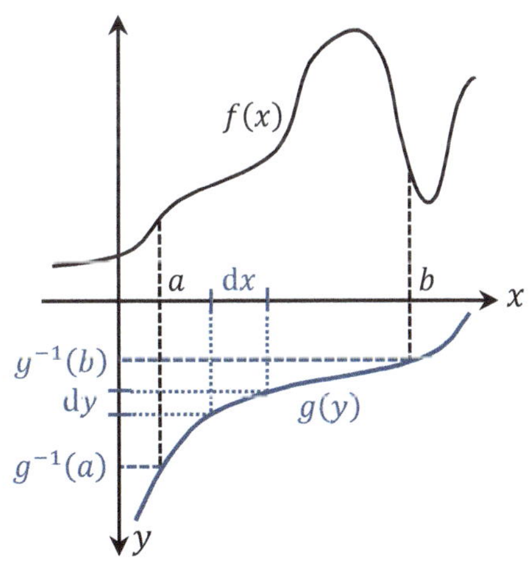

Figure 5.1. Changing the variable of integration from x to y, where $x = g(y)$, requires that the integrand $f(x)$ of $\int_a^b f(x)\mathrm{d}x$ be replaced by $f[g(y)]$, the increment $\mathrm{d}x$ of x be replaced by $g'(y)\mathrm{d}y$, and the limits of integration change from a and b to $g^{-1}(a)$ and $g^{-1}(b)$, respectively.

In two dimensions, one typically has access to a rich set of possible coordinate transforms. The well-known polar coordinates (r, ϕ), where $x = r\cos\phi$ and $y = r\sin\phi$, provide but one example.

Suppose that a coordinate transformation from the Cartesian (x, y) coordinates to a general curvilinear coordinate system (u, v) is defined by the functions $u(x, y)$ and $v(x, y)$, which may also be written in inverted form as $x(u, v)$ and $y(u, v)$. (In the aforementioned example of polar coordinates, the inverted functions are $r = \sqrt{x^2 + y^2}$ and $\phi = \tan^{-1}(y/x)$.) If you pick a point $(x_0, y_0) = (u_0, v_0)$ within a domain of integration, then move a short distance Δu along the u-coordinate while keeping v constant, you will have moved (within the xy-plane) along the infinitesimal vector $\boldsymbol{U} = [(\partial x/\partial u)\hat{\mathbf{x}} + (\partial y/\partial u)\hat{\mathbf{y}}]\Delta u$. Similarly, if starting at $(x_0, y_0) = (u_0, v_0)$, you move a short distance Δv along the v-coordinate while keeping u constant, you will have moved (within the xy-plane) along the infinitesimal vector $\boldsymbol{V} = [(\partial x/\partial v)\hat{\mathbf{x}} + (\partial y/\partial v)\hat{\mathbf{y}}]\Delta v$. The incremental area of the xy-plane enclosed between $\boldsymbol{U}$ and $\boldsymbol{V}$ is thus given by

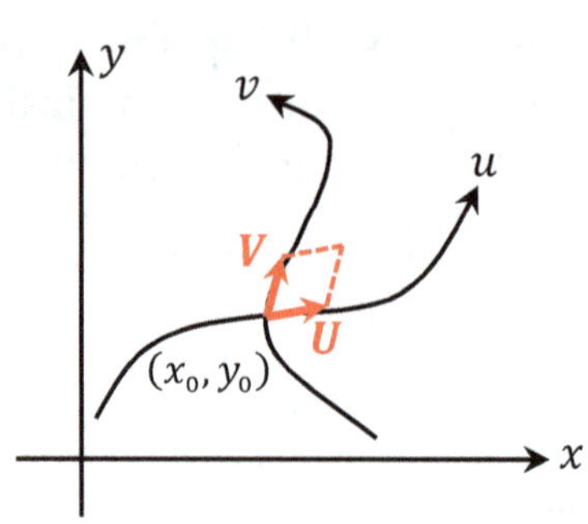

$$\text{Area increment} = |\boldsymbol{U} \times \boldsymbol{V}| = \left|\left(\frac{\partial x}{\partial u}\right)\left(\frac{\partial y}{\partial v}\right) - \left(\frac{\partial y}{\partial u}\right)\left(\frac{\partial x}{\partial v}\right)\right| \Delta u \Delta v$$

$$= \left\| \begin{matrix} \partial x/\partial u & \partial y/\partial u \\ \partial x/\partial v & \partial y/\partial v \end{matrix} \right\| \Delta u \Delta v. \qquad (4)$$

The absolute value of the 2×2 determinant appearing in Eq.(4) is the Jacobian of the transformation from (x, y) to (u, v) coordinates, which is often abbreviated as $\partial(x, y)/\partial(u, v)$. With the help of the Jacobian, the two-dimensional integral of a function such as $f(x, y)$ over some domain D within the xy-plane may be converted to an integral over the projection of D onto the uv-plane, as follows:

$$\iint_{D(x,y)} f(x, y)\mathrm{d}x\mathrm{d}y = \iint_{D(u,v)} f[x(u, v), y(u, v)] \frac{\partial(x,y)}{\partial(u,v)} \mathrm{d}u\mathrm{d}v. \qquad (5)$$

Example. In an elliptical coordinate system, $x = au\cos v$ and $y = bu\sin v$, where $a > 0$ and $b > 0$ are arbitrary constants. It is straightforward to show that

$$u^2 = (x/a)^2 + (y/b)^2, \qquad (6)$$

$$v = \tan^{-1}\left(\frac{ay}{bx}\right). \qquad (7)$$

The contours of constant u are ellipses whose semi-axes along the x and y directions are au and bu, respectively. The contours of constant v are rays having a constant slope $s = y/x$, emanating from the origin of the xy-coordinate system, with $v = \tan^{-1}(as/b)$. Figure 5.2 shows several contours of constant u and constant v in the xy-plane. The variable u ranges from zero (vanishingly small ellipse) to infinity (very large ellipse). The variable v ranges from zero (positive x-axis) to $\pi/2$ (positive y-axis), and from there to π (negative x-axis), to $3\pi/2$ (negative y-axis), and to 2π (back to positive x-axis).

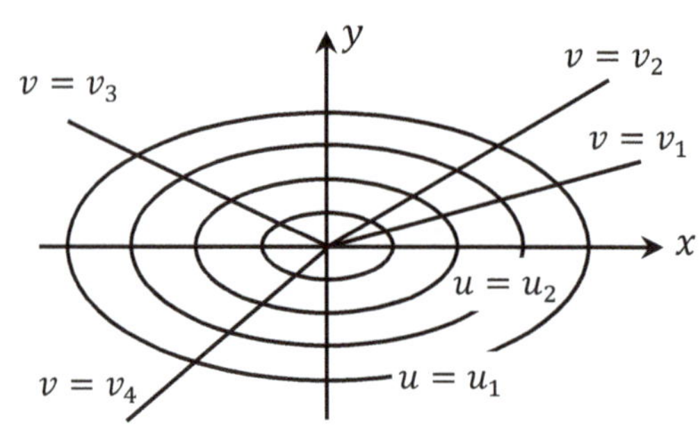

Figure 5.2. In an elliptical coordinate system, the contours of constant u are ellipses having a constant ratio a/b of the major to minor axis. The contours of constant v are rays emanating from the origin of the xy-plane and having slope $s = y/x$, where $v = \tan^{-1}(as/b)$.

The Jocabian of the transformation from the Cartesian xy-coordinates to the elliptical uv-coordinates is readily computed as follows:

$$\frac{\partial(x,y)}{\partial(u,v)} = \begin{Vmatrix} \partial x/\partial u & \partial y/\partial u \\ \partial x/\partial v & \partial y/\partial v \end{Vmatrix} = \begin{Vmatrix} a\cos v & b\sin v \\ -au\sin v & bu\cos v \end{Vmatrix}$$

$$= abu(\cos^2 v + \sin^2 v) = abu. \tag{8}$$

To find the area of an ellipse whose semi-axes along x and y are a and b, we write

$$\text{Area} = \int_{u=0}^{1} \int_{v=0}^{2\pi} 1 \cdot \frac{\partial(x,y)}{\partial(u,v)} \mathrm{d}u\mathrm{d}v = \int_{u=0}^{1} \int_{v=0}^{2\pi} (abu)\mathrm{d}u\mathrm{d}v = ab(\tfrac{1}{2}u^2|_0^1)(2\pi) = \pi ab.$$

Similarly, to find the moment of inertia for rotation around the x-axis of a thin, solid, elliptical plate having uniform mass-density $\rho(x,y) = 1.0$, we write

$$\text{Moment of inertia} = \iint_{\text{ellipse}} \rho(x,y)y^2\mathrm{d}x\mathrm{d}y = \int_{u=0}^{1} \int_{v=0}^{2\pi} (bu\sin v)^2 \frac{\partial(x,y)}{\partial(u,v)} \mathrm{d}u\mathrm{d}v$$

$$= \int_{u=0}^{1} \int_{v=0}^{2\pi} (bu\sin v)^2(abu)\mathrm{d}u\mathrm{d}v = ab^3(\tfrac{1}{4}u^4|_0^1)(\pi) = \tfrac{1}{4}\pi ab^3.$$

Digression: Perimeter of the ellipse. To compute the perimeter of the ellipse depicted in figure 5.3, we must first find an expression for the short segment $\Delta\ell = \sqrt{(\Delta x)^2 + (\Delta y)^2}$ located on the boundary of the ellipse, then integrate $\Delta\ell$ around the ellipse's perimeter.

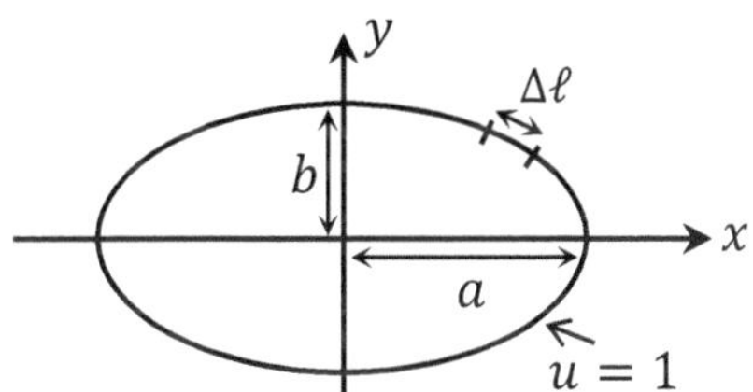

Figure 5.3. An ellipse, having major semi-axis a and minor semi-axis b, is centered at the origin of the Cartesian xy coordinates. In the curvilinear uv coordinates, we have $(x/a)^2 + (y/b)^2 = u^2$, indicating that the u-coordinate of the ellipse is fixed at $u = 1$. The v-coordinate, given by $v = \tan^{-1}(ay/bx)$, varies from 0 to 2π as one travels counterclockwise around the ellipse. The perimeter of the ellipse is found by integrating $\Delta\ell$ with respect to v, which ranges from 0 to 2π.

Switching to the curvilinear coordinates (u,v), specified by $x = au\cos v$ and $y = bu\sin v$, we find $(x/a)^2 + (y/b)^2 = u^2$, which indicates that the u-coordinate of the ellipse is fixed at $u = 1$. The v-coordinate, given by $v = \tan^{-1}(ay/bx)$, varies from 0 to 2π as one travels counterclockwise around the ellipse. We now write

$$\Delta\ell = \Delta v\sqrt{(\partial x/\partial v)^2 + (\partial y/\partial v)^2} = \Delta v\sqrt{a^2\sin^2 v + b^2\cos^2 v} = \Delta v\sqrt{a^2 - (a^2 - b^2)\cos^2 v},$$

which is readily integrated around the ellipse, as follows:

$$\text{Perimeter of ellipse} = \int_0^{2\pi} \sqrt{(\partial x/\partial v)^2 + (\partial y/\partial v)^2}\mathrm{d}v = 4\int_0^{\pi/2} \sqrt{a^2 - (a^2 - b^2)\cos^2 v}\,\mathrm{d}v$$

$$\boxed{k = \sqrt{1-(b/a)^2}} \rightarrow = 4a\int_0^{\pi/2} \sqrt{1 - k^2\cos^2 v}\,\mathrm{d}v = 4a\int_0^{\pi/2} \sqrt{1 - k^2\sin^2\theta}\,\mathrm{d}\theta \leftarrow \boxed{\theta = \tfrac{1}{2}\pi - v}$$

$$= 4aE(k). \leftarrow \boxed{E(k) \text{ is complete elliptic integral of the 2nd kind}}$$

In the special case of a circle, where $a = b$, we will have $k = 0$ and $E(0) = \pi/2$, which yields the perimeter as $4a(\pi/2) = 2\pi a$.

Note in the above derivation that there was no need to incorporate the Jacobian of the transformation into the integrand. The reason is that what is being calculated here is *not* an area, which would have required cross-multiplying $\boldsymbol{U} = (\partial x/\partial u)\hat{\boldsymbol{x}} + (\partial y/\partial u)\hat{\boldsymbol{y}}$ and $\boldsymbol{V} = (\partial x/\partial v)\hat{\boldsymbol{x}} + (\partial y/\partial v)\hat{\boldsymbol{y}}$ to arrive at the infinitesimal area $\|\boldsymbol{U} \times \boldsymbol{V}\| dudv$. In the absence of such infinitesimal areas, the need for introducing the Jacobian disappears.

The concept of Jacobian can be readily extended to higher-dimensional spaces. For instance, in the three-dimensional Cartesian coordinates (x, y, z), let a curvilinear coordinate system (u, v, w) be defined by $u(x, y, z)$, $v(x, y, z)$, and $w(x, y, z)$. The mapping from xyz to uvw being one to one, it is possible to invert the coordinate functions to arrive at $x(u, v, w)$, $y(u, v, w)$, and $z(u, v, w)$ as well. The short vectors representing small translations along the u, v, and w directions within the xyz system will then be

$$\boldsymbol{U} = [(\partial x/\partial u)\hat{\boldsymbol{x}} + (\partial y/\partial u)\hat{\boldsymbol{y}} + (\partial z/\partial u)\hat{\boldsymbol{z}}]\Delta u. \quad (9)$$

$$\boldsymbol{V} = [(\partial x/\partial v)\hat{\boldsymbol{x}} + (\partial y/\partial v)\hat{\boldsymbol{y}} + (\partial z/\partial v)\hat{\boldsymbol{z}}]\Delta v. \quad (10)$$

$$\boldsymbol{W} = [(\partial x/\partial w)\hat{\boldsymbol{x}} + (\partial y/\partial w)\hat{\boldsymbol{y}} + (\partial z/\partial w)\hat{\boldsymbol{z}}]\Delta w. \quad (11)$$

The volume of the parallelepiped defined by these three vectors can now be written as $\boldsymbol{U} \cdot (\boldsymbol{V} \times \boldsymbol{W})$, or, equivalently, as $\boldsymbol{W} \cdot (\boldsymbol{U} \times \boldsymbol{V})$ or as $\boldsymbol{V} \cdot (\boldsymbol{W} \times \boldsymbol{U})$. These triple vector products are identical and can be written in the form of a 3×3 determinant, as follows:

$$\text{Volume increment} = |\boldsymbol{U} \cdot (\boldsymbol{V} \times \boldsymbol{W})| = \left\| \begin{matrix} \partial x/\partial u & \partial y/\partial u & \partial z/\partial u \\ \partial x/\partial v & \partial y/\partial v & \partial z/\partial v \\ \partial x/\partial w & \partial y/\partial w & \partial z/\partial w \end{matrix} \right\| \Delta u \Delta v \Delta w. \quad (12)$$

As before, the 3×3 determinant appearing in Eq.(12) is the transformation Jacobian from xyz to uvw coordinates, which is often abbreviated as $\partial(x, y, z)/\partial(u, v, w)$. With the help of the Jacobian, the three-dimensional integral of a function such as $f(x, y, z)$ over some domain D within the xyz coordinates can be converted to an integral over the projection of D onto the uvw system, as follows:

$$\iint_{D(x,y,z)} f(x, y, z) \mathrm{d}x\mathrm{d}y\mathrm{d}z$$

$$= \iint_{D(u,v,w)} f[x(u, v, w), y(u, v, w), z(u, v, w)] \frac{\partial(x,y,z)}{\partial(u,v,w)} \mathrm{d}u\mathrm{d}v\mathrm{d}w. \quad (13)$$

The well-known spherical coordinate system (r, θ, ϕ) in which $x = r \sin\theta \cos\phi$, $y = r \sin\theta \sin\phi$, and $z = r \cos\theta$ provides a good example of a curvilinear coordinate system in three dimensions. As shown in problem 11, the Jacobian of transformation from (x, y, z) to (r, θ, ϕ) is given by $\partial(x, y, z)/\partial(r, \theta, \phi) = r^2 \sin\theta$.

Digression: In two-dimensional curvilinear coordinate transformations, small increments Δx and Δy of the xy-coordinates are related to small increments Δu and Δv of the uv-coordinates as follows:

$$(\Delta x \quad \Delta y) = (\Delta u \quad \Delta v) \begin{pmatrix} \partial x/\partial u & \partial y/\partial u \\ \partial x/\partial v & \partial y/\partial v \end{pmatrix}. \quad (14)$$

Let $(\Delta x, \Delta y)$ be a short vector in the Cartesian xy-plane and, similarly, $(\Delta u, \Delta v)$ be a short vector in the Cartesian uv-plane. The 2×2 matrix $\partial(x, y)/\partial(u, v)$ of Eq.(14) has two eigenvalues λ_1 and λ_2, as well as corresponding eigenvectors $\boldsymbol{\mathcal{E}}_1 = (\Delta u_1, \Delta v_1)$ and $\boldsymbol{\mathcal{E}}_2 = (\Delta u_2, \Delta v_2)$ in the uv-plane. When λ_1 and λ_2 are real-valued, the projections of their eigenvectors onto the xy-plane will be $(\Delta x_1, \Delta y_1) = \lambda_1 \boldsymbol{\mathcal{E}}_1$ and $(\Delta x_2, \Delta y_2) = \lambda_2 \boldsymbol{\mathcal{E}}_2$. Thus, aside from being scaled by λ_1 and λ_2, the eigenvectors in the uv-plane are the same as their projections onto the xy-plane. Upon multiplication by $|\lambda_1 \lambda_2|$, the area of the parallelogram formed by $\boldsymbol{\mathcal{E}}_1$ and $\boldsymbol{\mathcal{E}}_2$ in the uv-plane yields the area of the corresponding parallelogram in the xy-plane.

If λ_1 and λ_2 happen to be complex-valued, we will have $\lambda_1 = \lambda_2^* = |\lambda| \exp(\mathrm{i}\theta)$, whose pair of complex-conjugate eigenvectors will be $(\Delta u' \pm \mathrm{i}\Delta u'', \Delta v' \pm \mathrm{i}\Delta v'')$. The real-valued vectors in the uv-plane now become $\boldsymbol{\mathcal{E}}_1 = (\Delta u', \Delta v')$ and $\boldsymbol{\mathcal{E}}_2 = (\Delta u'', \Delta v'')$, which must be projected onto the xy-plane as $(\Delta x', \Delta y') = |\lambda|(\cos\theta\, \boldsymbol{\mathcal{E}}_1 - \sin\theta\, \boldsymbol{\mathcal{E}}_2)$ and $(\Delta x'', \Delta y'') = |\lambda|(\sin\theta\, \boldsymbol{\mathcal{E}}_1 + \cos\theta\, \boldsymbol{\mathcal{E}}_2)$. Once again, the infinitesimal area of the xy-plane occupied by these vectors is related to the corresponding area $\boldsymbol{\mathcal{E}}_1 \times \boldsymbol{\mathcal{E}}_2$ of the uv-plane as

$$(\Delta x', \Delta y') \times (\Delta x'', \Delta y'') = |\lambda|^2(\sin^2\theta + \cos^2\theta)\boldsymbol{\mathcal{E}}_1 \times \boldsymbol{\mathcal{E}}_2 = \lambda\lambda^*\, \boldsymbol{\mathcal{E}}_1 \times \boldsymbol{\mathcal{E}}_2. \quad (15)$$

In higher dimensions, the cross-product magnitude $|\boldsymbol{a} \times \boldsymbol{b}|$ is replaced by $\sqrt{(\boldsymbol{a}\cdot\boldsymbol{a})(\boldsymbol{b}\cdot\boldsymbol{b}) - (\boldsymbol{a}\cdot\boldsymbol{b})^2}$.

Considering that the determinant of any square matrix equals the product of its eigenvalues, we conclude that any infinitesimal area of the uv-plane must be multiplied by $|\lambda_1\lambda_2| = \|\partial(x,y)/\partial(u,v)\|$ to yield the corresponding (infinitesimal) area of the xy-plane. Since the present interpretation of the Jacobian of coordinate transformation between the xy and uv systems does *not* rely on the conventional notion of cross-multiplication in a two- or three-dimensional vector space, it is easily generalizable to curvilinear coordinate transformations in higher dimensions.

5.5. Geometry of a space curve. Consider a point P moving along a space curve as a function of time t. The position of the point in three-dimensional Euclidean space is specified as $\boldsymbol{P}(t)$, whose coordinates are $x(t)$, $y(t)$, and $z(t)$. We will have

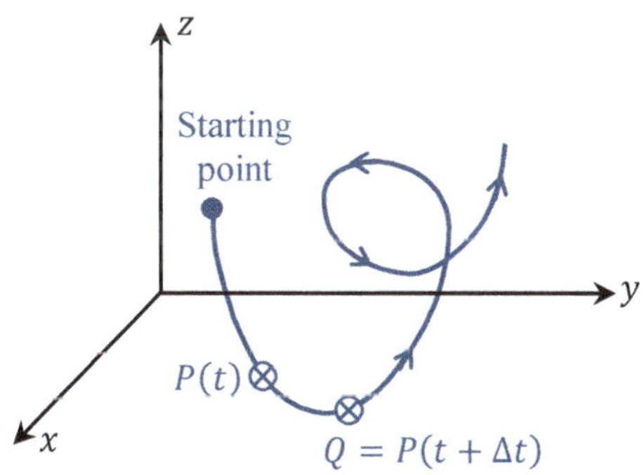

$$\frac{\Delta \boldsymbol{P}(t)}{\Delta t} = \frac{\Delta x(t)}{\Delta t}\hat{\boldsymbol{x}} + \frac{\Delta y(t)}{\Delta t}\hat{\boldsymbol{y}} + \frac{\Delta z(t)}{\Delta t}\hat{\boldsymbol{z}}$$
$$= \left(\frac{\Delta x}{\Delta s}\hat{\boldsymbol{x}} + \frac{\Delta y}{\Delta s}\hat{\boldsymbol{y}} + \frac{\Delta z}{\Delta s}\hat{\boldsymbol{z}}\right)\left(\frac{\Delta s}{\Delta t}\right). \quad (16)$$

In this equation, $s(t)$ is the arc length, measured from some arbitrary starting point along the curve. Note that, for sufficiently small Δt, we have $(\Delta s)^2 = (\Delta x)^2 + (\Delta y)^2 + (\Delta z)^2$. The unit vector $\hat{\boldsymbol{u}} = (\mathrm{d}x/\mathrm{d}s)\hat{\boldsymbol{x}} + (\mathrm{d}y/\mathrm{d}s)\hat{\boldsymbol{y}} + (\mathrm{d}z/\mathrm{d}s)\hat{\boldsymbol{z}}$ is tangent to the curve at each point along the path. We may obviously write the speed and the velocity of the particle at each point as follows (note that the velocity is along $\hat{\boldsymbol{u}}$):

$$\mathrm{d}s/\mathrm{d}t = \sqrt{(\mathrm{d}x/\mathrm{d}t)^2 + (\mathrm{d}y/\mathrm{d}t)^2 + (\mathrm{d}z/\mathrm{d}t)^2} \;\rightarrow\; \boldsymbol{V} = V\hat{\boldsymbol{u}} = \left(\frac{\mathrm{d}s}{\mathrm{d}t}\right)\hat{\boldsymbol{u}} = \frac{\mathrm{d}\boldsymbol{P}(t)}{\mathrm{d}t}. \quad (17)$$

The particle's acceleration $\boldsymbol{a}$ will then be

$$\boldsymbol{a} = \frac{\mathrm{d}\boldsymbol{V}}{\mathrm{d}t} = \frac{\mathrm{d}}{\mathrm{d}t}(V\hat{\boldsymbol{u}}) = \left(\frac{\mathrm{d}V}{\mathrm{d}t}\right)\hat{\boldsymbol{u}} + V\left(\frac{\mathrm{d}\hat{\boldsymbol{u}}}{\mathrm{d}t}\right) = \left(\frac{\mathrm{d}V}{\mathrm{d}t}\right)\hat{\boldsymbol{u}} + V^2\frac{\mathrm{d}\hat{\boldsymbol{u}}}{\mathrm{d}s}. \quad (18)$$

Now, $\mathrm{d}V/\mathrm{d}t = \mathrm{d}^2 s/\mathrm{d}t^2$ is the time rate of change of the particle's speed along the path. This component of the acceleration $\boldsymbol{a}$ is along the tangent vector $\hat{\boldsymbol{u}}$. The derivative $\mathrm{d}\hat{\boldsymbol{u}}/\mathrm{d}s$ is the rate of change of the direction of velocity along the path. Given that $\hat{\boldsymbol{u}}$ is a unit vector, $\mathrm{d}\hat{\boldsymbol{u}}/\mathrm{d}s$ must be orthogonal to $\hat{\boldsymbol{u}}$. We now write

$$\mathrm{d}\hat{\boldsymbol{u}}/\mathrm{d}s = (\mathrm{d}^2x/\mathrm{d}s^2)\hat{\boldsymbol{x}} + (\mathrm{d}^2y/\mathrm{d}s^2)\hat{\boldsymbol{y}} + (\mathrm{d}^2z/\mathrm{d}s^2)\hat{\boldsymbol{z}}. \tag{19}$$

The length of $\mathrm{d}\hat{\boldsymbol{u}}/\mathrm{d}s$ is the local curvature of the curve, which is the reciprocal of its radius of curvature ρ. Denoting the direction of $\mathrm{d}\hat{\boldsymbol{u}}/\mathrm{d}s$ by $\hat{\boldsymbol{n}}$, which is a unit vector orthogonal to the tangent vector $\hat{\boldsymbol{u}}$ at $\boldsymbol{P}$, we will have

$$\mathrm{d}\hat{\boldsymbol{u}}/\mathrm{d}s = (1/\rho)\hat{\boldsymbol{n}}; \qquad (\rho > 0 \text{ is the radius of curvature}). \tag{20}$$

The unit vector $\hat{\boldsymbol{n}}$ is commonly referred to as the *principal normal vector*.

$$1/\rho = \sqrt{(\mathrm{d}^2x/\mathrm{d}s^2)^2 + (\mathrm{d}^2y/\mathrm{d}s^2)^2 + (\mathrm{d}^2z/\mathrm{d}s^2)^2}. \tag{21}$$

The acceleration $\boldsymbol{a}$ given by Eq.(18) may, therefore, be expressed as follows:

$$\boldsymbol{a} = (\mathrm{d}V/\mathrm{d}t)\hat{\boldsymbol{u}} + (V^2/\rho)\hat{\boldsymbol{n}}. \tag{22}$$

Thus, when a particle moves along a space curve with speed $V(t)$, its acceleration can be resolved into a component $\mathrm{d}V/\mathrm{d}t$ along the curve and a component V^2/ρ along a direction that is perpendicular to the curve.

In addition to the unit vectors $\hat{\boldsymbol{u}}$ and $\hat{\boldsymbol{n}}$, which are orthogonal to each other, we define the *bi-normal* vector $\hat{\boldsymbol{b}}$, as follows:

$$\hat{\boldsymbol{b}} = \hat{\boldsymbol{u}} \times \hat{\boldsymbol{n}}. \tag{23}$$

Clearly, $\hat{\boldsymbol{b}}$ is also a unit vector, because $\hat{\boldsymbol{u}} \perp \hat{\boldsymbol{n}}$. If the space curve happens to be in a flat plane, $\hat{\boldsymbol{u}}$ and $\hat{\boldsymbol{n}}$ will be in that plane, and $\hat{\boldsymbol{b}}$ will be perpendicular to the plane at all times. Otherwise, $\hat{\boldsymbol{b}}$ will change direction in space as $\boldsymbol{P}$ moves along the curve. With each moving point $\boldsymbol{P}$ is thus associated a triad $(\hat{\boldsymbol{u}}, \hat{\boldsymbol{n}}, \hat{\boldsymbol{b}})$, which moves along the curve with $\boldsymbol{P}$. Differentiating $\hat{\boldsymbol{b}}$ along the path, we find

$$\frac{\mathrm{d}\hat{\boldsymbol{b}}}{\mathrm{d}s} = \frac{\mathrm{d}\hat{\boldsymbol{u}}}{\mathrm{d}s} \times \hat{\boldsymbol{n}} + \hat{\boldsymbol{u}} \times \frac{\mathrm{d}\hat{\boldsymbol{n}}}{\mathrm{d}s} = \frac{1}{\rho}\overset{0}{\cancel{\hat{\boldsymbol{n}} \times \hat{\boldsymbol{n}}}} + \hat{\boldsymbol{u}} \times \frac{\mathrm{d}\hat{\boldsymbol{n}}}{\mathrm{d}s} = \hat{\boldsymbol{u}} \times \frac{\mathrm{d}\hat{\boldsymbol{n}}}{\mathrm{d}s}. \tag{24}$$

Since $\hat{\boldsymbol{b}}$ is a unit vector, $\mathrm{d}\hat{\boldsymbol{b}}/\mathrm{d}s$ must be perpendicular to $\hat{\boldsymbol{b}}$. From Eq.(24), it is seen that $\mathrm{d}\hat{\boldsymbol{b}}/\mathrm{d}s$ is also perpendicular to $\hat{\boldsymbol{u}}$. Therefore, $\mathrm{d}\hat{\boldsymbol{b}}/\mathrm{d}s$ is aligned with $\hat{\boldsymbol{n}}$. Defining the length of $\mathrm{d}\hat{\boldsymbol{b}}/\mathrm{d}s$ as $-1/\tau$ (torsion of the curve), we will have

$$\mathrm{d}\hat{\boldsymbol{b}}/\mathrm{d}s = -(1/\tau)\hat{\boldsymbol{n}}. \tag{25}$$

The negative sign of $-1/\tau$ is introduced to make torsion positive when the vector triad $(\hat{\boldsymbol{u}}, \hat{\boldsymbol{n}}, \hat{\boldsymbol{b}})$ rotates in a right-handed sense about the tangent as it progresses along the curve. A curve is *twisted* if its torsion is nonzero. The length $|\tau|$ is the radius of torsion.

So far, we have expressions for $\mathrm{d}\hat{\boldsymbol{u}}/\mathrm{d}s$ and $\mathrm{d}\hat{\boldsymbol{b}}/\mathrm{d}s$. To determine $\mathrm{d}\hat{\boldsymbol{n}}/\mathrm{d}s$, we write $\hat{\boldsymbol{n}} = \hat{\boldsymbol{b}} \times \hat{\boldsymbol{u}}$, and proceed to differentiate, as follows:

$$\begin{aligned}\mathrm{d}\hat{\boldsymbol{n}}/\mathrm{d}s &= (\mathrm{d}\hat{\boldsymbol{b}}/\mathrm{d}s) \times \hat{\boldsymbol{u}} + \hat{\boldsymbol{b}} \times (\mathrm{d}\hat{\boldsymbol{u}}/\mathrm{d}s) = -(1/\tau)(\hat{\boldsymbol{n}} \times \hat{\boldsymbol{u}}) + (1/\rho)(\hat{\boldsymbol{b}} \times \hat{\boldsymbol{n}}) \\ &= (1/\tau)\hat{\boldsymbol{b}} - (1/\rho)\hat{\boldsymbol{u}}.\end{aligned} \tag{26}$$

The equations $\mathrm{d}\hat{\boldsymbol{u}}/\mathrm{d}s = \hat{\boldsymbol{n}}/\rho$, $\mathrm{d}\hat{\boldsymbol{b}}/\mathrm{d}s = -\hat{\boldsymbol{n}}/\tau$, and $\mathrm{d}\hat{\boldsymbol{n}}/\mathrm{d}s = (\hat{\boldsymbol{b}}/\tau) - (\hat{\boldsymbol{u}}/\rho)$ are known as Frenet's formulas. Dot multiplying $\mathrm{d}\hat{\boldsymbol{n}}/\mathrm{d}s$ into $\hat{\boldsymbol{b}}$ yields

$$\hat{\boldsymbol{b}} \cdot (\mathrm{d}\hat{\boldsymbol{n}}/\mathrm{d}s) = (1/\tau)\hat{\boldsymbol{b}} \cdot \hat{\boldsymbol{b}} - (1/\rho)\hat{\boldsymbol{u}} \cdot \hat{\boldsymbol{b}} = 1/\tau \quad \rightarrow \quad 1/\tau = (\hat{\boldsymbol{u}} \times \hat{\boldsymbol{n}}) \cdot (\mathrm{d}\hat{\boldsymbol{n}}/\mathrm{d}s)$$

$$\rightarrow \quad 1/\tau = \left(\hat{\boldsymbol{u}} \times \rho \frac{\mathrm{d}\hat{\boldsymbol{u}}}{\mathrm{d}s}\right) \cdot \left(\rho \frac{\mathrm{d}^2\hat{\boldsymbol{u}}}{\mathrm{d}s^2} + \frac{\mathrm{d}\rho}{\mathrm{d}s}\frac{\mathrm{d}\hat{\boldsymbol{u}}}{\mathrm{d}s}\right) = \rho^2 \left(\hat{\boldsymbol{u}} \times \frac{\mathrm{d}\hat{\boldsymbol{u}}}{\mathrm{d}s}\right) \cdot \frac{\mathrm{d}^2\hat{\boldsymbol{u}}}{\mathrm{d}s^2}$$

$$= \rho^2 \begin{vmatrix} \mathrm{d}x/\mathrm{d}s & \mathrm{d}y/\mathrm{d}s & \mathrm{d}z/\mathrm{d}s \\ \mathrm{d}^2x/\mathrm{d}s^2 & \mathrm{d}^2y/\mathrm{d}s^2 & \mathrm{d}^2z/\mathrm{d}s^2 \\ \mathrm{d}^3x/\mathrm{d}s^3 & \mathrm{d}^3y/\mathrm{d}s^3 & \mathrm{d}^3z/\mathrm{d}s^3 \end{vmatrix}. \qquad (27)$$

Equation (27) gives a compact expression for the torsion of the 3D space curve in terms of the radius of curvature ρ and the various derivatives of x, y, and z with respect to s. Note also that, in accordance with Eq.(21), ρ itself is a function of the second derivatives of x, y, and z with respect to s.

Suggestions for further reading

1. J.E. Marsden, A. Tromba, and A. Weinstein, *Basic Multivariable Calculus*, 3rd edition, Springer, New York (1993).
2. F.B. Hildebrand, *Advanced Calculus for Applications*, 2nd edition, Prentice-Hall, Upper Saddle River, NJ (1976).
3. G.B. Arfken and H.J. Weber, *Mathematical Methods for Physicists*, 6th edition, Academic Press, Cambridge, MA (2005).

Problems

Problem 1) A ray of light travels from an underwater object located at $(x_1, y_1, z_1) = (\ell, -h, 0)$ to an observer's eye at $(x_0, y_0, z_0) = (0, d, 0)$, as shown. The refractive index of the air, where the observer is located, is n_0, while that of the water is n_1. (The speed of light in a medium of refractive index n is given by $v = c/n$, where c is the speed of light in vacuum.)

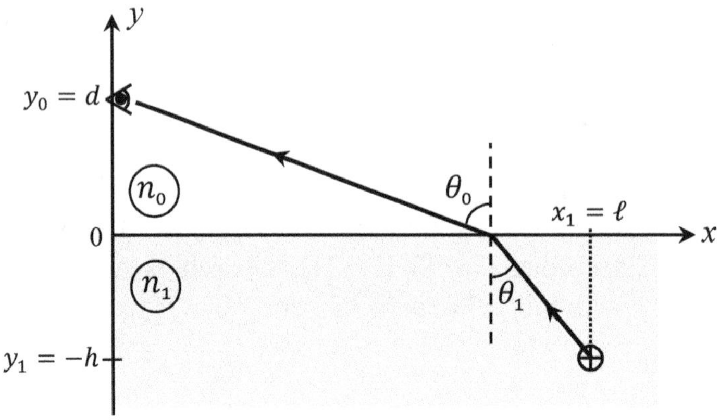

a) Locate the x-coordinate of the light ray where it emerges from the water in such a way as to *minimize* the time of travel from the object to the observer's eye.

Note: The answer is a root of a quartic (i.e., fourth degree) polynomial equation. Although such equations are generally solvable, you are being asked here only to find the equation — not its solution.

b) Use the result obtained in part (a) to verify Snell's law of refraction, namely, $n_1 \sin\theta_1 = n_0 \sin\theta_0$.

Problem 2) Consider the 2D function $g(x,y) = a + bx + cy + dx^2 + ey^2 + fxy$, which is defined on the entire xy-plane, with arbitrary real-valued constants a, b, c, d, e, and f. Find the point (or points) in the xy-plane at which the function is a local maximum, a local minimum, or a saddle point. In each case, determine the conditions that the coefficients a, b, c, d, e, f must satisfy for the specified point to be a maximum, a minimum, or a saddle-point.

Problem 3) Using the method of Lagrange multipliers, find two points in the xy-plane, one located on the straight line $ax + by = c$, the other on the circle $x^2 + y^2 = 1$, such that the distance between these two points is either a maximum or a minimum.

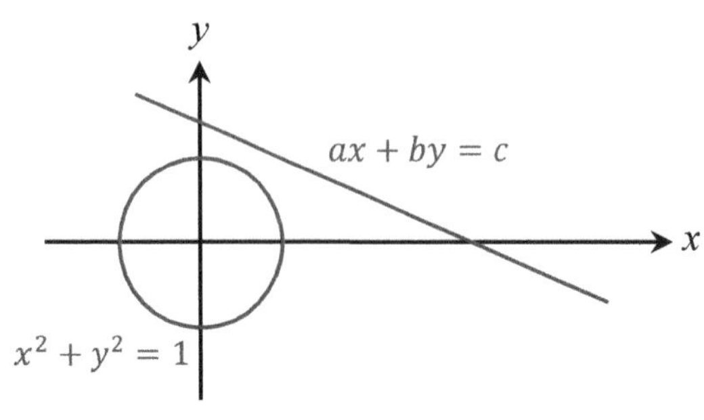

Problem 4) For a cylindrical can of radius r and height h, whose volume $V(r,h) = \pi r^2 h$ is fixed at V_0, use the method of Lagrange multipliers to determine the minimum surface area $S(r,h) = 2\pi r^2 + 2\pi rh$.

Problem 5) A house in the shape of a cuboid is to be built on a flat piece of land, as shown. The volume of the house $V = xyz$ is desired to be some constant value, V_0. However, the exposed surface area (that is, the exterior walls plus the roof) should be minimized. Use the method of Lagrange multipliers to determine the optimal dimensions of the house in terms of its desired volume V_0.

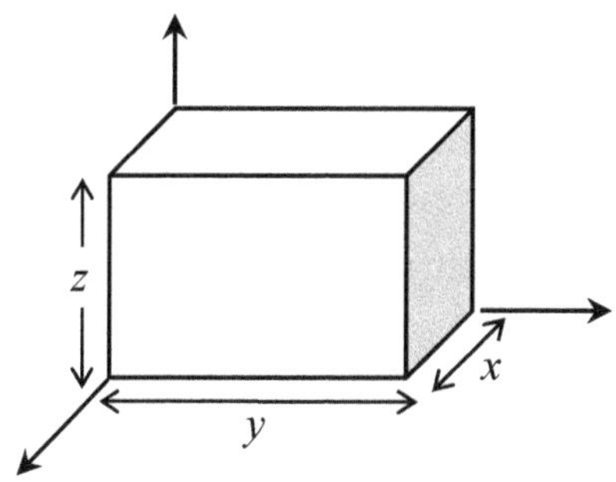

Problem 6) A cylindrical hole of radius r and height h is perforated within a solid sphere of radius R. Use the method of Lagrange multipliers to *maximize* the volume of the cylinder subject to the constraint that its round edges touch the surface of the sphere.

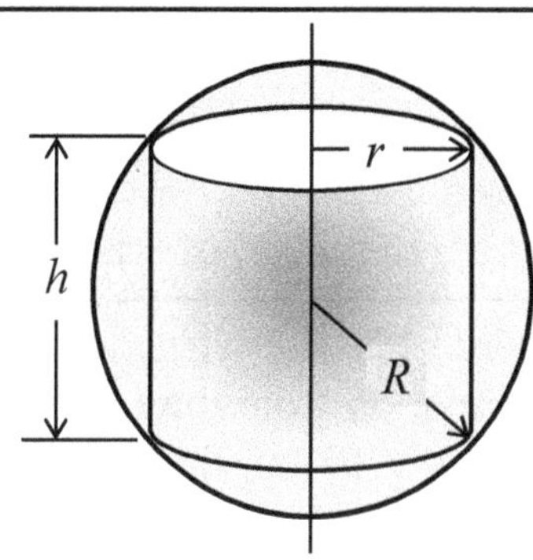

Problem 7) A straight stick of length L is to be cut into n pieces of lengths $x_1, x_2, \cdots, x_n$. The total length of the various pieces must obviously add up to L, that is, $x_1 + x_2 + \cdots + x_n = L$. We would like to devise a strategy for cutting the stick in such a way as to yield the maximum value for the product $x_1 x_2 \cdots x_n$. Use the method of Lagrange multipliers to show that $x_1 x_2 \cdots x_n$ is maximized when all the pieces are of equal length, that is, $x_1 = x_2 = \cdots = x_n = L/n$.

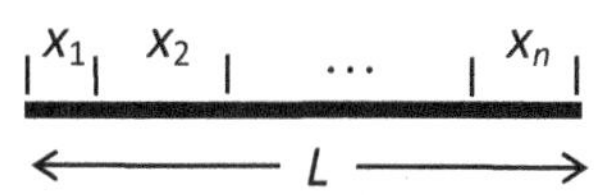

Problem 8) The outcome of an experiment is a random integer between 1 and N, each outcome having a probability $p_n \geq 0$. The Shannon entropy H of such an experiment is defined as

$$H(p_1, p_2, \dots, p_N) = -\sum_{n=1}^{N} p_n \ln(p_n).$$

Using the method of Lagrange multipliers, maximize this entropy subject to the constraint that the sum of the probabilities must be equal to unity, that is, $\sum_{n=1}^{N} p_n = 1$.

Problem 9) A paper cup has the shape of a hollow, right-circular cone, with height h, base radius ρ, and cone semi-angle θ, as shown. Clearly, $\rho = h \tan\theta$.

a) Determine the volume V of the cup as a function of h and θ.

b) Determine the surface area S of the cup (i.e., area of the flat sheet of paper used to make the cup) as a function of h and θ.

c) Use the method of Lagrange multipliers to minimize the surface area S subject to the constraint of a fixed volume $V = V_0$.

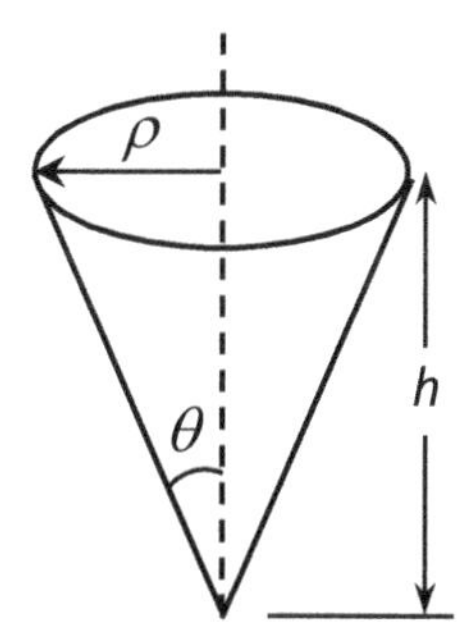

Problem 10) A flat, closed curve in the xy-plane is defined by its distance $r(\theta)$ from a central point, where $0 \leq \theta \leq 2\pi$ is the angular position on the curve, measured with respect to some arbitrary reference direction. The area enclosed by the curve is $A =$

½ $\int_0^{2\pi} r^2(\theta)d\theta$, while the perimeter of the closed curve is $P = \int_0^{2\pi} r(\theta)d\theta$. Use the method of Lagrange multipliers to find the particular function $r(\theta)$ that maximizes the area A of the curve for a given value of the perimeter $P = P_0$.

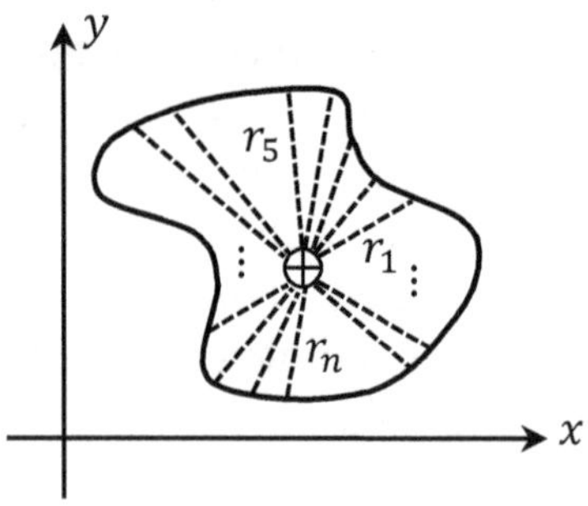

Hint: You may want to work with a discrete version of $r(\theta)$, where the radii $r_1, r_2, r_3, \cdots, r_n, \cdots, r_N$ evenly divide the available range of θ, creating an angular separation of $\Delta\theta$ between adjacent radii. The area and the perimeter of the curve may then be approximated as $A = ½ \sum r_n^2 \Delta\theta$ and $P = \sum r_n \Delta\theta$.

Problem 11) In standard spherical coordinates (r, θ, ϕ), the Cartesian coordinates are expressed as $x = r \sin\theta \cos\phi$, $y = r \sin\theta \sin\phi$, and $z = r \cos\theta$. What is the Jacobian $\partial(x, y, z)/\partial(r, \theta, \phi)$ of this transformation?

Problem 12) Let $[u(x, y), v(x, y)]$ represent an arbitrary curvilinear coordinate transformation from the Cartesian xy coordinates to the uv system. How is the Jacobian $\partial(x, y)/\partial(u, v)$ of this transformation related to that of the inverse transformation $[x(u, v), y(u, v)]$, which transforms the uv back to the xy coordinate system?

Problem 13) Within the Cartesian xy-plane, the uv coordinate system is defined by $xy = u$, $x^2 - y^2 = v$. Plot the curves of constant u and constant v in the xy-plane. Show that the vectors $\boldsymbol{U}$ and $\boldsymbol{V}$, which specify the directions of the local u and v coordinates, are orthogonal everywhere. What is the Jacobian $\partial(x, y)/\partial(u, v)$ of the transformation?

Problem 14) A transformation from the Cartesian xy coordinate system to the curvilinear uv system is defined by the equations $u = y/x^2$ and $v = x/y^2$.

a) Within the xy-plane, plot a set of curves that show the contours of constant u and another set that shows the contours of constant v. Aside from the singularity of the x-axis, where $y = 0$, and the singularity of the y-axis, where $x = 0$, show that the transformation uniquely associates every point (x, y) with a point (u, v) and vice versa.

b) Determine the Jacobian of the transformation from xy to uv, and also the Jacobian of the inverse transformation from uv to xy. Confirm that $\partial(x, y)/\partial(u, v)$ is the inverse of $\partial(u, v)/\partial(x, y)$.

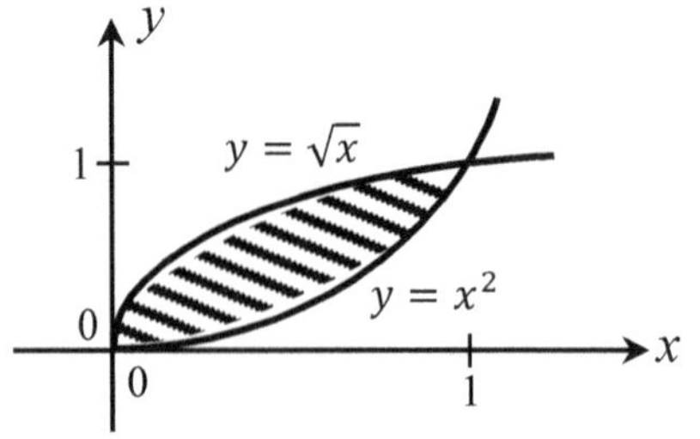

c) Find the area of the region of the xy-plane that is bounded by the curves $y = x^2$ and $y = \sqrt{x}$. (This is the shaded area in the figure.) Evaluate the relevant integral in both xy and uv coordinate systems and confirm that the results are identical.

Problem 15) The area of the ellipse $(x/a)^2 + (y/b)^2 = 1$ equals πab. Using the method of Lagrange multipliers, identify the largest area rectangle that fits inside this ellipse. Show that the ratio of the area of the maximal rectangle to that of the ellipse is $2/\pi$.

Chapter 6
Complex Analysis

"The opposite of a correct statement is a false statement. But the opposite of a profound truth may well be another profound truth."

Niels Bohr (1885–1962)

"In questions of science, the authority of a thousand is not worth the humble reasoning of a single individual."

Galileo Galilei (1564–1642)

6.1. Introduction. The theories of complex numbers and complex functions were originally developed during the eighteenth and nineteenth centuries, primarily by Leonhard Euler (1707–1783), Carl Friedrich Gauss (1777–1855), Augustin-Louis Cauchy (1789–1857), Bernhard Riemann (1826–1866), and Karl Weierstrass (1815–1897). Advances in theoretical physics during the twentieth century owed a great debt to the concepts and methods of complex function theory — also known as complex analysis. Modern physical and mathematical sciences would be inconceivable in the absence of complex numbers, complex functions, and complex plane techniques.[1-3†] Our goal in the present chapter is *not* a comprehensive coverage of complex analysis, which would require an entire volume by itself, but rather to present a few powerful tools and techniques that have found substantial applications in fundamental as well as applied physical sciences.

We will develop the concepts and methods of differentiation and integration for functions $f(z)$ of a complex variable $z = x + iy$. Complex number arithmetic and certain properties of functions of a complex variable were discussed in chapter 3. Here, we begin in section 2 by examining the derivative of a function at an arbitrary point in the complex plane and arrive at necessary and sufficient conditions for the existence of the derivative. These so-called Cauchy-Riemann conditions will then be employed to examine the differentiability of several elementary functions defined over the complex z-plane. The concept of analyticity, which requires the existence of the derivative of a function at a point z and also in the immediate neighborhood of z, is subsequently introduced.

Integration in the complex plane is the subject of section 3, where the significance of the integration path (or contour of integration) is emphasized. The most important result in complex analysis, the Cauchy-Goursat theorem, is stated and then proved in section 4. The Cauchy-Goursat theorem enables one to evaluate certain definite integrals of functions of a real variable, as well as those of functions of a complex variable, using complex plane techniques. The powerful methods of integration rooted in the Cauchy-Goursat theorem will be explored via several examples that follow our proof of the theorem in section 4. In almost all instances, evaluation of an integral in the complex plane reduces to the determination of so-called *residues* at one or more *poles* of the integrand. These notions will be further described and elaborated in section 5. In section 6 we demonstrate that functions of a complex variable can be expanded in a Taylor series in the vicinity of points at which they are analytic. Section 7 briefly describes the Liouville theorem, which is the statement of the fact that a function that is analytic throughout the entire complex plane cannot be bounded unless it is a constant.

In certain problems involving complex plane integration, it becomes necessary to establish that the integral over a semi-circle of radius R in the upper or lower half of the complex plane approaches zero when $R \to \infty$. A useful argument that sets the conditions

---

†Superscript numbers throughout the chapter identify references listed at the end of the chapter.

for the vanishing of the integral over such semi-circles is encapsulated in Jordan's lemma, which is the subject of section 8. In section 9 we invoke complex plane integration methods to arrive at the Kramers-Kronig relations, which are of immense significance in electrodynamics, in general, and in optics and photonics, in particular.

Section 10 provides a heuristic proof of the fundamental theorem of algebra, which states that any polynomial of order n defined over the complex plane will have n zeros in the complex plane. Finally, in section 11, we present a fairly simple and straightforward proof of Cauchy's argument principle, which relates the accumulated phase of a complex function during a single round trip around a closed contour to the difference between the number of poles and zeros of the function within the area enclosed by that contour.

6.2. Differentiating complex functions of a complex variable. Let the complex function $f(z) = u(x, y) + iv(x, y)$ be defined over an appropriately defined subset of the complex z-plane. The derivative $f'(z)$ of $f(z)$ with respect to z at a given point z_o is defined as follows:

$$f'(z)|_{z=z_0} = lim_{\Delta z\to 0}\,[f(z_o + \Delta z) - f(z_o)]/\Delta z. \tag{1}$$

In this equation, $\Delta z = \Delta x + \mathrm{i}\Delta y$ is a small complex number that approaches zero when both Δx and Δy become very small. For the derivative of $f(z)$ at $z = z_o$ to exist, it is imperative that the limiting value of the ratio $[f(z_o + \Delta z) - f(z_o)]/\Delta z$ be independent of the direction that Δz takes as it approaches zero. Thus, for instance, if Δz happens to be parallel to the x-axis, that is, $\Delta z = \Delta x$, we will find

$$f'(z)|_{z=z_0} = lim_{\Delta x\to 0}\frac{[u(x_0+\Delta x,y_0)+iv(x_0+\Delta x,y_0)] - [u(x_0,y_0)+iv(x_0,y_0)]}{\Delta x} = (\partial_x u + \mathrm{i}\partial_x v)_{x_0,y_0}. \tag{2a}$$

Alternatively, if Δz remains parallel to the y-axis as it approaches zero, that is, if $\Delta z = \mathrm{i}\Delta y$, then the limiting value of the ratio in Eq.(1) will be

$$f'(z)|_{z=z_0} = lim_{\Delta y\to 0}\frac{[u(x_0,y_0+\Delta y)+iv(x_0,y_0+\Delta y)]-[u(x_0,y_0)+iv(x_0,y_0)]}{\mathrm{i}\Delta y} = (\partial_y v - \mathrm{i}\partial_y u)_{x_0,y_0}. \tag{2b}$$

For the derivative of the function $f(z)$ to exist at $z = z_o$, it is necessary that the two values derived in Eqs.(2a) and 2(b) be identical; that is,

$$\partial_x u(x, y)|_{x_0,y_0} = \partial_y v(x, y)|_{x_0,y_0}, \tag{3a}$$

$$\partial_y u(x, y)|_{x_0,y_0} = -\partial_x v(x, y)|_{x_0,y_0}. \tag{3b}$$

These *necessary* conditions for the existence of the derivative of $f(z) = u(x, y) + iv(x, y)$ at $z = z_o = x_o + \mathrm{i}y_o$ are known as the Cauchy-Riemann conditions. It turns out that the Cauchy-Riemann conditions listed in Eqs.(3) are also *sufficient* for the existence of the derivative, the reason being that, for an arbitrary $\Delta z = \Delta x + \mathrm{i}\Delta y$, one may write

$$lim_{\Delta z\to 0}\frac{f(z_0+\Delta z) - f(z_0)}{\Delta z} = lim_{\Delta x,\Delta y\to 0}\frac{[u(x_0+\Delta x,y_0+\Delta y)+iv(x_0+\Delta x,y_0+\Delta y)] - [u(x_0,y_0)+iv(x_0,y_0)]}{\Delta x + \mathrm{i}\Delta y}$$

Cauchy-Riemann: → $\boxed{\partial_y u = -\partial_x v}$ $\boxed{\partial_y v = \partial_x u}$

$$= lim_{\Delta x,\Delta y\to 0}\frac{[(\partial_x u)\Delta x+(\partial_y u)\Delta y] + \mathrm{i}[(\partial_x v)\Delta x+(\partial_y v)\Delta y]}{\Delta x + \mathrm{i}\Delta y}$$

$$= lim_{\Delta x,\Delta y\to 0}\frac{(\partial_x u)(\Delta x+\mathrm{i}\Delta y) + \mathrm{i}(\partial_x v)(\Delta x+\mathrm{i}\Delta y)}{\Delta x + \mathrm{i}\Delta y}$$

$$= (\partial_x u + \mathrm{i}\partial_x v)_{x_0,y_0}. \tag{4}$$

The final result in Eq.(4) is the same as the derivative $f'(z)|_{z=z_0}$ obtained in Eq.(2a). Consequently, the Cauchy-Riemann conditions are sufficient for the limiting value of the ratio in Eq.(1) to be independent of the arbitrary orientation chosen for Δz. The function $f(z)$ is said to be differentiable at $z = z_0$ if and only if the Cauchy-Riemann conditions are satisfied at that point.[4-7]

Definitions. A function $f(z)$ is said to be *analytic* at a point $z = z_0$, if a small circle of radius $\varepsilon > 0$, centered at z_0, can be found such that $f(z)$ is differentiable at *every* point z within the circle—including z_0, of course. In other words, analyticity at z_0 requires not only the existence of the derivative $f'(z)$ at $z = z_0$, but also the existence of $f'(z)$ for all the points in the immediate vicinity of z_0.

A function is said to be analytic within a subset of the z-plane if the function is found to be analytic at each and every point belonging to that subset. When a function is analytic at *all* points in the complex z-plane, it is said to be an *entire* function.[6-10]

Example 1. Consider the quadratic polynomial function $f(z) = az^2 + bz + c$, where the a, b, and c coefficients are arbitrary complex numbers. Show that $f(z)$ is differentiable everywhere in the complex z-plane.

Proof: Let $a = a' + ia''$, $b = b' + ib''$, and $c = c' + ic''$. The real and imaginary parts of the function $f(z)$ can be identified as follows:

$$f(z) = (a' + ia'')(x + iy)^2 + (b' + ib'')(x + iy) + (c' + ic''). \tag{5a}$$

$$u(x, y) = a'x^2 - a'y^2 - 2a''xy + b'x - b''y + c'. \tag{5b}$$

$$v(x, y) = a''x^2 - a''y^2 + 2a'xy + b'y + b''x + c''. \tag{5c}$$

Consequently,

$$\partial_x u(x, y) = 2a'x - 2a''y + b'; \qquad \partial_y u(x, y) = -2a'y - 2a''x - b''. \tag{6a}$$

$$\partial_x v(x, y) = 2a''x + 2a'y + b''; \qquad \partial_y v(x, y) = -2a''y + 2a'x + b'. \tag{6b}$$

Clearly, $\partial_x u = \partial_y v$ and $\partial_y u = -\partial_x v$ at all points z. We conclude that the quadratic polynomial $f(z) = az^2 + bz + c$, being differentiable everywhere in the complex plane, is an *entire* function.

Example 2. The function $f(z) = z^*$ is *not* differentiable anywhere in the complex plane.

Discussion: Writing $f(z) = x - iy$, it is readily seen that $u(x, y) = x$ and $v(x, y) = -y$. Therefore, $\partial_x u = 1$, $\partial_y u = 0$, $\partial_x v = 0$, and $\partial_y v = -1$. While the second of the Cauchy-Riemann conditions is satisfied in this case, that is, $\partial_y u = -\partial_x v$, the first condition remains unsatisfied, namely, $\partial_x u \neq \partial_y v$ everywhere. Thus, the function $f(z) = z^*$ does *not* have a derivative at any point in the complex z-plane.

An alternative approach to showing the nonexistence of the derivative in the present example is to return to the original definition in Eq.(1) and express Δz in polar notation as $|\Delta z| \exp(i\varphi)$. We will have

$$[f(z_0 + \Delta z) - f(z_0)]/\Delta z = [(z_0 + \Delta z)^* - z_0^*]/\Delta z = (\Delta z)^*/\Delta z$$

$$= \frac{|\Delta z| \exp(-i\varphi)}{|\Delta z| \exp(i\varphi)} = \exp(-2i\varphi). \tag{7}$$

The above ratio is now seen to depend on the orientation angle φ of Δz, in clear violation of the requirements for the existence of the derivative.

Example 3. Show that the function $f(z) = \exp(az)$, where $a = a' + \mathrm{i}a''$ is an arbitrary constant, is differentiable everywhere in the complex plane.

Proof: Considering that $\exp(az) = \exp[(a'x - a''y) + \mathrm{i}(a''x + a'y)]$, we invoke the Euler identity to separate the real and imaginary parts of the function $f(z)$, as follows:

$$u(x, y) = \exp(a'x - a''y)\cos(a''x + a'y), \tag{8a}$$

$$v(x, y) = \exp(a'x - a''y)\sin(a''x + a'y). \tag{8b}$$

Consequently,

$$\partial_x u = [a'\cos(a''x + a'y) - a''\sin(a''x + a'y)]\exp(a'x - a''y), \tag{9a}$$

$$\partial_y u = -[a''\cos(a''x + a'y) + a'\sin(a''x + a'y)]\exp(a'x - a''y), \tag{9b}$$

$$\partial_x v = [a'\sin(a''x + a'y) + a''\cos(a''x + a'y)]\exp(a'x - a''y), \tag{9c}$$

$$\partial_y v = [-a''\sin(a''x + a'y) + a'\cos(a''x + a'y)]\exp(a'x - a''y). \tag{9d}$$

Clearly, $\partial_x u = \partial_y v$ and $\partial_y u = -\partial_x v$ at all points z. We conclude that the function $f(z) = \exp(az)$, being differentiable everywhere in the z-plane, is an *entire* function.

Example 4. Show that the function $f(z) = \ln(z)$ is differentiable everywhere in the complex plane except at $z = 0$ and on a branch-cut (e.g., the negative half of the real axis.)

Proof: Writing z in its polar representation as $z = \sqrt{x^2 + y^2}\exp[\mathrm{i}\tan^{-1}(y/x)]$, we will have $u(x, y) = ½\ln(x^2 + y^2)$ and $v(x, y) = \tan^{-1}(y/x)$. Therefore, $\partial_x u = x/(x^2 + y^2)$ and $\partial_y u = y/(x^2 + y^2)$. As for partial derivatives of v with respect to x and y, we write $\tan(v) = y/x$ and differentiate both sides of the equation to arrive at

$$(1 + \tan^2 v)\partial_x v = -y/x^2 \quad \rightarrow \quad \partial_x v(x, y) = -y/(x^2 + y^2), \tag{10a}$$

$$(1 + \tan^2 v)\partial_y v = 1/x \quad \rightarrow \quad \partial_y v(x, y) = x/(x^2 + y^2). \tag{10b}$$

Clearly, $\partial_x u = \partial_y v$ and $\partial_y u = -\partial_x v$ at all points z except at $z = 0$, where the function is not defined, and also on the branch-cut, where the imaginary part of $\ln(z)$ has a 2π discontinuity. The function $f(z) = \ln(z)$ is thus seen to be differentiable everywhere except at the origin ($z = 0$), and on its branch-cut, which could be chosen as the negative half of the real-axis.

Example 5. Show that the function $f(z) = 1/z$ is differentiable everywhere except at its singular point $z = 0$.

Discussion: We could continue to invoke the Cauchy-Riemann conditions to verify the existence of the derivative of $f(z)$. However, it is also possible to return to the original definition of the derivative given in Eq.(1) to decide the question of differentiability. In the present case we have

$$f'(z)|_{z=z_0} = lim_{\Delta z \to 0}\left(\frac{1}{z_0 + \Delta z} - \frac{1}{z_0}\right)/\Delta z = lim_{\Delta z \to 0}\left[-\frac{1}{z_0(z_0 + \Delta z)}\right]. \tag{11}$$

It must be clear that, when $\Delta z \to 0$, the term $z_0 + \Delta z$ in the denominator approaches z_0, irrespective of the path that Δz might take in its journey toward zero. The limit of the expression on the right-hand side of Eq.(11) is thus seen to be $-1/z_0^2$, unless z_0 happens to be zero, in which case neither the function $1/z$ nor its derivative at $z_0 = 0$ are defined.

We conclude that the derivative of $1/z$ is equal to $-1/z^2$ for all the points z in the complex plane with the exception of $z = 0$, where the function itself has a singularity. The function $f(z)$ is, therefore, analytic everywhere except at $z = 0$.

Example 6. The direct approach to deciding the question of differentiability used in the preceding example is often easier to pursue than verifying the Cauchy-Riemann conditions. For instance, in the case of the function $\exp(az)$ examined in Example 3, the direct approach would begin as follows:

$$\exp[a(z+\Delta z)] - \exp(az) = \exp(az)\,[\exp(a\Delta z) - 1] = \exp(az)\textstyle\sum_{n=1}^{\infty}(a\Delta z)^n/n!.$$

Dividing the expression on the right-hand side of this equation by Δz now yields

$$\{\exp[a(z+\Delta z)] - \exp(az)\}/\Delta z = a\exp(az)\,[1 + \textstyle\sum_{n=1}^{\infty}(a\Delta z)^n/(n+1)!]. \quad (12)$$

Given that $(n+1)! \geq 2^n$, the infinite sum appearing in Eq.(12) is seen to be less than the geometric series $\sum_{n=1}^{\infty}(a\Delta z/2)^n$, which, in the limit when $\Delta z \to 0$, approaches zero. The derivative of $f(z) = \exp(az)$ is thus seen to be independent of the orientation of Δz and is given by $f'(z) = a\exp(az)$.

In the case of the function $\ln(z)$, the direct approach to differentiation requires that at first we evaluate the change of the function between the point $z = re^{\mathrm{i}\varphi}$ and a neighboring point $z + \Delta z = (r + \Delta r)e^{\mathrm{i}(\varphi+\Delta\varphi)}$, as shown in figure 6.1. We will have

$$\ln(z+\Delta z) - \ln(z) = \ln(r+\Delta r) + \mathrm{i}(\varphi+\Delta\varphi) - \ln(r) - \mathrm{i}\varphi = \ln[1 + (\Delta r/r)] + \mathrm{i}\Delta\varphi. \quad (13)$$

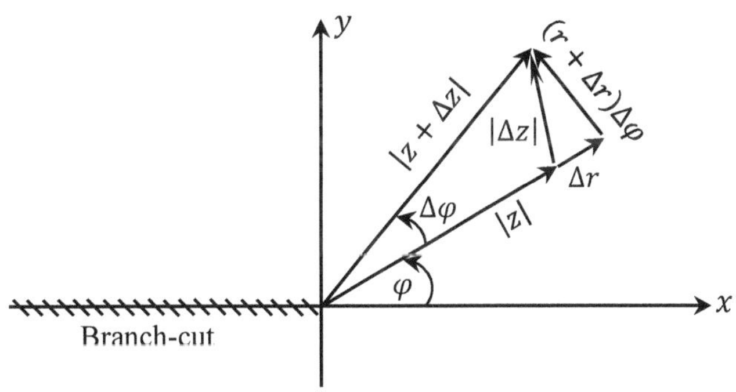

Figure 6.1. The point $z = r\exp(\mathrm{i}\varphi)$ is separated by Δz from a neighboring point $z + \Delta z = (r + \Delta r)\exp[\mathrm{i}(\varphi + \Delta\varphi)]$. For sufficiently small values of Δr and $\Delta\varphi$, one may write $\Delta z \cong (\Delta r + \mathrm{i}r\Delta\varphi)\exp(\mathrm{i}\varphi)$.

Thus, for small values of Δr, we have $\ln(z+\Delta z) - \ln(z) \cong (\Delta r/r) + \mathrm{i}\Delta\varphi$. Now, with reference to figure 6.1, and for sufficiently small values of Δr and $\Delta\varphi$, one may also write $\Delta z \cong (\Delta r + \mathrm{i}r\Delta\varphi)e^{\mathrm{i}\varphi}$. Consequently,

$$\frac{\mathrm{d}}{\mathrm{d}z}\ln z = lim_{\Delta z\to 0}\left[\frac{\ln(z+\Delta z) - \ln(z)}{\Delta z}\right] = lim_{\Delta r,\Delta\varphi\to 0}\left[\frac{(\Delta r/r) + \mathrm{i}\Delta\varphi}{(\Delta r + \mathrm{i}r\Delta\varphi)e^{\mathrm{i}\varphi}}\right] = \frac{1}{re^{\mathrm{i}\varphi}} = 1/z. \quad (14)$$

Needless to say, the above procedure cannot be carried out for points z on the branch-cut of $\ln(z)$, simply because a small change $\Delta\varphi$ in the phase angle φ could cause a jump across the branch-cut, thus resulting in a discontinuous change in the value of φ.

Brain teaser. Continue the following sequence: O T T F F S S E N T E T T ⋯.

6.3. Integrating complex functions of a complex variable. To integrate the complex function $f(z)$ we need to identify a starting point z_a, an end point z_b, and a path C that connects z_a to z_b in the complex plane. A typical path (or contour) of integration in the z-plane is shown in figure 6.2a. Subsequently, the path is divided into small segments, Δz_1, Δz_2, $\cdots$, Δz_N, as depicted in figure 6.2b. Each little arrow in the figure represents a complex number that is the difference between the terminal points of a short segment of the contour C. For each segment, we pick an arbitrary point somewhere on the segment, say, z_1 on the first segment, z_2 on the second segment, z_n on the nth segment, and so on. An approximation to the integral of $f(z)$ along the contour C, starting at z_a and ending at z_b, is then given by the sum of the complex numbers $f(z_n)\Delta z_n$ from $n = 1$ to N. In the limit when all the segments Δz_n become extremely small (i.e., $\Delta z_n \to 0$ for all n), the sum over n of $f(z_n)\Delta z_n$ approaches the desired integral.[4-8] In other words,

$$\int_{z_a}^{z_b} f(z)\mathrm{d}z = lim_{\Delta z_n \to 0} \sum_{n=1}^{N} f(z_n)\Delta z_n. \quad \leftarrow \boxed{\text{Integration path is } C} \qquad (15)$$

The integral is taken over the contour C and, in the absence of any other information about the function $f(z)$, we must assume that changing the contour (i.e., the path connecting z_a to z_b), will produce a different result for the integral.

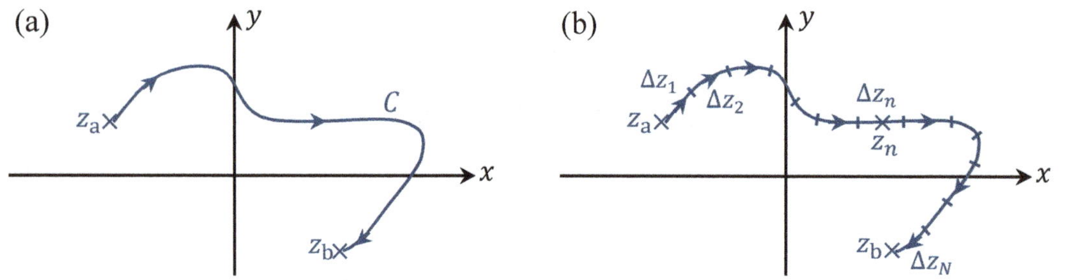

Figure 6.2. (a) The integration path (or contour) C, which lies entirely in the xy-plane, connects the beginning point z_a to the end point z_b of the integral of the function $f(z)$. (b) The contour C is divided into a large number of short segments Δz_n, and a point z_n is picked on each segment. The integral is the limit of the sum over n of the product $f(z_n)\Delta z_n$ in the limit when all $|\Delta z_n|$ become extremely small.

If the direction of integration is reversed, that is, if the integral starts at z_b and ends at z_a, then all the little arrows in figure 6.2b will flip over, changing each Δz_n into $-\Delta z_n$. Consequently, $\int_{z_b}^{z_a} f(z)\mathrm{d}z = -\int_{z_a}^{z_b} f(z)\mathrm{d}z$. In general, if the sum of $f(z_n)\Delta z_n$ fails to converge to a limiting value when the largest $|\Delta z_n|$ approaches zero, the integral is said to be nonexistent or divergent.

6.4. The Cauchy-Goursat theorem. Let $f(z)$ be an *analytic* function over a *simply connected* region of the complex z-plane. Then the integral $\oint_C f(z)\mathrm{d}z$ over any closed loop C that is fully contained within the simply connected region is exactly equal to zero.

By definition, $f(z)$ is analytic at $z = z_0$ if its first-order derivative $\mathrm{d}f(z)/\mathrm{d}z$ exists everywhere within a neighborhood of z_0—the neighborhood is a disk-shaped region centered at z_0 whose nonzero radius may be as small as desired. As for the requirement that the domain of $f(z)$ be simply connected, figure 6.3 shows one example of a simply connected region and two examples of regions that are *not* simply connected. A region is said to be simply connected if an arbitrary loop could be continually deformed and ultimately shrunk to zero size while remaining entirely within the region.[4-6]

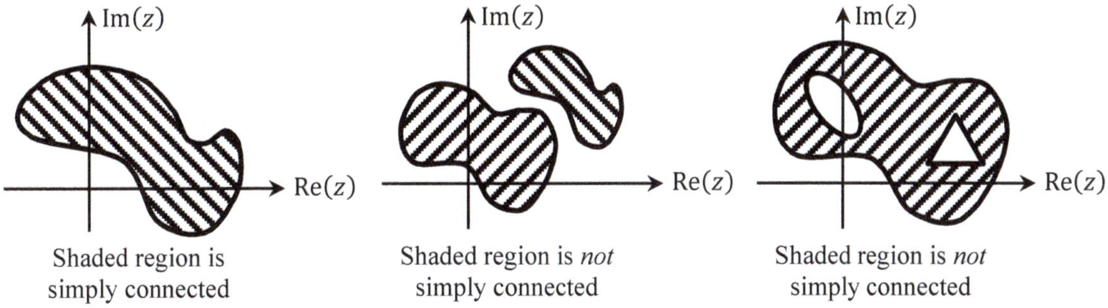

Figure 6.3. Examples of regions in the complex plane that are and are not simply connected.

Proof: Let the closed (and directed) loop C be broken into a large number of small loops, which are all in contact with one another, and which fully cover the area enclosed by the loop C. Since the boundaries of the smaller loops that are in contact with each other are traversed in opposite directions during integration, it is readily seen that adding up the integrals over the boundaries of the small loops is going to yield the integral over the boundary of the large loop C. Therefore, ideally, one would only need to prove that the integral of $f(z)$ around a very small loop c (contained within a simply connected region of analyticity) is zero. That way, adding up all the integrals around all the small loops will ensure that $\oint_C f(z)\mathrm{d}z = 0$.

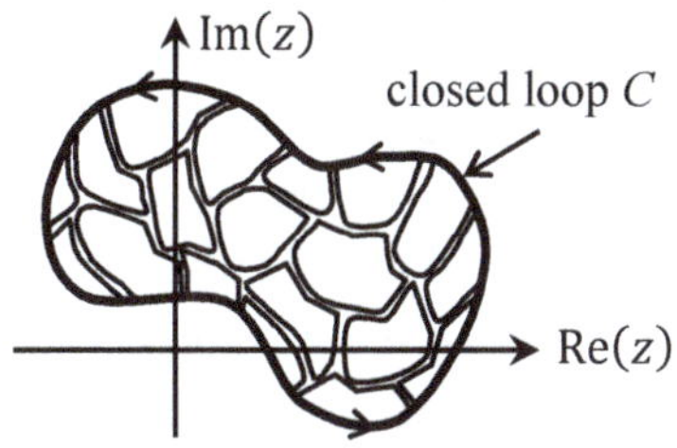

Consider a small loop c surrounding a fixed point z_0. Since $f(z)$ is analytic at z_0, its derivative $f'(z_0)$ exists. This means that $f'(z_0) \cong [f(z) - f(z_0)]/\Delta z$, irrespective of the orientation of Δz, provided that $|\Delta z|$ is sufficiently small. One may thus write $f(z) \cong f(z_0) + f'(z_0)(z - z_0)$ for all the points z located on the boundary of the small loop c that surrounds z_0. The smaller the loop becomes (i.e., the closer that all the points on c move toward z_0), the better this approximation will be. Note the importance of the fact that $f'(z_0)$ exists at the interior point z_0 of the loop, because the points z that comprise the small loop c surround z_0. This means that $z - z_0$ will assume all sorts of directions within the complex plane, yet, because $f'(z)|_{z_0}$ exists, $f(z)$ can be expressed as $f(z_0) + f'(z_0)(z - z_0)$ over the entire boundary of the small loop c. We thus write

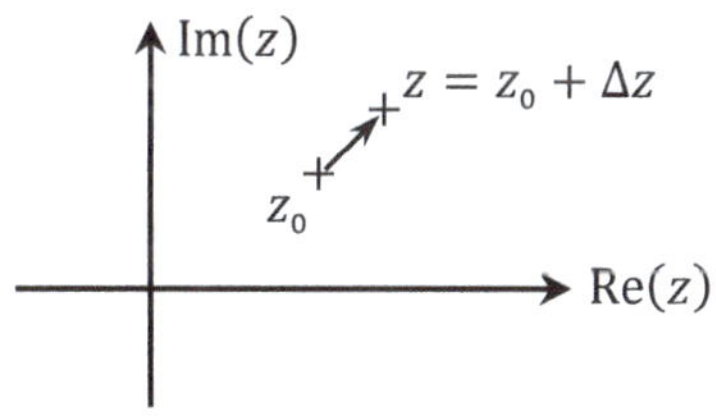

$$\oint_c f(z)\mathrm{d}z \cong \oint_c f(z_0)\mathrm{d}z + \oint_c f'(z_0)(z - z_0)\mathrm{d}z$$

$$= f(z_0)\cancelto{0}{\oint_c \mathrm{d}z} + f'(z_0)\oint_c z\mathrm{d}z - z_0 f'(z_0)\cancelto{0}{\oint_c \mathrm{d}z}. \qquad (16)$$

Obviously, $\oint_c \mathrm{d}z = 0$ over any closed loop. To see this, one might simply put the little complex vectors dz next to each other around the loop and observe that they add up to zero (i.e., the head of the last one will coincide with the tail of the first). As for the remaining term $f'(z_0)\oint_c z\mathrm{d}z$, note that $z\mathrm{d}z = ½\mathrm{d}(z^2)$; that is,

$(\Delta z)^2$ goes to 0 faster than $z\mathrm{d}z$

$$\Delta(z^2) = (z + \Delta z)^2 - z^2 = \cancel{z^2} + \cancelto{0}{(\Delta z)^2} + 2z\Delta z - \cancel{z^2} \quad \rightarrow \quad \mathrm{d}(z^2) = 2z\mathrm{d}z. \qquad (17)$$

Therefore, $\oint_c f(z)\mathrm{d}z \cong ½f'(z_0)\oint_c \mathrm{d}(z^2)$. But adding up all the $\mathrm{d}(z^2)$ around a closed loop causes the various terms in the summation to cancel each other out, as the diagram in figure 6.4 indicates. Consequently, in the limit of small loops, $\oint_c f(z)\mathrm{d}z = 0$.

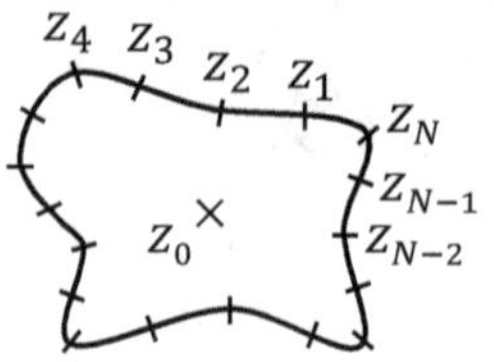

$$\oint_c \mathrm{d}(z^2) = (z_2^2 - z_1^2) + (z_3^2 - z_2^2) + \cdots + (z_N^2 - z_{N-1}^2) + (z_1^2 - z_N^2) = 0$$

Figure 6.4. Division of loop c into short segments $(\Delta z)_n = z_{n+1} - z_n$ demonstrates that $\oint_c \mathrm{d}(z^2) = 0$.

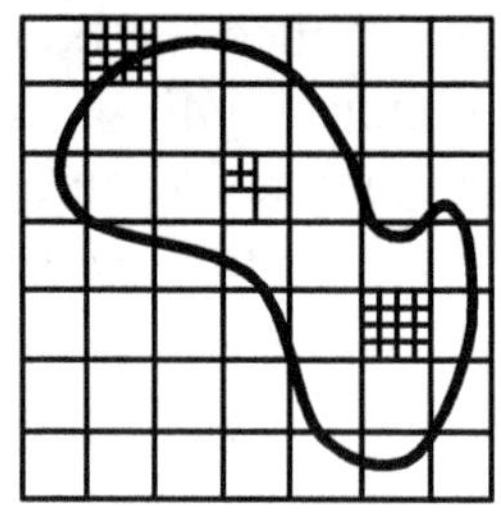

The proof requires one last step to ensure that the sum of the integrals over all the small loops also approaches zero.[6] To this end, we enclose the contour C inside a square, then break up the square into many small squares, as shown in the figure on the right. The boundary of the loop C must be considered as built up of short horizontal and vertical line segments, each belonging to a square within (i.e., inside) the loop C. Choose a small number ε and call the center of each square z_i. If, for a given square, the following inequality holds

$$\left| f'(z_i) - \frac{f(z) - f(z_i)}{z - z_i} \right| < \varepsilon, \tag{18}$$

then accept the square; otherwise, split it into four smaller squares and check the above inequality for each new square. Since the derivative exists at all points z_i inside the loop, sooner or later we will have divided the loop into squares that satisfy inequality (18). By approximating $f'(z_i)$ with $[f(z) - f(z_i)]/(z - z_i)$ for a given square, the absolute value of the error in evaluating the integral $\oint_\square f(z)\mathrm{d}z$ around that square will be less than $\varepsilon \oint_\square |z - z_i||\mathrm{d}z|$. Now, for each and every point on the square, $|z - z_i|$ is less than half the diameter of the square, and $\oint|\mathrm{d}z|$ is equal to the perimeter of the square. Therefore, the absolute value of the error in evaluating the integral for each square is less than $\varepsilon \times (\text{diameter}/2) \times \text{perimeter} = 2\sqrt{2}\varepsilon \times \text{area}$. Adding up the absolute values of the errors for all the squares will yield an entity less than $2\sqrt{2}\varepsilon \times$ (loop area). Thus, by choosing ε as small as desired, the absolute value of error can be made to approach zero.

Example 1. Find $\oint_{\text{unit circle}} z^n \mathrm{d}z$, where n is an arbitrary (positive, zero, or negative) integer, and the integration contour is a circle of radius 1 centered at the origin $z = 0$.

Discussion: In polar representation, $z = re^{\mathrm{i}\varphi}$ and $\mathrm{d}z = \mathrm{i}re^{\mathrm{i}\varphi}\mathrm{d}\varphi$. Considering that the integral is over the unit circle, we set $r = 1$ and rewrite the integral as follows:

$$\oint_{\text{unit circle}} z^n \mathrm{d}z = \int_{\varphi=0}^{2\pi} e^{\mathrm{i}n\varphi}(\mathrm{i}e^{\mathrm{i}\varphi})\mathrm{d}\varphi = \mathrm{i}\int_0^{2\pi} e^{\mathrm{i}(n+1)\varphi}\mathrm{d}\varphi$$

$$= \begin{cases} (n+1)^{-1}e^{\mathrm{i}(n+1)\varphi}\Big|_{\varphi=0}^{2\pi} = 0; & n \neq -1, \\ 2\pi\mathrm{i}; & n = -1. \end{cases} \tag{19}$$

It is seen that the integral vanishes for all integer values of n except for $n=-1$, where the integral equals $2\pi\mathrm{i}$. Of course, when $n \ge 0$, the function z^n is analytic throughout the entire complex plane and, in accordance with the Cauchy-Goursat theorem, one expects its integral over the unit circle to vanish. For $n=-1$, $f(z)=1/z$ has a singularity in the middle of the circle (at $z=0$) and, therefore, the integral of z^{-1} over the unit circle is not guaranteed to be zero. The fact that the integral turns out to be $2\pi\mathrm{i}$ does not contradict the Cauchy-Goursat theorem. Finally, for $n \le -2$, the function z^n has a singularity at $z=0$, which is inside the contour of integration. Once again, the Cauchy-Goursat theorem does not guarantee the vanishing of the integral, but the fact that the integral ends up being zero does not violate the theorem either.

Example 2. For an arbitrary closed contour C that encompasses the origin ($z=0$), show that $\oint_C z^{-1}\mathrm{d}z = -2\pi\mathrm{i}$, where the contour is traversed clockwise.

Proof: Since $\ln(z)$ is the anti-derivative of $1/z$, the integral may be written as $\oint_C \mathrm{d}(\ln z)$. The figure on the right shows the contour C that encompasses the origin. Any branch-cut of $\ln(z)$ must, of necessity, cut through the contour. Here, the chosen branch-cut is the negative half of the real axis, and the contour C is marked with the points $z_0, z_1, z_2, \cdots z_{N-2}, z_{N-1}, z_N$, with z_0 and z_N being on opposite sides of the branch-cut. Assuming the points z_n are sufficiently close to their immediate neighbors, a good approximation to the integral will be

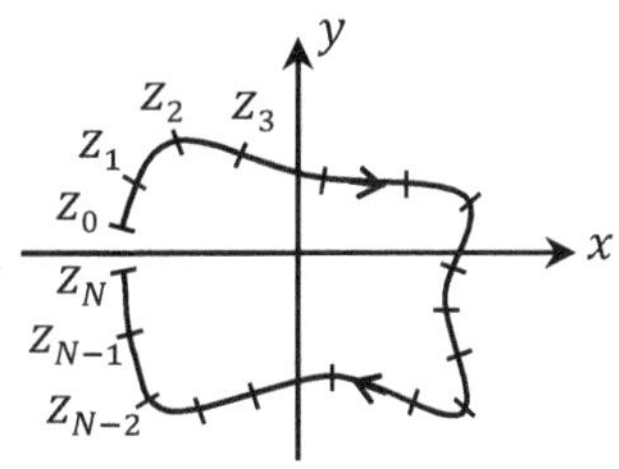

$$\oint_C z^{-1}\mathrm{d}z = \oint_C \mathrm{d}(\ln z) \cong (\ln z_1 - \ln z_0) + (\ln z_2 - \ln z_1) + \cdots + (\ln z_N - \ln z_{N-1})$$

$$= \ln z_N - \ln z_0 = (\ln|z_N| - \mathrm{i}\pi) - (\ln|z_0| + \mathrm{i}\pi) = -2\pi\mathrm{i}. \qquad (20)$$

Of course, in the limit when the points z_n become infinitesimally close to each other, the above result becomes exact. The result thus obtained is the same as that obtained in the preceding example, with the exception of the minus sign, which is due to the fact that the contours of integration in these two examples are traversed in opposite directions.

Example 3. Considering that $f(z)=\exp(z)$ is analytic throughout the entire z-plane, its integral around any closed loop is expected to be zero. Show, in particular, that the integral around a rectangular loop with corners at $z_{11}=x_1+\mathrm{i}y_1$, $z_{21}=x_2+\mathrm{i}y_1$, $z_{22}=x_2+\mathrm{i}y_2$, and $z_{12}=x_1+\mathrm{i}y_2$ vanishes.

Proof: Breaking up the integral around the rectangular loop into four integrals along the four legs of the loop, and noting that $\mathrm{d}z=\mathrm{d}x$ on the horizontal legs, whereas $\mathrm{d}z=\mathrm{i}\mathrm{d}y$ on the vertical legs, we will have

$$\oint_{\substack{\text{square}\\\text{loop}}} e^z\mathrm{d}z = e^{\mathrm{i}y_1}\int_{x_1}^{x_2} e^x\mathrm{d}x + \mathrm{i}e^{x_2}\int_{y_1}^{y_2} e^{\mathrm{i}y}\mathrm{d}y + e^{\mathrm{i}y_2}\int_{x_2}^{x_1} e^x\mathrm{d}x + \mathrm{i}e^{x_1}\int_{y_2}^{y_1} e^{\mathrm{i}y}\mathrm{d}y$$

$$= (e^{\mathrm{i}y_1} - e^{\mathrm{i}y_2})\int_{x_1}^{x_2} e^x\mathrm{d}x + \mathrm{i}(e^{x_2} - e^{x_1})\int_{y_1}^{y_2} e^{\mathrm{i}y}\mathrm{d}y$$

$$= (e^{\mathrm{i}y_1} - e^{\mathrm{i}y_2})(e^{x_2} - e^{x_1}) + (e^{x_2} - e^{x_1})(e^{\mathrm{i}y_2} - e^{\mathrm{i}y_1}) = 0. \qquad (21)$$

Example 4. Use the Cauchy-Goursat theorem to evaluate $\int_0^{2\pi}(1+a\cos\theta)^{-1}\mathrm{d}\theta$, where the parameter a could be either a real or a complex constant. Identify the range of values of a for which your final result is valid.

Discussion: Consider the unit circle centered at $z = 0$ in the complex z-plane depicted in figure 6.5. A point, z, on the perimeter of this circle has magnitude 1 and angle θ, where $0 \le \theta < 2\pi$. Writing $z = \exp(\mathrm{i}\theta)$, we will have $z^{-1} = \exp(-\mathrm{i}\theta)$ and $\mathrm{d}z = \mathrm{i}\exp(\mathrm{i}\theta)\,\mathrm{d}\theta$. Here, $\mathrm{d}z$ represents a short segment of the circle between the angles θ and $\theta + \mathrm{d}\theta$, oriented in the counterclockwise direction. Noting that $\cos\theta = ½(z + z^{-1})$ and $\mathrm{d}\theta = \mathrm{d}z/(\mathrm{i}z)$, we write the integral of interest as an integral around the unit circle, as follows:

$$\int_0^{2\pi}(1 + a\cos\theta)^{-1}\mathrm{d}\theta = \oint_{\text{unit circle}} \frac{\mathrm{d}z}{\mathrm{i}z[1 + ½a(z + z^{-1})]} = \left(\frac{2}{\mathrm{i}a}\right)\oint_{\text{unit circle}} \frac{\mathrm{d}z}{z^2 + (2/a)z + 1}. \quad (22)$$

The integrand is seen to have two poles at $z_{1,2} = (\pm\sqrt{1 - a^2} - 1)/a$. Since the product $z_1 z_2$ equals 1, we must have $|z_1| = 1/|z_2|$, which indicates that one of the poles is always inside the unit circle while the other one remains outside. Ambiguity arises only when both poles happen to reside on the unit circle, in which case we will have $(\pm\sqrt{1 - a^2} - 1)/a = \exp(\pm\mathrm{i}\varphi)$, yielding $\cos\varphi = -1/a$. Thus, for real-valued a, when $|a| \ge 1$, both poles z_1 and z_2 reside on the unit circle. This will *not* happen if $|a| < 1$ (for a real-valued a), or if a happens to be complex. Therefore, the only values of a that need to be excluded from the present analysis are those that are real and have absolute value greater than or equal to 1.

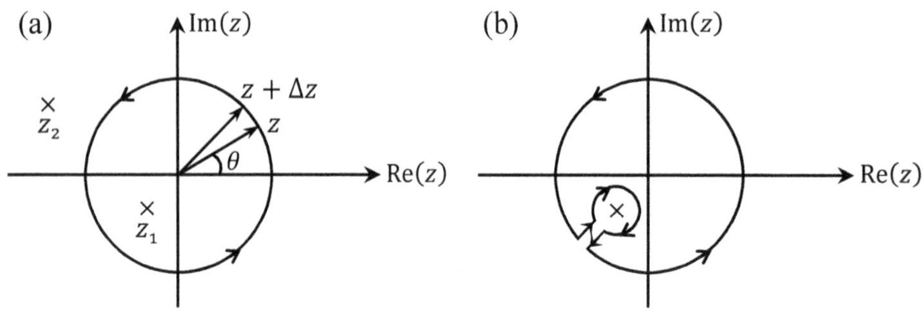

Figure 6.5. (a) The contour of integration is the unit circle in the complex z-plane. The integrand in Eq.(22) has two poles, one of which, z_1, is inside the unit circle while the other one, z_2, is outside. (b) By encircling and, therefore, removing the interior pole z_1 from the domain of integration, the modified contour falls within the domain of analyticity of the integrand, thus allowing the application of the Cauchy-Goursat theorem. The straight line segments of the modified contour can be made as close to each other as desired, thereby ensuring the cancellation of their contributions to the overall integral. The radius ε of the small circle can likewise be made sufficiently small.

Returning now to Eq.(22) and assuming that z_1 is the pole that resides inside the unit circle, we proceed to isolate this pole, as figure 6.5b indicates. The Cauchy-Goursat theorem now becomes applicable to the contour in figure 6.5b, and the integral around the unit circle becomes equal to the integral around the much smaller circle of radius ε surrounding the pole at z_1. (The two short segments that connect the unit circle to the small circle are infinitely close to each other and are traversed in opposite directions. Their contributions to the overall integral around the closed contour, therefore, cancel out.) For the small circle of radius ε, we can now write $z = z_1 + \varepsilon e^{\mathrm{i}\theta}$ and $\mathrm{d}z = \mathrm{i}\varepsilon e^{\mathrm{i}\theta}\mathrm{d}\theta$. Consequently,

$$\int_0^{2\pi}(1 + a\cos\theta)^{-1}\mathrm{d}\theta = \left(\frac{2}{\mathrm{i}a}\right)\oint_{\text{small circle}} \frac{\mathrm{d}z}{(z - z_1)(z - z_2)} = \left(\frac{2}{\mathrm{i}a}\right)\int_{\theta=0}^{2\pi} \frac{\mathrm{i}\varepsilon e^{\mathrm{i}\theta}\mathrm{d}\theta}{\varepsilon e^{\mathrm{i}\theta}(z_1 + \varepsilon e^{\mathrm{i}\theta} - z_2)}$$

$$\boxed{\text{In the limit when } \varepsilon \to 0} \rightarrow = \frac{2}{a(z_1 - z_2)}\int_0^{2\pi}\mathrm{d}\theta = 2\pi/\sqrt{1 - a^2}. \quad (23)$$

When a is complex, the sign of the square root appearing on the right-hand side of Eq.(23) must be chosen such that $|z_1| < |z_2|$.

Example 5. Use the Cauchy-Goursat theorem to evaluate $\int_{-\infty}^{\infty}(1+x^2)^{-1}\mathrm{d}x$.

Discussion: The integral under consideration is easily evaluated by means of the standard change of variable $x = \tan\theta$, which yields $\mathrm{d}x = (1+\tan^2\theta)\mathrm{d}\theta$, and, consequently,

$$\int_{-\infty}^{\infty}(1+x^2)^{-1}\mathrm{d}x = \int_{-½\pi}^{½\pi}\mathrm{d}\theta = \pi. \tag{24}$$

It is instructive to use a complex plane technique to determine the value of the same integral. The integrand thus becomes $f(z) = 1/(1+z^2)$, and our first choice for the contour of integration will be the one shown in figure 6.6a. However, one cannot apply the Cauchy-Goursat theorem to this contour, since the integrand has a singularity (or pole) at $z = \mathrm{i}$ inside the contour. In the modified contour depicted in figure 6.6b, we have isolated and removed the troublesome pole via a small circle of radius ε centered at $z = \mathrm{i}$. The overall integral of $f(z)$ around the contour of figure 6.6b must now be zero, and the desired integral along the x-axis, namely $\int_{-R}^{R}(1+x^2)^{-1}\mathrm{d}x$, is seen to be related to the integrals of $f(z)$ around the small circle, along the large semi-circle of radius R, and along the two vertical legs that connect the x-axis to the small circle.

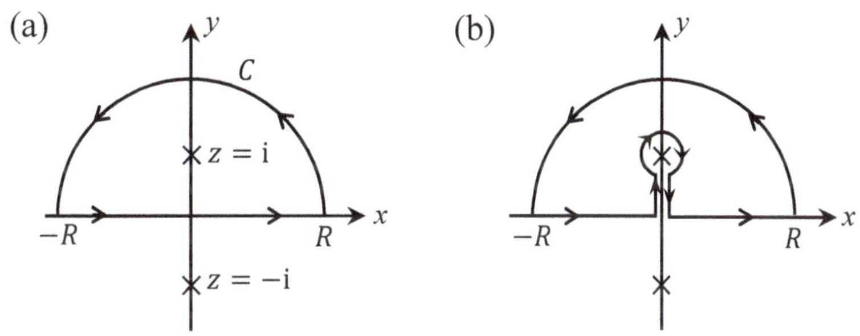

Figure 6.6. (a) A first attempt at choosing the integration contour for $\oint_C(1+z^2)^{-1}\mathrm{d}z$. The integral vanishes on the semi-circle of radius R in the limit when $R \to \infty$. However, the poles of the integrand are at $z = \pm\mathrm{i}$, and, considering that one of the poles is inside the closed contour C, the Cauchy-Goursat theorem cannot be applied. (b) The contour is modified by isolating the pole at $z = \mathrm{i}$. The modification involves breaking away from the x-axis, encircling the troublesome pole with a small circle of radius ε centered at $z = \mathrm{i}$, then returning to the x-axis. The two short vertical legs connecting the x-axis to the small circle are parallel and extremely close to each other.

In the limit when $R \to \infty$, the integral along the large semi-circle vanishes, while the contributions of the short vertical legs, being equal and opposite, cancel out. This leaves the small circle, on which $z = \mathrm{i} + \varepsilon e^{\mathrm{i}\theta}$, $\mathrm{d}z = \mathrm{i}\varepsilon e^{\mathrm{i}\theta}\mathrm{d}\theta$, and $f(z) = 1/[(z-\mathrm{i})(z+\mathrm{i})] = 1/[\varepsilon e^{\mathrm{i}\theta}(2\mathrm{i}+\varepsilon e^{\mathrm{i}\theta})]$. Thus, in the limit when $R \to \infty$ and $\varepsilon \to 0$, we will have

$$\int_{-\infty}^{\infty}\frac{\mathrm{d}x}{1+x^2} = lim_{\varepsilon\to 0}\int_{\theta=0}^{2\pi}\frac{1}{\cancel{\varepsilon e^{\mathrm{i}\theta}}(2\mathrm{i}+\varepsilon e^{\mathrm{i}\theta})}\mathrm{i}\cancel{\varepsilon e^{\mathrm{i}\theta}}\mathrm{d}\theta = \frac{2\pi\mathrm{i}}{2\mathrm{i}} = \pi. \tag{25}$$

The results of the two methods of integration given by Eqs.(24) and (25) are seen to be in complete agreement.

Example 6. Evaluate $\int_{-\infty}^{\infty}[e^{ax}/(1+e^x)]\,\mathrm{d}x$, where the real-valued parameter $a \in (0,1)$.

Discussion: The proposed contour of integration is shown in figure 6.7a. In the complex z-plane, the function $f(z) = e^{az}/(1+e^z)$ has singularities where its denominator vanishes, that is, where $\exp(z) = -1 = \exp[\mathrm{i}(2n+1)\pi]$, with n being any integer (positive, zero, or negative). Consequently, the singular points of $f(z)$ are $z_n = \mathrm{i}(2n+1)\pi$. The contour of integration in figure 6.7a encloses only one of these singularities, namely, that at $z_0 = \mathrm{i}\pi$. This singular point has been isolated by the modified contour depicted in figure 6.7b. The modified contour satisfies the requirements of the Cauchy-Goursat theorem and, therefore, the integral of $f(z)$ around the modified contour must equal zero.

On the lower leg of the rectangle, where $z = x$ and $\mathrm{d}z = \mathrm{d}x$, the function $f(z)$ is the same as the desired integrand, whereas on the upper leg, where $z = x + \mathrm{i}2\pi$ and $\mathrm{d}z = \mathrm{d}x$, we have $f(z) = e^{\mathrm{i}2\pi a}\, e^{ax}/(1+e^x)$. It is also easy to show that the vertical legs do not contribute to the integral in the limit when $L \to \infty$. Consequently, the integral around the rectangular loop shown in figure 6.7a is $(1 - e^{\mathrm{i}2\pi a})\int_{-\infty}^{\infty}[e^{ax}/(1+e^x)]\mathrm{d}x$, which must be equal to the integral around the small circle in figure 6.7b that encloses the pole at $z = z_0$.

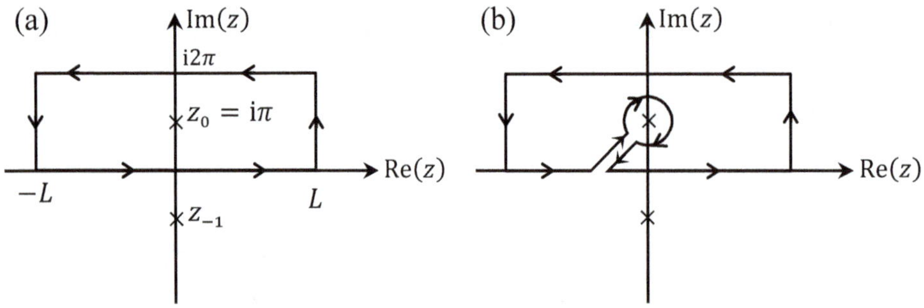

Figure 6.7. (a) A rectangular contour of width $2L$ and height 2π encloses only one of the singular points of the integrand, namely, $z_0 = \mathrm{i}\pi$. In the limit when $L \to \infty$, the vertical legs of the rectangle make no contribution to the integral, as the value of the integrand on both legs approaches zero. (b) The singularity at $z = z_0$ is cut out of the integration region by changing the integration contour such that it goes around z_0 via a small circle of radius ε.

Now, on the small circle of radius ε centered at z_0 we have $z = z_0 + \varepsilon e^{\mathrm{i}\theta}$. On this circle, $\mathrm{d}z = \mathrm{i}\varepsilon e^{\mathrm{i}\theta}\mathrm{d}\theta$, and the integral of $f(z)$ around the circle can be written as follows:

$$lim_{\varepsilon\to 0}\oint_{\text{small circle}} f(z)\mathrm{d}z = lim_{\varepsilon\to 0}\int_{\theta=0}^{2\pi}\frac{\exp[a(z_0+\varepsilon e^{\mathrm{i}\theta})]}{1+\exp(z_0+\varepsilon e^{\mathrm{i}\theta})}\mathrm{i}\varepsilon e^{\mathrm{i}\theta}\mathrm{d}\theta$$

$$\boxed{e^{\alpha} = 1+\alpha+½\alpha^2+\cdots} \rightarrow = lim_{\varepsilon\to 0}\int_0^{2\pi}\frac{e^{\mathrm{i}\pi a}(1+a\varepsilon e^{\mathrm{i}\theta}+½a^2\varepsilon^2 e^{2\mathrm{i}\theta}+\cdots)}{1+e^{\mathrm{i}\pi}(1+\varepsilon e^{\mathrm{i}\theta}+½\varepsilon^2 e^{2\mathrm{i}\theta}+\cdots)}\mathrm{i}\varepsilon e^{\mathrm{i}\theta}\mathrm{d}\theta$$

$$\boxed{e^{z_0} = e^{\mathrm{i}\pi} = -1} \rightarrow = -\mathrm{i}e^{\mathrm{i}\pi a}\, lim_{\varepsilon\to 0}\int_0^{2\pi}\frac{1+a\varepsilon e^{\mathrm{i}\theta}+½a^2\varepsilon^2 e^{2\mathrm{i}\theta}+\cdots}{\cancel{\varepsilon e^{\mathrm{i}\theta}}(1+½\varepsilon e^{\mathrm{i}\theta}+\cdots)}\cancel{\varepsilon e^{\mathrm{i}\theta}}\mathrm{d}\theta$$

$$= -\mathrm{i}2\pi e^{\mathrm{i}\pi a}. \tag{26}$$

Finally, equating the integral around the rectangle with that around the small circle, we find

$$\int_{-\infty}^{\infty}\frac{\exp(ax)}{1+\exp(x)}\mathrm{d}x = \frac{-\mathrm{i}2\pi\exp(\mathrm{i}\pi a)}{1-\exp(\mathrm{i}2\pi a)} = \frac{2\pi\mathrm{i}}{\exp(\mathrm{i}\pi a)-\exp(-\mathrm{i}\pi a)} = \frac{\pi}{\sin(\pi a)}. \tag{27}$$

The integral diverges at $a = 0$ and $a = 1$ but is convergent within the $(0, 1)$ interval.

Brain teaser. A staircase has 100 steps. Standing on each step is a man wearing either a white or a black hat. None of the men knows the color of his own hat. All the men are facing downstairs, so that the man on the highest step can see the colors of the hats on the 99 men below him. Similarly, the man on the ninety-ninth step can see the colors of the 98 hats below him, and so on. Before the game begins, the men are allowed to strategize among themselves. Once the game is started, each man is expected to guess the color of his own hat, then announce it in a loud and clear voice that the other men can hear. What strategy should the men employ to maximize the total number of their correct guesses?

6.5. Poles and residues. So far, we have seen that the integral of $f(z)/(z-z_0)$, where $f(z)$ is analytic at $z=z_0$, taken around a small circle of radius ε centered at z_0, approaches $2\pi i f(z_0)$ in the limit when $\varepsilon \to 0$. In such cases, z_0 is referred to as a *first-order pole* of the function $f(z)/(z-z_0)$, while $f(z_0)$ is known as the *residue* of the function at that pole.[4-9]

The question naturally arises as to the poles of higher order and their corresponding residues. For instance, the function $f(z)/(z-z_0)^2$, where $f(z)$ is analytic at $z=z_0$, has a second-order pole at z_0. What is the integral of this function around a small circle of radius ε centered at z_0, in the limit when $\varepsilon \to 0$? One way to answer this question is to consider $(z-z_0)^2$ as the product of $(z-z_1)$ and $(z-z_2)$, in the limit when both z_1 and z_2 approach z_0. The function $f(z)/[(z-z_1)(z-z_2)]$ now has two first-order poles, one at z_1 with a residue of $f(z_1)/(z_1-z_2)$, the other at z_2 with a residue of $f(z_2)/(z_2-z_1)$. The integral around an infinitesimal circle centered at z_0 thus approaches the sum of the integrals around z_1 and z_2 in the limit when both z_1 and z_2 approach z_0; that is,

$$lim_{\varepsilon\to 0}\oint_{\text{small circle}} \frac{f(z)}{(z-z_0)^2}\mathrm{d}z = (2\pi\mathrm{i})\, lim_{z_1,z_2\to z_0}\left[\frac{f(z_1)}{z_1-z_2}+\frac{f(z_2)}{z_2-z_1}\right]$$

$$= (2\pi\mathrm{i})\, lim_{z_1,z_2\to z_0}\frac{f(z_1)-f(z_2)}{z_1-z_2} = (2\pi\mathrm{i}) f'(z)|_{z=z_0}. \quad (28)$$

The residue of $f(z)/(z-z_0)^2$ at the second-order pole z_0 is thus seen to be *not* the value of $f(z)$ at $z=z_0$, but rather the derivative of $f(z)$ at the location z_0 of the pole.

A more general approach to the problem of finding residues at high-order poles is to resort to a Taylor series expansion of $f(z)$ around the pole at $z=z_0$. In general, for an nth-order pole at $z=z_0$, we write

$$\oint_{\text{small circle}} \frac{f(z)}{(z-z_0)^n}\mathrm{d}z$$

$$= \oint_{\text{small circle}} \frac{f(z_0)+f'(z_0)(z-z_0)+(1/2!)f''(z_0)(z-z_0)^2+\cdots+[1/(n-1)!]f^{(n-1)}(z_0)(z-z_0)^{n-1}+\cdots}{(z-z_0)^n}\mathrm{d}z. \quad (29)$$

Now, it is easy to show, for integer values of m (positive, zero, and negative), that the integral of $(z-z_0)^m$ around a circle of arbitrary radius vanishes, unless $m=-1$, in which case the integral equals $2\pi\mathrm{i}$ (see section 4, Example 1). Consequently, the sole surviving term in Eq.(29) will be the one associated with $(z-z_0)^{n-1}$ in the numerator. We thus arrive at

$$\oint_{\text{small circle}}[f(z)/(z-z_0)^n]\mathrm{d}z = (2\pi\mathrm{i})\, f^{(n-1)}(z_0)/(n-1)!. \quad (30)$$

The residue of $f(z)/(z-z_0)^n$, where n is a positive integer, is therefore given by the $(n-1)^{\text{st}}$ derivative of $f(z)$ at $z=z_0$, divided by the factorial of $(n-1)$. Needless to say, for a first-order pole, $n=1$, $(n-1)!=0!=1$, and the residue simply becomes

$f(z_0)$. For a second-order pole, $n = 2$, $(n-1)! = 1! = 1$, and the residue becomes $f'(z_0)$, as demonstrated earlier. For a third-order pole, $n = 3$, $(n-1)! = 2! = 2$, and the residue becomes $\tfrac{1}{2}f''(z_0)$, and so on.[4-11]

Example 1. Show that $\displaystyle\int_0^\infty \frac{\sin(\alpha x)}{\sinh(\beta x)}\mathrm{d}x = (\pi/2\beta)\tanh(\pi\alpha/2\beta);\quad \mathrm{Re}(\beta) > 0,\ \alpha > 0.$

Proof: The first step involves extending the integration domain from $(0,\infty)$ to $(-\infty,\infty)$, by recognizing that the integrand is an even function of x. In the second step, we use the identity $\sin(\alpha x) = [\exp(\mathrm{i}\alpha x) - \exp(-\mathrm{i}\alpha x)]/(2\mathrm{i})$ to split the integral into two, as follows:

$$\int_0^\infty \frac{\sin(\alpha x)}{\sinh(\beta x)}\mathrm{d}x = \frac{1}{4\mathrm{i}}\int_{-\infty}^\infty \frac{\exp(\mathrm{i}\alpha x)}{\sinh(\beta x)}\mathrm{d}x - \frac{1}{4\mathrm{i}}\int_{-\infty}^\infty \frac{\exp(-\mathrm{i}\alpha x)}{\sinh(\beta x)}\mathrm{d}x. \tag{31}$$

To find the poles of these integrands, we set $\sinh(\beta z)$ equal to zero, which yields $\exp(2\beta z) = 1 = \exp(\mathrm{i}2n\pi)$. Therefore, the poles are at $z_n = \mathrm{i}n\pi/\beta$, where $n = 0, \pm 1, \pm 2, \cdots$. The integration contour for the two integrals on the right-hand side of Eq.(31) is shown on the right. Next, we express $\sinh(\beta z)$ on the upper leg of the rectangular contour by writing $z = x + \mathrm{i}\eta$, where η is an arbitrary complex constant, as follows:

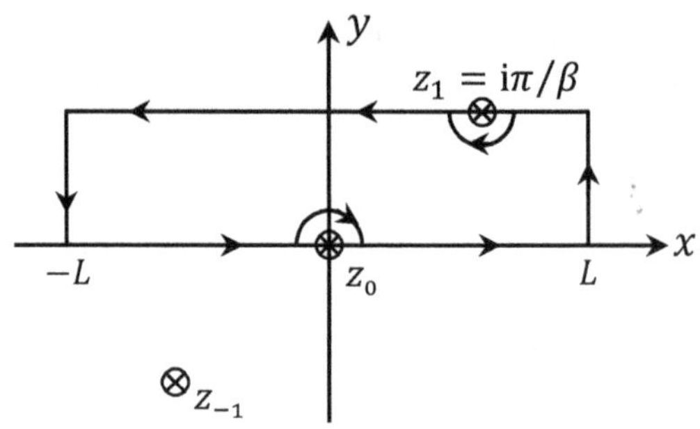

$$\begin{aligned}\sinh(\beta z) &= \tfrac{1}{2}\{\exp[\beta(x+\mathrm{i}\eta)] - \exp[-\beta(x+\mathrm{i}\eta)]\}\\ &= \tfrac{1}{2}\exp(\beta x)\,[\cos(\beta\eta) + \mathrm{i}\sin(\beta\eta)] - \tfrac{1}{2}\exp(-\beta x)\,[\cos(\beta\eta) - \mathrm{i}\sin(\beta\eta)]\\ &= \sinh(\beta x)\cos(\beta\eta) + \mathrm{i}\cosh(\beta x)\sin(\beta\eta). \end{aligned}\tag{32}$$

Therefore, on the upper leg of the rectangular contour, where $z = x + \mathrm{i}(\pi/\beta)$, we have $\sinh(\beta z) = -\sinh(\beta x)$. On the vertical legs, it is not difficult to show that the integrands approach zero when $L \to \infty$. We thus have

Residues at z_0 and z_1

$$\begin{aligned}\int_{-\infty}^\infty \frac{\exp(\mathrm{i}\alpha x)}{\sinh(\beta x)}\mathrm{d}x + \int_{-\infty}^\infty \frac{\exp[\mathrm{i}\alpha(x+\mathrm{i}\pi/\beta)]}{\sinh(\beta x)}\mathrm{d}x &= \mathrm{i}\pi\left[\frac{\exp(\mathrm{i}\alpha z_0)}{\beta\cosh(\beta z_0)} + \frac{\exp(\mathrm{i}\alpha z_1)}{\beta\cosh(\beta z_1)}\right]\\ &= \left(\frac{\mathrm{i}\pi}{\beta}\right)[1 - \exp(-\pi\alpha/\beta)]. \end{aligned}\tag{33}$$

This equation may be simplified as follows:

$$\int_{-\infty}^\infty \frac{\exp(\mathrm{i}\alpha x)}{\sinh(\beta x)}\mathrm{d}x = \left(\frac{\mathrm{i}\pi}{\beta}\right)\frac{1-\exp(-\pi\alpha/\beta)}{1+\exp(-\pi\alpha/\beta)} = \left(\frac{\mathrm{i}\pi}{\beta}\right)\tanh\left(\frac{\pi\alpha}{2\beta}\right). \tag{34}$$

As for the second half of the integral in Eq.(31), a similar procedure yields

Residues at z_0 and z_1

$$\begin{aligned}\int_{-\infty}^\infty \frac{\exp(-\mathrm{i}\alpha x)}{\sinh(\beta x)}\mathrm{d}x + \int_{-\infty}^\infty \frac{\exp[-\mathrm{i}\alpha(x+\mathrm{i}\pi/\beta)]}{\sinh(\beta x)}\mathrm{d}x &= \mathrm{i}\pi\left[\frac{\exp(-\mathrm{i}\alpha z_0)}{\beta\cosh(\beta z_0)} + \frac{\exp(-\mathrm{i}\alpha z_1)}{\beta\cosh(\beta z_1)}\right]\\ &= \left(\frac{\mathrm{i}\pi}{\beta}\right)[1 - \exp(\pi\alpha/\beta)]. \end{aligned}\tag{35}$$

Consequently,

$$\int_{-\infty}^{\infty} \frac{\exp(-\mathrm{i}\alpha x)}{\sinh(\beta x)} \mathrm{d}x = \left(\frac{\mathrm{i}\pi}{\beta}\right) \frac{1 - \exp(\pi\alpha/\beta)}{1 + \exp(\pi\alpha/\beta)} = -\left(\frac{\mathrm{i}\pi}{\beta}\right) \tanh\left(\frac{\pi\alpha}{2\beta}\right). \tag{36}$$

Following a final substitution into Eq.(31), we arrive at

$$\int_{0}^{\infty} \frac{\sin(\alpha x)}{\sinh(\beta x)} \mathrm{d}x = \left(\frac{\pi}{2\beta}\right) \tanh\left(\frac{\pi\alpha}{2\beta}\right). \tag{37}$$

Example 2. The function $f(z) = z^n \exp(az)$ is analytic everywhere in the complex z-plane except possibly at the origin, where $z = 0$ could be a pole of the function. Here, n is a positive, zero, or negative integer and $a = a' + \mathrm{i}a''$ is an arbitrary constant. Express the loop integral $\oint f(z)\mathrm{d}z$ around the unit circle centered at the origin as an integral over the polar angle θ from 0 to 2π, then use the residue theorem to evaluate the integral.

Discussion: On the unit circle, we have $z = \exp(\mathrm{i}\theta)$ and $\mathrm{d}z = \mathrm{i}\exp(\mathrm{i}\theta)\,\mathrm{d}\theta$. Therefore,

$$\begin{aligned} \oint f(z)\mathrm{d}z &= \int_0^{2\pi} \exp(\mathrm{i}n\theta) \exp[(a' + \mathrm{i}a'')(\cos\theta + \mathrm{i}\sin\theta)]\,[\mathrm{i}\exp(\mathrm{i}\theta)]\mathrm{d}\theta \\ &= \mathrm{i}\int_0^{2\pi} \exp(a'\cos\theta - a''\sin\theta) \exp\{\mathrm{i}[a'\sin\theta + a''\cos\theta + (n+1)\theta]\}\,\mathrm{d}\theta \\ &= -\int_0^{2\pi} \exp(a'\cos\theta - a''\sin\theta) \sin[a'\sin\theta + a''\cos\theta + (n+1)\theta]\,\mathrm{d}\theta \\ &\quad +\mathrm{i}\int_0^{2\pi} \exp(a'\cos\theta - a''\sin\theta) \cos[a'\sin\theta + a''\cos\theta + (n+1)\theta]\,\mathrm{d}\theta \\ &= \begin{cases} 0; & n \geq 0, \\ \mathrm{i}2\pi\, a^{|n+1|}/|n+1|!; & n < 0. \quad \leftarrow \text{Residue at } z = 0 \end{cases} \end{aligned} \tag{38}$$

The final result is somewhat simplified if we define a new integer $m = n + 1$. We will have

$$\begin{aligned} &\int_0^{2\pi} \exp(a'\cos\theta - a''\sin\theta) \sin[a'\sin\theta + a''\cos\theta + m\theta]\,\mathrm{d}\theta \\ &= \begin{cases} 0; & m \geq 0, \\ \frac{2\pi}{|m|!}(a'^2 + a''^2)^{|m|/2} \sin[|m| \arctan(a''/a')]; & m < 0. \end{cases} \end{aligned} \tag{39}$$

$$\begin{aligned} &\int_0^{2\pi} \exp(a'\cos\theta - a''\sin\theta) \cos[a'\sin\theta + a''\cos\theta + m\theta]\,\mathrm{d}\theta \\ &= \begin{cases} 0; & m > 0, \\ \frac{2\pi}{|m|!}(a'^2 + a''^2)^{|m|/2} \cos[|m| \arctan(a''/a')]; & m \leq 0. \end{cases} \end{aligned} \tag{40}$$

The preceding integrals for the case of $m \leq 0$ are listed as Eqs.(**3.937**-3 and 4) in Gradshteyn and Ryzhik's *Table of Integrals, Series, and Products*.[12]

6.6. Analytic functions have a Taylor series expansion. The notion of analyticity for functions defined on the complex plane is much more powerful than the corresponding notion for functions of a real variable. Consider, for instance, the function $f(x) = \text{step}(x)\exp(-1/x)$ of the real variable x. Not only is this function continuous, but it also permits differentiation to all orders everywhere on the x-axis, including at $x = 0$. However, all derivatives of $f(x)$ at $x = 0$ turn out to be zero, indicating that the function cannot be expanded in a Taylor series around the origin.

In contrast, a complex function such as $f(z)$, if analytic at $z = z_0$—that is, if first-order differentiable at z_0 as well as within a small neighborhood of z_0—turns out to be differentiable to all orders at z_0. Moreover, the function can be expanded in a Taylor series at least within a circle centered at $z = z_0$ and contained in the domain of analyticity surrounding z_0. The proof of the preceding statement, due to Cauchy, begins by selecting, within the domain of analyticity of $f(z)$, a circular contour C centered at $z = z_0$. For an arbitrary point z' inside the circle C, application of the Cauchy-Goursat theorem yields

$$
\begin{aligned}
f(z') &= \frac{1}{2\pi i}\oint_C \frac{f(z)}{z - z'}\,dz \\
&= \frac{1}{2\pi i}\oint_C \frac{f(z)}{(z - z_0) - (z' - z_0)}\,dz = \frac{1}{2\pi i}\oint_C \frac{f(z)}{z - z_0} \times \frac{1}{1 - (z' - z_0)/(z - z_0)}\,dz \\
&= \frac{1}{2\pi i}\oint_C \frac{f(z)}{z - z_0} \times \sum_{n=0}^{\infty}\left(\frac{z' - z_0}{z - z_0}\right)^n dz \leftarrow \boxed{|z' - z_0| < |z - z_0| \text{ for all } z' \text{ inside } C.} \\
&= \frac{1}{2\pi i}\sum_{n=0}^{\infty}(z' - z_0)^n \oint_C \frac{f(z)}{(z - z_0)^{n+1}}\,dz = f(z_0) + \sum_{n=1}^{\infty}\frac{f^{(n)}(z_0)}{n!}(z' - z_0)^n. \quad (41)
\end{aligned}
$$

In other words, when z' is within the circle C, the function $f(z')$ can be expanded in a Taylor series around $z = z_0$. Completing the proof requires demonstrating that $f(z)$ has derivatives of all orders at z_0. To this end, note that, within the domain of analyticity of $f(z) = f(x + iy) = u(x, y) + iv(x, y)$, the following Cauchy-Riemann conditions are satisfied:

$$
\frac{\partial u}{\partial x} = \frac{\partial v}{\partial y} \quad \text{and} \quad \frac{\partial u}{\partial y} = -\frac{\partial v}{\partial x}. \qquad (42)
$$

Now, the first derivative of $f(z)$ being $f'(z) = u_1(x, y) + iv_1(x, y) = (\partial u/\partial x) + i(\partial v/\partial x)$, the corresponding Cauchy-Riemann conditions may be written

$$
\frac{\partial u_1}{\partial x} = \frac{\partial v_1}{\partial y} \quad \rightarrow \quad \frac{\partial^2 u}{\partial x^2} = \frac{\partial^2 v}{\partial x \partial y}, \qquad (43a)
$$

$$
\frac{\partial u_1}{\partial y} = -\frac{\partial v_1}{\partial x} \quad \rightarrow \quad \frac{\partial^2 u}{\partial x \partial y} = -\frac{\partial^2 v}{\partial x^2}. \qquad (43b)
$$

Substitution from Eqs.(42) into Eqs.(43) reveals that the Cauchy-Riemann conditions for $f'(z)$ are automatically satisfied. Having thus shown that $f'(z)$ is analytic within the domain of analyticity of $f(z)$, the same argument can be repeated for the second- and higher-order derivatives of $f(z)$.

6.7. Liouville's theorem. The Liouville's theorem of complex analysis, named after the French mathematician Joseph Liouville (1809–1882), states that a bounded entire function (i.e., a function that is first-order differentiable throughout the entire complex plane) must be equal to a constant.

Consider the entire function $f(z)$ and assume that $|f(z)| < F_0$ for all z. The nth derivative of the function at $z = 0$, namely, $f^{(n)}(0)$, may be written as a contour integral over a circle C of radius R centered at the origin; that is,

$$
f^{(n)}(0) = \frac{n!}{2\pi i}\oint_C \frac{f(z)}{z^{n+1}}\,dz. \qquad (44)
$$

Using the fact that the absolute value of an integral is less than or equal to the integral of the absolute value of the integrand, we write

$$\left|f^{(n)}(0)\right| \le \frac{n!}{2\pi}\int_0^{2\pi} \frac{|f(z)|}{|z|^{n+1}} R\mathrm{d}\theta = n!\, F_0/R^n. \tag{45}$$

The radius R of the integration contour C may now be allowed to go to infinity, confirming that all the derivatives of $f(z)$ at $z = 0$ indeed vanish. Consequently, the bounded entire function $f(z)$ must be a constant.

6.8. Jordan's Lemma. Consider the integral $F(s) = \int_{-\infty}^{\infty} f(x)\exp(-\mathrm{i}2\pi sx)\,\mathrm{d}x$ applied to an arbitrary function $f(x)$. We would like to evaluate $F(s)$—generally known as the Fourier transform of $f(x)$—in the complex z-plane by closing the integration path on a semi-circle either in the lower-half or the upper-half of the complex plane (see figure 6.8a). We assume that $f(z)$ is analytic everywhere except at a few poles, say, $z_1, z_2, \cdots z_n$, and also that $f(z) \to 0$ when $|z| \to \infty$. The question then becomes, "Can one set the integral over the semi-circle (in the limit when $R \to \infty$) equal to zero to evaluate $F(s)$ using the Cauchy-Goursat theorem?"

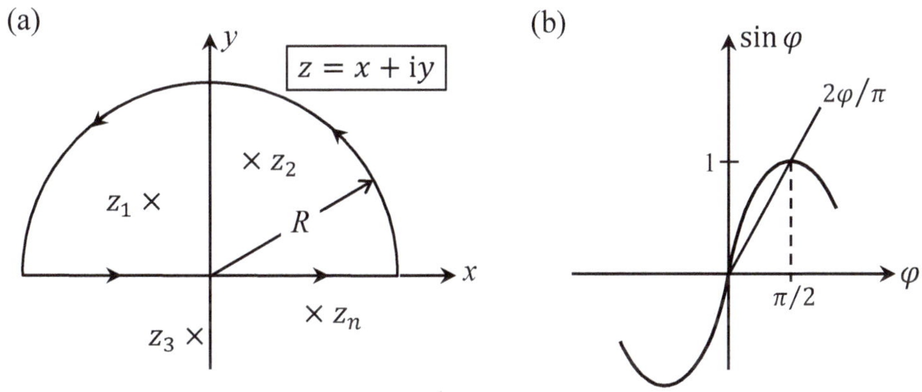

Figure 6.8. (a) Integration contour in the upper half of the z-plane. (b) $\sin\varphi \ge 2\varphi/\pi$ when $0 \le \varphi \le \pi/2$.

To answer this question, first observe that the choice of the semi-circle depends on the sign of s. For negative s, the semi-circle must be in the upper-half of the complex plane, because that is where $\exp(-\mathrm{i}2\pi sz) = \exp(2\pi sy)\exp(-\mathrm{i}2\pi sx) \to 0$ when $R \to \infty$. However, for positive s, the contour must close on the semi-circle in the lower-half of the z-plane. Once the contour is specified, we must check to see if the integral over the corresponding semi-circle would approach zero when $R \to \infty$.

Considering that $f(z) \to 0$ in the limit when $|z| \to \infty$, we can always find, for an arbitrarily small positive ε, a radius R such that $|f(z)| < \varepsilon$ when $|z| > R$. We will then have (for $s < 0$)

$\left|\int_C g(z)\mathrm{d}z\right| \le \int_C |g(z)\mathrm{d}z|$

$$\left|\int_{\substack{\text{semi}\\ \text{circle}}} f(z)\exp(-\mathrm{i}2\pi sz)\mathrm{d}z\right| \le \int_0^{\pi} \left|f(z)\exp[-\mathrm{i}2\pi s\underbrace{R\exp(\mathrm{i}\varphi)}_{z}]\underbrace{\mathrm{i}R\exp(\mathrm{i}\varphi)\mathrm{d}\varphi}_{\mathrm{d}z}\right|$$

$R\exp(\mathrm{i}\varphi)$

$|\exp(-\mathrm{i}2\pi sR\cos\varphi)| = 1$; $|\mathrm{i}R\exp(\mathrm{i}\varphi)| = R$

$$\le R\varepsilon\int_0^{\pi}\exp(2\pi sR\sin\varphi)\,\mathrm{d}\varphi = 2R\varepsilon\int_0^{\pi/2}\exp(2\pi sR\sin\varphi)\,\mathrm{d}\varphi$$

$\sin\varphi \ge 2\varphi/\pi$; see Fig.1(b)

$$\le 2R\varepsilon\int_0^{\pi/2}\exp(4sR\varphi)\,\mathrm{d}\varphi = \frac{\varepsilon}{2s}[\exp(2\pi sR) - 1] \le \frac{\varepsilon}{2|s|}. \tag{46}$$

$s < 0$

It is thus seen that, in the end, the arc length of the semi-circle (which is proportional to R) is eliminated and also, since $\varepsilon \to 0$ when $R \to \infty$, the entire integral over the semi-circle approaches zero.[4-9] A similar argument applies to the case of $s > 0$, albeit with the semi-circle being in the lower-half of the z-plane. Jordan's lemma is also applicable to the inverse Fourier integral, $f(x) = \int_{-\infty}^{\infty} F(s)\exp(\mathrm{i}2\pi sx)\,\mathrm{d}s$, where the roles of s and x are reversed. The contour will then be in the upper-half of the plane when $x > 0$, and in the lower-half of the plane when $x < 0$.

Example 1. Find the area under the function $\operatorname{sinc}(x) = \sin(\pi x)/(\pi x)$.

Discussion: Considering that $\sin(\pi x) = \operatorname{Imag}(e^{\mathrm{i}\pi x})$, at first we evaluate $\int_{-\infty}^{\infty} \frac{\exp(\mathrm{i}\pi x)}{\pi x}\,\mathrm{d}x$, then take its imaginary part as the area under $\operatorname{sinc}(x)$. For the integration contour shown in the figure on the right, Jordan's lemma ensures that the integral of $f(z) = \exp(\mathrm{i}\pi z)/(\pi z)$ on the large semi-circle in the upper-half of the z-plane vanishes. The only pole of the integrand at $z = 0$ is excluded from the region of integration by the small semi-circle of radius ε centered at the origin. The desired integral on the x-axis is thus equal to the integral over this small semi-circle; that is,

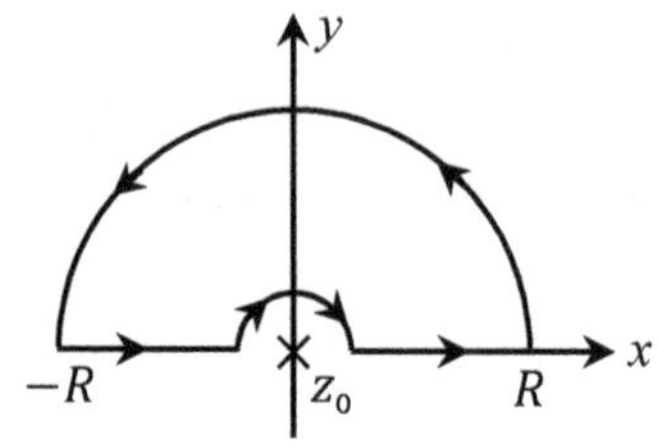

$$\int_{-\infty}^{-\varepsilon} \frac{\exp(\mathrm{i}\pi x)}{\pi x}\,\mathrm{d}x + \int_{\varepsilon}^{\infty} \frac{\exp(\mathrm{i}\pi x)}{\pi x}\,\mathrm{d}x = \int_{\theta=0}^{\pi} \frac{\exp(\mathrm{i}\pi\varepsilon e^{\mathrm{i}\theta})}{\pi\cancel{\varepsilon e^{\mathrm{i}\theta}}}\mathrm{i}\cancel{\varepsilon e^{\mathrm{i}\theta}}\,\mathrm{d}\theta \underset{(\varepsilon \to 0)}{\rightarrow} \mathrm{i}\pi/\pi = \mathrm{i}. \quad (47)$$

The imaginary part of the integral of $\exp(\mathrm{i}\pi x)/(\pi x)$ is thus seen to be 1. Therefore, the area under $\operatorname{sinc}(x)$ is also equal to 1.

Example 2. Show that $\int_{-\infty}^{\infty} \frac{x\exp(\mu x)}{\exp(\nu x) - 1}\,\mathrm{d}x = \left[\frac{\pi/\nu}{\sin(\pi\mu/\nu)}\right]^2; \qquad \operatorname{Re}(\nu) > \operatorname{Re}(\mu) > 0.$

Proof: The poles of the integrand in the z-plane are obtained by setting the denominator to zero (i.e., $\exp(\nu z) - 1 = 0 \rightarrow z_n = \mathrm{i}2\pi n/\nu$ for $n = \pm 1, \pm 2, \pm 3, \cdots$). Note that $n = 0$ does *not* represent a pole, as z in the integrand's numerator cancels the corresponding zero in the denominator. The integration contour has a semi-circular portion that is either in the upper- or the lower-half of the z-plane, depending on the ratio μ/ν.

$$\int_{-\infty}^{\infty} \frac{x\exp(\mu x)}{\exp(\nu x) - 1}\,\mathrm{d}x = 2\pi\mathrm{i}\sum_{n=1}^{\infty} \frac{z_n\exp(\mu z_n)}{\nu\exp(\nu z_n)} = -\left(\frac{2\pi}{\nu}\right)^2 \sum_{n=1}^{\infty} n\exp(\mathrm{i}2\pi n\mu/\nu). \quad (48)$$

To evaluate the infinite sum appearing on the right-hand side of Eq.(48), we note that $\sum_{n=1}^{\infty}\exp(\alpha n)$, where $\operatorname{Re}(\alpha) < 0$, is given by

Geometric series → $\sum_{n=1}^{\infty}\exp(\alpha n) = \exp(\alpha)/[1 - \exp(\alpha)] = [\exp(-\alpha) - 1]^{-1}$. (49)

Differentiating this identity with respect to α then yields

$$\sum_{n=1}^{\infty} n\exp(\alpha n) = \frac{\mathrm{d}}{\mathrm{d}\alpha}\sum_{n=1}^{\infty}\exp(\alpha n) = \frac{\mathrm{d}}{\mathrm{d}\alpha}[\exp(-\alpha) - 1]^{-1}$$

$$= \exp(-\alpha)\,[\exp(-\alpha) - 1]^{-2} = \frac{1}{[\exp(-\alpha/2) - \exp(\alpha/2)]^2} = \frac{1}{[2\sinh(\alpha/2)]^2}. \quad (50)$$

We now substitute Eq.(50) into Eq.(48), using $\alpha = \mathrm{i}2\pi\mu/\nu$, to obtain

$$\int_{-\infty}^{\infty} \frac{x\exp(\mu x)}{\exp(\nu x)-1}\mathrm{d}x = -\frac{(2\pi/\nu)^2}{[2\sinh(\mathrm{i}\pi\mu/\nu)]^2} = \left[\frac{\pi/\nu}{\sin(\pi\mu/\nu)}\right]^2. \quad (51)$$

6.9. The Kramers-Kronig relations. In the Lorentz oscillator model of classical electrodynamics,[13-15] barring the Clausius-Mossotti correction, the susceptibility of a linear, homogeneous, isotropic dielectric medium is given by

$$\chi(\omega) = \sum_{k=1}^{K} \frac{f_k\omega_p^2}{\omega_{0k}^2 - \omega^2 - \mathrm{i}\gamma_k\omega}. \quad (52)$$

In this equation, K is the number of independent elementary oscillators, ω_p is the plasma frequency, and the parameters f_k, ω_{0k}, and γ_k are, respectively, the oscillator strength, the resonance frequency, and the damping coefficient associated with the kth oscillator. The susceptibility associated with each oscillator thus has two poles at $\omega = \pm(\omega_{0k}^2 - \tfrac{1}{4}\gamma_k^2)^{\frac{1}{2}} - \tfrac{1}{2}\mathrm{i}\gamma_k$. The fact that these poles are in the lower-half of the complex ω-plane renders the function $\chi(\omega)$ analytic in the upper-half of the ω-plane.

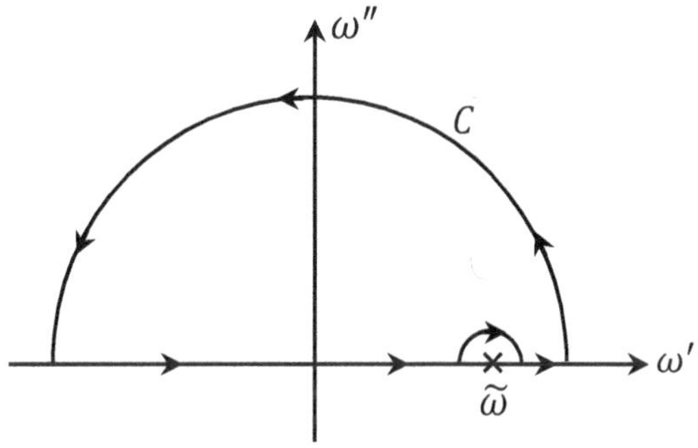

Figure 6.9. When a function such as $\chi(\omega)$ is analytic in the upper-half of the complex ω-plane, the new function $\chi(\omega)/(\omega-\tilde{\omega})$ will have no singularities within the contour shown. Consequently, when the large semi-circle expands to infinity, the integral of $\chi(\omega)/(\omega-\tilde{\omega})$ over the real axis becomes equal to its integral over the small semi-circle (albeit traversed in the opposite direction to that shown in the figure). Note that the integral over the large semi-circle will not vanish unless $\chi(\omega)$ approaches zero sufficiently rapidly as $|\omega| \to \infty$.

The Kramers-Kronig relations apply to *any* complex function, such as $\chi(\omega)$, which is analytic in the upper-half of the complex plane and approaches zero sufficiently rapidly as $|\omega| \to \infty$ within the upper-half of the complex plane. To see this, consider the function $\chi(\omega)/(\omega-\tilde{\omega})$, where $\tilde{\omega}$ is an arbitrary real-valued constant. This function is analytic in the upper-half of the plane, but has a pole on the real axis at $\omega = \tilde{\omega}$. Application of the Cauchy-Goursat theorem to the closed contour C depicted in figure 6.9 reveals that the integral of $\chi(\omega)/(\omega-\tilde{\omega})$ over the entire contour is zero. Now, on the large semi-circle the function approaches zero as the semi-circle expands to infinity, while the integral over the infinitesimal semi-circle is given by the residue of the function at $\omega = \tilde{\omega}$, namely $-\mathrm{i}\pi\chi(\tilde{\omega})$. Consequently,

$$\int_{-\infty}^{\infty}[\chi(\omega)/(\omega-\tilde{\omega})]\,\mathrm{d}\omega = \mathrm{i}\pi\chi(\tilde{\omega}). \quad (53)$$

Denoting the real and imaginary parts of $\chi(\omega)$ by $\chi'(\omega)$ and $\chi''(\omega)$, respectively, it is now easy to see from Eq.(53) that the following Kramers-Kronig relations hold:

$$\chi'(\tilde{\omega}) = (1/\pi)\int_{-\infty}^{\infty}[\chi''(\omega)/(\omega-\tilde{\omega})]\,\mathrm{d}\omega, \quad (54a)$$

$$\chi''(\tilde{\omega}) = -(1/\pi)\int_{-\infty}^{\infty}[\chi'(\omega)/(\omega-\tilde{\omega})]\,\mathrm{d}\omega. \quad (54b)$$

In general $\chi(\omega)$ is Hermitian, as its inverse Fourier transform must be a real-valued function of time representing the impulse-response of the material medium's polarization $\boldsymbol{P}(\boldsymbol{r},t)$ to an applied electric field $\boldsymbol{E}(\boldsymbol{r})\exp(-\mathrm{i}\omega t)$. Consequently, $\chi'(\omega)$ is an even function of ω, while $\chi''(\omega)$ is an odd function. Given that $\chi(\omega)$ is analytic in the upper-half of the complex ω-plane, and also that $\chi(\omega)\to 0$ sufficiently rapidly as $|\omega|\to\infty$, The Cauchy-Goursat theorem guarantees that $\int_{-\infty}^{\infty}\chi(\omega)\mathrm{d}\omega=0$. However, since $\chi''(\omega)$ is odd, its integral is automatically zero, which means that, for the even function $\chi'(\omega)$, we must have $\int_0^{\infty}\chi'(\omega)\mathrm{d}\omega=0$. This identity may, of course, be confirmed by a direct integration of $\mathrm{Re}[\chi(\omega)]$ given by Eq.(52), but the result obtained with the aid of the Cauchy-Goursat theorem is more general and applies to a large class of functions.

In a similar vein, let us consider the function $\omega\chi(\omega)$, which is analytic in the upper-half of the complex ω-plane but does *not* vanish sufficiently rapidly when $|\omega|\to\infty$. If one assumes that, on a large semi-circle of radius R (such as that depicted in figure 6.9), $\chi(\omega)\to\alpha/\omega^2$ as $R\to\infty$, then the contribution of the large semi-circle to the contour integral becomes

$$\int_{\text{semi-circle}}\omega\chi(\omega)\mathrm{d}\omega \;\to\; \int_{\text{semi-circle}}(\alpha/\omega)\mathrm{d}\omega=\alpha\int_{\theta=0}^{\pi}\frac{\mathrm{i}R\exp(\mathrm{i}\theta)}{R\exp(\mathrm{i}\theta)}\mathrm{d}\theta=\mathrm{i}\alpha\pi. \tag{55}$$

In the absence of singularities in the upper-half of the plane, the integral of $\omega\chi(\omega)$ over a closed contour consisting of the real axis and the large semi-circle in the upper-half of the plane goes to zero. Given that $\omega\chi'(\omega)$ is an odd function of ω, whereas $\omega\chi''(\omega)$ is even, we will have

$$\int_{-\infty}^{\infty}\omega\chi(\omega)\mathrm{d}\omega=2\mathrm{i}\int_0^{\infty}\omega\chi''(\omega)\mathrm{d}\omega=-\mathrm{i}\alpha\pi. \tag{56}$$

A comparison with Eq.(52) reveals that $\int_0^{\infty}\omega\chi''(\omega)\mathrm{d}\omega=\tfrac{1}{2}\pi\omega_p^2\sum_k f_k$. This is one of the so-called sum rules described in J. D. Jackson's *Classical Electrodynamics.*[13]

An alternative form of the Kramers-Kronig relations is obtained by breaking up Eqs.(54a) and (54b) into integrals over positive and negative halves of the ω-axis, then invoking the Hermitian character of $\chi(\omega)$, namely, $\chi(-\omega)=\chi^*(\omega)$, to simplify the resulting equations.[15] We find

$$\chi'(\tilde{\omega})=\frac{1}{\pi}\int_{-\infty}^{0}\frac{\chi''(\omega)}{\omega-\tilde{\omega}}\mathrm{d}\omega+\frac{1}{\pi}\int_0^{\infty}\frac{\chi''(\omega)}{\omega-\tilde{\omega}}\mathrm{d}\omega=\frac{1}{\pi}\int_0^{\infty}\frac{\chi''(-\omega)}{-\omega-\tilde{\omega}}\mathrm{d}\omega+\frac{1}{\pi}\int_0^{\infty}\frac{\chi''(\omega)}{\omega-\tilde{\omega}}\mathrm{d}\omega$$

$$=\frac{1}{\pi}\int_0^{\infty}\left[\frac{\chi''(\omega)}{\omega+\tilde{\omega}}+\frac{\chi''(\omega)}{\omega-\tilde{\omega}}\right]\mathrm{d}\omega=\frac{2}{\pi}\int_0^{\infty}\frac{\omega\chi''(\omega)}{\omega^2-\tilde{\omega}^2}\mathrm{d}\omega. \tag{57a}$$

$$\chi''(\tilde{\omega})=-\frac{1}{\pi}\int_{-\infty}^{0}\frac{\chi'(\omega)}{\omega-\tilde{\omega}}\mathrm{d}\omega-\frac{1}{\pi}\int_0^{\infty}\frac{\chi'(\omega)}{\omega-\tilde{\omega}}\mathrm{d}\omega=-\frac{1}{\pi}\int_0^{\infty}\frac{\chi'(-\omega)}{-\omega-\tilde{\omega}}\mathrm{d}\omega-\frac{1}{\pi}\int_0^{\infty}\frac{\chi'(\omega)}{\omega-\tilde{\omega}}\mathrm{d}\omega$$

$$=\frac{1}{\pi}\int_0^{\infty}\left[\frac{\chi'(\omega)}{\omega+\tilde{\omega}}-\frac{\chi'(\omega)}{\omega-\tilde{\omega}}\right]\mathrm{d}\omega=-\frac{2\tilde{\omega}}{\pi}\int_0^{\infty}\frac{\chi'(\omega)}{\omega^2-\tilde{\omega}^2}\mathrm{d}\omega. \tag{57b}$$

We mention in passing that the singularity of the integrands in Eqs.(54a) and (54b), as well as those in Eqs.(57a) and (57b), requires that the "principal value" of each integral be evaluated.

6.10. Fundamental theorem of algebra. Let $P_n(z) = c_0 + c_1 z + c_2 z^2 + \cdots + c_{n-1} z^{n-1} + c_n z^n$, where the coefficients $c_0, c_1, c_2, \cdots, c_{n-1}, c_n$ are complex constants, be an nth-order polynomial function of the complex variable z. The polynomial equation $P_n(z) = 0$ of degree $n \geq 1$ defined over the entire complex plane z has a total of n roots $z_1, z_2, \cdots, z_n$, some of which might be repetitive (i.e., overlap each other).

Proof: For $n = 1$, $P_1(z) = 0$ has a single root, $z_1 = -c_0/c_1$. For $n = 2$, $P_2(z) = 0$ has two (possibly overlapping) roots $z_{1,2} = -½(c_1/c_2) \pm \sqrt{¼(c_1/c_2)^2 - (c_0/c_2)}$. Assuming that the nth-order equation $P_n(z) = 0$ has n roots, we would like to show that $P_{n+1}(z) = 0$ has $n + 1$ roots. This task can be accomplished provided that, within the complex z-plane, a point $\tilde{z}$ can be located that satisfies $P_{n+1}(\tilde{z}) = 0$. To prove the assertion, note that, upon locating $\tilde{z}$, the polynomial $P_{n+1}(z)$ may be factored as follows:

$$P_{n+1}(z) = (z - \tilde{z})(c_0' + c_1' z + c_2' z^2 + \cdots + c_{n-1}' z^{n-1} + c_n' z^n). \tag{58}$$

The unknown coefficients $c_0', c_1', \cdots, c_{n-1}', c_n'$ may be obtained from the coefficients $c_0, c_1, \cdots, c_n, c_{n+1}$ of $P_{n+1}(z)$ using the recursion relations

$$\begin{aligned} c_{n+1} &= c_n', \\ c_n &= c_{n-1}' - c_n' \tilde{z}, \\ c_{n-1} &= c_{n-2}' - c_{n-1}' \tilde{z}, \\ &\vdots \\ c_1 &= c_0' - c_1' \tilde{z}, \\ c_0 &= -c_0' \tilde{z}. \end{aligned} \tag{59}$$

Thus, in accordance with Eq.(58), the remaining n roots of $P_{n+1}(z)$ are the roots of the nth-order polynomial $P_n(z) = c_0' + c_1' z + c_2' z^2 + \cdots + c_{n-1}' z^{n-1} + c_n' z^n$, which have already been assumed to exist. The proof by induction will then be complete.

It remains to prove that $P_{n+1}(z)$ has at least one root $\tilde{z}$ in the z-plane. To this end, we rely on the fact that $P_{n+1}(z)$ is an analytic function everywhere in the z-plane. Since the derivative with respect to z of $P_{n+1}(z)$ is an nth-order polynomial, and since it is already assumed that any n^{th}-order polynomial has, at most, n distinct roots in the z-plane, there exist no more than n points $\tilde{z}_1, \tilde{z}_2, \cdots, \tilde{z}_n$, at which the derivative $P_{n+1}'(z)$ of $P_{n+1}(z)$ vanishes. With reference to figure 6.10, the projected zeros of $P_{n+1}'(z)$ onto the complex w-plane, where $w = P_{n+1}(z)$, are denoted by $\tilde{w}_1, \tilde{w}_2, \cdots, \tilde{w}_n$.

Our goal is to locate a point $\tilde{z}$ in the z-plane, whose projection coincides with the origin of the w-plane, namely $\tilde{w} = P_{n+1}(\tilde{z}) = 0$. We begin by selecting an arbitrary point z_0 in the z-plane and find its projection $w_0 = P_{n+1}(z_0)$ in the w-plane. Next we draw, as shown in figure 6.10, a continuous path C_w from w_0 to the origin of the w-plane, taking care to avoid crossing the points $\tilde{w}_1, \tilde{w}_2, \cdots, \tilde{w}_n$, which are projections into the w-plane of the roots of $P_{n+1}'(z) = 0$. To construct the corresponding path C_z within the z-plane, we compute $P_{n+1}'(z_0)$, then move a short distance Δz away from z_0 in the z-plane, such that $P_{n+1}(z_0 + \Delta z)$ becomes a close neighbor of w_0 along the C_w path. For sufficiently small Δz, this requires the satisfaction of the following equation:

$$\Delta z = \frac{P_{n+1}(z_0 + \Delta z) - P_{n+1}(z_0)}{P_{n+1}'(z_0)}. \tag{60}$$

In this way, $z_o + \Delta z$ will end up being on the desired path C_z, which is the path from z_o to $\tilde{z}$ in the z-plane. By selecting the next adjacent point along C_w and repeating the above procedure, one constructs the desired path C_z in small increments Δz, until one arrives at $\tilde{z}$.

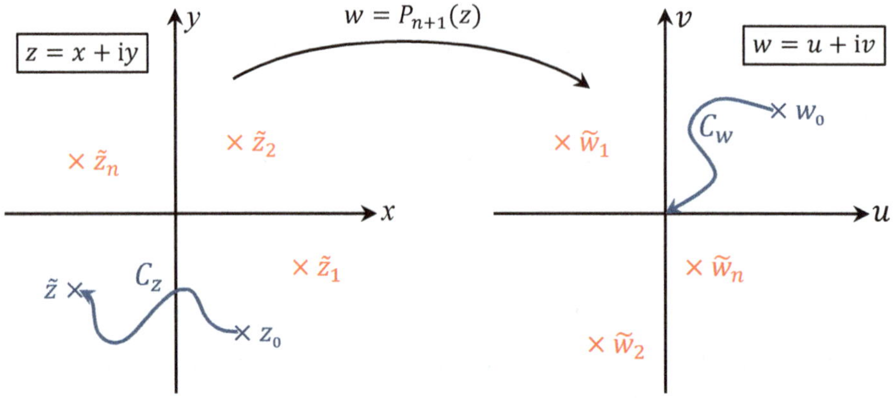

Figure 6.10. The $(n+1)^{\text{st}}$-order polynomial function $P_{n+1}(z)$ projects the points of the complex z-plane into the w-plane. The zeros of $P'_{n+1}(z)$ are denoted by $\tilde{z}_1, \tilde{z}_2, \cdots, \tilde{z}_n$, while their respective projections are identified as $\tilde{w}_1, \tilde{w}_2, \cdots, \tilde{w}_n$. In searching for a zero $\tilde{z}$ of $P_{n+1}(z)$, the arbitrarily selected z_o serves as a starting point; w_o is the projection of z_o into the w-plane. The continuous path C_w, which begins at w_o and terminates at $w = 0$, must avoid the points $\tilde{w}_1, \tilde{w}_2, \cdots, \tilde{w}_n$, but is otherwise arbitrary. The continuous path C_z is constructed (in small incremental steps) such that, when a point z travels along C_z from z_o to $\tilde{z}$, its projection moves along C_w from w_o to $w = 0$.

Note that, by selecting a trajectory C_w which avoids the points $\tilde{w}_1, \tilde{w}_2, \cdots, \tilde{w}_n$, we have ensured that $P'_{n+1}(z) \neq 0$ everywhere along the corresponding path C_z. Therefore, progress toward z along C_z cannot come to a premature halt before $\tilde{z}$ is reached. It is also worth mentioning that $\tilde{z}$, the sought-after zero of $P_{n+1}(z)$, cannot be infinitely far away from z_0, because, when $|z| \to \infty$, the function $P_{n+1}(z)$, which rapidly approaches $c_{n+1}z^{n+1}$, can no longer vanish. This indicates that $P_{n+1}(z)$ does not have a zero outside a sufficiently large circle centered at $z = 0$.

Digression: In no way does the computational accuracy needed to construct the trajectory C_z in a practical setting impact the validity of the preceding proof of the fundamental theorem. For numerical calculations, however, although Eq.(60) offers a simple recipe for computing successive points z along C_z, there exist other, more efficient ways of moving along this trajectory toward the desired destination $\tilde{z}$. For instance, suppose that, in addition to $P'_{n+1}(z)$, one also evaluates $P''_{n+1}(z)$ at each point z along the trajectory. Moreover, assume that the zeros $\bar{\bar{z}}_1, \bar{\bar{z}}_2, \cdots, \bar{\bar{z}}_{n-1}$ of $P''_{n+1}(z)$, an $(n-1)^{\text{st}}$-order polynomial, are identified, and that the chosen trajectory C_w properly avoids their corresponding projections $\bar{\bar{w}}_1, \bar{\bar{w}}_2, \cdots, \bar{\bar{w}}_{n-1}$ onto the w-plane. One may then choose Δz in such a way as to satisfy the identity

$$P_{n+1}(z+\Delta z) = P_{n+1}(z) + P'_{n+1}(z)\Delta z + \tfrac{1}{2}P''_{n+1}(z)(\Delta z)^2. \tag{61}$$

The two solutions of this quadratic equation are readily obtained as follows:

$$\Delta z = -\frac{P'_{n+1}(z)}{P''_{n+1}(z)} \pm \left\{\left[\frac{P'_{n+1}(z)}{P''_{n+1}(z)}\right]^2 + \frac{2[P_{n+1}(z+\Delta z) - P_{n+1}(z)]}{P''_{n+1}(z)}\right\}^{1/2}. \tag{62}$$

Considering that $\Delta w = P_{n+1}(z+\Delta z) - P_{n+1}(z)$, the incremental step along C_w, is a small complex number, of the two solutions for Δz given by Eq.(62), only the smaller solution will be acceptable. One may thus use Eq.(62) instead of Eq.(60) to construct C_z in a more efficient way.

Example. Let $f(z) = P_m(z)/P_n(z)$ be the ratio of two polynomials, one of order m, the other of order n, where $0 \le m \le n-2$. Show that the sum of the residues of $f(z)$ is equal to zero.

Proof: When $|z| \to \infty$, the absolute value of the function $f(z)$ approaches zero in proportion to $|z|^{m-n}$, which is faster than $1/|z|$. Consequently, the integral of $f(z)$ around a circle of radius R centered at the origin of the z-plane will approach zero when $R \to \infty$. The poles of $f(z)$, which are the n roots of the polynomial equation $P_n(z) = 0$, will all be inside the circle. In what follows, we shall assume for the sake of simplicity that the n roots of the polynomial are distinct from one another. Thus, in the immediate vicinity of the kth root, z_k, we may write

$$P_n(z) \cong P_n(z_k) + P_n'(z_k)(z - z_k) = P_n'(z_k)(z - z_k). \tag{63}$$

The residue associated with z_k is thus found to be $P_m(z_k)/P_n'(z_k)$, and, in accordance with the residue theorem, since the integral of $f(z)$ around the large circle is zero, we will have

$$\sum_{k=1}^{n} \frac{P_m(z_k)}{P_n'(z_k)} = 0. \tag{64}$$

As a simple example, consider the function $f(z) = z/[(z-z_1)(z-z_2)(z-z_3)]$, which has three simple poles at $z = z_1$, $z = z_2$, and $z = z_3$. We will have,

$$\begin{aligned}\text{sum of residues} &= \frac{z_1}{(z_1-z_2)(z_1-z_3)} + \frac{z_2}{(z_2-z_1)(z_2-z_3)} + \frac{z_3}{(z_3-z_1)(z_3-z_2)}\\ &= \frac{z_1(z_2-z_3)-z_2(z_1-z_3)+z_3(z_1-z_2)}{(z_1-z_2)(z_1-z_3)(z_2-z_3)}\\ &= \frac{z_1z_2-z_1z_3-z_2z_1+z_2z_3+z_3z_1-z_3z_2}{(z_1-z_2)(z_1-z_3)(z_2-z_3)} = 0.\end{aligned} \tag{65}$$

The above theorem is far more general than stated here for the ratios of polynomials, and applies to *any* complex function $f(z)$ which is analytic everywhere except at its isolated poles, and which declines sufficiently rapidly when $|z| \to \infty$ that its integral over a large circle centered at the origin would vanish. Under these circumstances, the sum of the residues of $f(z)$ at its points of singularity will always add up to zero.

6.11. Cauchy's Argument Principle. Consider the simple analytic function $f_1(z) = (z - z_0)^n$, where z_0 is an arbitrary point in the z-plane, while $n > 0$ is an arbitrary positive integer. Let the closed contour C, which encloses the point z_0, be traversed counterclockwise. Clearly, the phase of $f_1(z)$ changes by $2\pi n$ during each round trip around C. Thus, dividing the accumulated phase (in a single round trip around z_0) by 2π, yields the order n of the zero of the function $f_1(z)$. Note that the accumulated phase (during a round trip) would be zero if the contour C did not enclose z_0. Similarly, the function $f_2(z) = (z - z_0)^n(z - z_1)^m$, where both n and m are positive integers, has the property that, its accumulated phase in a single round trip along a contour that encloses both z_0 and z_1, is equal to $2\pi(n + m)$. Generalization of this concept to polynomial functions that possess an arbitrary number of zeros in the complex plane should be obvious.

Next, consider the function $f_3(z) = 1/(z - z_0)^n$, which is analytic everywhere except at its single n^{th}-order pole at $z = z_0$. Let the closed contour C surround the pole z_0

of $f_3(z)$. The accumulated phase during a single counterclockwise round trip along C would then be equal to $-2\pi n$. Needless to say, if the pole happens to be outside the closed contour C, then the accumulated phase would be zero. The argument can be readily extended to any function that can be expressed as the ratio of two arbitrary polynomials, namely $f(z) = P(z)/Q(z)$. Explicitly, a closed contour that contains some or all of the poles and zeros of $f(z)$, when traversed counterclockwise, would accumulate (in a single round trip) a total phase equal to 2π times the difference between the number of enclosed poles and zeros—with pole and zero multiplicities properly taken into account.

Suppose now that an analytic function with no poles and zeros, say, $f(z) = \exp(az)$, is under consideration. If we write $a = a' + \mathrm{i}a''$ and $z = x + \mathrm{i}y$, then the phase of $f(z)$ could be written as $\varphi(z) = a''x + a'y$. In a single round trip around any closed contour C, the shadow on the x-axis of the traveling point will move to the right and to the left, but will eventually return to its starting point; therefore, $a''x$ will have a cumulative value of zero at the end of the round trip. Similarly, the projection on the y-axis of the traveling point will move up and down but will eventually return to its starting point, so the accumulated change in $a'y$ will also be zero. We conclude that the overall phase of $f(z)$ does not change at all after a round trip along C.

In contrast to the preceding examples, let $f(z) = \ln(z)$, which is *not* analytic on a branch-cut, say, $\mathrm{Re}(z) \le 0$, $\mathrm{Im}(z) = 0$. For a simple contour, such as a circle of radius $R > 1$ centered at the origin of the z-plane, the value of the function at $z = R\exp(\mathrm{i}\theta)$ will be $f(z) = \ln R + \mathrm{i}\theta$. Traversing the circle counterclockwise from slightly below the branch-cut all the way to the opposite point located slightly above (or just on) the branch-cut, the value of θ would change continuously from $-\pi$ to π, causing the phase of $f(z)$ to rise from $-\tan^{-1}(\pi/\ln R)$ to $\tan^{-1}(\pi/\ln R)$. Not only does the accumulated phase in this case depend on the shape and size of the chosen contour, but also the resulting accumulated phase conveys no information about the poles and zeros of the function. In general, Cauchy's *argument principle* does *not* apply to functions that have branch-cuts or, excepting functions that possess isolated poles, fail the test of analyticity on and within the closed contour C.

The general proof of the *argument principle* for $f(z) = |f(z)|\exp[\mathrm{i}\varphi(z)]$ that is analytic not only on the closed contour C but also, aside from isolated poles, inside the contour, relies on the properties of the function $f'(z)/f(z)$, whose singularities are the poles and zeros of $f(z)$. Considering that $f'(z)/f(z) = \mathrm{d}\ln[f(z)]/\mathrm{d}z$, the closed-contour integral $\oint_C [f'(z)/f(z)]\mathrm{d}z$ is the sum of the incremental changes in $\ln[f(z)] = \ln|f(z)| + \mathrm{i}\varphi(z)$ during a single round trip along C. Since $|f(z)|$ is finite and continuous everywhere on C, the total round trip change in $\ln|f(z)|$ will be zero. Therefore, the overall integral will be i times the cumulative change in the phase $\varphi(z)$ of the function $f(z)$.

Now, in accordance with the Cauchy-Goursat theorem, the contour integral is also equal to $2\pi\mathrm{i}$ times the sum of the residues of $f'(z)/f(z)$ inside C. All that is left to prove, therefore, is that the function $f'(z)/f(z)$ has a residue equal to n at an nth-order zero of $f(z)$, and a residue equal to $-m$ at an mth-order pole of $f(z)$. To this end, we write $f(z) = (z - z_0)^n g(z)$ in the vicinity of an nth-order zero z_0, and also $f(z) = h(z)/(z - z_1)^m$ in the vicinity of an mth-order pole z_1, where the remainder functions $g(z)$ and $h(z)$ are analytic at and in the immediate neighborhood of z_0 and z_1, respectively. We now write

$$\text{at and around } z_0: \quad \frac{f'(z)}{f(z)} = \frac{n(z-z_0)^{n-1}g(z) + (z-z_0)^n g'(z)}{(z-z_0)^n g(z)} = \frac{n}{z-z_0} + \frac{g'(z)}{g(z)}, \tag{66}$$

at and around z_1: $$\frac{f'(z)}{f(z)} = \frac{-m(z-z_1)^{-m-1}h(z) + (z-z_1)^{-m}h'(z)}{(z-z_1)^{-m}h(z)} = -\frac{m}{z-z_1} + \frac{h'(z)}{h(z)}. \qquad (67)$$

Clearly, the residue of $f'(z)/f(z)$ at z_0 is n, while that at z_1 is $-m$. The proof is thus complete.

References

1. Stephen Hawking, *God Created the Integers: The Mathematical Breakthroughs That Changed History*, 2nd edition, Running Press, Philadelphia (2007).
2. J. Gray, *The Real and the Complex: A History of Analysis in the 19th Century*, Springer, Switzerland (2015).
3. Eric Temple Bell, *Men of Mathematics*, Simon and Schuster, New York (1986).
4. R.V. Churchill and J.W. Brown, *Complex Variables and Applications*, 6th edition, McGraw-Hill, New York (1995).
5. G.B. Arfken and H.J. Weber, *Mathematical Methods for Physicists*, 6th edition, Academic Press, Burlington, MA (2005).
6. F.B. Hildebrand, *Advanced Calculus for Applications*, 2nd edition, Prentice-Hall, New Jersey (1976).
7. J.H. Mathews and R.W. Howell, *Complex Analysis for Mathematics and Engineering*, 5th edition, Jones & Bartlett, Sudbury, Massachusetts (2006).
8. J. Mathews and R.L. Walker, *Mathematical Methods of Physics*, 2nd edition, Benjamin/Cummings Publishing, California (1970).
9. G. Stephenson and P.M. Radmore, *Advanced Mathematical Methods for Engineering and Science Students*, Cambridge University Press, United Kingdom (1990).
10. W. Kaplan, *Advanced Mathematics for Engineers*, Addison-Wesley Publishing Co., Reading, Massachusetts (1981).
11. R. Courant and D. Hilbert, *Methods of Mathematical Physics*, volume I, Interscience, New York (1937).
12. I.S. Gradshteyn and I.M. Ryzhik, *Table of Integrals, Series, and Products*, 7th edition, Academic Press, Burlington, MA (2007).
13. J.D. Jackson, *Classical Electrodynamics*, 3rd edition, Wiley, New York (1999); see Eq.(7.122).
14. M. Born and E. Wolf, *Principles of Optics*, 7th (expanded) edition, Cambridge University Press, Cambridge, United Kingdom (1999).
15. M. Mansuripur, *Field, Force, Energy and Momentum in Classical Electrodynamics* (Revised Edition), Bentham Science Publishers, Sharjah, UAE (2017); see chapter 6.

Problems

Problem 1) Show that the third-order polynomial $f(z) = az^3 + bz^2 + cz + d$, where a, b, c, and d are arbitrary complex coefficients, is *analytic* at all points z in the complex plane. (In other words, show that the Cauchy-Riemann conditions are satisfied at all z.)

Problem 2) Use the Cauchy-Riemann conditions to determine the domain of analyticity of the function $f(z) = 1/(z^2 + 1)$.

Problem 3) Using the definition of the derivative for functions of a complex variable z, show that $f(z) = z^n$, where n is an arbitrary positive integer, is analytic throughout the complex plane.

Problem 4) Using the method of proof by induction, show that any N^{th} degree polynomial, namely $f(z) = \sum_{n=0}^{N} a_n z^n$, where N is a positive integer and a_n are arbitrary complex coefficients, is analytic throughout the entire complex plane.

Problem 5) Show that the following functions satisfy the Cauchy-Riemann conditions:

a) $f(z) = 1/(z - z_0)$; $z \neq z_0$, where z_0 is a fixed point in the complex plane.

b) $g(z) = \exp(\alpha z)$; where α is an arbitrary complex constant.

c) $h(z) = \ln(\beta z)$; where β is an arbitrary complex constant and $z = |z| \exp(\mathrm{i}\phi)$, with $-\pi < \phi \leq \pi$.

In each case, find the derivative of the function in the region where it exists.

Problem 6) Writing $f(re^{\mathrm{i}\phi}) = R(r,\phi) \exp[\mathrm{i}\Phi(r,\phi)]$, where $R(r,\phi)$ and $\Phi(r,\phi)$ are real valued and differentiable functions of r and ϕ, show that the Cauchy-Riemann conditions in polar coordinates become $\partial_r R = (R/r)\partial_\phi \Phi$ and $(1/r)\partial_\phi R = -R\partial_r \Phi$.

Problem 7) Find the analytic function $f(z) = u(x,y) + \mathrm{i}v(x,y)$, where

a) $u(x,y) = x^3 - 3xy^2$;

b) $v(x,y) = \exp(-y) \sin x$.

Problem 8) a) The function $f(z)$ is differentiable in the complex plane along a given path that starts at z_a and ends at z_b. Show that the integral of $f'(z)$ along the path, namely, $\int_{z_a}^{z_b} f'(z)dz$, is equal to $f(z_b) - f(z_a)$. In particular, if the path is a closed loop C, show that $\oint_C f'(z)dz = 0$.

b) Let $f(z)$ and $g(z)$ be differentiable functions along a given path from z_a to z_b. Show that the method of integration by parts applies along the given path; that is,

$$\int_{z_a}^{z_b} f(z)g'(z)dz = [f(z_b)g(z_b) - f(z_a)g(z_a)] - \int_{z_a}^{z_b} f'(z)g(z)dz.$$

In particular, if the path is the closed loop C, show that the following equality will hold:

$$\oint_C f(z)g'(z)dz = -\oint_C f'(z)g(z)dz.$$

c) Verify the result obtained in part (b) by evaluating closed-loop integrals of the functions $f(z)g'(z)$ and $f'(z)g(z)$ where $f(z) = \exp(az)$ and $g(z) = 1/(z - z_0)$.

Problem 9) Show that $\oint_C (z^2 + z)^{-1} dz = 0$, where C is a circle of radius $R > 1$, centered at the origin.

Problem 10) Evaluate $\oint_C (z^2 - 1)^{-1} dz$, where C is a circle of radius 2 centered at the origin.

Problem 11) For real-valued parameters a and b, where $a > |b|$, show that

$$\int_0^{2\pi} \frac{d\theta}{a + b\cos\theta} = \int_0^{2\pi} \frac{d\theta}{a + b\sin\theta} = 2\pi/\sqrt{a^2 - b^2}.$$

Problem 12) Show that $\int_0^{2\pi} (1 - 2t\cos\theta + t^2)^{-1} d\theta = 2\pi/(1 - t^2)$ when $|t| < 1$. What happens if $|t| > 1$? What happens if $|t| = 1$?

Problem 13) Show that $(a > 0)$:

a) $$\int_{-\infty}^{\infty} \frac{\cos x}{x^2 + a^2} dx = (\pi/a)\exp(-a).$$

b) $$\int_{-\infty}^{\infty} \frac{x\sin x}{x^2 + a^2} dx = \pi\exp(-a).$$

How is the right-hand side of each equation modified if $\cos x$ in (a) is replaced by $\cos(kx)$, and $\sin x$ in (b) is replaced by $\sin(kx)$?

Problem 14) Show that $\int_0^{\infty} \sin(x^2)\, dx = \int_0^{\infty} \cos(x^2)\, dx = \sqrt{\pi}/(2\sqrt{2})$.

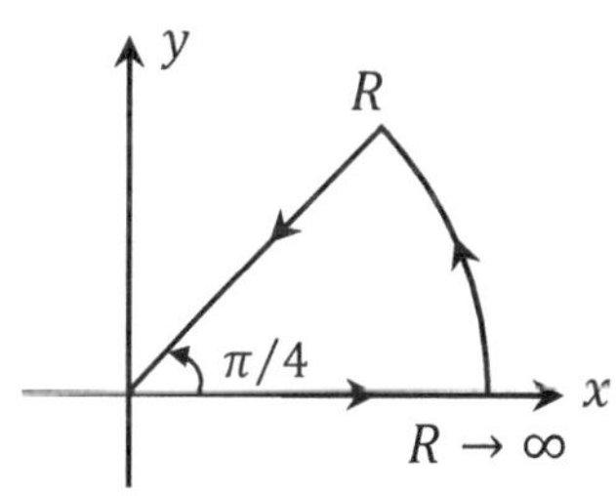

Hint: Try the contour shown in the figure.

Problem 15) Show that $\int_{-\infty}^{\infty} \frac{x^2}{1+x^4} dx = \pi/\sqrt{2}$.

Problem 16) Let $f(z)$ be analytic at $z = z_0$. Assuming that z_0 is a simple zero of $f(z)$, that is, $f(z) = (z - z_0)g(z)$ and $g(z_0) \neq 0$, evaluate $(z - z_0)/f(z)$ at $z = z_0$ *without* an explicit determination of $g(z)$. Check your result for $f(z) = z^3 + 1$ at $z_0 = -1$.

Problem 17) Show that $\int_0^{\pi} \frac{d\theta}{(a + \cos\theta)^2} = \frac{\pi a}{(a^2-1)^{3/2}}$; $a > 1$.

Problem 18) Show that $\int_0^{\infty} \frac{dx}{(x^2 + a^2)^2} = \frac{\pi}{4a^3}$; $a > 0$.

Problem 19) Show that $\int_{-\infty}^{\infty} \frac{\cos(bx) - \cos(ax)}{x^2} dx = \pi(a - b)$; $a > b > 0$.

Problem 20) Show that $\int_0^\infty \frac{dx}{1+x^n} = \frac{\pi/n}{\sin(\pi/n)}$, where $n \geq 2$ is an integer.

Hint: Try the contour shown in the figure. Also use the result obtained in problem 16 to determine the residue at the single pole located inside the contour.

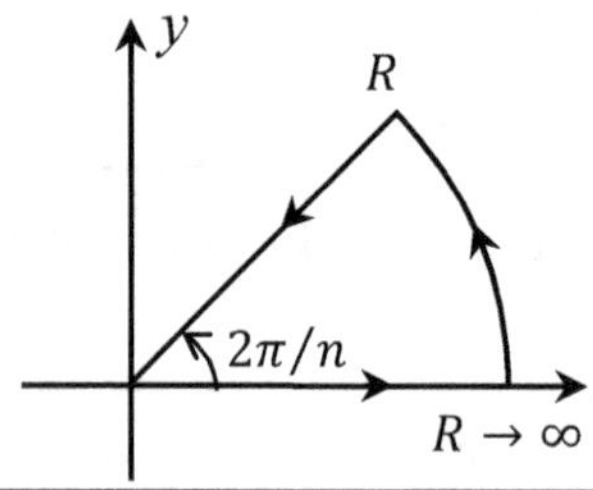

Problem 21) Show that the function $f(z) = z^4 - 2\cos(2\theta)z^2 + 1$ has zeros $z_{1,2} = \pm\exp(i\theta)$ and $z_{3,4} = \pm\exp(-i\theta)$. Use this result to confirm that

a) $$\int_{-\infty}^{\infty} \frac{dx}{x^4 - 2\cos(2\theta)x^2 + 1} = \frac{\pi}{2\sin\theta},$$

b) $$\int_{-\infty}^{\infty} \frac{x^2 dx}{x^4 - 2\cos(2\theta)x^2 + 1} = \frac{\pi}{2\sin\theta}.$$

Problem 22) Evaluate $\int_0^\infty [(\ln t)^2/(1+t^2)]dt$ using an appropriate series expansion of the integrand and show that the integral is equal to $4\sum_{n=0}^\infty [(-1)^n/(2n+1)^3]$. Next, make a change of variable $x = \ln(t)$ to rewrite the integral as ½ $\int_{-\infty}^\infty (x^2/\cosh x)dx$. Using the contour shown in the figure, confirm that the integral is equal to $\pi^3/8$.

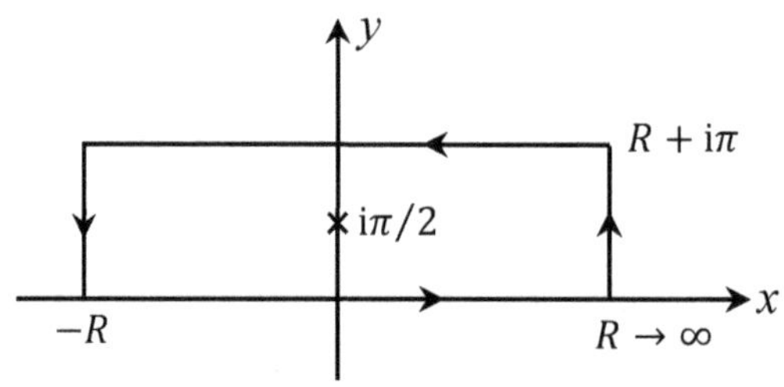

Problem 23) Using the contour shown in the figure, and noting that $z = 0$ is a branch point while the positive x-axis is a cut line, verify the following integrals:

a) $$\int_0^\infty \frac{x^{-a}}{x+1}dx = \frac{\pi}{\sin(\pi a)}; \qquad 0 < a < 1,$$

b) $$\int_0^\infty \frac{x^a}{(x+1)^2}dx = \frac{\pi a}{\sin(\pi a)}; \quad -1 < a < 1.$$

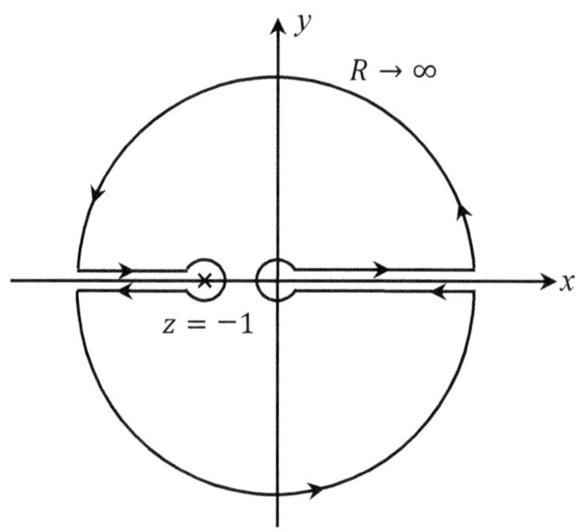

Problem 24) Using the contour shown, verify that $\int_0^\infty [x/(x^4+1)]dx = \pi/4$.

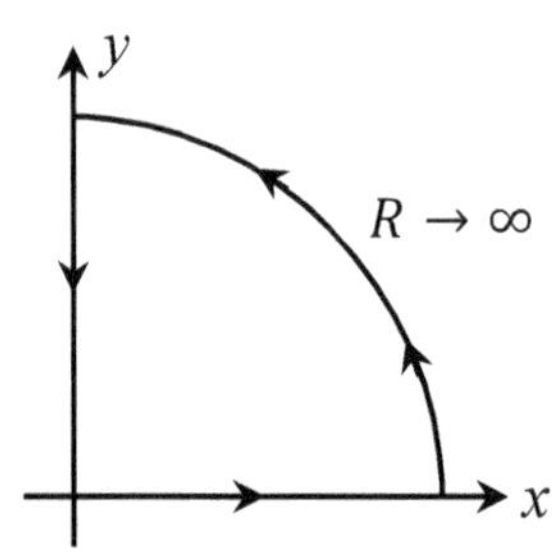

Problem 25) Use residue calculus to show the following:

a) $\int_0^{\pi/2} \sin^4\theta \, d\theta = \int_0^{\pi/2} \cos^4\theta \, d\theta = 3\pi/16.$

b) $\displaystyle\int_0^{2\pi} \frac{\sin^2\theta}{5+4\cos\theta} d\theta = \pi/4.$

c) $\displaystyle\int_0^{2\pi} \frac{d\theta}{a^2+\sin^2\theta} = \int_0^{2\pi} \frac{d\theta}{a^2+\cos^2\theta} = \frac{2\pi}{a\sqrt{a^2+1}}; \quad a > 0.$

d) $\displaystyle\int_0^{\infty} \frac{dx}{x^4+4a^4} = \frac{\pi}{8a^3}; \quad a > 0.$

e) $\displaystyle\int_0^{\infty} \frac{dx}{(x^2+a^2)^2} = \frac{\pi}{4a^3}; \quad a > 0.$

Problem 26) Use the residue calculus to show that

$$\int_0^{\pi} \cos^{2n}\theta \, d\theta = \pi \frac{(2n)!}{2^{2n}(n!)^2} = \pi \frac{(2n-1)!!}{(2n)!!}, \qquad n = 0, 1, 2, \cdots.$$

Hint: Use the fact that $\cos\theta = (z + z^{-1})/2$ on the unit-circle. Also helpful is the binomial expansion.

Problem 27) Show that $\displaystyle\int_{-\infty}^{\infty} \frac{\cos x}{\pi^2 - 4x^2} dx = ½$. (Note that the integrand does *not* diverge at $x = \pm\pi/2$, because the numerator, $\cos x$, also vanishes at these points.)

Problem 28) Use residue calculus to show that

$$\int_0^{\infty} \frac{\cos(ax)}{x^4+4b^4} dx = \frac{\pi}{8b^3} \exp(-ab)\,[\cos(ab) + \sin(ab)]; \quad a \ge 0, \;\; b > 0.$$

Use this result to derive a formula for the integral $\displaystyle\int_0^{\infty} \frac{\cos(ax)}{(x^4+4b^4)^2} dx$.

Problem 29) Use residue calculus to show that

$$\int_{-\infty}^{\infty} \frac{x^2 dx}{(x^2+1)[x^2 - 2x\cos(\omega) + 1]} = \frac{\pi}{2|\sin\omega|}; \qquad \omega \text{ real}, \;\; \sin(\omega) \ne 0.$$

Problem 30) Use residue calculus to show that

$$\int_{-\infty}^{\infty} \frac{\cos(ax)}{(x+b)^2+c^2} dx = (\pi/c)\exp(-ac)\cos(ab); \quad a \ge 0, \;\; c > 0.$$

Differentiating both sides of this identity with respect to *a* and *c*, deduce that

$$\int_{-\infty}^{\infty} \frac{x\sin(ax)}{(x+b)^2+c^2} dx = (\pi/c)\exp(-ac)\,[c\cos(ab) + b\sin(ab)]; \quad a > 0, \;\; c > 0.$$

$$\int_{-\infty}^{\infty} \frac{\cos(ax)}{[(x+b)^2+c^2]^2} dx = \left(\frac{\pi}{2c^3}\right)(1+ac)\exp(-ac)\cos(ab); \qquad a \ge 0, \;\; c > 0.$$

Problem 31) The function $f(z)$ is analytic within the complex plane except at a finite number of poles, none of which is located at the origin or on the positive real axis. Also, the function $f(z)\ln(z)$ declines sufficiently rapidly with increasing z that its integral over a circle of radius R approaches zero when $R \to \infty$. Using the positive real axis as a branch-cut for $f(z)\ln(z)$, show that

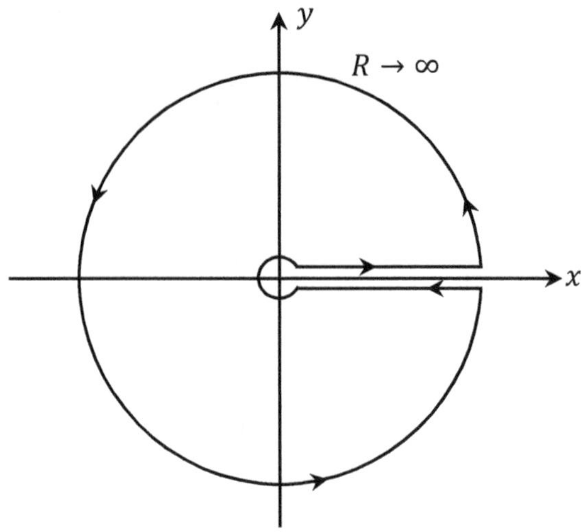

$$\int_0^\infty f(x)dx = -\sum_{\text{all singularities}} \text{Residues of } f(z)\ln(z).$$

Use the above result to verify the following:

a) $$\int_0^\infty \frac{dx}{x^3+a^3} = \frac{2\pi\sqrt{3}}{9a^2}; \qquad a > 0.$$

b) $$\int_0^\infty \frac{dx}{(x+a)(x^2+b^2)} = \frac{\pi a + 2b\ln(b/a)}{2b(a^2+b^2)}; \qquad a > 0,\ b > 0.$$

Problem 32) Hyperbolic sine and cosine functions are defined in the complex z-plane as follows:

$$\cosh(z) = ½[\exp(z) + \exp(-z)]; \qquad \sinh(z) = ½[\exp(z) - \exp(-z)].$$

a) Writing $z = x + \mathrm{i}y$, show that $\cosh(z) = \cosh(x)\cos(y) + \mathrm{i}\sinh(x)\sin(y)$.

b) Find all the zeros of $\cosh(z)$, namely, all the points in the z-plane where $\cosh(z) = 0$.

c) On the horizontal line $z = x + \mathrm{i}\pi$, show that $\cosh(z) = \cosh(x + \mathrm{i}\pi) = -\cosh(x)$.

d) Using complex plane techniques based on the Cauchy-Goursat theorem, prove the validity of the following definite integrals:

(i) $\displaystyle\int_{-\infty}^{\infty} \frac{1}{\cosh(x)}dx = \pi.$ (ii) $\displaystyle\int_{-\infty}^{\infty} \frac{x}{\cosh(x)}dx = 0.$ (iii) $\displaystyle\int_{-\infty}^{\infty} \frac{x^2}{\cosh(x)}dx = \pi^3/4.$

Problem 33) Use the complex plane integration techniques to verify the stated value of the following integral:

$$\int_0^\infty \frac{\cos(\alpha x)}{\cosh(\beta x)}dx = \frac{\pi/2\beta}{\cosh(\pi\alpha/2\beta)}; \qquad \text{Re}(\beta) > 0,\ \alpha \text{ any real number.}$$

Hint: Considering that the integrand $\cos(\alpha x)/\cosh(\beta x)$ is an even function of x, the domain of the integral can be extended from $(0,\infty)$ to $(-\infty,\infty)$. Also, the identity $\cos(\alpha x) = ½[\exp(\mathrm{i}\alpha x) + \exp(-\mathrm{i}\alpha x)]$ may be useful.

Problem 34) Use the Cauchy-Riemann conditions to determine if and where the function $f(z) = f(x+\mathrm{i}y) = (x-y)^2 + 2\mathrm{i}(x+y)$ is differentiable. What is the region, if any, within which $f(z)$ is analytic?

Problem 35) The complex function $f(z) = u(r,\theta) + \mathrm{i}v(r,\theta)$ is defined in a proper region within the complex plane $z = r\exp(\mathrm{i}\theta)$.

a) Derive the Cauchy-Riemann conditions in terms of the partial derivatives of $u(r,\theta)$ and $v(r,\theta)$ with respect to the polar coordinates r and θ.

b) Use the result in part (a) to show that the function $f(z) = z^{1/2}$ is analytic everywhere in the complex plane except on a branch-cut.

c) Identify an appropriate branch-cut for the function $f(z) = z^{1/2}$.

Problem 36) Use residue calculus to show that

$$\int_0^\infty \frac{dx}{(x^2+a^2)(x^2+b^2)} = \frac{\pi}{2ab(a+b)}; \qquad a > 0, \quad b > 0.$$

Problem 37) Use the Cauchy-Riemann conditions to determine the domain of analyticity of each of the following functions:

a) $f_1(z) = 1/(z - z_0)$;

b) $f_2(z) = \exp(z^2)$.

Problem 38) Use the Cauchy-Riemann conditions to determine the domain of analyticity of each of the following functions:

a) $f_1(z) = \cos z$.

b) $f_2(z) = 1/[1 + \exp(z)]$.

Problem 39) a) Invoking the Cauchy-Riemann conditions, demonstrate that the function $f(z) = \exp(-z^2)$ is analytic throughout the entire complex z-plane.

b) Find the derivative $f'(z)$ of $f(z)$ at the arbitrarily chosen point $z = z_0$.

Problem 40) Let $f(z) = p(x,y) + \mathrm{i}q(x,y)$ and $g(z) = r(x,y) + \mathrm{i}s(x,y)$ be complex functions of the complex variable $z = x + \mathrm{i}y$. Both functions are defined at $z = z_0$ and are differentiable at that point. Show that the function $h(z) = f(z)g(z)$ is also differentiable at $z = z_0$ by confirming that $h(z)$ satisfies the Cauchy-Riemann conditions.

Problem 41) Confirm the following integral (tabulated in Gradshteyn & Ryzhik as **3.471**-15):

$$\int_0^\infty x^{-1/2}\exp(-px - q/x)\,\mathrm{d}x = \sqrt{\pi/p}\,\exp\!\left(-2\sqrt{pq}\right); \qquad \mathrm{Re}(p) > 0,\ \mathrm{Re}(q) \geq 0.$$

Hint: You may want to start by symmetrizing the integrand using complex-plane techniques, so that the integral would reduce to $2(q/p)^{1/4}\exp(-2\sqrt{pq})\int_0^\infty \exp\!\left[-\sqrt{pq}(x - x^{-1})^2\right]\mathrm{d}x$.

Problem 42) Use complex-plane integration methods to evaluate the following integral:

$$\int_{-\infty}^\infty \frac{\mathrm{d}x}{(x^2-1)(x^2-2\mathrm{i}x-2)}.$$

The contour of integration may be closed in either the upper-half or the lower-half of the complex z-plane. Confirm that the result of integration is independent of the half-plane chosen to close the contour.

Chapter 7

Fourier Transformation

"A new scientific truth does not triumph by convincing its opponents and making them see the light, but rather because its opponents eventually die, and a new generation grows up that is familiar with it."

Max Planck (1858–1947)

7.1. Introduction. On December 21, 1807, Jean-Baptiste Joseph Fourier (1768–1830) presented to the French Academy the theory that, decades later, would become synonymous with his name. Fourier's paper was criticized by the likes of Lagrange, Poisson, Legendre, and Laplace for lack of mathematical rigor and generality and was rejected for publication in the *Memoirs of the Academy*. Fourier eventually published his groundbreaking results in his book, *Théorie Analytique de la Chaleur* (The Analytic Theory of Heat), in 1822—a few years after Lagrange had passed away.

The principal idea behind Fourier's theory is that essentially all functions of practical interest in physical sciences can be represented by superpositions of sines and cosines of differing amplitudes and frequencies. Complex analysis allows Fourier transform theory to be simplified and presented in terms of integrals of functions that are modulated (i.e., multiplied) by complex exponentials of the form $\exp(\pm \mathrm{i}2\pi sx)$. A complex function of a real variable, say, $f(x)$, has a Fourier transform that is, in general, another complex function of a different real variable, say, $F(s)$. One may then treat $f(x)$ as a superposition of complex exponentials $\exp(\mathrm{i}2\pi sx)$, with the parameter s being the frequency of oscillations along the x-axis. The complex amplitude associated with each frequency s is given by $F(s)$. Every function $f(x)$ has its own unique Fourier transform $F(s)$, and every function $F(s)$ uniquely identifies the corresponding function $f(x)$ via an inverse Fourier transformation.

Many natural phenomena can be readily understood and explained in terms of the Fourier transform theory. For example, when white light breaks up into its rainbow colors (with the aid of a prism, a diffraction grating, or a water droplet), each emerging color represents a Fourier component (i.e., sine, cosine, or complex exponential) of the white light whose overall amplitude as a function of time may be obtained by adding up the sinusoidal waveforms associated with the individual colors. In other words, these sinusoids of differing amplitude and frequency—which the human visual system perceives as colors—collectively constitute the white light. The use of Fourier transformation simplifies the analysis of many physical problems, as it is often the case that solving the governing equations of a linear system is much easier for individual sines and cosines (or for complex exponentials in general), than it is for arbitrary functions.

The present chapter begins with the introduction, in section 2, of certain special functions that find frequent application in the theory of Fourier transforms. Of particular interest, here, is Dirac's delta-function, $\delta(x)$, which will be treated in greater detail in section 3. The Fourier integral and its inverse are introduced in section 4, where we provide a few examples in which the transforms of several important functions are derived. Section 5 is devoted to a brief proof of Fourier's fundamental theorem, which asserts that the inverse Fourier integral does in fact reconstitute the original function, $f(x)$, after its initial conversion, via Fourier transformation, to $F(s)$. In section 6, we state, and then prove, several interesting as well as useful theorems that shed light on important characteristics of the Fourier transformation. In addition to providing insight into the inner workings of the Fourier transform operation, these theorems enable one to determine the transforms of certain functions without the need to invoke the full machinery of evaluating complex integrals.

A especially useful function in the Fourier transform theory is the comb function, $\text{comb}(x)$, which is a periodic array of unit delta functions. We derive the Fourier transform of the comb function in section 7, then use it to arrive at the Fourier series expansion of periodic functions in section 8, and also to examine the sampling theorem and the corresponding Nyquist criterion in section 9.

Generalization of the Fourier transform theory to higher-dimensional spaces is the subject of section 10. Then, in section 11, we rely on Fourier transformation in a two-dimensional space to describe a rudimentary version of the classical theory of optical diffraction—the theory that forms a basis for studies of Fourier optics. Although this book does not delve into Fourier optics to any great depth, we find it useful to discuss, in section 12, the method of stationary-phase approximation. It turns out that, in many situations arising in diffraction theory and in Fourier optics, the requisite integrals are difficult to evaluate—even when numerical techniques and powerful computers are summoned. It is fortunate that the relevant integrals in such circumstances can be evaluated, albeit approximately, with relative ease and with hardly any computation at all. Section 12 presents the essential ideas and the general approach to stationary-phase approximation, first in a one-dimensional setting, and, subsequently, in the case of two-dimensional spaces, such as those encountered in Fourier optics. A related—although different in many respects—method of approximating certain integrals is the saddle-point technique, which is discussed in the final section of the chapter.

7.2. Special functions. The functions $\text{step}(x)$, $\text{sign}(x)$, $\text{rect}(x)$, $\text{tri}(x)$, $\text{sinc}(x)$, and the Dirac delta-function $\delta(x)$ are defined as follows:

Step function: $$\text{step}(x) = \begin{cases} 0, & x < 0; \\ 1, & x > 0. \end{cases} \qquad (1)$$

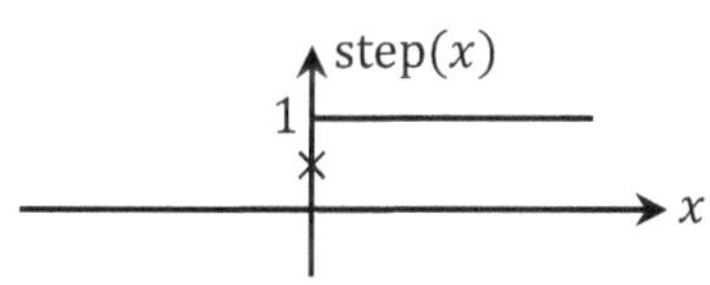

At $x = 0$, the step function may be assigned any arbitrary value without affecting its fundamental properties. It is customary to define the value of the step function at $x = 0$ as ½, although, strictly speaking, this is not necessary.

Sign function: $$\text{sign}(x) = \begin{cases} -1, & x < 0; \\ +1, & x > 0. \end{cases} \qquad (2)$$

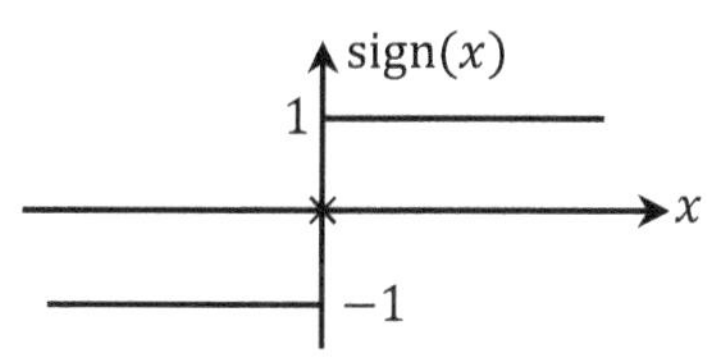

At $x = 0$, the sign function is usually assigned the value of zero, although, strictly speaking, this is not necessary and any arbitrary value assigned to this point of discontinuity of the function will be acceptable.

Rectangular pulse: $$\text{rect}(x) = \begin{cases} 0, & |x| > ½; \\ 1, & |x| < ½. \end{cases} \qquad (3)$$

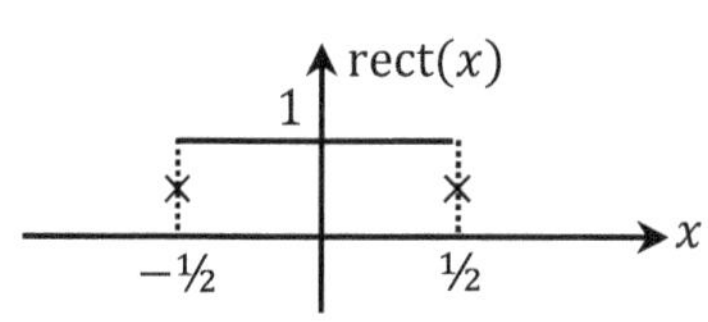

The area under the rect function is equal to 1. At the points of discontinuity, the function may be assigned arbitrary values, although it is customary to define the value of the function at these points as $\text{rect}(\pm½) = ½$.

Triangular pulse: $\mathrm{tri}(x) = \begin{cases} 0, & |x| > 1; \\ 1 - |x|, & |x| < 1. \end{cases}$ (4)

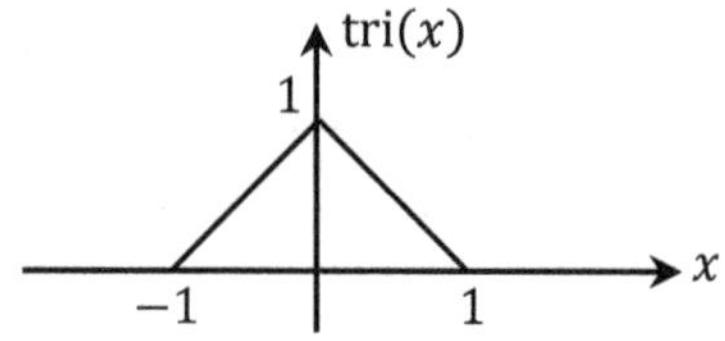

The area under $\mathrm{tri}(x)$ is equal to 1. The function is continuous and well-defined everywhere, although its derivative has discontinuities at $x = \pm 1$ and $x = 0$.

The sinc function: $\mathrm{sinc}(x) = \sin(\pi x)/(\pi x)$. (5)

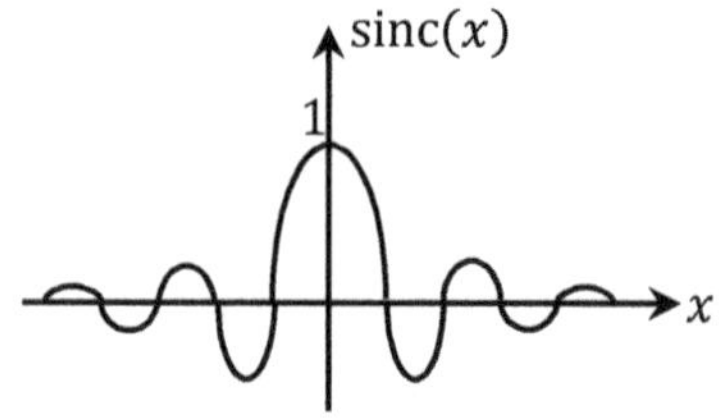

The function is equal to 1.0 at $x = 0$. Also, the area under $\mathrm{sinc}(x)$ is 1.0. The zero-crossings occur at integer values of x, namely at $x = \pm 1, \pm 2, \pm 3$, etc.

Dirac's delta-function: The delta-function $\delta(x)$ does *not* have a unique shape but *does* have unique properties. One may think of $\delta(x)$ as the limiting form of $\alpha^{-1}\mathrm{rect}(x/\alpha)$, where the real and positive parameter α becomes exceedingly small.[1†] Thus, $\delta(x)$ is a tall, narrow, even function of x, that is centered at the origin of the coordinates, $x = 0$, and has an area equal to 1; that is, $\int_{-\infty}^{\infty} \delta(x)\mathrm{d}x = 1$.

Alternative definitions of $\delta(x)$, which are frequently found to be useful as well, include $lim_{\alpha \to 0}\, \alpha^{-1}\mathrm{tri}(x/\alpha)$, or $\lim_{\alpha \to 0}\, \alpha^{-1}\mathrm{sinc}(x/\alpha)$, or $\lim_{\alpha \to 0}\, \alpha^{-1} \exp[-\pi(x/\alpha)^2]$ (see figure 7.1). Essentially any well-behaved even function $f(x)$ with an area equal to 1, when scaled in accordance with $\alpha^{-1}f(x/\alpha)$ for sufficiently small values of α, becomes a good approximation to $\delta(x)$.

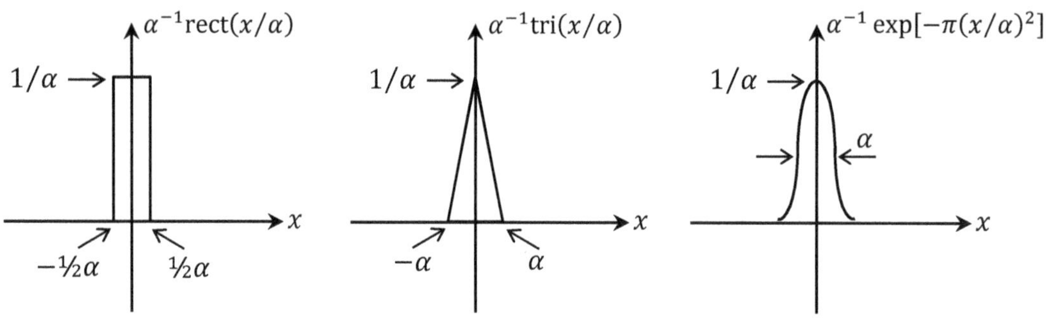

Figure 7.1. Any well-behaved function $f(x)$ that is even and has area equal to 1, when scaled as $\alpha^{-1}f(x/\alpha)$, becomes a good approximation to $\delta(x)$, provided that α is sufficiently small.

Another way of viewing the δ-function is as the derivative of the unit step function; that is, $\delta(x) = (\mathrm{d}/\mathrm{d}x)\mathrm{step}(x)$. The sharp transition at the midpoint of the step function (i.e., the jump from 0 at $x = 0^-$ to 1 at $x = 0^+$) can be smoothed out if one modifies the function ever so slightly by introducing a narrow transition region, say, from $x = -\alpha$ to $x = \alpha$, within which the step function acquires a large but finite slope (see the figure on the right). The derivative of the modified step function, which, for sufficiently small α, provides a good approximation to the δ-function, is now seen to be a tall, narrow, and even function of x, which has an area equal to 1.

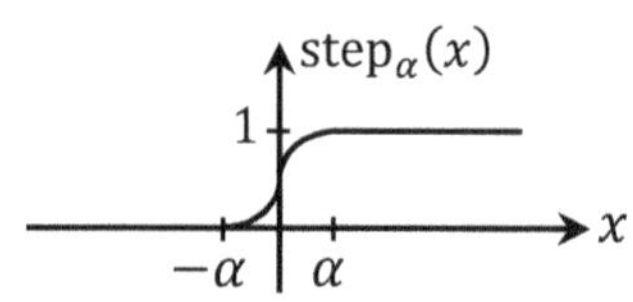

The most important property of $\delta(x)$ is its sifting property.[1] Suppose $g(x)$ is an arbitrary function of x that is continuous at $x = 0$ and has the value $g(0)$ at $x = 0$. If we

[†] Superscript numbers throughout the chapter identify references listed at the end of the chapter.

multiply $g(x)$ and $\delta(x)$ and proceed to determine the area under the product function, we will find $\int_{-\infty}^{\infty} g(x)\delta(x)\mathrm{d}x = g(0)$. If $g(x)$ happens to be discontinuous at $x = 0$, its multiplication into $\delta(x)$ yields a function whose area is given by $\int_{-\infty}^{\infty} g(x)\delta(x)\mathrm{d}x = ½[g(0^+) + g(0^-)]$. The sifting property of $\delta(x)$ is a direct consequence of its definition as the limiting form of a function that is tall, narrow, even, and has an area equal to 1.0.

7.3. Properties of $\boldsymbol{\delta(x)}$ and its derivatives. The properties of Dirac's δ-function are extremely important not only for the theory of Fourier transforms, but also in other areas of physical and mathematical sciences. This section is devoted to an in-depth analysis of some of these properties. Starting with the definition $\delta(x) = lim_{\alpha\to 0}\, \alpha^{-1}\mathrm{rect}(x/\alpha)$, the sifting property of the δ-function in conjunction with an arbitrary function $f(x)$ is written

$$\int_{-\infty}^{\infty} f(x)\delta(x - x_0)\mathrm{d}x = \begin{cases} f(x_0); & \text{if } f(\cdot) \text{ is continuous at } x_0, \\ ½[f(x_0^-) + f(x_0^+)]; & \text{if } f(\cdot) \text{ is discontinuous at } x_0. \end{cases} \quad (6)$$

Alternatively, suppose $f(x)$ is analytic and $\delta(x) = lim_{\alpha\to 0}(\sqrt{\pi}\alpha)^{-1}\exp[-(x/\alpha)^2]$. Invoking the definite integral $\int_{-\infty}^{\infty} x^{2n}\exp(-x^2)\,\mathrm{d}x = \sqrt{\pi}(2n-1)!!/2^n$ (G&R 3.461-3), we find, on expanding $f(x)$ into a Taylor series around x_0,

over-dot represents differentiation

odd terms integrate to zero

$$\begin{aligned}\int_{-\infty}^{\infty} f(x)\delta(x - x_0)\mathrm{d}x \\ &= lim_{\alpha\to 0}(\sqrt{\pi}\alpha)^{-1}\int_{-\infty}^{\infty}[f(x_0) + \dot{f}(x_0)(x - x_0) + \tfrac{1}{2!}\ddot{f}(x_0)(x - x_0)^2 + \cdots] \\ &\qquad\qquad \times \exp\{-[(x - x_0)/\alpha]^2\}\,\mathrm{d}x \\ &= lim_{\alpha\to 0}(1/\sqrt{\pi})\int_{-\infty}^{\infty}[f(x_0) + \alpha\dot{f}(x_0)y + \tfrac{1}{2!}\alpha^2\ddot{f}(x_0)y^2 + \cdots]\exp(-y^2)\,\mathrm{d}y \\ &= lim_{\alpha\to 0}\left[f(x_0) + \tfrac{1}{2}\times\tfrac{\alpha^2}{2!}\ddot{f}(x_0) + \tfrac{3}{4}\times\tfrac{\alpha^4}{4!}f^{(4)}(x_0) + \tfrac{15}{8}\times\tfrac{\alpha^6}{6!}f^{(6)}(x_0) + \cdots\right] = f(x_0). \quad (7)\end{aligned}$$

The sifting property of $\delta(x)$, expressed in Eq.(6), is generally valid, irrespective of the specific form that one may employ to describe the δ-function, provided that the chosen function is tall, narrow, symmetric around the peak, has unit area, and declines rapidly when the argument of the function moves away from the peak's location.

7.3.1. Sifting property of $\boldsymbol{\dot{\delta}(x)}$. Let $\delta(x) = lim_{\alpha\to 0}\,\alpha^{-1}\mathrm{tri}(x/\alpha)$. Then $\dot{\delta}(x)$ assumes the limiting form of a function confined to the interval $(-\alpha, \alpha)$, its value being $1/\alpha^2$ when $-\alpha < x < 0$, and $-1/\alpha^2$ when $0 < x < \alpha$. Thus, the sifting property of $\dot{\delta}(x)$ is

$$\begin{aligned}\int_{-\infty}^{\infty} f(x)\dot{\delta}(x - x_0)\mathrm{d}x &= lim_{\alpha\to 0}\frac{\alpha}{\alpha^2}[f(x_0 - ½\alpha) - f(x_0 + ½\alpha)] \\ &= \begin{cases} -\dot{f}(x_0); & \text{if } \dot{f}(\cdot) \text{ is continuous at } x_0, \\ -½[\dot{f}(x_0^-) + \dot{f}(x_0^+)]; & \text{if } \dot{f}(\cdot) \text{ is discontinuous at } x_0. \end{cases} \quad (8)\end{aligned}$$

Alternatively, we may use the method of integration by parts in conjunction with the sifting property of $\delta(x)$ given in Eq.(6) to arrive at the sifting property of $\dot{\delta}(x)$; that is,

$$\begin{aligned}\int_{-\infty}^{\infty} f(x)\dot{\delta}(x - x_0)\mathrm{d}x &= \underbrace{f(x)\delta(x - x_0)|_{-\infty}^{+\infty}}_{0} - \int_{-\infty}^{\infty}\dot{f}(x)\delta(x - x_0)\,\mathrm{d}x \\ &= \begin{cases} -\dot{f}(x_0); & \text{if } \dot{f}(\cdot) \text{ is continuous at } x_0, \\ -½[\dot{f}(x_0^-) + \dot{f}(x_0^+)]; & \text{if } \dot{f}(\cdot) \text{ is discontinuous at } x_0. \end{cases} \quad (9)\end{aligned}$$

As was the case with $\delta(\cdot)$, the sifting property of $\dot{\delta}(\cdot)$ is a universal property that does *not* depend on the limiting functional form chosen to represent $\delta(\cdot)$, provided that the chosen function is tall, narrow, symmetric around the peak, has unit area, and declines rapidly when the argument of the function moves away from the peak's location. For instance, if $f(x)$ is analytic and $\delta(x) = lim_{\alpha\to 0}(\sqrt{\pi}\alpha)^{-1}\exp[-(x/\alpha)^2]$, we find

$$\int_{-\infty}^{\infty} f(x)\dot{\delta}(x-x_0)dx$$
$$= lim_{\alpha\to 0}\left(\frac{-2}{\sqrt{\pi}\alpha^3}\right)\int_{-\infty}^{\infty}[f(x_0)+\dot{f}(x_0)(x-x_0)+\tfrac{1}{2!}\ddot{f}(x_0)(x-x_0)^2+\cdots]$$
$$\times (x-x_0)\exp\{-[(x-x_0)/\alpha]^2\}\,dx$$

odd terms integrate to zero

$$= -\,lim_{\alpha\to 0}\frac{2}{\sqrt{\pi}}\int_{-\infty}^{\infty}[\dot{f}(x_0)y^2+\frac{\alpha^2}{3!}\dddot{f}(x_0)y^4+\frac{\alpha^4}{5!}f^{(5)}(x_0)y^6+\cdots]\exp(-y^2)\,dy$$
$$= -\,lim_{\alpha\to 0}\left[\dot{f}(x_0)+\tfrac{3}{2}\times\frac{\alpha^2}{3!}\dddot{f}(x_0)+\tfrac{15}{4}\times\frac{\alpha^4}{5!}f^{(5)}(x_0)+\tfrac{105}{8}\times\frac{\alpha^6}{7!}f^{(7)}(x_0)+\cdots\right]$$
$$= -\dot{f}(x_0). \tag{10}$$

Similar sifting properties can be readily proven for higher derivatives of the δ-function.

7.3.2. Derivative of a continuous function of a discontinuous function. Let the function $\varphi(x)$ have a discontinuity equal to φ_0 at $x = x_0$. One may write $\varphi(x) = \tilde{\varphi}(x) + \varphi_0\text{step}(x-x_0)$, where $\tilde{\varphi}(x)$ is smooth and well behaved at $x = x_0$. Let us form the composite function $f(x) = \sin[\varphi(x)]$ and consider its derivative $f'(x)$ with respect to x. The discontinuity of $f(x)$ produces, on differentiation, a δ-function at $x = x_0$. We would like to apply the chain rule of differentiation and verify that the magnitude of the δ-function thus produced is equal to the discontinuity of $f(x)$ at $x = x_0$; in other words, the goal is to show that $f'(x) = \varphi'(x)\cos[\varphi(x)]$ contains a δ-function whose magnitude equals the discontinuity of $f(x)$ at $x = x_0$. To this end, we split $f(x)$ into two simpler functions, as follows:

$$f(x) = \sin[\varphi(x)] = \sin[\tilde{\varphi}(x) + \varphi_0\text{step}(x-x_0)]$$
$$= \sin[\tilde{\varphi}(x)]\cos[\varphi_0\text{step}(x-x_0)] + \cos[\tilde{\varphi}(x)]\sin[\varphi_0\text{step}(x-x_0)]. \tag{11}$$

It is clear from Eq.(11) that the analysis of discontinuity should be confined to the simpler functions $\sin[\varphi_0\text{step}(x-x_0)]$ and $\cos[\varphi_0\text{step}(x-x_0)]$. In what follows, we shall analyze only $g(x) = \sin[\varphi_0\text{step}(x-x_0)]$, which is zero when $x < x_0$, and equals $\sin\varphi_0$ when $x > x_0$. Straightforward differentiation of $g(x)$ now yields

$$g'(x) = (\sin\varphi_0)\delta(x-x_0). \tag{12}$$

If one insists on applying the chain rule of differentiation to $g(x)$, a naïve application of the chain rule to the discontinuity at $x = x_0$ would yield

$$g'(x) = \varphi_0\delta(x-x_0)\cos[\varphi_0\text{step}(x-x_0)] = ½(1+\cos\varphi_0)\varphi_0\delta(x-x_0). \tag{13}$$

In this equation, the coefficient $½(1+\cos\varphi_0)$ is the mid-value of the discontinuity of $\cos[\varphi_0\text{step}(x-x_0)]$ at $x = x_0$, which is what one would expect based on the sifting property of $\delta(x-x_0)$ when it is multiplied by a discontinuous function at $x = x_0$. The derivative of $g(x)$ in Eq.(13) is clearly different from that obtained by direct differentiation in Eq.(12). So, what went wrong?

The problem here is that the discontinuity of $\cos[\varphi_0 \text{step}(x-x_0)]$ at $x=x_0$ is *not* a simple jump from 1 to $\cos\varphi_0$. This is a *structured* discontinuity, which is better understood if $\text{step}(x-x_0)$ is treated as the limiting form of a continuous function. Suppose the sharp transition from 0 to 1 of $\text{step}(x-x_0)$ is replaced with the straight line segment $½+½(x-x_0)/\varepsilon$ in the interval $x_0-\varepsilon \le x \le x_0+\varepsilon$. We will have

$$\frac{\mathrm{d}}{\mathrm{d}x}[\varphi_0 \text{step}(x-x_0)] \rightarrow \left(\frac{\varphi_0}{2\varepsilon}\right)\text{rect}\left(\frac{x-x_0}{2\varepsilon}\right). \tag{14}$$

$$\cos[\varphi_0 \text{step}(x-x_0)] \rightarrow \cos[½\varphi_0+½(x-x_0)\varphi_0/\varepsilon]; \qquad (|x-x_0|\le\varepsilon). \tag{15}$$

$$\int_{x_0-\varepsilon}^{x_0+\varepsilon}\cos[½\varphi_0+½(x-x_0)\varphi_0/\varepsilon]\,\mathrm{d}x = (2\varepsilon/\varphi_0)\sin[½\varphi_0+½(x-x_0)\varphi_0/\varepsilon]_{x_0-\varepsilon}^{x_0+\varepsilon}$$

$$= (2\varepsilon/\varphi_0)\sin\varphi_0. \tag{16}$$

When the $\text{rect}(\cdot)$ function appearing in Eq.(14) is multiplied by $\cos[½\varphi_0 + ½(x-x_0)\varphi_0/\varepsilon]$ of Eq.(15), the area under the product function will be $\sin\varphi_0$, as seen in Eq.(16). Thus, in the limit when $\varepsilon\rightarrow 0$, the derivative of $g(x)$ will be a δ-function of magnitude $\sin\varphi_0$, in agreement with the result obtained in Eq.(12).

7.3.3. A peculiar behavior of the δ-function. When working with δ-functions, one must remain alert and watch for pitfalls. Consider the real-valued function $f(x)$, which is defined as an integral of the product of an arbitrary (real-valued) function $g(x')$ and a δ-function whose argument depends on yet another arbitrary (real-valued) function $h(x')$:

$$f(x)=\int_{-\infty}^{\infty} g(x')\,\delta[x-h(x')]\mathrm{d}x'. \tag{17}$$

Suppose the argument of the δ-function has a zero at $x'=x'_0(x)$; that is, $h(x'_0)=x$. Given that the slope of the δ-function's argument at $x'=x'_0(x)$ is $S_0=-\dot{h}(x'_0)$, the sifting property of $\delta(\cdot)$ enables one to write $f(x)$ as follows:

$$f(x)=g(x'_0)/|\dot{h}(x'_0)|. \tag{18}$$

Differentiation with respect to x now entails the following steps:

a) $\quad h(x'_0)=x \quad\rightarrow\quad \dot{h}(x'_0)\dot{x}'_0(x)=1 \quad\rightarrow\quad \dot{x}'_0(x)=1/\dot{h}(x'_0). \tag{19}$

b) $\quad \frac{\mathrm{d}}{\mathrm{d}x'}|\dot{h}(x'_0)| = \frac{\mathrm{d}}{\mathrm{d}x'}\sqrt{\dot{h}^2(x'_0)} = \dot{h}(x'_0)\ddot{h}(x'_0)/|\dot{h}(x'_0)|. \tag{20}$

c) $$\dot{f}(x)=\left[\frac{\mathrm{d}g(x'_0)}{\mathrm{d}x}\right]|\dot{h}(x'_0)|^{-1}-g(x'_0)\left[\frac{\mathrm{d}|\dot{h}(x'_0)|}{\mathrm{d}x}\right]|\dot{h}(x'_0)|^{-2}$$

$$=\left[\frac{\mathrm{d}g(x'_0)}{\mathrm{d}x'}\right]\left[\frac{\mathrm{d}x'_0(x)}{\mathrm{d}x}\right]|\dot{h}(x'_0)|^{-1}-g(x'_0)\left[\frac{\mathrm{d}|\dot{h}(x'_0)|}{\mathrm{d}x'}\right]\left[\frac{\mathrm{d}x'_0(x)}{\mathrm{d}x}\right]|\dot{h}(x'_0)|^{-2}. \tag{21}$$

Consequently,

$$\dot{f}(x)=\frac{\dot{g}(x'_0)}{\dot{h}(x'_0)|\dot{h}(x'_0)|}-\frac{g(x'_0)\ddot{h}(x'_0)}{|\dot{h}(x'_0)|^3}=\frac{\dot{g}(x'_0)\dot{h}(x'_0)-g(x'_0)\ddot{h}(x'_0)}{|\dot{h}(x'_0)|^3}. \tag{22}$$

This is the *correct* expression for $\dot{f}(x)$. An *incorrect* formula is obtained if we begin by differentiating the δ-function under the integral sign in Eq.(17), then invoke the sifting property of $\dot{\delta}(\cdot)$, as follows:

$$\dot{f}(x)=\int_{-\infty}^{\infty} g(x')\,\dot{\delta}[x-h(x')]\mathrm{d}x' = \dot{g}(x'_0)\dot{h}(x'_0)/|\dot{h}(x'_0)|^3. \tag{23}$$

To understand this failure of reversing the differentiation-integration sequence, consider the 2D Gaussian function $G(x,x') = (\sqrt{\pi}/\alpha)\exp\{-[x-h(x')]^2/\alpha^2\}$ over the $x'x$-plane, depicted in figure 7.2. For any finite value of $\alpha > 0$, the function $f(x) = \int_{-\infty}^{\infty} g(x')G(x,x')\mathrm{d}x'$ is well-behaved, and its derivative may be found by moving the differential operator $\partial/\partial x$ under the integral; that is, $\dot{f}(x) = \int_{-\infty}^{\infty} g(x')\partial_x G(x,x')\mathrm{d}x'$. Nevertheless, this rearranging of the order of differentiation and integration becomes impermissible in the limit $\alpha \to 0$, when the function $G(x,x')$ approaches a tall, thin wall erected over the curved line represented by the function $h(x')$ within the $x'x$-plane.

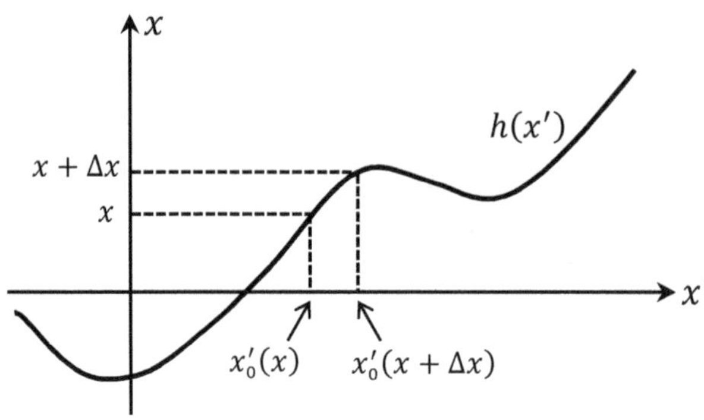

Figure 7.2. The 2D Gaussian function $G(x,x') = (\sqrt{\pi}/\alpha)\exp\{-[x-h(x')]^2/\alpha^2\}$ represents a δ-function along the x'-axis when the value of x is kept constant. The peak of the δ-function occurs at $x'_0(x)$, and its area varies with the slope $\dot{h}(x'_0)$ of the function $h(x')$ at the location of the peak. For any finite value of $\alpha > 0$, the 2D Gaussian function resembles a mountain ridge raised over the curve represented by $h(x')$. When $\alpha \to 0$, the ridge becomes a tall, thin wall erected over a curved line within the $x'x$-plane.

Brain teaser. There exist a total of 25 horses and you are allowed to race only 5 horses at a time. Each race determines the relative speeds of the 5 horses involved, but not their absolute speeds. How many races are needed to find the three fastest horses?

7.4. The Fourier transform. Fourier transformation is an operation that is carried out on a generally complex function $f(x)$ of the real variable x. The Fourier integral operates on $f(x)$ and produces a new function, $F(s)$, which is, once again, a generally complex function, but this time, of the real variable s. The Fourier integral is defined as follows:

$$F(s) = \int_{-\infty}^{\infty} f(x)\exp(-\mathrm{i}2\pi sx)\,\mathrm{d}x. \tag{24}$$

We will show in the next section that the function $f(x)$ can be recovered from its Fourier transform $F(s)$ by means of the following inverse Fourier transform operation:

$$f(x) = \int_{-\infty}^{\infty} F(s)\exp(\mathrm{i}2\pi sx)\,\mathrm{d}s. \tag{25}$$

The script letter $\mathcal{F}$ is sometimes used to describe the Fourier operator and its inverse; that is, $\mathcal{F}\{f(x)\} = F(s)$ and $\mathcal{F}^{-1}\{F(s)\} = f(x)$.

Example 1. The Fourier transform of $f(x) = \operatorname{rect}(x)$ is readily determined as follows:

$$F(s) = \mathcal{F}\{\operatorname{rect}(x)\} = \int_{-\infty}^{\infty} \operatorname{rect}(x)\exp(-\mathrm{i}2\pi sx)\,\mathrm{d}x = \int_{-1/2}^{1/2} \exp(-\mathrm{i}2\pi sx)\,\mathrm{d}x$$

$$= \frac{1}{-\mathrm{i}2\pi s}\exp(-\mathrm{i}2\pi sx)\Big|_{x=-1/2}^{1/2} = \frac{\exp(-\mathrm{i}\pi s) - \exp(\mathrm{i}\pi s)}{-\mathrm{i}2\pi s} = \frac{\sin(\pi s)}{\pi s} = \operatorname{sinc}(s). \tag{26}$$

Example 2. The Fourier transform of $f(x) = \operatorname{tri}(x)$ is obtained by first invoking the even symmetry of the function, then using the method of integration by parts, as follows:

$$F(s) = \mathcal{F}\{\operatorname{tri}(x)\} = \int_{-\infty}^{\infty} \operatorname{tri}(x)\exp(-\mathrm{i}2\pi sx)\,\mathrm{d}x = \int_{-1}^{1}(1-|x|)\cos(2\pi sx)\,\mathrm{d}x$$

$$= 2\int_0^1 (1-x)\cos(2\pi sx)\,\mathrm{d}x = \frac{\sin(2\pi sx)}{\pi s}\bigg|_{x=0}^{1} - \frac{x\sin(2\pi sx)}{\pi s}\bigg|_{x=0}^{1} + \int_0^1 \frac{\sin(2\pi sx)}{\pi s}\mathrm{d}x$$

$$= \frac{\sin(\cancel{2\pi s})}{\pi s} - \frac{\sin(\cancel{2\pi s})}{\pi s} - \frac{\cos(2\pi sx)}{2\pi^2 s^2}\bigg|_{x=0}^{1} = \frac{1-\cos(2\pi s)}{2\pi^2 s^2} = \frac{\sin^2(\pi s)}{\pi^2 s^2} = \mathrm{sinc}^2(s). \qquad (27)$$

Example 3. The Fourier transform of $f(x) = \exp(-\alpha|x|)$, where α is a real and positive constant is given by

$$F(s) = \mathcal{F}\{\exp(-\alpha|x|)\} = \int_{-\infty}^{\infty} \exp(-\alpha|x|)\exp(-\mathrm{i}2\pi sx)\,\mathrm{d}x$$

$$= \int_{-\infty}^{0} \exp(\alpha x)\exp(-\mathrm{i}2\pi sx)\,\mathrm{d}x + \int_0^{\infty} \exp(-\alpha x)\exp(-\mathrm{i}2\pi sx)\,\mathrm{d}x$$

$$= \frac{1}{\alpha-\mathrm{i}2\pi s}\exp[(\alpha-\mathrm{i}2\pi s)x]\bigg|_{x=-\infty}^{0} - \frac{1}{\alpha+\mathrm{i}2\pi s}\exp[-(\alpha+\mathrm{i}2\pi s)x]\bigg|_{x=0}^{\infty}$$

$$= \frac{1}{\alpha-\mathrm{i}2\pi s} + \frac{1}{\alpha+\mathrm{i}2\pi s} = \frac{2\alpha}{\alpha^2+4\pi^2 s^2}. \qquad (28)$$

Example 4. The function $f(x) = 1.0$ can be considered the limit of $f_\alpha(x) = \exp(-\alpha|x|)$ when $\alpha \to 0$. Based on the result of the preceding example, one could argue that $F(s) = \mathcal{F}\{1.0\} = lim_{\alpha\to 0} F_\alpha(s) = lim_{\alpha\to 0}[2\alpha/(\alpha^2+4\pi^2 s^2)]$. Now, the plot of $F_\alpha(s)$ depicted on the right shows that, in the limit when $\alpha \to 0$, the function becomes tall (height $= 2/\alpha$) and narrow (width at half-height $= \alpha/\pi$), while remaining symmetric (i.e., even) and retaining its area of 1. The function $F_\alpha(s)$ thus approaches $\delta(s)$ when $\alpha \to 0$. The Fourier transform of the constant function $f(x) = 1.0$ is seen to be $F(s) = \delta(s)$.

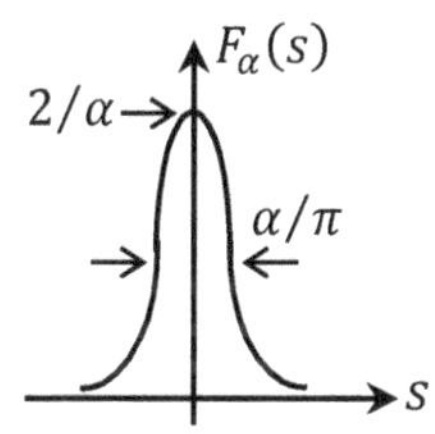

Example 5. The Fourier transform (also known as the frequency spectrum) of Dirac's δ-function is readily obtained with the aid of the sifting property of $\delta(x)$, as follows:

$$\mathcal{F}\{\delta(x)\} = \int_{-\infty}^{\infty} \delta(x)\exp(-\mathrm{i}2\pi sx)\,\mathrm{d}x = 1.0. \qquad (29)$$

The inverse transform of $\mathcal{F}\{\delta(x)\} = F(s) = 1.0$ thus yields

$$\delta(x) = \mathcal{F}^{-1}\{1.0\} = \int_{-\infty}^{\infty} \exp(\mathrm{i}2\pi sx)\,\mathrm{d}s \;\to\; \delta(x) = 2\int_0^{\infty} \cos(2\pi sx)\,\mathrm{d}s. \qquad (30)$$

Upon differentiating Eq.(30) with respect to x, we find

$$\dot{\delta}(x) = \frac{\mathrm{d}\delta(x)}{\mathrm{d}x} = \mathrm{i}2\pi\int_{-\infty}^{\infty} s\exp(\mathrm{i}2\pi sx)\,\mathrm{d}s \;\to\; \dot{\delta}(x) = -4\pi\int_0^{\infty} s\sin(2\pi sx)\,\mathrm{d}s. \qquad (31)$$

It is now easy to verify some of the scaling properties of $\delta(x)$ and $\dot{\delta}(x)$, such as

$$\delta(\alpha x) = 2\int_0^{\infty} \cos(2\pi s\alpha x)\,\mathrm{d}s = |\alpha|^{-1}\delta(x). \qquad (32)$$

$$\dot{\delta}(\alpha x) = -4\pi\int_0^{\infty} s\sin(2\pi s\alpha x)\,\mathrm{d}s = (\alpha/|\alpha|^3)\dot{\delta}(x). \qquad (33)$$

The parameter α appearing in Eqs.(32) and (33) is real-valued and nonzero, but is otherwise arbitrary.

Example 6. The inverse Fourier transform of $F(s) = 1/(s^2 + 1)$ is found by evaluating the inverse Fourier integral using complex plane techniques, as follows:

$$f(x) = \mathcal{F}^{-1}\{1/(s^2 + 1)\} = \int_{-\infty}^{\infty}[\exp(\mathrm{i}2\pi xs)/(s^2 + 1)]\mathrm{d}s. \quad (34)$$

Moving to the complex z-plane, we note that $f(z) = \exp(\mathrm{i}2\pi xz)/(z^2 + 1)$ has two poles at $z_{1,2} = \pm\mathrm{i}$. (To avoid confusing the parameter x of the inverse Fourier integral with the real part of z, in place of our usual $z = x + \mathrm{i}y$, we now write $z = \eta + \mathrm{i}\zeta$.) The function $f(z)$ satisfies the requirements of Jordan's lemma (see chapter 6, section 8) provided that we use, as part of the integration contour, a semi-circle of radius R in the upper-half of the plane when $x > 0$, and in the lower-half of the plane when $x < 0$. The integral along the semi-circle then vanishes and the desired (Fourier) integral along the real axis becomes $2\pi\mathrm{i}$ times the residue of $f(z)$ at the pole that is located inside the contour of integration. For $x > 0$, the enclosed pole is $z_1 = \mathrm{i}$ and the residue is $\exp(\mathrm{i}2\pi xz_1)/(z_1 + \mathrm{i}) = \exp(-2\pi x)/(2\mathrm{i})$. As for $x < 0$, the enclosed pole is $z_2 = -\mathrm{i}$ and the residue is $-\exp(\mathrm{i}2\pi xz_2)/(z_2 - \mathrm{i}) = \exp(2\pi x)/(2\mathrm{i})$. We thus find the inverse transform of $F(s)$ as $f(x) = \pi\exp(-2\pi|x|)$.

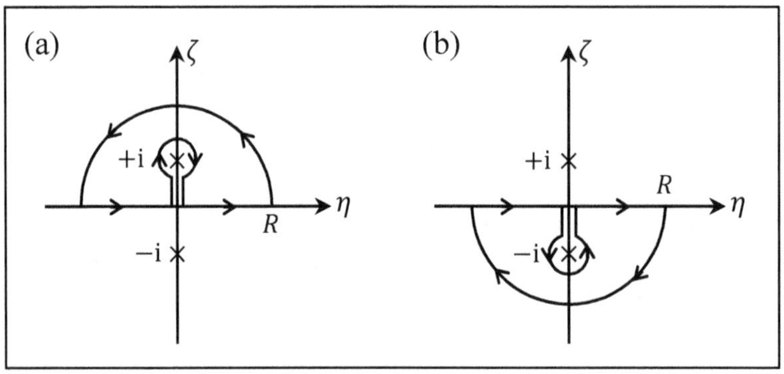

Figure 7.3. Contours of integration in the complex plane $z = \eta + \mathrm{i}\zeta$. In (a) the contour closes in the upper-half plane because $x > 0$. In (b), which corresponds to $x < 0$, the contour closes in the lower-half plane. Jordan's lemma ensures that, in each case, the integral over the large semi-circle vanishes in the limit when $R \to \infty$. Note that the small circles surrounding the poles at $z_1 = \mathrm{i}$ and $z_2 = -\mathrm{i}$ are traversed in opposite directions. Consequently, the residue at $z_2 = -\mathrm{i}$ must be multiplied by -1 to account for its clockwise direction of integration.

Example 7. The Fourier transform of $f(x) = \mathrm{step}(x)$ may be obtained as the limit of the Fourier transform of $f_\alpha(x) = \mathrm{step}(x)\exp(-\alpha x)$ when the real-valued and positive parameter α approaches zero. We will have

$$F(s) = \mathcal{F}\{\mathrm{step}(x)\} = lim_{\alpha\to 0}\int_0^{\infty}\exp(-\alpha x)\exp(-\mathrm{i}2\pi sx)\,\mathrm{d}x$$

$$= lim_{\alpha\to 0}\left(\frac{1}{\alpha + \mathrm{i}2\pi s}\right) = lim_{\alpha\to 0}\left(\frac{\alpha}{\alpha^2+4\pi^2s^2} - \frac{\mathrm{i}2\pi s}{\alpha^2+4\pi^2s^2}\right). \quad (35)$$

For sufficiently small values of α, the real part of the function appearing on the right-hand side of Eq.(35) is a tall, narrow, even function of s, whose area remains constant at ½, irrespective of the value of α. We conclude that the real part of $F(s)$ equals ½$\delta(s)$. As for its imaginary part, the plot of $2\pi s/(\alpha^2 + 4\pi^2 s^2)$ for small values of α, shown on the right, indicates that the function approaches $1/(2\pi s)$ when $\alpha \to 0$. All in all, the Fourier transform of $f(x) = \mathrm{step}(x)$ is found to be $F(s) = ½\delta(s) - \mathrm{i}/(2\pi s)$.

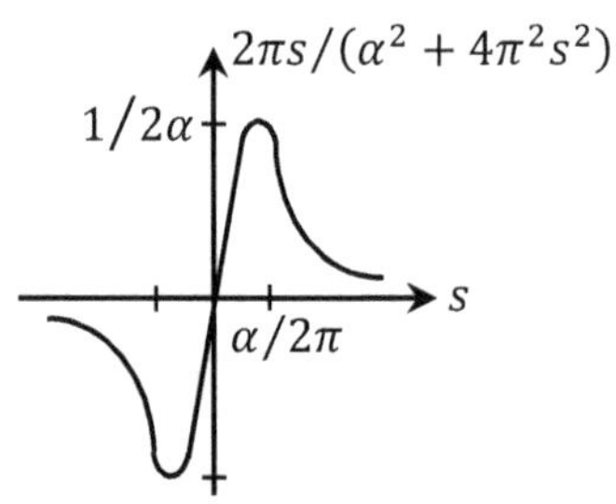

Example 8. The Fourier transform of $f(x) = \text{sign}(x) = x/|x|$ may be found straightforwardly as the limit of the Fourier transform of $f_\alpha(x) = \text{sign}(x)\exp(-\alpha|x|)$, when the real-valued and positive parameter α approaches zero. We will find

$$F(s) = \mathcal{F}\{\text{sign}(x)\}$$

$$= lim_{\alpha\to 0}\left[\int_0^\infty \exp(-\alpha x)\exp(-\text{i}2\pi s x)\,\text{d}x - \int_{-\infty}^0 \exp(\alpha x)\exp(-\text{i}2\pi s x)\,\text{d}x\right]$$

$$= lim_{\alpha\to 0}\left(\frac{1}{\alpha + \text{i}2\pi s} - \frac{1}{\alpha - \text{i}2\pi s}\right) = lim_{\alpha\to 0}\left(-\frac{\text{i}4\pi s}{\alpha^2 + 4\pi^2 s^2}\right) = -\text{i}/(\pi s). \qquad (36)$$

A quick comparison with Examples 4 and 7 now reveals the consistency of the above results; that is,

$$\mathcal{F}\{\text{sign}(x)\} = \mathcal{F}\{2\text{step}(x) - 1.0\} = 2\mathcal{F}\{\text{step}(x)\} - \mathcal{F}\{1.0\}$$

$$= 2[½\delta(s) - \text{i}/(2\pi s)] - \delta(s) = -\text{i}/(\pi s). \qquad (37)$$

Example 9. The Fourier transform of the Gaussian function $f(x) = \exp(-\pi x^2)$ is $F(s) = \exp(-\pi s^2)$. The proof involves completing the square in the exponent of the integrand, followed by switching the integration path from the upper leg to the lower leg of the rectangular contour in the complex plane, as shown in figure 7.4. We will have

$$\mathcal{F}\{\exp(-\pi x^2)\} = \int_{-\infty}^{\infty} \exp(-\pi x^2)\exp(-\text{i}2\pi s x)\,\text{d}x$$

$$= \int_{-\infty}^{\infty} \exp[-\pi(x + \text{i}s)^2]\exp(-\pi s^2)\,\text{d}x$$

$$= \exp(-\pi s^2)\int_{-\infty}^{\infty} \exp[-\pi(x + \text{i}s)^2]\,\text{d}x$$

$$= \exp(-\pi s^2)\int_{-\infty}^{\infty} \exp(-\pi x^2)\,\text{d}x = \exp(-\pi s^2). \qquad (38)$$

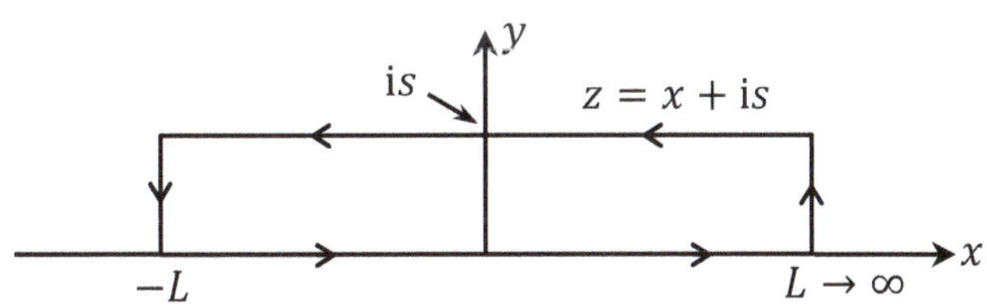

Figure 7.4. The integral appearing in the penultimate line of Eq.(38) is $\int \exp(-\pi z^2)\,\text{d}z$ along the upper leg of the rectangular contour, where $z = x + \text{i}s$ and $\text{d}z = \text{d}x$. Since $\exp(-\pi z^2)$ is analytic everywhere, and considering that the vertical legs of the rectangle, in the limit when $L \to \infty$, do not contribute, the loop integral must vanish. This implies that the integrals along the upper and lower legs of the rectangle are equal in magnitude and opposite in sign. In the final line of Eq.(38), the integral along the upper leg has been replaced with that along the lower leg.

Example 10. The Fourier transform of $f(x) = \exp(\text{i}\pi x^2)$ is $F(s) = \exp[-\text{i}\pi(s^2 - ¼)]$. The proof requires completing the square in the exponent of the integrand, then switching to the complex plane and using the contour shown in figure 7.5. We will have

$$\mathcal{F}\{\exp(\text{i}\pi x^2)\} = \int_{-\infty}^{\infty} \exp(\text{i}\pi x^2)\exp(-\text{i}2\pi s x)\,\text{d}x$$

$$= \int_{-\infty}^{\infty} \exp[\text{i}\pi(x - s)^2 - \text{i}\pi s^2]\,\text{d}x$$

$$= \exp(-\mathrm{i}\pi s^2) \int_{-\infty}^{\infty} \exp[\mathrm{i}\pi(x-s)^2]\, \mathrm{d}x$$

change of variable: $\eta = x - s$ → $$= 2\exp(-\mathrm{i}\pi s^2) \int_0^{\infty} \exp(\mathrm{i}\pi\eta^2)\, \mathrm{d}\eta. \qquad (39)$$

The integration contour in the complex $\eta\zeta$-plane shown in figure 7.5 allows one to replace the last integral in Eq.(39) with an integral along the 45° line, where $z = r\exp(\mathrm{i}\pi/4)$. Along the 45° line, the integrand $\exp(\mathrm{i}\pi z^2)$ equals $\exp(-\pi r^2)$, and $\mathrm{d}z = \exp(\mathrm{i}\pi/4)\, \mathrm{d}r$. Consequently,

$$\mathcal{F}\{\exp(\mathrm{i}\pi x^2)\} = 2\exp(-\mathrm{i}\pi s^2)\exp(\mathrm{i}\pi/4) \int_0^{\infty} \exp(-\pi r^2)\, \mathrm{d}r = \exp[-\mathrm{i}\pi(s^2 - ¼)]. \qquad (40)$$

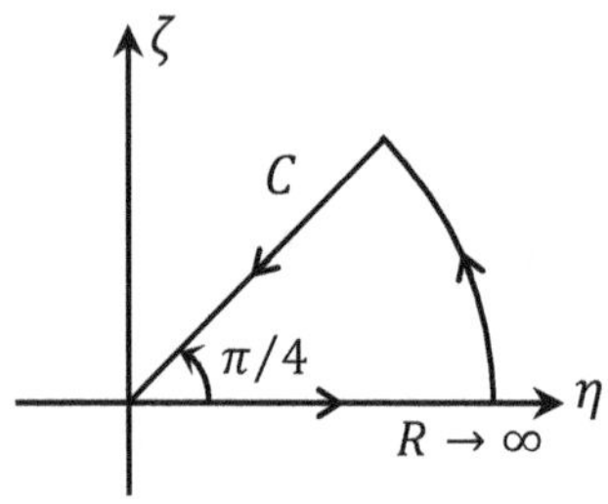

Figure 7.5. The integral on the last line of Eq.(40) is taken on the real-axis η of the complex $\eta\zeta$-plane. Considering that $\exp(\mathrm{i}\pi z^2)$ is analytic everywhere across the z-plane, the integral around the closed contour C is zero. The desired integral may thus be evaluated on the 45° line plus the arc of the circle of radius R in the limit when $R \to \infty$.

Digression: The contribution of the arc depicted in figure 7.5 to the loop integral approaches zero when $R \to \infty$. To show this, we use an argument that closely parallels Jordan's lemma described in chapter 6, section 8. Considering that $z = R\exp(\mathrm{i}\theta)$ on the circle of radius R, we will have

$$\left|\int_{\mathrm{arc}} \exp(\mathrm{i}\pi z^2)\, \mathrm{d}z\right| = \left|\int_{\theta=0}^{\pi/4} \exp(\mathrm{i}\pi R^2 e^{\mathrm{i}2\theta})\mathrm{i}Re^{\mathrm{i}\theta}\, \mathrm{d}\theta\right| \le R\int_{\theta=0}^{\pi/4} \exp[-\pi R^2 \sin(2\theta)]\, \mathrm{d}\theta$$

$$\le R\int_0^{\pi/4} \exp(-4R^2\theta)\, \mathrm{d}\theta = [1 - \exp(-\pi R^2)]/(4R) \to 0 \text{ when } R \to \infty.$$

In general, it is *not* true that the Fourier transform of the real (or imaginary) part of a function is the real (or imaginary) part of its Fourier transform. In the case of $f(x) = \exp(\mathrm{i}\pi x^2)$, however, since both its real and imaginary parts are even functions of x, we are allowed to equate the real and imaginary parts of the function with the corresponding parts of its Fourier transform. We thus obtain

$$\mathcal{F}\{\cos(\pi x^2)\} = \cos[\pi(s^2 - ¼)]. \qquad (41)$$

$$\mathcal{F}\{\sin(\pi x^2)\} = -\sin[\pi(s^2 - ¼)] = \cos[\pi(s^2 + ¼)]. \qquad (42)$$

Example 11. In this example, we use the gamma function $\Gamma(\nu) = \int_0^{\infty} x^{\nu-1}\exp(-x)\, \mathrm{d}x$ to determine the Fourier transform of the function $f(x) = 1/|x|^{\nu}$, where $0 < \nu < 1$. Since the complex function $z^{\nu} = |z|\exp(\mathrm{i}\nu\theta)$ is multivalued, we must specify a branch-cut in the complex plane. A convenient branch-cut is along the negative imaginary axis, which confines the phase angle θ to the range $(-90°, 270°]$. The Fourier transform of $f(x)$ may now be written as follows:

$$F(s) = \int_{-\infty}^{\infty} f(x)\exp(-\mathrm{i}2\pi sx)\, \mathrm{d}x$$

$$= \int_{-\infty}^{0} [\exp(-\mathrm{i}\pi)x]^{-\nu}\exp(-\mathrm{i}2\pi sx)\, \mathrm{d}x + \int_0^{\infty} x^{-\nu}\exp(-\mathrm{i}2\pi sx)\, \mathrm{d}x. \qquad (43)$$

Note that we have set $|x| = \exp(-i\pi)x$ along the negative x-axis. This choice ensures that, when x is replaced by the complex variable z, the correct $f(z)$ is obtained at $z = |z|\exp(i\pi)$.

The two integrals in Eq.(43) must be evaluated on different contours, as shown in the following diagram. The contributions of the large quarter-circles to the corresponding integrals vanish in the limit of large R, in accordance with Jordan's lemma. Also, the small quarter-circles of radius ε do not contribute to the integrals because $dz = i\varepsilon\exp(i\theta)\,d\theta$ approaches zero faster than $z^{\nu} = \varepsilon^{\nu}\exp(i\nu\theta)$ in the limit when $\varepsilon \to 0$. Consequently, each integral along the real axis is replaced by one along the imaginary axis, as follows.

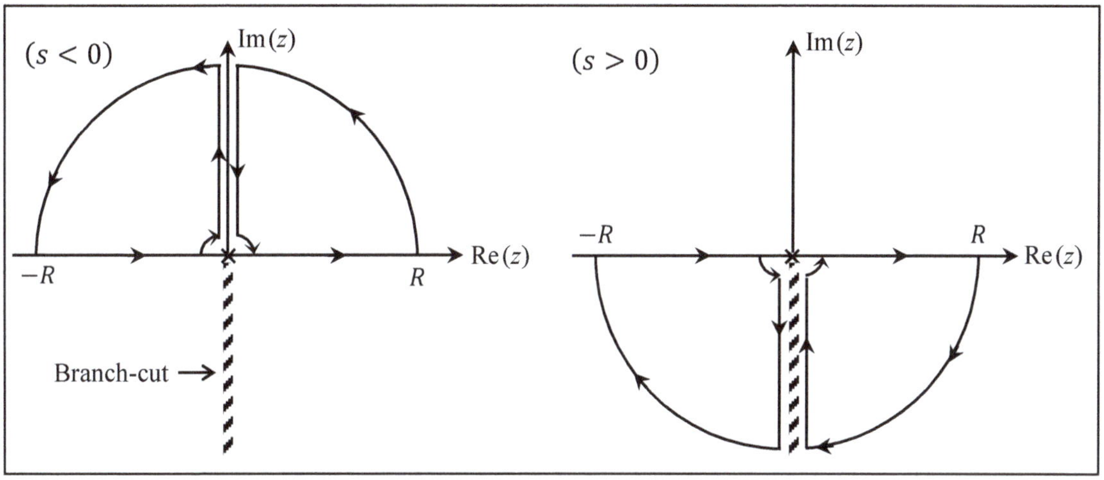

For $s < 0$, both integrals of $F(s)$ are replaced by integrals along the positive imaginary axis where $z = iy$. We will have

$$F(s) = -i\int_0^{\infty}[\exp(-i\pi)iy]^{-\nu}\exp(2\pi sy)\,dy + i\int_0^{\infty}(iy)^{-\nu}\exp(2\pi sy)\,dy$$

$$= -i\exp(+\tfrac{1}{2}i\nu\pi)\int_0^{\infty}y^{-\nu}\exp(2\pi sy)\,dy + i\exp(-\tfrac{1}{2}i\nu\pi)\int_0^{\infty}y^{-\nu}\exp(2\pi sy)\,dy$$

$$= 2\sin(\tfrac{1}{2}\nu\pi)\,(2\pi|s|)^{\nu-1}\int_0^{\infty}\zeta^{-\nu}\exp(-\zeta)\,d\zeta \quad \leftarrow \text{change of variable: } \zeta = 2\pi|s|y$$

$$= \frac{2\Gamma(1-\nu)\sin(\tfrac{1}{2}\nu\pi)}{(2\pi|s|)^{1-\nu}}. \tag{44}$$

For $s > 0$, the two integrals of $F(s)$ are replaced by integrals along the negative imaginary axis, where, in the third quadrant, $z = i^3|y|$, whereas in the fourth quadrant $z = -i|y|$. Since we are going to choose the range of integration along the negative imaginary axis from zero to $+\infty$, we shall ignore the absolute-value sign and write $|y|$ simply as y. We will have

$$F(s) = -i^3\int_0^{\infty}[\exp(-i\pi)i^3y]^{-\nu}\exp(-2\pi sy)\,dy - i\int_0^{\infty}(-iy)^{-\nu}\exp(-2\pi sy)\,dy$$

$$= i\exp(-\tfrac{1}{2}i\nu\pi)\int_0^{\infty}y^{-\nu}\exp(-2\pi sy)\,dy - i\exp(+\tfrac{1}{2}i\nu\pi)\int_0^{\infty}y^{-\nu}\exp(-2\pi sy)\,dy$$

$$= 2\sin(\tfrac{1}{2}\nu\pi)\,(2\pi s)^{\nu-1}\int_0^{\infty}\zeta^{-\nu}\exp(-\zeta)\,d\zeta \quad \leftarrow \text{change of variable: } \zeta = 2\pi sy$$

$$= \frac{2\Gamma(1-\nu)\sin(\tfrac{1}{2}\nu\pi)}{(2\pi|s|)^{1-\nu}}. \tag{45}$$

It is seen that the expressions of $F(s)$ thus obtained for positive and negative s are identical. This should not come as a surprise considering that $f(x)$ is a real and even function of x, whose Fourier transform must also be a real and even function of s.

7.5. Proof of Fourier's fundamental theorem. The Fourier transform preserves the information content of the function on which it operates. It is thus possible to recover the original function by performing an inverse Fourier transformation. This is a major property of the Fourier transform operation, without which the transformation would be rather useless. The proof of the Fourier theorem is relatively easy if, following a change in the order of integration, one invokes the properties of the δ-function, as follows:

$$\begin{aligned}\mathcal{F}^{-1}\{F(s)\} &= \int_{-\infty}^{\infty} F(s)\exp(\mathrm{i}2\pi sx)\,\mathrm{d}s \\ &= \int_{-\infty}^{\infty}\overbrace{\left[\int_{-\infty}^{\infty} f(x')\exp(-\mathrm{i}2\pi sx')\,\mathrm{d}x'\right]}^{F(s)}\exp(\mathrm{i}2\pi sx)\,\mathrm{d}s \\ &\overset{\text{changing the order of integration}}{=} \int_{-\infty}^{\infty} f(x')\underbrace{\int_{-\infty}^{\infty}\exp[-\mathrm{i}2\pi s(x'-x)]\,\mathrm{d}s}_{\delta(x'-x)}\,\mathrm{d}x' \\ &= \int_{-\infty}^{\infty} f(x')\delta(x'-x)\mathrm{d}x' = f(x). \quad \leftarrow \text{sifting property of } \delta(x')\end{aligned} \tag{46}$$

To the extent that the integrals in Eq.(46) exist, and the conditions for changing the order of integration are satisfied, the inverse Fourier integral reproduces the original function $f(x)$.

7.6. Fourier transform theorems. There exist several theorems that prove various properties of the Fourier integral and its inverse.[1] These theorems occasionally provide a simple way to determine Fourier transforms of some unknown functions in terms of other (known) Fourier transforms. Also, the theorems often provide a shortcut to extracting information from the Fourier integral and its inverse, which might otherwise require extra work. Listed below are a few useful theorems, along with their proofs and some examples of their applications.

7.6.1. Area under a function. It is easy to show that the area under the function $f(x)$ is equal to the value of its Fourier transform $F(s)$ at $s = 0$; that is, $\int_{-\infty}^{\infty} f(x)\mathrm{d}x = F(0)$. To prove this result, simply set $s = 0$ in Eq.(24). Likewise, setting $x = 0$ in Eq.(25) yields the area under the Fourier transform $F(s)$ as the value of the function $f(x)$ at the origin; that is, $\int_{-\infty}^{\infty} F(s)\mathrm{d}s = f(0)$.

Thus, the area under $F(s) = \mathrm{sinc}(s)$ must be 1.0, because $f(0) = \mathrm{rect}(0) = 1$. Similarly, the area under $g(x) = \mathrm{sinc}^2(x)$ equals 1.0, because its Fourier transform $G(s) = \mathrm{tri}(s)$ is equal to 1 at $s = 0$.

7.6.2. Scaling theorem. The Fourier transform of $f(\alpha x)$ is $|\alpha|^{-1}F(s/\alpha)$, where α is a real-valued constant parameter.

Proof: A change of variable of the Fourier integral from x to $y = \alpha x$ yields

$$\mathcal{F}\{f(\alpha x)\} = \int_{-\infty}^{\infty} f(\alpha x)\exp(-\mathrm{i}2\pi sx)\,\mathrm{d}x = \alpha^{-1}\int_{\mp\infty}^{\pm\infty} f(y)\exp(-\mathrm{i}2\pi sy/\alpha)\,\mathrm{d}y. \tag{47}$$

Note that the domain $(-\infty,\infty)$ of integration remains intact if $\alpha > 0$, but must be switched to $(\infty,-\infty)$ if $\alpha < 0$. In the latter case, the integration domain is restored to $(-\infty,\infty)$ if the integral is multiplied by -1. This change of sign may then be incorporated into the coefficient α if we write it as $-|\alpha|$ when α happens to be negative. We will have

$$\mathcal{F}\{f(\alpha x)\} = |\alpha|^{-1} \int_{-\infty}^{\infty} f(y) \exp(-\mathrm{i}2\pi s y/\alpha)\, \mathrm{d}y = |\alpha|^{-1} F(s/\alpha). \quad (48)$$

For example, knowing that the Fourier transform of $\mathrm{rect}(x)$ is $\mathrm{sinc}(s)$, immediately implies that $\mathrm{rect}(2x)$ transforms to ½$\mathrm{sinc}(s/2)$. Similarly, flipping the function $\mathrm{step}(x)$ around the vertical axis produces $\mathrm{step}(-x)$, whose Fourier transform is seen from Example 7 of section 4 in conjunction with the scaling theorem to be ½$\delta(s) + \mathrm{i}/(2\pi s)$.

7.6.3. Linear superposition theorem. Based on the definitions given in Eqs.(24) and (25), it must be obvious that the Fourier transform and its inverse are linear operators, meaning that $\mathcal{F}\{\alpha f(x) + \beta g(x)\} = \alpha F(s) + \beta G(s)$, where α and β are arbitrary (complex) constants, while $f(x)$ and $g(x)$ are arbitrary functions of x whose Fourier transforms are given by $F(s)$ and $G(s)$, respectively.

As an example, consider the function depicted in figure 7.6, which may be expressed as $f(x) = 2\mathrm{rect}(x/8) - \mathrm{rect}(x/6) + 4\mathrm{tri}(x/2)$. The Fourier transform of this function is readily seen to be $F(s) = 16\,\mathrm{sinc}(8s) - 6\,\mathrm{sinc}(6s) + 8\,\mathrm{sinc}^2(2s)$.

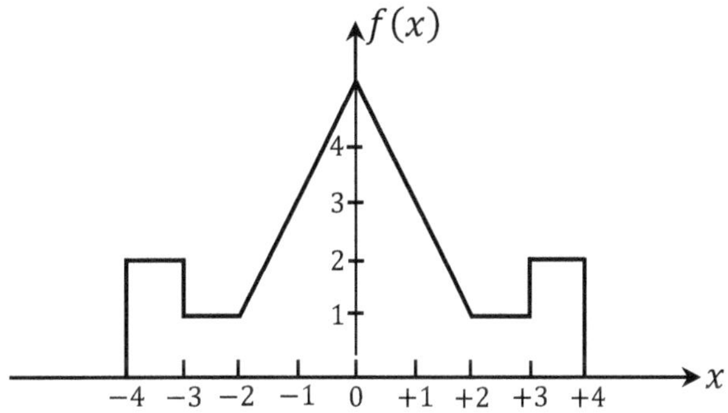

Figure 7.6. The function $f(x)$ is the superposition of a rectangular function of width 8 and height 2, another rectangular function of width 6 and height -1, and a triangular function of width 4 (at the base) and height 4. The Fourier transform $F(s)$ is obtained with the aid of the linear superposition theorem and the scaling theorem.

7.6.4. Parseval's theorem. For two arbitrary functions $f(x)$ and $g(x)$, whose Fourier transforms are given by $F(s)$ and $G(s)$, respectively, the following identity always holds:

$$\int_{-\infty}^{\infty} f(x) g^*(x) \mathrm{d}x = \int_{-\infty}^{\infty} F(s) G^*(s) \mathrm{d}s. \quad (49)$$

Proof: We replace one of the two functions, say, $g(x)$, with its inverse Fourier integral, then change the order of integration and rearrange the terms to arrive at the final result. We will have

$$\begin{aligned}\int_{-\infty}^{\infty} f(x) g^*(x) \mathrm{d}x &= \int_{-\infty}^{\infty} f(x) \left[\int_{-\infty}^{\infty} G(s) \exp(\mathrm{i}2\pi s x)\, \mathrm{d}s\right]^* \mathrm{d}x \\ &= \iint_{-\infty}^{\infty} f(x) G^*(s) \exp(-\mathrm{i}2\pi s x)\, \mathrm{d}s \mathrm{d}x \\ &= \int_{-\infty}^{\infty} \left[\int_{-\infty}^{\infty} f(x) \exp(-\mathrm{i}2\pi s x)\, \mathrm{d}x\right] G^*(s) \mathrm{d}s \\ &= \int_{-\infty}^{\infty} F(s) G^*(s) \mathrm{d}s. \end{aligned} \quad (50)$$

In the special case when $f(x) = g(x)$, the Parseval theorem reduces to

$$\int_{-\infty}^{\infty} |f(x)|^2 \mathrm{d}x = \int_{-\infty}^{\infty} |F(s)|^2 \mathrm{d}s. \quad (51)$$

This result is sometimes referred to as the theorem that asserts the "conservation of energy" of a function, since $|f(x)|^2$ might represent the energy-density associated with x, while $|F(s)|^2$ is the energy-density of the spectrum (or frequency content) of $f(x)$.

7.6.5. Shift theorem. The Fourier transform of $f(x - x_0)$ is given by $\exp(-\mathrm{i}2\pi x_0 s)\, F(s)$.

Proof: A change of variable of the Fourier integral from x to $y = x - x_0$ yields

$$\mathcal{F}\{f(x - x_0)\} = \int_{-\infty}^{\infty} f(x - x_0) \exp(-\mathrm{i}2\pi s x)\, \mathrm{d}x = \int_{-\infty}^{\infty} f(y) \exp[-\mathrm{i}2\pi s(y + x_0)]\, \mathrm{d}y$$

$$= \exp(-\mathrm{i}2\pi s x_0) \int_{-\infty}^{\infty} f(y) \exp(-\mathrm{i}2\pi s y)\, \mathrm{d}y$$

$$= \exp(-\mathrm{i}2\pi s x_0)\, F(s). \qquad (52)$$

It is also easy to show that multiplication of $f(x)$ by the single-frequency phase-factor $\exp(\mathrm{i}2\pi s_0 x)$ causes the Fourier transform of the function to shift to $s = s_0$; that is,

$$\mathcal{F}\{\exp(\mathrm{i}2\pi s_0 x)\, f(x)\} = F(s - s_0). \qquad (53)$$

As an example, consider the function $\cos(2\pi s_0 x) = [\exp(\mathrm{i}2\pi s_0 x) + \exp(-\mathrm{i}2\pi s_0 x)]/2$. Since the Fourier transform of $f(x) = 1$ is $F(s) = \delta(s)$, the two exponential functions in the expression for the cosine function will have a pair of shifted δ-functions for their Fourier transforms; that is,

$$\mathcal{F}\{\cos(2\pi s_0 x)\} = \tfrac{1}{2}[\delta(s - s_0) + \delta(s + s_0)]. \qquad (54)$$

Similarly, one can easily show that

$$\mathcal{F}\{\sin(2\pi s_0 x)\} = \tfrac{1}{2\mathrm{i}}[\delta(s - s_0) - \delta(s + s_0)]. \qquad (55)$$

7.6.6. Differentiation theorem. The Fourier transform of the derivative $f'(x)$ of $f(x)$ is obtained by multiplying $\mathrm{i}2\pi s$ into $F(s)$.

Proof: Differentiating both sides of Eq.(25) with respect to x yields

$$f'(x) = \frac{\mathrm{d}f(x)}{\mathrm{d}x} = \frac{\mathrm{d}}{\mathrm{d}x}\int_{-\infty}^{\infty} F(s)\exp(\mathrm{i}2\pi s x)\, \mathrm{d}s = \int_{-\infty}^{\infty} F(s) \frac{\mathrm{d}}{\mathrm{d}x} \exp(\mathrm{i}2\pi s x)\, \mathrm{d}s$$

$$= \int_{-\infty}^{\infty} F(s)(\mathrm{i}2\pi s) \exp(\mathrm{i}2\pi s x)\, \mathrm{d}s = \mathcal{F}^{-1}\{\mathrm{i}2\pi s F(s)\}. \qquad (56)$$

The same procedure, of course, may be applied to the Fourier integral in Eq.(24), which would then yield $\mathcal{F}\{-\mathrm{i}2\pi x f(x)\} = F'(s)$. Alternatively, one could use the method of integration by parts to arrive at the same result, as shown below.

$$\mathcal{F}^{-1}\{F'(s)\} = \int_{-\infty}^{\infty} F'(s) \exp(\mathrm{i}2\pi s x)\, \mathrm{d}s$$

$$= \overbrace{F(s) \exp(\mathrm{i}2\pi s x)|_{s=-\infty}^{\infty}}^{\to 0} - \int_{-\infty}^{\infty} F(s)(\mathrm{i}2\pi x) \exp(\mathrm{i}2\pi s x)\, \mathrm{d}s$$

$$= -\mathrm{i}2\pi x \int_{-\infty}^{\infty} F(s) \exp(\mathrm{i}2\pi s x)\, \mathrm{d}s = -\mathrm{i}2\pi x f(x). \qquad (57)$$

As an example, consider the function $f(x) = \mathrm{step}(x)$, whose derivative is $f'(x) = \delta(x)$. Since $F(s) = ½\delta(s) - \mathrm{i}/(2\pi s)$, the Fourier transform of $\delta(x)$, in accordance with the differentiation theorem, must be $\mathrm{i}2\pi s F(s) = \mathrm{i}\pi s \delta(s) + 1$. Considering that $s\delta(s) = 0$

(see, for example, the sifting property of the δ-function), we conclude that $\mathcal{F}\{\delta(x)\} = 1$, in agreement with Eq.(29).

7.6.7. Convolution theorem. When two functions are multiplied together to produce a new function, say, $h(x) = f(x)g(x)$, the Fourier transform of the product will be the result of a convolution operation between the Fourier transforms of the original functions; that is, $H(s) = F(s) * G(s)$. The asterisk placed between $F(s)$ and $G(s)$ represents the convolution operation, defined as $H(s) = \int_{-\infty}^{\infty} F(s')G(s - s')\mathrm{d}s'$.

Similarly, if the convolution operation is performed on two functions to produce a new function, say, $r(x) = p(x) * q(x) = \int_{-\infty}^{\infty} p(x')q(x - x')\mathrm{d}x'$, then the Fourier transform of the resulting function will be the product of the Fourier transforms of the original functions; that is, $R(s) = P(s)Q(s)$. We provide the proof for only one of the preceding statements, as the other statement can be proven in very much the same way.

Proof: $\mathcal{F}\{f(x)g(x)\} = \int_{-\infty}^{\infty} f(x)g(x)\exp(-\mathrm{i}2\pi sx)\,\mathrm{d}x.$

$$= \int_{-\infty}^{\infty}\left[\int_{-\infty}^{\infty} F(s')\exp(\mathrm{i}2\pi s'x)\,\mathrm{d}s'\right]g(x)\exp(-\mathrm{i}2\pi sx)\,\mathrm{d}x$$

$$= \int_{-\infty}^{\infty} F(s')\left\{\int_{-\infty}^{\infty} g(x)\exp[-\mathrm{i}2\pi(s - s')x]\,\mathrm{d}x\right\}\mathrm{d}s'$$

$$= \int_{-\infty}^{\infty} F(s')G(s - s')\mathrm{d}s'. \tag{58}$$

As an example, the function $\mathrm{tri}(x)$ is the convolution of $\mathrm{rect}(x)$ with itself; that is, $\mathrm{tri}(x) = \mathrm{rect}(x) * \mathrm{rect}(x)$. Consequently, $\mathcal{F}\{\mathrm{tri}(x)\} = \mathrm{sinc}^2(s)$.

Convolution between the δ-function and any other function $f(x)$, reproduces the function $f(x)$. This is readily inferred from the sifting property of the δ-function, which yields

$$f(x) * \delta(x) = \int_{-\infty}^{\infty} f(x')\delta(x - x')\mathrm{d}x' = f(x). \tag{59}$$

The convolution theorem confirms this property of the δ-function by virtue of the fact that $\mathcal{F}\{\delta(x)\} = 1.0$, whose multiplication with $F(s)$ does not modify $F(s)$.

The convolution of any function $f(x)$ with a shifted δ-function, say, $\delta(x - x_0)$, where x_0 is an arbitrary constant, causes a shift of $f(x)$ to the location of the δ-function, that is

$$f(x) * \delta(x - x_0) = \int_{-\infty}^{\infty} f(x')\delta(x - x' - x_0)\mathrm{d}x' = f(x - x_0). \tag{60}$$

This result is consistent with the convolution theorem, since $\mathcal{F}\{\delta(x - x_0)\} = \exp(-\mathrm{i}2\pi x_0 s)$, whose multiplication with $\mathcal{F}\{f(x)\}$ yields $\exp(-\mathrm{i}2\pi x_0 s)\,F(s)$, which, in accordance with the shift theorem, is the Fourier transform of $f(x - x_0)$.

7.7. The comb function and its Fourier transform. The standard comb function is defined as $\mathrm{comb}(x) = \sum_{n=-\infty}^{\infty}\delta(x - n)$. It is a periodic function (period = 1) consisting of individual δ-functions of unit area. A plot of the comb function is shown on the right. An interesting (and extremely useful) property of $f(x) = \mathrm{comb}(x)$ is that its Fourier transform is given by $F(s) = \mathrm{comb}(s)$, as demonstrated below.[1]

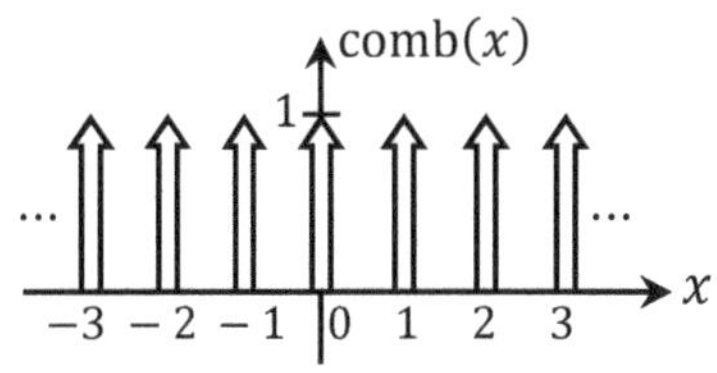

$$\begin{aligned}\mathcal{F}\{\mathrm{comb}(x)\} &= \int_{-\infty}^{\infty} \mathrm{comb}(x)\exp(-\mathrm{i}2\pi sx)\,\mathrm{d}x \\ &= lim_{N\to\infty}\int_{-\infty}^{\infty}\left[\sum_{n=-N}^{N}\delta(x-n)\right]\exp(-\mathrm{i}2\pi sx)\,\mathrm{d}x \\ &= lim_{N\to\infty}\sum_{n=-N}^{N}\exp(-\mathrm{i}2\pi ns) \\ &= lim_{N\to\infty}\exp(\mathrm{i}2\pi Ns)\sum_{n=0}^{2N}\exp(-\mathrm{i}2\pi ns) \\ &= lim_{N\to\infty}\exp(\mathrm{i}2\pi Ns)\left\{\frac{1-\exp[-\mathrm{i}2\pi(2N+1)s]}{1-\exp(-\mathrm{i}2\pi s)}\right\} \\ &= lim_{N\to\infty}\frac{\exp[\mathrm{i}\pi(2N+1)s]-\exp[-\mathrm{i}\pi(2N+1)s]}{\exp(\mathrm{i}\pi s)-\exp(-\mathrm{i}\pi s)} \\ &= lim_{N\to\infty}\frac{\sin[\pi(2N+1)s]}{\sin(\pi s)}. \end{aligned} \tag{61}$$

Geometric series: $\sum_{n=0}^{2N} x^n = \frac{1-x^{2N+1}}{1-x}$

In the last line of Eq.(61), the numerator $\sin[\pi(2N+1)s]$ is a rapidly oscillating function of s, which may be ignored, as it "chops up" any function that it multiplies into positive and negative pieces, which, on integration, end up cancelling each other out. The only time that the behavior of the function on the right-hand side of Eq.(61) is nontrivial is when the denominator, $\sin(\pi s)$, approaches zero, which happens when s approaches an integer value $(0, \pm1, \pm2, \cdots)$. In the immediate vicinity of $s = n$, an integer, the denominator may be approximated as $\sin(\pi s) \cong \pi(-1)^n(s-n)$. Now, the numerator can always be written as $(-1)^n \sin[\pi(2N+1)(s-n)]$ because the product $n(2N+1)$ is either an odd or an even integer—depending on whether n is odd or even. When n is even, the argument of the modified sine function is augmented by an integer multiple of 2π, which has no effect on the value the function, and when n is odd, the argument is augmented by an odd multiple of π, which has the effect of multiplying the value of the sine function by -1. All in all, for sufficiently large N, one may replace the ratio $\sin[\pi(2N+1)s]/\sin(\pi s)$ in Eq.(61) with zero, when s is not too close to an integer, and with $\sin[\pi(2N+1)(s-n)]/[\pi(s-n)]$ in the vicinity of an integer n. Thus,

$$\begin{aligned}\mathcal{F}\{\mathrm{comb}(x)\} &= lim_{N\to\infty}\sum_{n=-\infty}^{\infty}\frac{\sin[\pi(2N+1)(s-n)]}{\pi(s-n)} \\ &= lim_{N\to\infty}\sum_{n=-\infty}^{\infty}(2N+1)\,\mathrm{sinc}[(2N+1)(s-n)] \\ &= \sum_{n=-\infty}^{\infty}\delta(s-n) = \mathrm{comb}(s). \end{aligned} \tag{62}$$

The scaling property of the δ-function, namely, $\delta(\alpha x) = |\alpha|^{-1}\delta(x)$, can be used when one tries to construct a comb function that has a different period. For instance, the function $\mathrm{comb}(x/p)$ has a period p, which is a positive real number. We will have

$$\mathrm{comb}(x/p) = \sum_{n=-\infty}^{\infty}\delta[(x/p)-n] = \sum_{n=-\infty}^{\infty}p\delta(x-np). \tag{63}$$

In words, the period of $\mathrm{comb}(x/p)$ is p, but each "tooth of the comb" is a δ-function whose area equals p. When p is greater than 1, the δ-functions constituting $\mathrm{comb}(x/p)$ become larger, and when p is less than 1, the δ-functions become smaller. The period and the height of the stretched (or compressed) comb function are thus tied together. As for the Fourier transform of $\mathrm{comb}(x/p)$, the scaling theorem of Fourier transforms yields

$$\mathcal{F}\{\mathrm{comb}(x/p)\} = p\,\mathrm{comb}(ps). \tag{64}$$

Keep in mind that the Fourier transform of $\mathrm{comb}(x/p)$ now has period $1/p$, but, because of the coefficient p that multiplies $\mathrm{comb}(ps)$ in Eq.(64), each tooth of the comb in the Fourier domain has an area of 1. Alternatively, the function $(1/p)\text{comb}(x/p)$, which has period p and teeth consisting of unit δ-functions, is Fourier transformed into $\text{comb}(ps)$, which has period $1/p$, with each tooth having an area of $1/p$.

7.8. Expanding periodic functions into a Fourier series. Let $f_p(x)$ be a (generally complex) periodic function of the real variable x, having period p. Denoting by $f(x)$ a single period of this function, say, the part of $f_p(x)$ between $x = x_0$ and $x = x_0 + p$, we can write

$$f_p(x) = (1/p)\text{comb}(x/p) * f(x). \tag{65}$$

The Fourier transform of $f_p(x)$ is thus the product of $F(s)$ and $\text{comb}(ps)$; that is,

$$F_p(s) = \text{comb}(ps)F(s) = (1/p)\sum_{n=-\infty}^{\infty} F(n/p)\delta[s-(n/p)]. \tag{66}$$

In words, the Fourier transform of the periodic function $f_p(x)$ is an array of δ-functions of magnitude $(1/p)F(n/p)$ located at regular intervals along the s-axis at $s_n = n/p$, where $n = 0, \pm 1, \pm 2, \cdots$ and $F(n/p)$ is the sampled value of the Fourier transform $F(s)$ of the single period $f(x)$ at the location s_n of the nth δ-function. The function $f_p(x)$ is reconstructed from $F_p(s)$ using inverse Fourier transformation, as follows:

$$\begin{aligned} f_p(x) &= \int_{-\infty}^{\infty} F_p(s)\exp(\mathrm{i}2\pi sx)\,\mathrm{d}s \\ &= \int_{-\infty}^{\infty}(1/p)\sum_{n=-\infty}^{\infty} F(n/p)\delta[s-(n/p)]\exp(\mathrm{i}2\pi sx)\,\mathrm{d}s \\ &= \sum_{n=-\infty}^{\infty}(1/p)F(n/p)\exp(\mathrm{i}2\pi nx/p). \end{aligned} \tag{67}$$

It is thus seen that $f_p(x)$ can be expressed as a superposition of complex exponential functions $\exp(\mathrm{i}2\pi nx/p)$, each with a coefficient $(1/p)F(n/p)$, obtained by sampling the Fourier transform $F(s)$ of a single period of the function at an integer-multiple n/p of the inverse of the period p. The coefficient is further normalized by p. Equation (67) is generally referred to as the Fourier series expansion of the periodic function $f_p(x)$.

If $f_p(x)$ happens to be a real function, the Fourier transform $F(s)$ of $f(x)$ will be Hermitian, that is, $F(-s) = F^*(s)$, in which case the Fourier series in Eq.(67) may be further simplified, as follows:

$$f_p(x) = (1/p)F(0) + (2/p)\sum_{n=1}^{\infty}\text{Real}\{F(n/p)\exp(\mathrm{i}2\pi nx/p)\}. \tag{68}$$

The Fourier series of a periodic real function, given by Eq.(68), is sometimes written as

$$f_p(x) = A_0 + \sum_{n=1}^{\infty} A_n\cos[2\pi(n/p)x + \varphi_n]. \tag{69}$$

The Fourier series coefficients are then given by

$$A_0 = (1/p)F(0) = (1/p)\int_{x_0}^{x_0+p} f_p(x)\mathrm{d}x. \tag{70a}$$

$$A_n\exp(\mathrm{i}\varphi_n) = (2/p)F(n/p) = (2/p)\int_{x_0}^{x_0+p} f_p(x)\exp(-\mathrm{i}2\pi nx/p)\,\mathrm{d}x. \tag{70b}$$

Occasionally, the Fourier series of a periodic real function is expressed in terms of sines and cosines, as follows:

$$f_p(x) = A_0 + \sum_{n=1}^{\infty}[B_n \cos(2\pi nx/p) + C_n \sin(2\pi nx/p)]. \tag{71}$$

In this case, the constant term A_0 will be given by Eq.(70a), whereas the coefficients B_n and C_n of the cosine and sine functions will be given by

$$B_n = (2/p)\text{Real}\{F(n/p)\} = (2/p)\int_{x_0}^{x_0+p} f_p(x)\cos(2\pi nx/p)\,dx. \tag{72a}$$

$$C_n = -(2/p)\text{Imag}\{F(n/p)\} = (2/p)\int_{x_0}^{x_0+p} f_p(x)\sin(2\pi nx/p)\,dx. \tag{72b}$$

If, in addition to being real valued, $f_p(x)$ happens to be an even function, then $C_n = 0$ and the Fourier series in Eq.(71) will consist of the constant term A_0 and the cosine functions only. In contrast, if $f_p(x)$ is an odd and real function, then $A_0 = 0$ and $B_n = 0$, in which case the Fourier series in Eq.(71) will consist solely of sine functions.

7.9. Sampling theorem and the Nyquist rate. Suppose the Fourier transform $F(s)$ of the function $f(x)$ is band-limited, that is, outside a bandwidth W, the spectral distribution of $f(x)$ is negligible, namely $F(s) \cong 0$ for $|s| \geq W/2$. If the function is now sampled at regular intervals p by multiplying $f(x)$ with $(1/p)\text{comb}\,(x/p)$, the Fourier transform of the sampled function can be written as $\text{comb}(ps) * F(s)$, as shown in figure 7.7. Considering that $\text{comb}(ps)$ is a regular array of δ-functions of magnitude $1/p$ located at $1/p$ intervals, the sampled function's Fourier transform becomes an array of shifted copies of $(1/p)F(s)$, similarly replicated at $1/p$ intervals.

To prevent overlap among the tails of various copies $(1/p)F(s - n/p)$ of the shifted spectral distribution function, it is necessary that $1/p$ be greater than W. Thus, if the function $f(x)$ is sampled at regular intervals p, with $p < 1/W$, the Fourier transform of the sampled function will consist of replica of $F(s)$ at regular $1/p$ intervals without any overlap (or aliasing) whatsoever. The condition $p < 1/W$ is known as the Nyquist criterion for sampling a band-limited function.

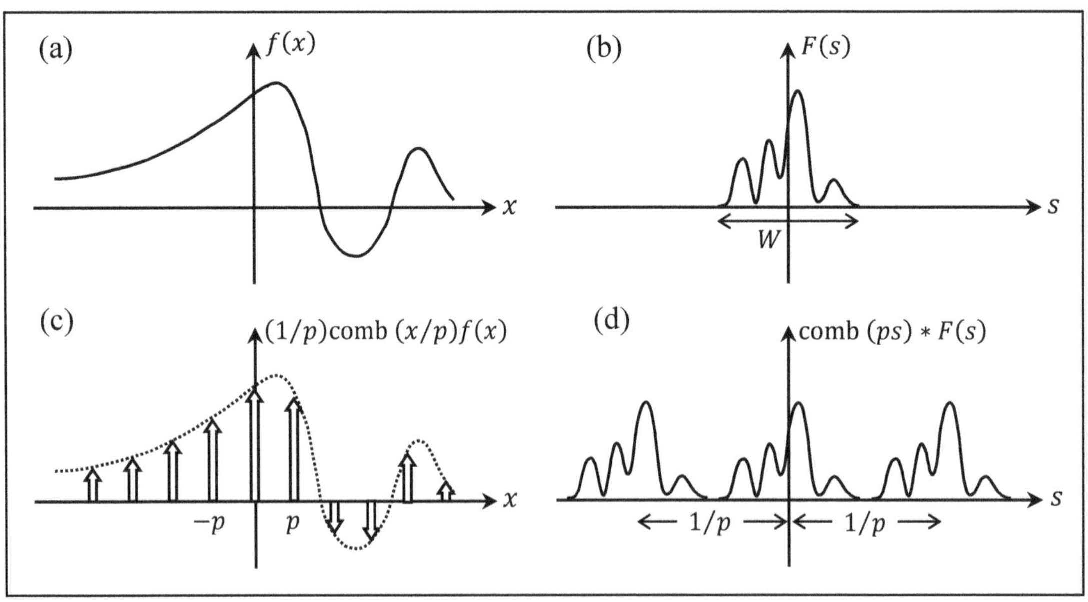

Figure 7.7. (a) The function $f(x)$, which may be real or complex, is an arbitrary function of the real variable x. (b) The nonzero values of the Fourier transform $F(s)$ of $f(x)$ are confined to the frequency range $|s| < W/2$. (c) Sampling of $f(x)$ is accomplished by multiplying with a comb function whose teeth, repeated at regular intervals p, have unit magnitude (i.e., the area under each δ-function is 1.0). (d) The Fourier transform of the sampled function is obtained by replicating the function $(1/p)F(s)$ at regular $1/p$ intervals along the s-axis.

Assuming the function is sampled at a rate that satisfies Nyquist's criterion, it is always possible to reconstruct the original $f(x)$ from a knowledge of its sample values. In fact, to reconstruct $f(x)$, all one needs to do is to apply a low-pass filter to the sampled function. Based on the depiction in figure 7.7d, it is trivial to see that, on multiplying the transfer function $H(s) = p\,\mathrm{rect}(ps)$ of a low-pass filter into $\mathrm{comb}(ps) * F(s)$, all the redundant copies of $F(s)$ will be eliminated and only the central (unshifted) copy survives. Subsequently, the original $f(x)$ is obtained following an inverse Fourier transformation. Given that multiplication with $H(s)$ in the Fourier domain is equivalent to convolution with $h(x) = \mathcal{F}^{-1}\{H(s)\} = \mathrm{sinc}(x/p)$ in the x-domain, the output of the low-pass filter will be

$$[(1/p)\mathrm{comb}\,(x/p)f(x)] * \mathrm{sinc}(x/p) = \textstyle\sum_{n=-\infty}^{\infty} f(np)\mathrm{sinc}[(x/p) - n] = f(x). \quad (73)$$

In practice, a pre-filter is typically applied to $f(x)$ to truncate its frequency content and/or to clean it up before sampling the function at (or slightly above) the Nyquist rate. Also, during reconstruction, the transfer function $H(s)$ of the low-pass filter may not be as simple as a $\mathrm{rect}(\cdot)$ function but could contain additional features to smooth out the transition near the cut-off frequency and/or to boost or attenuate certain spectral features of the reconstructed function.

7.10. Fourier transformation in higher dimensions. The idea of Fourier transformation can be readily extended to functions of higher dimensions. For instance, in a two-dimensional (2D) space, the Fourier transform and its inverse are defined as follows:

$$F(s_x, s_y) = \iint_{-\infty}^{\infty} f(x, y) \exp[-\mathrm{i}2\pi(s_x x + s_y y)]\, \mathrm{d}x\mathrm{d}y. \quad (74)$$

$$f(x, y) = \iint_{-\infty}^{\infty} F(s_x, s_y)\exp[\mathrm{i}2\pi(s_x x + s_y y)]\, \mathrm{d}s_x\mathrm{d}s_y. \quad (75)$$

If $f(x, y)$ happens to be a separable function, that is, $f(x, y) = g(x)h(y)$, then its Fourier transform will be the product of the one-dimensional transforms of its two constituents; that is, $F(s_x, s_y) = G(s_x)H(s_y)$. In general, however, the functions of interest are not separable, and the 2D integrals in Eqs.(74) and (75) cannot be further simplified. Similar concepts and methods apply to higher-dimensional functions.

The special functions introduced in section 1 can be similarly generalized and extended to two or higher dimensions. For instance, one can construct a 2D rectangular function in the form of $\mathrm{rect}(x/L_x)\mathrm{rect}(y/L_y)$, which equals 1.0 over an $L_x \times L_y$ rectangular area centered at the origin of the xy-plane and equals zero outside that rectangle. It is also possible to define an inseparable 2D function, $\mathrm{circ}(x, y)$, which equals 1.0 over a circle of radius $r = \sqrt{x^2 + y^2} = 1.0$ centered at the origin of the xy-plane and equals zero outside that circle. In the same vein, one may define the separable sinc function in two dimensions, $f(x, y) = \mathrm{sinc}(x)\mathrm{sinc}(y)$, as well as the inseparable sinc function, $g(x, y) = \mathrm{sinc}(r)$, where $r = \sqrt{x^2 + y^2}$ is the distance from the origin of coordinates.

Delta-functions in higher-dimensional spaces may be constructed as products of lower-dimensional δ-functions. For instance,

$$\delta(x, y) = \delta(x)\delta(y) = lim_{\alpha,\beta\to 0}(1/\alpha\beta)\, \mathrm{rect}(x/\alpha)\mathrm{rect}(y/\beta). \quad (76)$$

Alternatively, one may choose to define such higher-dimensional δ-functions in terms of the limits of non-separable functions, say,

$$\delta(x,y) = lim_{R\to 0}[(\pi R^2)^{-1}\text{circ}(r/R)]. \quad (77)$$

These and many similar definitions for the δ-function in 2D space are equivalent and may be used interchangeably, provided that the adopted function in the proper limit approaches something tall, narrow, and symmetric (with respect to the origin of coordinates), and with a volume of 1.0 under the function. The sifting property of a two-dimensional δ-function will then be expressed as follows:

$$\iint_{-\infty}^{\infty} f(x,y)\delta(x-x_0, y-y_0)\text{d}x\text{d}y = f(x_0, y_0). \quad (78)$$

Similar generalizations can be made straightforwardly for other special functions, such as the various derivatives of the δ-function.

7.11. Elementary theory of diffraction. Theories of optical diffraction are intimately tied to (and deeply rooted in) the Fourier representation of the electromagnetic field.[2-5] As a simple example, consider, with reference to figure 7.8, a monochromatic planewave of wavelength λ and amplitude a_0 illuminating a phase-amplitude object. The object is located in the xy-plane at $z = 0$, and its transmission coefficient is specified by the complex function $t(x,y)$. The complex amplitude of the transmitted beam immediately after the object is thus given by $a(x,y,z=0^+) = a_0 t(x,y)$. Using a pair of Fourier and inverse Fourier transforms, we write

$$A(s_x, s_y) = \iint_{-\infty}^{\infty} a(x,y,z=0^+)\exp[-\text{i}2\pi(s_x x + s_y y)]\,\text{d}x\text{d}y, \quad (79a)$$

$$a(x,y,z=0^+) = \iint_{-\infty}^{\infty} A(s_x, s_y)\exp[\text{i}2\pi(s_x x + s_y y)]\,\text{d}s_x\text{d}s_y. \quad (79b)$$

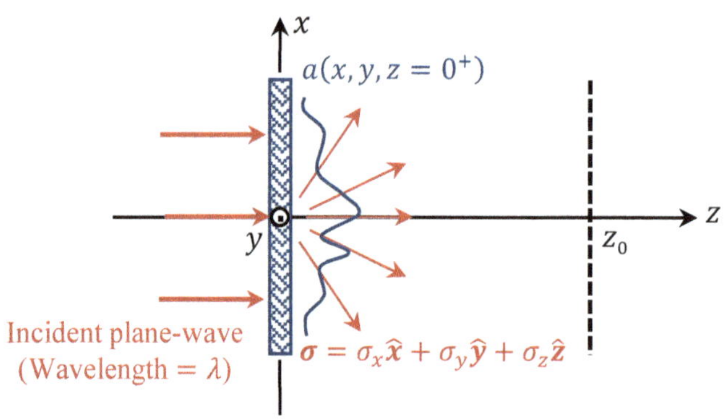

Figure 7.8. A plane, monochromatic, electromagnetic wave, having wavelength λ and amplitude a_0, is incident on an object with a transmission coefficient $t(x,y)$. The object is located in the xy-plane at $z = 0$. The transmitted wave, whose amplitude $a(x,y,z=0^+)$ is $a_0 t(x,y)$, is mathematically decomposed into a superposition of plane waves propagating along various directions that are specified by unit vectors $\boldsymbol{\sigma} = \sigma_x\hat{\boldsymbol{x}} + \sigma_y\hat{\boldsymbol{y}} + \sigma_z\hat{\boldsymbol{z}}$. The goal of diffraction calculations is to determine the field amplitude distribution $a(x,y,z=z_0)$ in the destination plane, which is an xy-plane located at $z = z_0$.

It is well known from Maxwellian electrodynamics[6,7] that a plane electromagnetic wave propagating in free space along the unit vector $\boldsymbol{\sigma} = \sigma_x\hat{\boldsymbol{x}} + \sigma_y\hat{\boldsymbol{y}} + \sigma_z\hat{\boldsymbol{z}}$, having frequency ω and wavelength $\lambda = 2\pi c/\omega$, has the following electric field distribution:

$$\boldsymbol{E}(x,y,z,t) = \boldsymbol{E}_0\exp[\text{i}(2\pi/\lambda)(x\sigma_x + y\sigma_y + z\sigma_z)]\exp(-\text{i}\omega t). \quad (80)$$

Here, $\boldsymbol{E}_0$ is a constant vector that is perpendicular to $\boldsymbol{\sigma}$ and carries information about the magnitude, phase, and direction of the E-field oscillations. In many situations of practical interest, the orientation of $\boldsymbol{E}_0$ remains more or less invariant, and it is the magnitude and phase of the E-field that are of primary concern. The scalar theory of diffraction has been developed to address situations where, to a good approximation, the vector nature of $\boldsymbol{E}_0$ can be ignored.[2] A comparison between Eqs.(79b) and (80) reveals that the (complex) E-field amplitude immediately after the object shown in figure 7.8 is a superposition of plane waves of amplitude $A(s_x, s_y)$, where $s_x = \sigma_x/\lambda$ and $s_y = \sigma_y/\lambda$. Note that, in the $z = 0^+$ plane, the term $z\sigma_z$ vanishes. However, given that $\boldsymbol{\sigma}$ is a unit vector, the value of σ_z is readily determined, as follows:

$$\sigma_z = \sqrt{1 - \sigma_x^2 - \sigma_y^2} = \sqrt{1 - (\lambda s_x)^2 - (\lambda s_y)^2}. \tag{81}$$

If we now choose to propagate the beam to the xy-plane located at $z = z_0$, we must add the term $z_0\sigma_z$ to the exponential factor under the integral sign in Eq.(79b), in which case we will have

$$a(x, y, z = z_0)$$
$$= \iint_{-\infty}^{\infty} A(s_x, s_y) \exp\left[\mathrm{i}2\pi z_0\sqrt{(1/\lambda)^2 - s_x^2 - s_y^2}\right] \exp[\mathrm{i}2\pi(s_x x + s_y y)]\, \mathrm{d}s_x \mathrm{d}s_y. \tag{82}$$

It is thus seen that electromagnetic wave propagation in free space between the xy-planes at $z = 0^+$ and $z = z_0$ requires that we first compute the Fourier transform $A(s_x, s_y)$ of the scalar field distribution at $z = 0^+$, then multiply this Fourier transform with the propagation factor $\exp\left[\mathrm{i}2\pi z_0\sqrt{(1/\lambda)^2 - s_x^2 - s_y^2}\right]$, and, finally, carry out an inverse Fourier transformation in accordance with Eq.(82) to arrive at the distribution of the field in the destination plane.[5]

In practice, the integral in Eq.(82) need only be evaluated within the $s_x s_y$-plane over a circle of radius $1/\lambda$. This is because the square root appearing in the exponential factor becomes imaginary outside the circle of radius $1/\lambda$. The rapid decay with distance z_0 of the resulting exponential factor indicates that the plane waves whose s_x, s_y values fall outside the circle of radius $1/\lambda$—the so-called evanescent waves[2-7]—need not be taken into account if the destination plane happens to be more than a few wavelengths away from the object.

For distances z_0 that are not too close, but also not too far away from the object (i.e., in the so-called near-field or Fresnel diffraction regime), both forward and inverse transforms in Eqs.(79a) and (82) must be evaluated. Typically, these Fourier integrals do not submit to analytic methods and their evaluation requires numerical computations.

As the destination plane moves further away from the object, z_0 becomes larger, and one eventually moves into the far-field (or Fraunhofer) diffraction regime. In the far field, the oscillations of the phase-factor $\exp\left[\mathrm{i}2\pi z_0\sqrt{(1/\lambda)^2 - s_x^2 - s_y^2}\right]$ with the changing s_x, s_y become too rapid for numerical evaluation of the inverse Fourier integral of Eq.(82). Fortunately, under such circumstances, the integral in Eq.(82) yields to an approximate calculation with the aid of a powerful approximation technique known as the stationary-phase method. This approximation technique will be described in the following section, first in the context of one-dimensional integrals, and, subsequently, in a more general form suitable for two-dimensional problems.

7.12. Method of stationary-phase approximation. Consider the following one-dimensional integral in which $f(x)$ is a generally complex function of the real variable x, whereas the function $g(x)$ is real and the parameter η is real and large. The domain of integration is the interval $[a, b]$ on the real axis.

$$I = \int_a^b f(x) \exp[\mathrm{i}\eta g(x)]\, \mathrm{d}x. \tag{83}$$

As the point x moves along the x-axis from a to b, any small change in the value of $g(x)$ will be amplified due to multiplication by the large parameter η, thus resulting in rapid oscillations of the complex exponential function. Assuming $f(x)$ does not vary too rapidly with small changes of the variable x, the integral over several (rapid) oscillations of the integrand would vanish. The net contribution to the integral thus appears to be zero whenever these circumstances prevail. In contrast, consider a point, say, $x = x_0$, around which $g(x)$ remains more or less flat. In the immediate vicinity of x_0, the complex exponential does not vary all that much and, therefore, the contribution of the integrand around $x = x_0$ to the overall integral could become significant. The point x_0 at which $g(x)$ is locally flat is generally referred to as a stationary point of $g(x)$. To find all such stationary points, one must solve the equation $g'(x) = 0$. The domain of integration can then be broken up into short segments of the x-axis, each covering a small neighborhood of one of the stationary points.

Writing the Taylor series expansion of $g(x)$ up to and including the second-order term around a stationary point x_n, and recognizing that, by definition, the first derivative of $g(x)$ at $x = x_n$ vanishes, we will have

$$g(x) \cong g(x_n) + \cancelto{0}{g'(x_n)}(x - x_n) + \tfrac{1}{2} g''(x_n)(x - x_n)^2. \tag{84}$$

Denoting the interval in the immediate vicinity of x_n by $[x_n - \varepsilon, x_n + \varepsilon]$, where ε is a small number, we may write

$$I \cong \sum_{x_n} \int_{x_n-\varepsilon}^{x_n+\varepsilon} f(x) \exp\{\mathrm{i}\eta[g(x_n) + \tfrac{1}{2} g''(x_n)(x - x_n)^2]\}\, \mathrm{d}x$$

$$\boxed{\text{change of variable: } y = x - x_n} \rightarrow \cong \sum_{x_n} f(x_n) \exp[\mathrm{i}\eta g(x_n)] \int_{x_n-\varepsilon}^{x_n+\varepsilon} \exp[\tfrac{1}{2}\mathrm{i}\eta g''(x_n)(x - x_n)^2]\, \mathrm{d}x$$

$$\cong \sum_{x_n} f(x_n) \exp[\mathrm{i}\eta g(x_n)] \int_{-\varepsilon}^{\varepsilon} \exp[\tfrac{1}{2}\mathrm{i}\eta g''(x_n) y^2]\, \mathrm{d}y. \tag{85}$$

The domain of integration around each x_n may now be expanded from $(-\varepsilon, \varepsilon)$ to $(-\infty, \infty)$, as the major contribution to each integral comes primarily from the immediate vicinity of $y = 0$. Another change of variable $x = \sqrt{\tfrac{1}{2}|\eta g''(x_n)|}\, y$ now yields

$$I \cong \sum_{x_n} \sqrt{2/|\eta g''(x_n)|}\, f(x_n) \exp[\mathrm{i}\eta g(x_n)] \underbrace{\int_{-\infty}^{+\infty} \exp(\overset{\text{sign of } \eta g''(x_n)}{\pm} \mathrm{i}x^2)\, \mathrm{d}x.}_{\sqrt{\pi}\exp(\pm \mathrm{i}\pi/4)} \tag{86}$$

The end result of the stationary-phase approximation thus turns out to be

$$I \cong \sum_{x_n} \sqrt{2\pi/|\eta g''(x_n)|} \exp(\pm \mathrm{i}\pi/4) f(x_n) \exp[\mathrm{i}\eta g(x_n)]. \tag{87}$$

Needless to say, the approximation improves for larger values of η, as the oscillations of the integrand away from the stationary point(s) intensify. If a stationary point happens to coincide with either end point a or b of the integration domain, its contribution to the overall integral will reduce by a factor of 2. These results become inapplicable if $g''(x_n)$ for some stationary point turns out to be vanishingly small; in such cases, one must revisit Eq.(84) and consider the effects of higher-order terms in the Taylor series expansion of $g(x)$ around x_n.

7.12.1. Stationary-phase approximation in two-dimensional space. Let $f(x,y)$ be an arbitrary complex-valued function, $g(x,y)$ an arbitrary real-valued function, and η a large, real-valued, positive constant. The integral

$$I = \iint_{\mathcal{D}} f(x,y)\exp[\mathrm{i}\eta g(x,y)]\,\mathrm{d}x\mathrm{d}y \tag{88}$$

is defined over an appropriate domain $\mathcal{D}$ within the xy-plane. A stationary point of this integral is a point at which $\partial_x g(x,y) = \partial_y g(x,y) = 0$. Assuming the stationary point (x_0, y_0) is inside the domain $\mathcal{D}$, in what follows we shall determine the contribution of this stationary point to the integral I.

When η is sufficiently large, the complex exponential oscillates rapidly, causing the integral over any region containing a large number of such oscillations to average out to zero. The only regions for which this argument does *not* hold are the neighborhoods of stationary points, where $g(x,y)$ does not vary much, thus allowing the integral to acquire a significant value. A Taylor series expansion of $g(x,y)$ around the stationary point (x_0, y_0) yields

$$g(x,y) \cong g(x_0,y_0) + \cancelto{0}{\partial_x g|_{(x_0,y_0)}}(x-x_0) + \cancelto{0}{\partial_y g|_{(x_0,y_0)}}(y-y_0)$$
$$+\tfrac{1}{2}\partial_{xx}^2 g|_{(x_0,y_0)}(x-x_0)^2 + \partial_{xy}^2 g|_{(x_0,y_0)}(x-x_0)(y-y_0) + \tfrac{1}{2}\partial_{yy}^2 g|_{(x_0,y_0)}(y-y_0)^2. \tag{89}$$

Simplifying the notation by writing the second derivatives at (x_0, y_0) as g_{xx}, g_{xy}, and g_{yy}, the integral I may now be approximated as follows:

$$I \cong f(x_0,y_0)\exp[\mathrm{i}\eta g(x_0,y_0)] \iint_{-\infty}^{\infty} \exp[\tfrac{1}{2}\mathrm{i}\eta(g_{xx}x^2 + 2g_{xy}xy + g_{yy}y^2)]\,\mathrm{d}x\mathrm{d}y. \tag{90}$$

Note that the domain of integration in Eq. (90) has been expanded from a small neighborhood of the stationary point (x_0, y_0) to the entire xy-plane, the reason being that the only place where the integral in Eq. (90) has a significant value is the neighborhood of (x_0, y_0); everywhere else the integrand oscillates rapidly and the integral averages out to zero, so extending the domain of integration to the entire plane does not make much of a difference. Note also that a change of variables has been made in Eq.(90) so that $x - x_0 \to x$ and $y - y_0 \to y$.

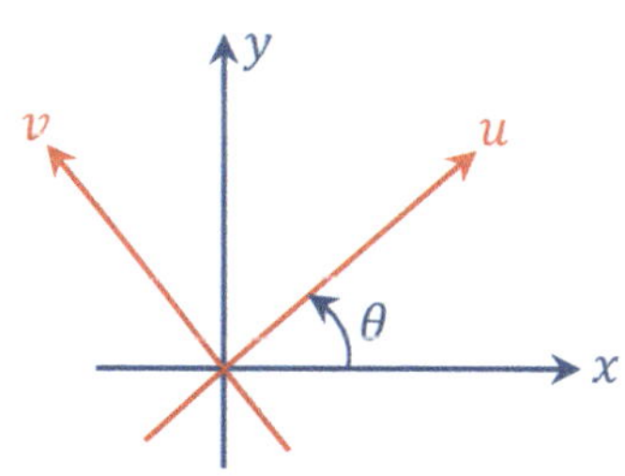

If we now rotate the coordinates through an angle θ in the xy-plane, such that $x = u\cos\theta - v\sin\theta$ and $y = u\sin\theta + v\cos\theta$, the exponent of the integrand becomes

$$g_{xx}x^2 + 2g_{xy}xy + g_{yy}y^2 = g_{xx}(u\cos\theta - v\sin\theta)^2$$
$$+2g_{xy}(u\cos\theta - v\sin\theta)(u\sin\theta + v\cos\theta) + g_{yy}(u\sin\theta + v\cos\theta)^2$$
$$= (g_{xx}\cos^2\theta + g_{xy}\sin 2\theta + g_{yy}\sin^2\theta)u^2 - [(g_{xx} - g_{yy})\sin 2\theta - 2g_{xy}\cos 2\theta]uv$$
$$+(g_{xx}\sin^2\theta - g_{xy}\sin 2\theta + g_{yy}\cos^2\theta)v^2. \tag{91}$$

Here, we have used the trigonometric identities $\sin 2\theta = 2\sin\theta\cos\theta$ and $\cos 2\theta = \cos^2\theta - \sin^2\theta$. Choosing the angle θ to eliminate the cross term uv, that is, $\tan 2\theta = 2g_{xy}/(g_{xx} - g_{yy})$, the remaining terms on the right-hand side of Eq. (91) become

$$\tfrac{1}{2}\left[g_{xx} + g_{yy} + \frac{(g_{xx}-g_{yy})^2 + 4g_{xy}^2}{g_{xx} - g_{yy}}\cos 2\theta\right]u^2 + \tfrac{1}{2}\left[g_{xx} + g_{yy} - \frac{(g_{xx}-g_{yy})^2 + 4g_{xy}^2}{g_{xx} - g_{yy}}\cos 2\theta\right]v^2. \tag{92}$$

Invoking the identity

$$\cos 2\theta = \frac{1}{\sqrt{1+\tan^2 2\theta}} = (g_{xx} - g_{yy})/\sqrt{(g_{xx} - g_{yy})^2 + 4g_{xy}^2}, \tag{93}$$

the exponent of the integrand of Eq.(90), which is given in Eq. (92), simplifies as follows:

$$\tfrac{1}{2}\left[g_{xx} + g_{yy} + \sqrt{(g_{xx} - g_{yy})^2 + 4g_{xy}^2}\right]u^2 + \tfrac{1}{2}\left[g_{xx} + g_{yy} - \sqrt{(g_{xx} - g_{yy})^2 + 4g_{xy}^2}\right]v^2. \tag{94}$$

The integral over the xy-plane in Eq. (90) now becomes a separable integral over the uv-plane, namely

$$\iint_{-\infty}^{\infty} \exp[½\mathrm{i}\eta(g_{xx}x^2 + 2g_{xy}xy + g_{yy}y^2)]\,\mathrm{d}x\mathrm{d}y$$

$$= \int_{-\infty}^{\infty} \exp\left\{¼\mathrm{i}\eta\left[g_{xx} + g_{yy} + \sqrt{(g_{xx} - g_{yy})^2 + 4g_{xy}^2}\right]u^2\right\}\mathrm{d}u$$

$$\times \int_{-\infty}^{\infty} \exp\left\{¼\mathrm{i}\eta\left[g_{xx} + g_{yy} - \sqrt{(g_{xx} - g_{yy})^2 + 4g_{xy}^2}\right]v^2\right\}\mathrm{d}v$$

± depending on the sign of the coefficient of u^2 or that of v^2.

$$= \frac{2\int_{-\infty}^{\infty} \exp(\pm\mathrm{i}u'^2)\mathrm{d}u'}{\left|\eta\left[g_{xx}+g_{yy}+\sqrt{(g_{xx}-g_{yy})^2+4g_{xy}^2}\right]\right|^{½}} \times \frac{2\int_{-\infty}^{\infty} \exp(\pm\mathrm{i}v'^2)\mathrm{d}v'}{\left|\eta\left[g_{xx}+g_{yy}-\sqrt{(g_{xx}-g_{yy})^2+4g_{xy}^2}\right]\right|^{½}}$$

$$= \frac{4\int_{-\infty}^{\infty} \exp(\pm\mathrm{i}u^2)\mathrm{d}u \times \int_{-\infty}^{\infty} \exp(\pm\mathrm{i}v^2)\mathrm{d}v}{2|\eta||g_{xx}g_{yy} - g_{xy}^2|^{½}} = \frac{2\pi\exp(\pm\mathrm{i}\pi/4)\exp(\pm\mathrm{i}\pi/4)}{|\eta||g_{xx}g_{yy} - g_{xy}^2|^{½}}$$

$$= \frac{\mathrm{i}2\pi\nu}{|\eta||g_{xx}g_{yy} - g_{xy}^2|^{½}}; \qquad \nu = \begin{cases} \pm 1 & \text{if } g_{xx}g_{yy} > g_{xy}^2 \text{ and } g_{xx} \gtrless 0, \\ -\mathrm{i} & \text{if } g_{xx}g_{yy} < g_{xy}^2. \end{cases} \tag{95}$$

The identity $\int_{-\infty}^{\infty} \exp(\pm\mathrm{i}x^2)\,\mathrm{d}x = \sqrt{\pi}\exp(\pm\mathrm{i}\pi/4)$ has been used in the above derivation. The stationary-phase approximation to the integral I is thus found to be

$$I \cong \frac{\mathrm{i}2\pi\nu}{|\eta|\,|g_{xx}g_{yy} - g_{xy}^2|^{½}} f(x_0, y_0)\exp[\mathrm{i}\eta g(x_0, y_0)]. \tag{96}$$

7.13. Saddle point approximation. Let $g(z) = g^{(\mathrm{r})}(z) + \mathrm{i}g^{(\mathrm{i})}(z)$ be a complex function of the complex variable $z = x + \mathrm{i}y$. Assuming that in the vicinity of $z = z_0$ the function is analytic and that $g'(z_0) = 0$, we will have

$$g^{(\mathrm{r})}(x,y) = g^{(\mathrm{r})}(x_0,y_0) + ½g_{xx}^{(\mathrm{r})}(x - x_0)^2 + ½g_{yy}^{(\mathrm{r})}(y - y_0)^2 + g_{xy}^{(\mathrm{r})}(x - x_0)(y - y_0) + \cdots. \tag{97}$$

Here, $g_{xx}^{(\mathrm{r})} = (\partial^2/\partial x^2)g^{(\mathrm{r})}(x_0, y_0)$, with a similar notation used for $g_{yy}^{(\mathrm{r})}$ and $g_{xy}^{(\mathrm{r})}$. The Cauchy-Riemann conditions for analyticity at z_0 require that $g_x^{(\mathrm{r})} = g_y^{(\mathrm{i})}$ and $g_y^{(\mathrm{r})} = -g_x^{(\mathrm{i})}$. Consequently, $g_{xx}^{(\mathrm{r})} = g_{xy}^{(\mathrm{i})}$ and $g_{yy}^{(\mathrm{r})} = -g_{xy}^{(\mathrm{i})}$, which results in $g_{xx}^{(\mathrm{r})} = -g_{yy}^{(\mathrm{r})}$. Needless to say, parallel relations exist for the imaginary part $g^{(\mathrm{i})}(x,y)$ of $g(z)$ and its second-order partial derivatives at $z = z_0$.

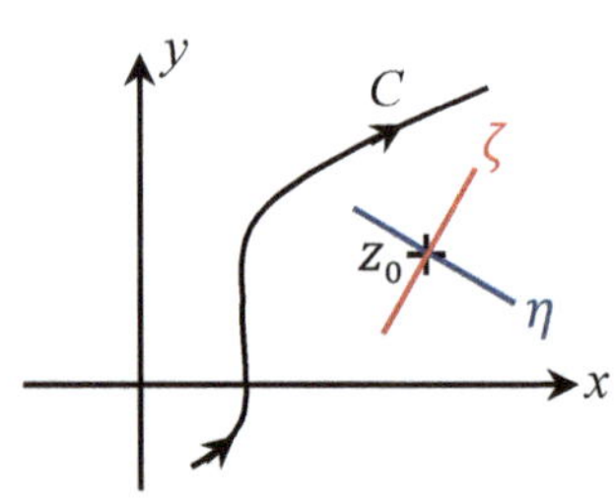

We now rotate the xy coordinates through an angle φ around the point (x_0, y_0) so that the new coordinates (η, ζ) are related to $(x - x_0, y - y_0)$ via the unitary transformation $(x - x_0) = \eta\cos\varphi - \zeta\sin\varphi$ and $(y - y_0) = \eta\sin\varphi + \zeta\cos\varphi$. The Taylor series expansion of $g^{(\mathrm{r})}(x, y)$ around (x_0, y_0) is now written

$$g^{(\mathrm{r})}(\eta,\zeta) = g^{(\mathrm{r})}(x_0, y_0) + ½g_{xx}^{(\mathrm{r})}(\eta\cos\varphi - \zeta\sin\varphi)^2 - ½g_{xx}^{(\mathrm{r})}(\eta\sin\varphi + \zeta\cos\varphi)^2$$

$$+g_{xy}^{(\mathrm{r})}(\eta\cos\varphi - \zeta\sin\varphi)(\eta\sin\varphi + \zeta\cos\varphi) + \cdots$$

$$= g^{(\mathrm{r})}(x_0, y_0) + ½\left[g_{xx}^{(\mathrm{r})}\cos 2\varphi + g_{xy}^{(\mathrm{r})}\sin 2\varphi\right](\eta^2 - \zeta^2)$$

$$+\left[g_{xy}^{(\mathrm{r})}\cos 2\varphi - g_{xx}^{(\mathrm{r})}\sin 2\varphi\right]\eta\zeta + \cdots. \quad (98)$$

Setting $\tan(2\varphi) = g_{xy}^{(\mathrm{r})}/g_{xx}^{(\mathrm{r})}$ ensures that the $\eta\zeta$ term in Eq.(98) disappears. The local curvature of $g^{(\mathrm{r})}(\eta,\zeta)$ along the η direction is now seen to be the opposite of that along ζ, indicating that the point z_0 is a *saddle point* of the function $g^{(\mathrm{r})}(x,y)$. In other words, if at $z = z_0$ the real part $g^{(\mathrm{r})}(z)$ of $g(z)$ has a peak along the η direction, then it will have a valley along the ζ direction or vice versa. It is not difficult to show that the above choice of φ results in $g^{(\mathrm{r})}(\eta,\zeta) = g^{(\mathrm{r})}(x_0, y_0) + ½\left([g_{xx}^{(\mathrm{r})}]^2 + [g_{xy}^{(\mathrm{r})}]^2\right)^{½}(\eta^2 - \zeta^2) + \cdots$. A similar argument applies to the imaginary part $g^{(\mathrm{i})}(x,y)$ of the function $g(z)$. This time, however, the coordinates must rotate through an angle φ', such that $\tan(2\varphi') = g_{xy}^{(\mathrm{i})}/g_{xx}^{(\mathrm{i})}$, which is the inverse of $\tan(2\varphi)$. Therefore, the principal axes of curvature of $g^{(\mathrm{i})}(x,y)$ at $z = z_0$ are rotated by 45° relative to those of $g^{(\mathrm{r})}(x,y)$ within the xy-plane. Stated differently, the saddle of the real part of $g(z)$ at $z = z_0$ is rotated around z_0 by 45° relative to the saddle of the imaginary part of the function. A simple example is provided by the function $g(z) = z^2$, whose derivative vanishes at $z = 0$. Whereas $g^{(\mathrm{r})}(x,y) = x^2 - y^2$ has its principal axes of curvature along the x and y axes, the function $g^{(\mathrm{i})}(x,y) = 2xy$ has its principal axes along the $\pm 45°$ lines within the xy-plane.

Following this preliminary introduction, we now turn to the subject of saddle point approximation, which involves the approximate evaluation of $\int_C f(z)\exp[\eta g(z)]\,\mathrm{d}z$ along a continuous path C in the complex z-plane—the integration path C may be an open or a closed contour. The functions $f(z)$ and $g(z)$ are analytic, except possibly at a few singular points, and the constant parameter η is real valued and large (although it could be positive or negative). One can deform the integration path C to go through any desired point in the z-plane, provided that no singularities of the integrand are crossed while the path is being continuously deformed. (If singularities are crossed, then, in accordance with the Cauchy-Goursat theorem, their contributions to the integral—in the form of residues—must be accounted for.) The path must be deformed until it goes through a point z_0 where (i) $g'(z_0) = 0$; (ii) the curvature of $\mathrm{Re}[g(z)]$ along the deformed path at $z = z_0$ is large and opposite in sign to that of η; and (iii) along the deformed path at $z = z_0$, the function $\mathrm{Re}[g(z)]$ acquires either its maximum (when $\eta > 0$) or its minimum (when $\eta < 0$) value among all points on the deformed path C'. (For a detailed description of the path of steepest ascent or descent see the digression at the end of the section.) To simplify the notation, we continue to refer to the real and imaginary parts of $g(z)$ as $g^{(\mathrm{r})}(z)$ and $g^{(\mathrm{i})}(z)$, respectively.

Assuming that an appropriate saddle point $z = z_0$ of $g^{(\mathrm{r})}(z)$ has been identified, and that the integration contour has been properly deformed to wind its way through z_0 (in the correct direction), the logic of saddle point approximation boils down to the following proposition: The magnitude of the integrand $f(z)\exp[\eta g(z)]$ is so much greater at and around z_0 than everywhere else along the deformed contour C' that our efforts must be concentrated solely at and around $z = z_0$, ignoring the contributions to the integral along the rest of the path. We thus proceed to expand $f(z)$ and $g(z)$ into their respective Taylor series at $z = z_0$, using a simplified notation for the various derivatives at z_0, and also writing $z - z_0 = s\hat{\zeta}$, where the real number s (which could be positive or negative) is the distance from z_0 to z along the (deformed) contour C', while $\hat{\zeta}$ is a fixed, unit-magnitude complex number directed along C' at $z = z_0$. We will have

$$\begin{aligned} f(z) &= f(z_0) + f'(z_0)(z - z_0) + \tfrac{1}{2!} f''(z_0)(z - z_0)^2 + \tfrac{1}{3!} f'''(z_0)(z - z_0)^3 + \cdots \\ &= f_0 + f_1\hat{\zeta}s + \tfrac{1}{2} f_2\hat{\zeta}^2 s^2 + \tfrac{1}{6} f_3\hat{\zeta}^3 s^3 + \cdots . \end{aligned} \tag{99}$$

$$\begin{aligned} g(z) &= g(z_0) + \cancelto{0}{g'(z_0)}(z - z_0) + \tfrac{1}{2!} g''(z_0)(z - z_0)^2 + \tfrac{1}{3!} g'''(z_0)(z - z_0)^3 + \cdots \\ &= g_0 + \tfrac{1}{2} g_2\hat{\zeta}^2 s^2 + \tfrac{1}{6} g_3\hat{\zeta}^3 s^3 + \tfrac{1}{24} g_4\hat{\zeta}^4 s^4 + \cdots . \end{aligned} \tag{100}$$

Considering that $g''(z_0) = g_{xx}^{(\mathrm{r})} + \mathrm{i}g_{xx}^{(\mathrm{i})} = g_{xx}^{(\mathrm{r})} - \mathrm{i}g_{xy}^{(\mathrm{r})} = |g''(z_0)|\exp(-\mathrm{i}2\varphi)$, where the fixed angle $\varphi = ½\tan^{-1}[g_{xy}^{(\mathrm{r})}/g_{xx}^{(\mathrm{r})}]$, and that $z - z_0 = s\hat{\zeta} = s\exp(\mathrm{i}\theta)$, the quadratic term in the above expansion of $g(z)$ may be written as $½|g''(z_0)|s^2\exp[2\mathrm{i}(\theta - \varphi)]$. If the (deformed) integration path at z_0 is oriented such that $\theta = \varphi$ when $\eta < 0$, or $\theta = ½\pi + \varphi$ when $\eta > 0$, then the product of η and the quadratic term of $g(z)$ will be a large negative number, in which case the integral under consideration may be simplified as follows:

$$\int_{C'} f(z)\exp[\eta g(z)]\,\mathrm{d}z \cong \int_{-\infty}^{\infty}\left(f_0 + f_1\hat{\zeta}s + \tfrac{1}{2}f_2\hat{\zeta}^2 s^2 + \tfrac{1}{6}f_3\hat{\zeta}^3 s^3 + \cdots\right)$$

Note: the integral's limits have changed to $\pm\infty$.

change of variable $x = \sqrt{½|\eta g_2|}\,s$

$$\times \exp(\eta g_0)\exp(-½|\eta g_2|s^2)\exp\left(\tfrac{1}{6}\eta g_3\hat{\zeta}^3 s^3 + \tfrac{1}{24}\eta g_4\hat{\zeta}^4 s^4 + \cdots\right)\hat{\zeta}\,\mathrm{d}s$$

$$\cong \frac{f_0\exp(\eta g_0)\hat{\zeta}}{\sqrt{½|\eta g_2|}}\int_{-\infty}^{\infty}\exp(-x^2)\left(1 + \frac{f_1/f_0}{\sqrt{½|\eta g_2|}}\hat{\zeta}x + \frac{f_2/f_0}{2(½|\eta g_2|)}\hat{\zeta}^2x^2 + \frac{f_3/f_0}{6(½|\eta g_2|)^{3/2}}\hat{\zeta}^3x^3 + \cdots\right)$$

$e^{\alpha} = 1 + \alpha + ½\alpha^2 + \cdots$

$$\times\left(1 + \frac{\eta g_3}{6(½|\eta g_2|)^{3/2}}\hat{\zeta}^3x^3 + \frac{\eta g_4}{24(½|\eta g_2|)^2}\hat{\zeta}^4x^4 + \frac{(\eta g_3)^2}{72(½|\eta g_2|)^3}\hat{\zeta}^6x^6 + \cdots\right)\mathrm{d}x$$

arranged in ascending powers of $1/|\eta|$. sign of η

$$= \frac{f_0\exp(\eta g_0)\hat{\zeta}}{\sqrt{½|\eta g_2|}}\int_{-\infty}^{\infty}\left\{1 + \frac{1}{|\eta|}\left[\frac{(f_2/f_0)}{|g_2|}\hat{\zeta}^2x^2 \pm \left(\frac{4(f_1/f_0)g_3 + g_4}{6|g_2|^2}\right)\hat{\zeta}^4x^4 + \frac{(g_3)^2}{9|g_2|^3}\hat{\zeta}^6x^6\right] + \cdots\right\}$$

$$\times \exp(-x^2)\,\mathrm{d}x$$

odd powers of x are removed because they integrate to zero.

$$= \sqrt{\frac{2\pi}{|\eta g_2|}}f_0\exp(\eta g_0)\hat{\zeta}\left\{1 + \frac{1}{|\eta|}\left[\frac{(f_2/f_0)}{2|g_2|}\hat{\zeta}^2 \pm \left(\frac{4(f_1/f_0)g_3 + g_4}{8|g_2|^2}\right)\hat{\zeta}^4 + \frac{5(g_3)^2}{24|g_2|^3}\hat{\zeta}^6\right] + \cdots\right\}. \tag{101}$$

The following identities have been used in the last line of the above derivation:

$$\int_{-\infty}^{\infty}\exp(-x^2)\,\mathrm{d}x = \sqrt{\pi}, \tag{102a}$$

$$\int_{-\infty}^{\infty}x^2\exp(-x^2)\,\mathrm{d}x = \sqrt{\pi}/2, \tag{102b}$$

$$\int_{-\infty}^{\infty}x^4\exp(-x^2)\,\mathrm{d}x = 3\sqrt{\pi}/4, \tag{102c}$$

$$\int_{-\infty}^{\infty}x^6\exp(-x^2)\,\mathrm{d}x = 15\sqrt{\pi}/8. \tag{102d}$$

The identities in Eq.(102) have been derived by successive application of the method of integration by parts.

Digression: Within its domain of analyticity, the function $g(z) = g(x + \mathrm{i}y) = g^{(\mathrm{r})}(x, y) + \mathrm{i}g^{(\mathrm{i})}(x, y)$ satisfies the Cauchy-Riemann conditions, namely, $g_x^{(\mathrm{r})} = g_y^{(\mathrm{i})}$ and $g_y^{(\mathrm{r})} = -g_x^{(\mathrm{i})}$. This indicates that the gradient of the real part of $g(z)$, namely,

$$\boldsymbol{\nabla} g^{(\mathrm{r})}(x, y) = g_x^{(\mathrm{r})}\hat{\boldsymbol{x}} + g_y^{(\mathrm{r})}\hat{\boldsymbol{y}} = g_x^{(\mathrm{r})}\hat{\boldsymbol{x}} - g_x^{(\mathrm{i})}\hat{\boldsymbol{y}}, \tag{103}$$

which is a two-dimensional vector in the xy-plane, is orthogonal to the gradient of the imaginary part of $g(z)$, namely,

$$\boldsymbol{\nabla} g^{(\mathrm{i})}(x, y) = g_x^{(\mathrm{i})}\hat{\boldsymbol{x}} + g_y^{(\mathrm{i})}\hat{\boldsymbol{y}} = g_x^{(\mathrm{i})}\hat{\boldsymbol{x}} + g_x^{(\mathrm{r})}\hat{\boldsymbol{y}}. \tag{104}$$

As an aside, note that, at any point $z = x + \mathrm{i}y$ within the domain of analyticity of $g(z)$ in the complex plane, the vector $\boldsymbol{\nabla} g^{(\mathrm{r})}(x, y)$ coincides with the local arrow representing $g^{*\prime}(z) = g_x^{(\mathrm{r})} - \mathrm{i}g_x^{(\mathrm{i})}$ (i.e., the conjugate of the derivative $g'(z)$ of $g(z)$). Similarly, the vector $\boldsymbol{\nabla} g^{(\mathrm{i})}(x, y)$ is locally coincident with the arrow associated with the complex number $\mathrm{i}g^{*\prime}(z) = g_x^{(\mathrm{i})} + \mathrm{i}g_x^{(\mathrm{r})}$.

The orthogonality of $\boldsymbol{\nabla} g^{(\mathrm{r})}$ to $\boldsymbol{\nabla} g^{(\mathrm{i})}$ is of great significance as it ensures that, along any path in the z-plane where $g^{(\mathrm{r})}(x, y)$ is maximally ascendant (or maximally descendant), the function $g^{(\mathrm{i})}(x, y)$ remains constant — and vice-versa. In other words, the contours of constant $g^{(\mathrm{i})}(x, y)$ are exactly the paths of steepest ascent (or descent) for $g^{(\mathrm{r})}(x, y)$. Similarly, the contours of constant $g^{(\mathrm{r})}(x, y)$ are precisely the paths of steepest ascent (or descent) for $g^{(\mathrm{i})}(x, y)$. If, while moving along such a path, one arrives at a saddle point (i.e., a point z_0 where $g'(z_0) = 0$), then as was shown in the beginning of the present section, the continuation of a path of steepest ascent beyond the saddle becomes a path of steepest descent, whereas the continuation of a path of steepest descent beyond the saddle becomes a path of steepest ascent.

A saddle point for $g^{(\mathrm{r})}$ is also a saddle point for $g^{(\mathrm{i})}$, and vice-versa. Four contours meet at each and every saddle point. On two of these contours, which are orthogonal to each other at the saddle point, $g^{(\mathrm{r})}$ remains constant. Along one of these two, $g^{(\mathrm{i})}$ reaches a peak at the saddle point, whereas along the other one $g^{(\mathrm{i})}$ reaches a valley. On the remaining two contours (which are also orthogonal to each other at the saddle point), $g^{(\mathrm{i})}$ remains constant, while $g^{(\mathrm{r})}$ acquires a maximum along one, and a minimum along the other (again, both occurring at the same saddle point). These two sets of curves are rotated by 45° relative to each other.

Returning to our previous example, in the case of the analytic function $g(z) = z^2$, the contours of constant $g^{(\mathrm{r})}$, namely, $x^2 - y^2 = c'$, are everywhere orthogonal to contours of constant $g^{(\mathrm{i})}$, namely, $xy = c''$. (Here, c' and c'' are a pair of nonzero but otherwise arbitrary constants.) The only saddle point of $g(z) = z^2$ is at $z_0 = 0$, where the pair of constant $g^{(\mathrm{r})}$ contours, $y = \pm x$, meets with the pair of constant $g^{(\mathrm{i})}$ contours, $x = 0$ and $y = 0$. While the two contours belonging to each pair are clearly orthogonal to one another at the saddle point $z_0 = 0$, each pair is seen to be rotated by 45° relative to the other one.

References

1. R.N. Bracewell, *The Fourier Transform and Its Applications*, 2nd edition, McGraw-Hill, New York (1978).
2. J.W. Goodman, *Introduction to Fourier Optics*, 3rd edition, Roberts & Company, Englewood, Colorado (2005).
3. M. Born and E. Wolf, *Principles of Optics*, 7th edition, Cambridge University Press, London (1999).
4. M.V. Klein, *Optics*, Wiley, New York (1970).
5. M. Mansuripur, *Classical Optics and Its Applications*, 2nd edition, Cambridge University Press, London (2009).
6. J.D. Jackson, *Classical Electrodynamics*, 3rd edition, Wiley, New York (1998).
7. M. Mansuripur, *Field, Force, Energy and Momentum in Classical Electrodynamics*, revised edition, Bentham Science Publishers, Sharjah, UAE (2017).

Problems

Problem 1) Convolution (denoted by $*$) is an operation performed on two functions, $f(x)$ and $g(x)$ yielding a third function $h(x)$, where $h(x) = f(x) * g(x) = \int_{-\infty}^{\infty} f(x')g(x - x')\mathrm{d}x'$. Show that the Fourier transforms of these functions are related as follows: $H(s) = F(s)G(s)$. The latter result indicates that the order of $f(x)$ and $g(x)$ is irrelevant as far as $h(x)$ is concerned. Use a simple change of the integration variable to verify that $f(x) * g(x) = g(x) * f(x)$.

Problem 2) Show that $\mathrm{tri}(x) = \mathrm{rect}(x) * \mathrm{rect}(x)$. Conclude that $\mathcal{F}\{\mathrm{tri}(x)\} = \mathrm{sinc}^2(s)$.

Problem 3) a) Plot the function $f(x) = (\mathrm{d}/\mathrm{d}x)\mathrm{tri}(x)$. Use the definition of the Fourier transform, namely $F(s) = \int_{-\infty}^{\infty} f(x)\exp(-\mathrm{i}2\pi sx)\,\mathrm{d}x$, to determine $F(s)$ by direct integration.

b) Considering that $\mathcal{F}\{g'(x)\} = \mathrm{i}2\pi sG(s)$ for an arbitrary function $g(x)$, whose derivative with respect to x is denoted by $g'(x)$, and whose Fourier transform by $G(s)$, use the result of (a) to obtain the Fourier transform of the function $\mathrm{tri}(x)$.

Problem 4) Prove the sifting property of $\delta'(x) = (\mathrm{d}/\mathrm{d}x)\delta(x)$, namely, assuming that at $x = 0$ the function $f(x)$ is continuous and differentiable, show that

$$\int_{-\infty}^{\infty} f(x)\delta'(x)\mathrm{d}x = -f'(0).$$

Problem 5) The convolution between an arbitrary function $f(x)$ and a delta function located at $x = x_0$ would result in a shift along the x-axis of $f(x)$ by x_0; that is,

$$f(x) * \delta(x - x_0) = f(x - x_0).$$

Prove the above statement using (a) direct integration, and (b) the convolution theorem of the Fourier transform theory.

Problem 6) Use contour integration to find the Fourier transform of the function $\mathrm{sinc}(x) = \sin(\pi x)/(\pi x)$; that is, $\mathcal{F}\{\mathrm{sinc}(x)\} = \int_{-\infty}^{\infty}[\sin(\pi x)/(\pi x)]\exp(-\mathrm{i}2\pi sx)\,\mathrm{d}x$.

Problem 7) a) The function $f(x)$ is real and even; that is, $f(x) = f(-x)$. Show that its Fourier transform $F(s)$ is also real and even.

b) The function $g(x)$ is real and odd; that is, $g(x) = -g(-x)$. Show that its Fourier transform $G(s)$ is purely imaginary and odd.

c) The real function $h(x)$ is neither odd nor even. Express $h(x)$ as the superposition of a pair of even and odd functions, $h_{\mathrm{e}}(x)$ and $h_{\mathrm{o}}(x)$. Show that the real and imaginary parts of the Fourier transform of $h(x)$, namely $H(s) = H_{\mathrm{R}}(s) + \mathrm{i}H_{\mathrm{I}}(s)$, are directly related to the Fourier transforms of $h_{\mathrm{e}}(x)$ and $h_{\mathrm{o}}(x)$.

Problem 8) a) Use direct integration to show that the Fourier transform of the function $f(x) = \exp(\mathrm{i}2\pi s_0 x)$, where s_0 is a fixed frequency, is $F(s) = \delta(s - s_0)$.

b) With the aid of the result obtained in part (a), find the Fourier transforms of the following functions: $\cos(2\pi s_0 x)$, $\sin(2\pi s_0 x)$, $\cos^2(2\pi s_0 x)$, and $\sin^2(2\pi s_0 x)$.

Problem 9) Plot the function $1/\cosh(\pi x)$, then show that its Fourier transform is $1/\cosh(\pi s)$.

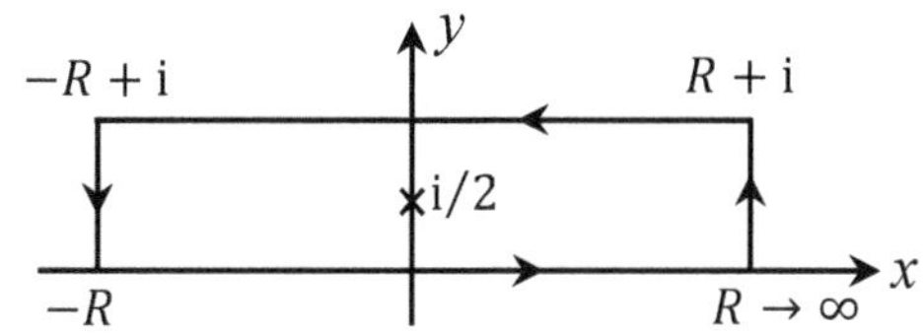

Hint: Use the contour shown. To find the residue of the integrand at the simple pole $z_0 = \mathrm{i}/2$, use the first few terms in the Taylor series expansion of $\cosh(\pi x)$.

Problem 10) Plot the function $f(x) = 1/\cosh^2(\pi x)$, then show that its Fourier transform is $F(s) = 2s/\sinh(\pi s)$.

Hint: This problem is similar to problem 9, except for the pole at $z_0 = \mathrm{i}/2$ being a second-order pole.

Problem 11) Without using problem 10 in the reverse direction, show that the Fourier transform of $f(x) = 2x/\sinh(\pi x)$ is $F(s) = 1/\cosh^2(\pi s)$.

Hint: One way to solve this problem is to use an infinitely large semi-circular contour in the upper-half of the complex plane, then calculate a closed form for the sum of all residues within the semi-circle.

Problem 12) With the help of the results obtained in chapter 6, problem 14, prove the following identities:

a) $\mathcal{F}\{\sin(ax^2)\} = \sqrt{\pi/a}\cos\left(\frac{\pi^2 s^2}{a} + \frac{\pi}{4}\right), \quad a > 0;$

b) $\mathcal{F}\{\cos(ax^2)\} = \sqrt{\pi/a}\cos\left(\frac{\pi^2 s^2}{a} - \frac{\pi}{4}\right), \quad a > 0.$

Problem 13) Find the Fourier transforms of $f(x) = \operatorname{sinc}(ax)$ and $g(x) = \exp(-b|x|)$, where both a and b are real, positive constants. Use the convolution theorem to show that the Fourier transform of $h(x) = f(x)g(x) = \operatorname{sinc}(ax)\exp(-b|x|)$ is

$$H(s) = \left(\frac{1}{\pi a}\right)\tan^{-1}\left[\frac{\pi a b}{2\pi^2 s^2 + ½(b^2 - \pi^2 a^2)}\right].$$

This result simplifies to $\mathcal{F}\{\exp(-|x|)\sin(x)/x\} = \tan^{-1}(1/2\pi^2 s^2)$ in the special case when $\pi a = b = 1$.

Problem 14) a) Show that the Fourier transform of $1/|x|^{1/2}$ is $1/|s|^{1/2}$.

b) Show that the Fourier transform of $\operatorname{sign}(x)/|x|^{1/2}$ is $-\mathrm{i}\operatorname{sign}(s)/|s|^{1/2}$.

Hint: Start by splitting the domain of integration into positive and negative x, convert the integral over negative x to one over positive x, then change the variable from x to y^2.

Problem 15) Plot the function $\tanh(\pi x)$, then show that its Fourier transform is given by $F(s) = -\mathrm{i}/\sinh(\pi s)$.

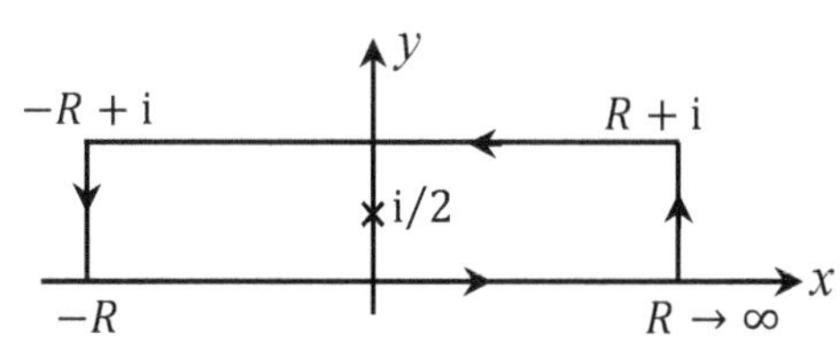

Hint: Use the rectangular contour shown. Pay particular attention to the vertical legs of the rectangle. To determine the residue of the integrand at $z_0 = \mathrm{i}/2$, the first couple of terms in the Taylor series expansion of $\cosh(\pi x)$ would suffice.

Problem 16) Let $f(x)$ and $g(x)$ be two arbitrary functions of the real variable x whose Fourier transforms are given by $F(s)$ and $G(s)$, respectively. In general, $f(x)$, $g(x)$, $F(s)$, and $G(s)$ are complex-valued functions of their respective real variables.

a) Using the defining integrals of direct and inverse Fourier transformation, prove the following identity, which is commonly referred to as Parseval's theorem:

$$\int_{-\infty}^{\infty} f(x)g^*(x)\mathrm{d}x = \int_{-\infty}^{\infty} F(s)G^*(s)\mathrm{d}s.$$

Considering that $\mathcal{F}\{\mathrm{rect}(x)\} = \mathrm{sinc}(s)$, $\mathcal{F}\{\mathrm{tri}(x)\} = \mathrm{sinc}^2(s)$, and $\mathcal{F}\{\exp(-|x|)\} = 2/[1+(2\pi s)^2]$, use Parseval's theorem to evaluate the following definite integrals:

b) $\int_{-\infty}^{\infty} \mathrm{sinc}^3(s)\mathrm{d}s$, c) $\int_{-\infty}^{\infty} \mathrm{sinc}^4(s)\mathrm{d}s$, d) $\int_{0}^{\infty} \exp(-x)\,\mathrm{sinc}(x)\mathrm{d}x$.

Problem 17) Use two different methods to determine the Fourier transform $F(s)$ of the depicted function $f(x)$. The function equals $\cos(2\pi x)$ when $|x| < ¾$, and is zero otherwise. The first method is based on direct integration. For the second method, express $f(x)$ as the product of a periodic function and a rectangular pulse, then use the convolution theorem. Confirm that the two methods yield precisely the same result.

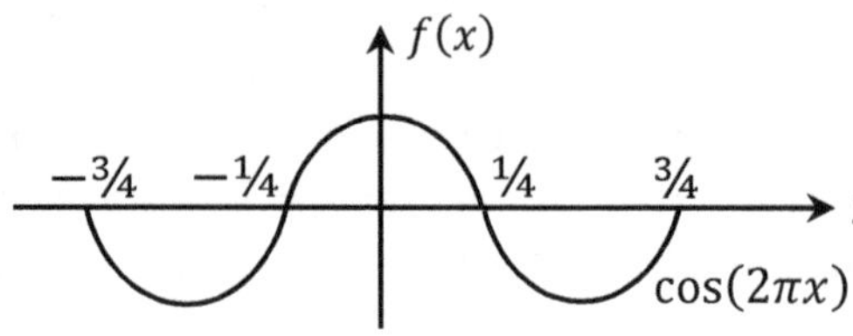

Problem 18) Find the Fourier transforms of the functions $f(x)$ and $g(x)$ shown below.

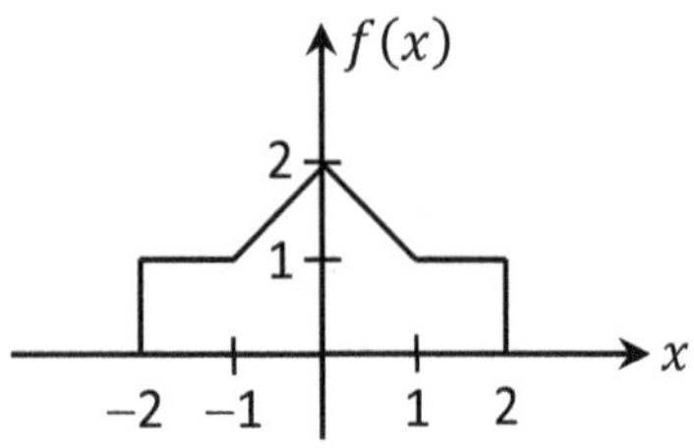

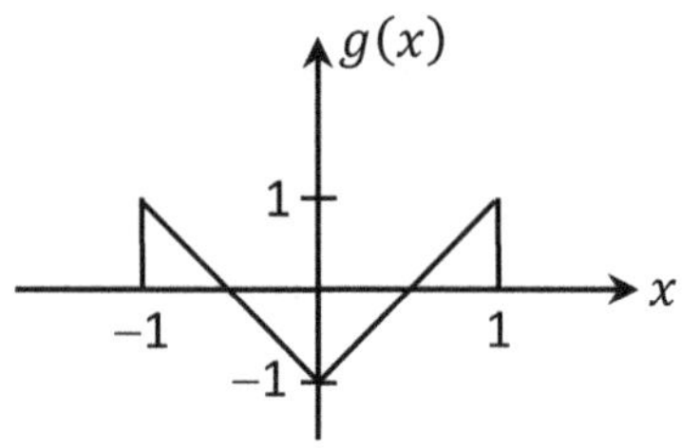

Problem 19) a) Show that the Fourier transform of the function $f(x) = \cos(2\pi f_o x)$ is given by $F(s) = ½[\delta(s-f_o)+\delta(s+f_o)]$.

b) Show that the Fourier transform of $\sin(2\pi f_o x)$ is $[\delta(s-f_o)-\delta(s+f_o)]/(2\mathrm{i})$.

c) Using the result of part (a), find the Fourier transform of $\cos^2(\pi f_o x)$.

d) Use the differentiation theorem to find the Fourier transform of the derivative of $\cos^2(\pi f_o x)$. Show that your final result is consistent with the result obtained in part (b).

Problem 20) A superposition of a large number of rectangular functions of variable width β and infinitesimal height $\mathrm{d}\beta$ produces the following function:

$$f(x) = ½\int_0^2 \mathrm{rect}(x/\beta)\,\mathrm{d}\beta.$$

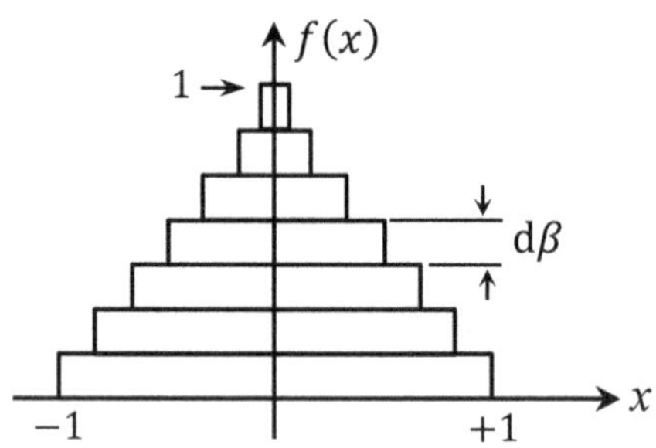

a) What well-known, standard function does $f(x)$ represent?

b) Find the Fourier transform $F(s)$ of $f(x)$ and confirm that $F(s)$ does in fact correspond to the standard function that you associated with $f(x)$ in part (a).

Problem 21) a) Using the scaling and differentiation theorems, find the Fourier transform of $f(x) = x\,\text{rect}(x/2)$.

b) Verify that the result obtained in (a) is in agreement with the function $F(s) = \mathcal{F}\{f(x)\}$ obtained by direct integration.

Problem 22) a) Find the Fourier transform $F_L(s)$ of the function $f_L(x)$ depicted below. (Hint: Use integration by parts.)

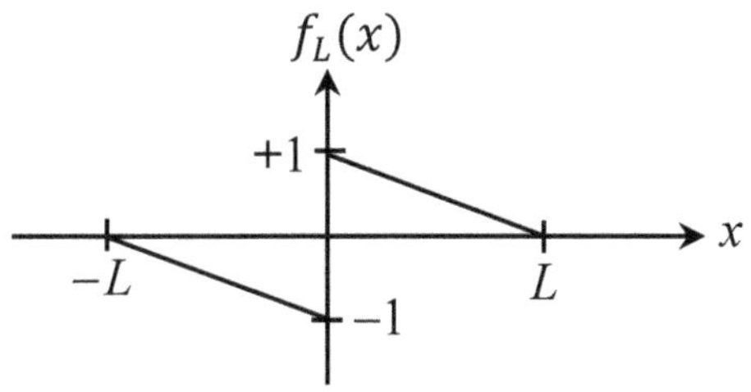

$$f_L(x) = \begin{cases} -1-(x/L), & -L \le x < 0; \\ 1-(x/L), & 0 < x \le L. \end{cases}$$

b) The function $f_L(x)$ approaches a well-known, standard function in the limit when $L \to \infty$. Identify that function.

c) Determine the Fourier transform of the standard function that you identified in part (b) by finding the limit when $L \to \infty$ of $F_L(s)$ obtained in part (a).

d) Use complex plane integration to find the inverse Fourier transform of the function $F(s) = 1/(i\pi s)$. Specify the contours of integration for both $x > 0$ and $x < 0$, and explain why certain segments of the contours do not contribute to the inverse Fourier integral.

Problem 23) Find the Fourier series coefficients of the following periodic functions:

(a) $f(x) = \sum_{n=-\infty}^{\infty} \text{rect}[(x-np)/a]$; $\quad p > a > 0.$

(b) $g(x) = \sum_{n=-\infty}^{\infty} \text{tri}[(x-np)/a]$; $\quad p > 2a > 0.$

Problem 24) Using the function $f(x) = \text{sinc}(x/2)\text{comb}(x)$ and its Fourier transform, show that the series $1 - \frac{1}{3} + \frac{1}{5} - \frac{1}{7} + \cdots$ is equal to $\pi/4$.

Hint: Use the fact that the area under $f(x)$ is equal to $F(s=0)$.

Problem 25) a) Using the function $f(x) = \text{sinc}^2(x/2)\text{comb}(x)$ and its Fourier transform, show that $\sum_{n=0}^{\infty}[1/(2n+1)^2] = \pi^2/8$.

b) Use the result obtained in (a) to verify that $\zeta(2) = \sum_{n=1}^{\infty}(1/n^2) = \pi^2/6$, where $\zeta(\cdot)$ is Riemann's zeta function.

Hint: Use the fact that the area under $f(x)$ is equal to $F(s=0)$.

Problem 26) a) Assuming that a is a positive real constant, use the function

$$f(x) = \left[\frac{2a}{a^2+(2\pi x)^2}\right]\text{comb}(x)$$

and its Fourier transform to show that

$$\sum_{n=1}^{\infty} 1/[1+(2\pi n/a)^2] = \frac{a}{4}\left[\frac{1+\exp(-a)}{1-\exp(-a)}\right] - \frac{1}{2}.$$

b) In the limit when $a \to 0$, show that the result obtained in part (a) yields $\zeta(2) = \sum_{n=1}^{\infty}(1/n^2) = \pi^2/6$.

Hint: Use the fact that the area under $f(x)$ is equal to $F(s = 0)$.

Problem 27) In general, the Fourier series of a periodic function $f(x)$, whose period is the real and positive constant p, may be written as

$$f(x) = \sum_{n=-\infty}^{\infty} c_n \exp(\mathrm{i}2\pi nx/p).$$

Let $x = a$, an arbitrary point on the x-axis, denote the beginning of a period; the point $x = a + p$ will then indicate the end of that period. Proceed to define a single-period of $f(x)$ over the interval $(a, a + p)$ as $\tilde{f}(x)$, namely $\tilde{f}(x) = f(x)$ if $a \le x < a + p$, and $\tilde{f}(x) = 0$ otherwise. The Fourier transform of the single period, $\tilde{f}(x)$, will then be denoted by $\tilde{F}(s)$.

a) Show that $c_n = (1/p)\tilde{F}(n/p) = (1/p)\int_a^{a+p} f(x)\exp(-\mathrm{i}2\pi nx/p)\,\mathrm{d}x$.

b) Show that $c_0 = (1/p)\int_a^{a+p} f(x)\mathrm{d}x$, which is the average value of the function over one period.

c) If $f(x)$ happens to be real valued, prove that $c_{-n} = c_n^*$. Writing $c_n = a_n - \mathrm{i}b_n$, verify that the Fourier series may subsequently be written as

$$f(x) = c_0 + 2\sum_{n=1}^{\infty} a_n \cos(2\pi nx/p) + 2\sum_{n=1}^{\infty} b_n \sin(2\pi nx/p),$$

where $a_n = p^{-1}\int_a^{a+p} f(x)\cos(2\pi nx/p)\,\mathrm{d}x$ and $b_n = p^{-1}\int_a^{a+p} f(x)\sin(2\pi nx/p)\,\mathrm{d}x$.

Problem 28) a) Show that any function $\tilde{f}(x)$, irrespective of its width, when replicated at regular intervals p along the x-axis, yields a periodic function $f(x) = \sum_{n=-\infty}^{\infty} \tilde{f}(x - np)$.

b) The function $\tilde{f}(x) = \mathrm{rect}(x)$, whose width is 1.0, is replicated along the x-axis at regular intervals $p = ¾$, thus producing the periodic function

$$f(x) = \sum_{n=-\infty}^{\infty} \mathrm{rect}(x - ¾n).$$

Plot the function $f(x)$, then derive its Fourier series coefficients, namely,

$$c_n = (1/p)\int_0^p f(x)\exp(-\mathrm{i}2\pi nx/p)\,\mathrm{d}x.$$

Problem 29) Consider the function $f(x)$, whose Fourier transform $F(s)$ is band-limited, having a bandwidth of W along the s-axis. To be specific, we assume that $F(s) = 0$ when $|s| > W/2$. The function $f(x)$ is sampled at regular intervals p, then each sample is replaced with a rectangular pulse of height $f(np)$ and width a, where $a \le p$. The sampled function may thus be written as $\hat{f}(x) = \sum_{n=-\infty}^{\infty} f(np)\mathrm{rect}[(x - np)/a]$. Find the Fourier transform $\hat{F}(s)$ of the sampled function and show that the original $f(x)$ may be recovered from $\hat{f}(x)$with the aid of low-pass filtering.

Problem 30) a) Plot the function $f(x) = \mathrm{comb}(x)\cos(\pi x)$, find its Fourier transform $F(s)$, then plot the function $F(s)$.

b) Plot the function $g(x) = \mathrm{comb}(x - ½)$, find its Fourier transform $G(s)$, then plot the function $G(s)$.

Problem 31) The real-valued function $h(x)$ is defined over the interval $0 \le x \le L$. It is desired to expand this function in a Fourier cosine series, as follows:

$$h(x) = \sum_{n=0}^{\infty} c_n \cos(\pi n x/L).$$

Find the coefficients c_n.

Problem 32) The function $f(x) = |\cos(\pi x)|$ defined on the real x-axis is periodic, with a period $p = 1$, as shown.

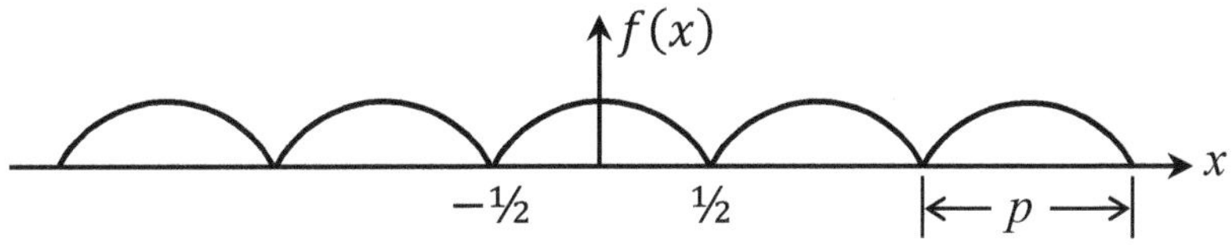

a) Using a combination of the functions $\text{rect}(x)$, $\text{comb}(x)$, and $\cos(\pi x)$, express $f(x)$ *without* invoking the absolute value operator $|\cdot|$.

b) Find the Fourier series expansion of $f(x)$.

Problem 33) Show that the periodic function $\phi_1(x)$, having period $p = 2\pi$ and defined as $-\frac{1}{2}(\pi + x)$ when $-\pi \le x < 0$, and $\frac{1}{2}(\pi - x)$ when $0 < x \le \pi$, is represented by the following Fourier series:

$$\phi_1(x) = \sum_{n=1}^{\infty} \frac{\sin(nx)}{n}.$$

Show, by integrating $\phi_1(x)$ over an appropriate range of x, that

$$\phi_2(x) = \sum_{n=1}^{\infty} \frac{\cos(nx)}{n^2} = \begin{cases} \frac{1}{4}(\pi + x)^2 - (\pi^2/12); & -\pi \le x \le 0, \\ \frac{1}{4}(\pi - x)^2 - (\pi^2/12); & 0 \le x \le \pi. \end{cases}$$

Note that, by raising the power of n in the denominator, integration has improved the rate of convergence of the series.

Problem 34) Given $\psi_{2s}(x) = \sum_{n=1}^{\infty} \sin(nx)/n^{2s}$ and $\psi_{2s+1}(x) = \sum_{n=1}^{\infty} \cos(nx)/n^{2s+1}$, develop the following recurrence relations:

a) $\psi_{2s}(x) = \int_0^x \psi_{2s-1}(y)\text{d}y$,

b) $\psi_{2s+1}(x) = \zeta(2s + 1) - \int_0^x \psi_{2s}(y)\text{d}y$.

Here, $\zeta(s) = \sum_{n=1}^{\infty}(1/n^s)$ is the Riemann zeta function.

Problem 35) a) Find the Fourier series representation of the periodic function $f(x)$, having period $p = 1$, and defined over the interval $0 \le x < 1$ as $f(x) = \frac{1}{2} - x$.

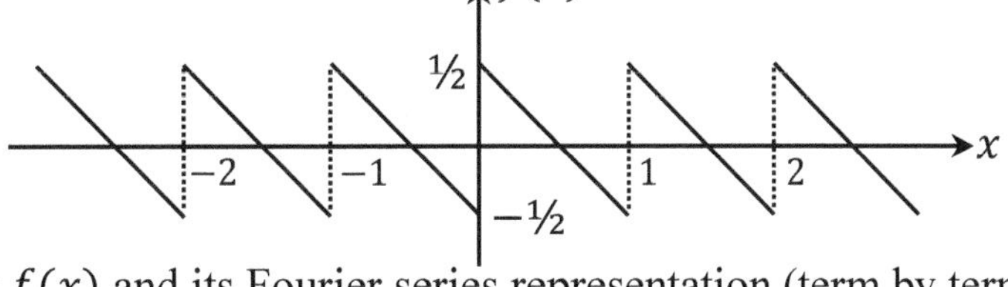

b) Differentiating $f(x)$ and its Fourier series representation (term by term) with respect to x, show that the periodic function $f'(x)$ is indeed equal to the derivative of the Fourier series representation of $f(x)$.

Problem 36) Use two different methods to find the Fourier transform of the periodic function $f(x)$ shown below. Confirm that both methods yield the same answer.

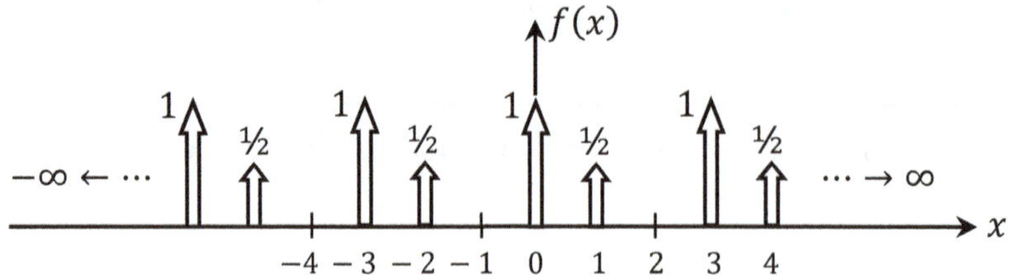

Problem 37) A monochromatic plane wave of wavelength λ illuminates a flat, partially transparent object at normal incidence, as shown. The complex amplitude distribution of the transmitted beam immediately after the object is given by the function $f(x, z = 0^+)$. In this problem, we shall ignore the vectorial nature of the electromagnetic field, treating it instead as a scalar waveform.

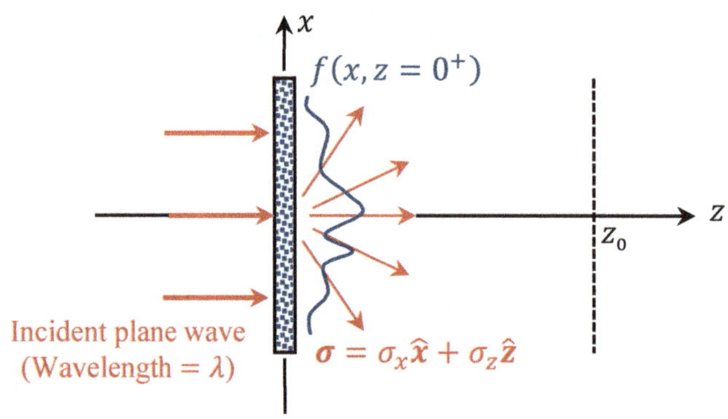

a) Expressing $f(x, z = 0^+)$ as the inverse transform of its own Fourier transform, $F(s, z = 0^+)$, find the complex amplitude of the emergent plane wave that propagates in the $z > 0$ half space along the unit vector $\boldsymbol{\sigma} = \sigma_x \hat{\mathbf{x}} + \sigma_z \hat{\mathbf{z}}$.

b) What is the complex amplitude distribution $f(x, z = z_0)$ of the light field in a plane perpendicular to the z-axis at $z = z_0$?

c) Use the method of stationary-phase approximation to obtain the field profile in the far field; in other words, find an approximate expression for $f(x, z = z_0)$ when z_0 is sufficiently far away from the object.

Problem 38) An integral representation of $J_0(x)$, the Bessel function of first kind, 0th order, is

$$J_0(x) = (1/2\pi) \int_0^{2\pi} \exp(\mathrm{i}x \sin\theta)\, \mathrm{d}\theta.$$

Use the method of stationary-phase approximation to show that the asymptotic form of $J_0(x)$ for large values of x is $J_0(x) \sim \sqrt{2/(\pi x)} \cos(x - \pi/4)$.

Problem 39) Determine an approximate value for the following integral, assuming that the real-valued parameter η is large and positive:

$$I = \int_{-1}^{+1} \exp\left(-x^2 + \mathrm{i}\eta\sqrt{1 - x^2}\right) \mathrm{d}x.$$

Problem 40) Plot the functions $f(x) = \sinh(ax)/\sinh(bx)$ and $g(x) = \cosh(ax)/\sinh(bx)$, where $b > a > 0$, then show that their Fourier transforms are given by

$$F(s) = \frac{(\pi/b)\sin(\pi a/b)}{\cos(\pi a/b) + \cosh(2\pi^2 s/b)},$$

$$G(s) = -\frac{\mathrm{i}(\pi/b)\sinh(2\pi^2 s/b)}{\cos(\pi a/b) + \cosh(2\pi^2 s/b)}.$$

Hint: Start by Fourier transforming $\exp(\pm ax)/\sinh(bx)$, using an infinitely large semi-circular contour in the upper-half or lower-half of the complex plane, depending on the sign of s. The poles of the integrand are the zeros of $\sinh(bz)$, namely, $z_n = \mathrm{i}n\pi/b$, where the integer n could be positive, zero, or negative. In the vicinity of each pole, $\sinh(bz)$ may be approximated by the first couple of terms in its Taylor series expansion; that is, $\sinh(bz) \cong \sinh(bz_n) + b\cosh(bz_n)(z - z_n) = (-1)^n b(z - z_n)$.

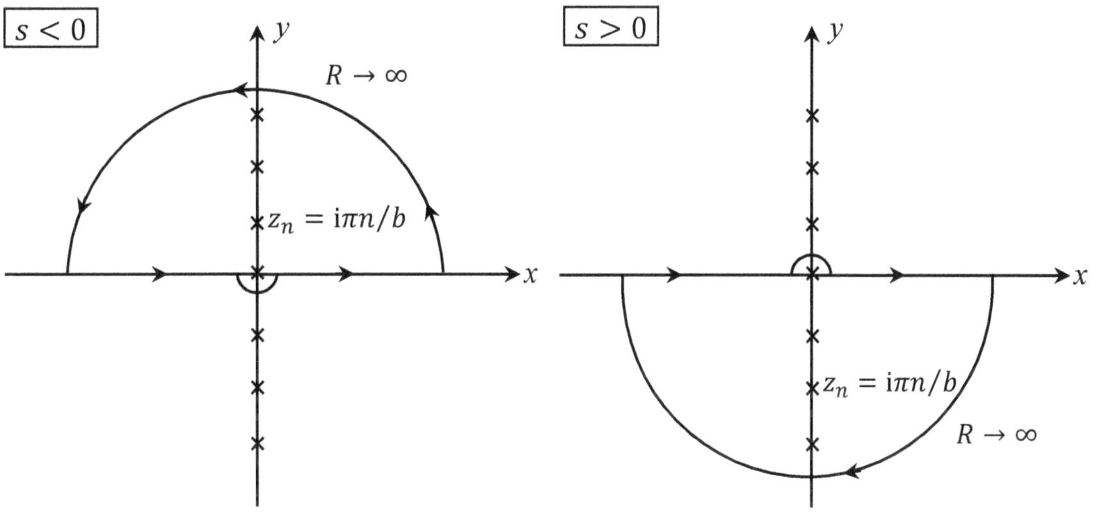

Problem 41) Plot the function $f(x) = \exp(-ax)\ln[1 + \exp(x)]$, where $0 < \mathrm{Re}(a) < 1$, then show that its Fourier transform is given by

$$F(s) = \frac{\pi}{(a + \mathrm{i}2\pi s)\sin[\pi(a + \mathrm{i}2\pi s)]}.$$

Hint: Use the contour shown. The function $\ln(1 + e^z)$ has no poles but has branch points at $z_n = \mathrm{i}(2n + 1)\pi$, where its argument vanishes. The branch-cut shown connects $z_1 = \mathrm{i}\pi$ to a branch point at infinity. Going around the small circle in the counterclockwise direction does not contribute anything to the integral, but adds the constant $2\pi\mathrm{i}$ to $\ln(1 + e^z)$. On the lower segment of the branch-cut, therefore, $\ln(1 + e^z)$ differs from that on the upper segment by $2\pi\mathrm{i}$.

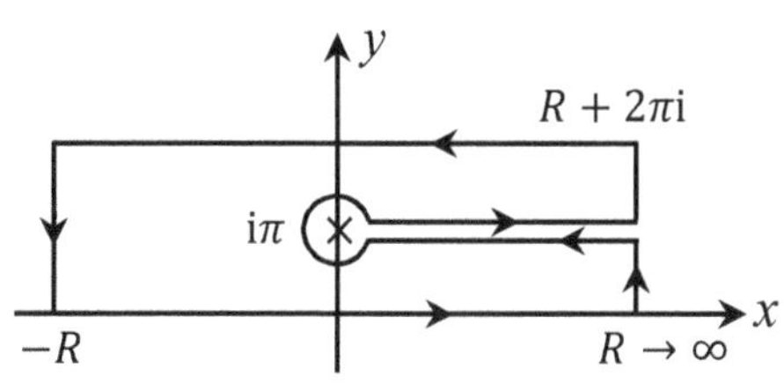

Problem 42) The goal of the present problem is to demonstrate that the Fourier transform $F(s)$ of the function $f(x) = \cos(2\pi s_0 x)$ is a pair of delta-functions located at $s = \pm s_0$. Consider the function $f_\alpha(x) = \exp(-\alpha x^2)\cos(2\pi s_0 x)$, where α is a positive real parameter which will eventually be made to approach zero, thus allowing $f_\alpha(x)$ to approach $f(x)$.

a) Use the methods of complex-plane integration to determine the Fourier transform of $e^{-\alpha x^2} e^{\mathrm{i}2\pi s_0 x}$.

b) Repeat part (a) for the function $e^{-\alpha x^2} e^{-\mathrm{i}2\pi s_0 x}$.

c) Combine the results obtained in parts (a) and (b) to show that the Fourier transform $F_\alpha(s)$ of the function $f_\alpha(x)$ approaches a pair of δ-functions in the limit when $\alpha \to 0$.

Hint: The Fourier transform of $g(x) = \exp(-\pi x^2)$ is $G(s) = \exp(-\pi s^2)$, indicating that the area under the function $\exp(-\pi x^2)$ is equal to 1, namely, $\int_{-\infty}^{\infty} \exp(-\pi x^2)\,\mathrm{d}x = G(0) = 1$.

Problem 43) a) Invoking Euler's Gamma function, $\Gamma(x) = \int_0^\infty y^{x-1}e^{-y}\mathrm{d}y$, defined here for $x > 0$, show that $\int_0^\infty \exp(-x^3)\,\mathrm{d}x = ⅓\Gamma(⅓)$.

b) In the method of stationary-phase approximation, the following integral is evaluated by focusing attention at and around stationary points such as x_0, where $g'(x_0) = 0$.

$$\int_a^b f(x) \exp[\mathrm{i}\eta g(x)]\,\mathrm{d}x.$$

Here, $f(x)$ is a generally complex-valued function of the real variable x, η is a large (positive or negative) real number, and $g(x)$ is a real-valued function of x. In section 12, we assumed that $a < x_0 < b$, and $g''(x_0) \neq 0$. We then evaluated the contribution of the region in the immediate vicinity of x_0 to the above integral in terms of $f(x_0)$, $g(x_0)$, η, and $g''(x_0)$. In the present problem, you are asked to assume that $g''(x_0) = 0$ but $g'''(x_0) \neq 0$. Applying to method of stationary-phase approximation, estimate the value of the above integral in terms of $f(x_0)$, $g(x_0)$, η, and $g'''(x_0)$. (**Hint**: $\cos(\pi/6) = \sqrt{3}/2$.)

Problem 44) The Fourier transform of $f(x)$ is given by $F(s) = \int_{-\infty}^{\infty} f(x) \exp(-\mathrm{i}2\pi sx)\,\mathrm{d}x$. Extending the function $f(x)$ to the complex z-plane, assuming that $f(z)$ is analytic within the horizontal strip between the real axis x and the straight-line $z = x + \mathrm{i}y_0$ (which parallels the x-axis), and also assuming that $f(z) \to 0$ as $L \to \infty$ on the vertical lines $z = \pm L + \mathrm{i}y$ where $0 \le y \le y_0$, show that the Fourier transform of $f(x + \mathrm{i}y_0)$ is given by $\exp(-2\pi y_0 s)\,F(s)$.

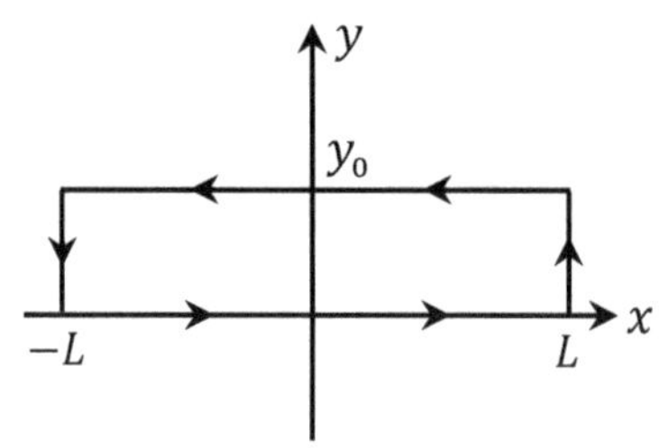

Problem 45) The function $f(x)$ equals 1 when $1 \le x \le 3$, and zero otherwise, as shown.

a) Find the Fourier transform $F(s)$ of $f(x)$ by direct integration.

b) Express $f(x)$ in terms of the elementary function $\mathrm{rect}(x)$. Use the shift and scaling theorems of the Fourier transform theory to determine $F(s)$ for the function thus obtained. Confirm that your result agrees with that obtained in part (a).

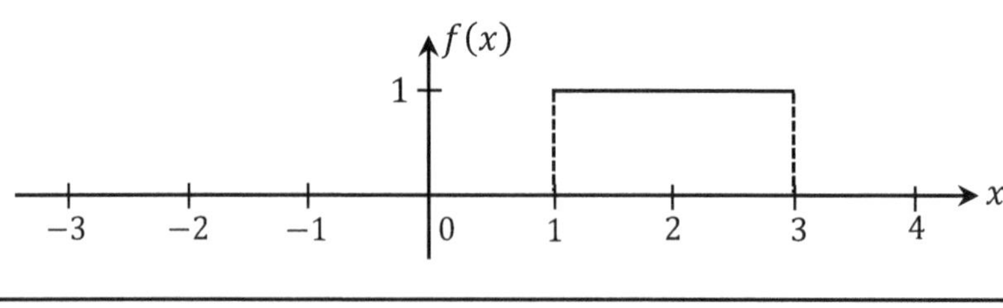

Problem 46) The cross-correlation of the functions $f(x)$ and $g(x)$ is defined as follows:

$$f(x) \otimes g(x) = \int_{-\infty}^{\infty} f(x')g(x' - x)\mathrm{d}x'.$$

a) In what way does cross-correlation differ from the convolution operation?

b) Denoting the Fourier transforms of $f(x)$ and $g(x)$ by $F(s)$ and $G(s)$, respectively, show that the Fourier transform of the cross-correlation function is given by

$$\mathcal{F}\{f(x) \otimes g(x)\} = F(s)G(-s).$$

Chapter 8

Ordinary Differential Equations

"We ought to regard the present state of the universe as the effect of its antecedent state and as the cause of the state that is to follow. An intelligence knowing all the forces acting in nature at a given instant, as well as the momentary positions of all things in the universe, would be able to comprehend in one single formula the motions of the largest bodies as well as the lightest atoms in the world, provided that its intellect were sufficiently powerful to subject all data to analysis; to it nothing would be uncertain, the future as well as the past would be present to its eyes. The perfection that the human mind has been able to give to astronomy affords but a feeble outline of such an intelligence."

Pierre-Simon Laplace (1749–1827)

8.1. Introduction. When a physical system's behavior happens to depend on a function of a single variable, say, $f(x)$, and also on some of the derivatives of that function, namely $\dot{f}(x) = f'(x) = \mathrm{d}f(x)/\mathrm{d}x$, $\ddot{f}(x) = f''(x) = \mathrm{d}^2f(x)/\mathrm{d}x^2$, etc., the equation that governs the evolution of the system is typically an ordinary differential equation (ODE). The simplest ODE is a first-order equation with constant coefficients such as $a\dot{f}(x) + bf(x) = g(x)$. Here, the constant coefficients a and b are the system parameters, whereas $g(x)$ is an external excitation that drives the evolution of the system.

The highest derivative of $f(x)$ appearing in a differential equation determines the order of the equation. Thus, a second-order ODE contains $\ddot{f}(x)$, and possibly $\dot{f}(x)$ and $f(x)$, but it cannot contain $\dddot{f}(x)$ or higher derivatives. A differential equation is said to be linear if excitation by $g(x) = \alpha g_1(x) + \beta g_2(x)$ elicits the response $f(x) = \alpha f_1(x) + \beta f_2(x)$ from the system. Here, α and β are arbitrary constants, while $f_1(x)$ and $f_2(x)$ are the response of the system to individual excitations $g_1(x)$ and $g_2(x)$, respectively. A differential equation is shift invariant if, for any arbitrary x_0, the response to $g(x - x_0)$ turns out to be $f(x - x_0)$. For example, the differential equation $p(x)\ddot{f}(x) + q(x)\dot{f}(x) + r(x)f(x) = g(x)$, where $p(x)$, $q(x)$, and $r(x)$ are arbitrary functions of the variable x, is a second-order ODE that is linear but *not* shift invariant. In contrast, the first-order ODE $a\dot{f}(x) + bf^2(x) = g(x)$ is nonlinear but shift invariant.

A differential equation can have one or more solutions in the absence of external excitations. Setting $g(x) = 0$ and solving the equation yields the so-called *homogeneous* solutions, which may be identified as $f_{h1}(x), f_{h2}(x), \cdots, f_{hn}(x)$. A solution that satisfies the differential equation in the presence of an excitation function $g(x)$ is commonly referred to as a *particular* solution and may be denoted as $f_p(x)$. For linear ODEs, the general solution is the sum of the particular solution and a linear combination of the homogeneous solutions; that is, $f(x) = a_1f_{h1}(x) + a_2f_{h2}(x) + \cdots + a_nf_{hn}(x) + f_p(x)$. Here, $a_1, a_2, \cdots, a_n$ are arbitrary constants that must be determined by requiring the solution to satisfy certain *initial* conditions. For instance, a second-order linear ODE generally has two independent homogeneous solutions, $f_{h1}(x)$ and $f_{h2}(x)$. The constant coefficients a_1, a_2 of the general solution to such an equation can be determined if, for example, the values of $f(x)$ and $\dot{f}(x)$ at an arbitrary point $x = x_0$ are specified a priori.

The present chapter is primarily concerned with linear ODEs. We will present examples of the first-, second-, and third-order differential equations and proceed to describe their homogeneous as well as particular solutions using standard methods of solving ODEs. Section 2, which is devoted to an analysis of first-order linear equations, draws on several examples from passive electrical circuits. In section 3 we introduce the quintessential harmonic oscillator, the mass-and-spring system, then solve its governing second-order differential equation by means of the Fourier transform method. Another case involving a second-order ODE is treated in section 4, where we first derive the homogeneous and particular solutions, then check the result against the solution obtained via Fourier transformation. Analysis of a degenerate third-order ODE is the subject of section 5. Here, we learn how to treat degeneracies by first solving the problem for the nondegenerate case, then resorting to a limit operation as the degeneracy sets in.

Another powerful method of finding the solutions of linear ODEs, the Frobenius method, is described in section 6, in the context of solving Bessel's differential equation. A linear, second-order ODE whose coefficients of $f(x)$, $\dot{f}(x)$, and $\ddot{f}(x)$ are functions of the variable x, Bessel's equation appears frequently in mathematical physics and engineering sciences in conjunction with diffusion or wave propagation in systems of spherical or cylindrical symmetry. (You will encounter several examples of such problems in chapter 9, which is devoted to a treatment of partial differential equations.) The method of Frobenius is applicable to other important differential equations of mathematical physics, such as those of Hermite, Legendre, and Airy. Solution of these equations, which are fairly straightforward, are left as end-of-chapter problems for the student. Finally, parametric oscillation, which has found applications in electronic amplifiers and nonlinear optics, forms the subject of the last section of the chapter.

8.2. First-order linear differential equations. Consider the following first-order linear differential equation whose coefficients are functions of the variable x:

$$p(x)\dot{f}(x) + q(x)f(x) = r(x). \tag{1}$$

Here, $p(x)$, $q(x)$, and $r(x)$ are arbitrary functions of the real variable x, while $f(x)$ is the desired function, which is required to satisfy Eq.(1). Upon multiplying both sides of Eq.(1) with an arbitrary function $g(x)$, we will have

$$g(x)p(x)\dot{f}(x) + g(x)q(x)f(x) = g(x)r(x). \tag{2}$$

The left-hand side of Eq.(2) may be expressed as the derivative of the function $g(x)p(x)f(x)$, provided that $g(x)$ is chosen to satisfy the following identity:

$$\frac{\mathrm{d}}{\mathrm{d}x}[g(x)p(x)] = g(x)q(x) \quad \rightarrow \quad \frac{\mathrm{d}}{\mathrm{d}x}\ln[g(x)p(x)] = q(x)/p(x)$$

$$\rightarrow \ln[g(x)p(x)] = \int [q(x)/p(x)]\mathrm{d}x \quad \rightarrow \quad g(x) = \frac{1}{p(x)}\exp\left[\int \frac{q(x)}{p(x)}\mathrm{d}x\right]. \tag{3}$$

With this choice for $g(x)$, Eq.(2) can now be integrated to yield

$$\frac{\mathrm{d}}{\mathrm{d}x}[g(x)p(x)f(x)] = g(x)r(x) \quad \rightarrow \quad f(x) = [g(x)p(x)]^{-1}\int g(x)r(x)\mathrm{d}x. \tag{4}$$

The function $f(x)$ given by Eq.(4) is the complete solution of our ordinary first-order differential equation. In the special case of a homogeneous equation (i.e., when $r(x) = 0$), we will have $\int g(x)r(x)\mathrm{d}x = C$, where C is the integration constant. The solution to the homogeneous equation is thus $f(x) = C[g(x)p(x)]^{-1} = C\exp\{-\int [q(x)/p(x)]\mathrm{d}x\}$.

Example 1. A first-order differential equation with constant coefficients has $p(x) = a$, $q(x) = b$, and $r(x) = c$, where a, b, and c are arbitrary constants. Consequently,

$$g(x) = a^{-1}\exp(bx/a) \;\rightarrow\; f(x) = \exp(-bx/a)\int (c/a)\exp(bx/a)\,\mathrm{d}x$$
$$\rightarrow\; f(x) = \exp(-bx/a)\,[(c/b)\exp(bx/a) + C]$$
$$\rightarrow\; f(x) = (c/b) + C\exp(-bx/a). \quad (5)$$

The solution thus obtained consists of two parts: the particular solution, c/b, and the homogeneous solution, $C\exp(-bx/a)$. The constant of integration C will be determined once an initial condition, say, $f(x_0)$, has been specified.

Example 2. The differential equation $x\dot{f}(x) + f(x) = x^2$ has $p(x) = x$, $q(x) = 1$, and $r(x) = x^2$. Consequently,

$$g(x) = x^{-1}\exp(\int \mathrm{d}x/x) = x^{-1}\exp(\ln x) = x^{-1}x = 1. \quad (6)$$

$$f(x) = x^{-1}\int x^2\mathrm{d}x = \tfrac{1}{3}x^2 + Cx^{-1}. \quad (7)$$

Once again, the general solution is seen to consist of a particular solution, $\tfrac{1}{3}x^2$, and a homogeneous solution, C/x. The constant of integration C will be determined once an initial condition has been specified.

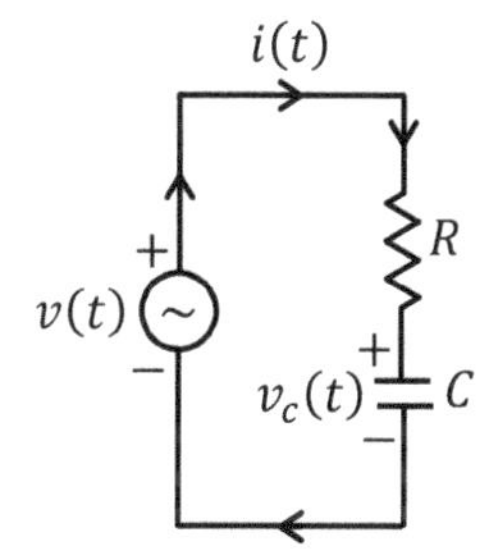

Example 3. In the circuit shown on the right-hand side, the time derivative of the voltage across the capacitor determines the current flowing through the circuit in accordance with $i(t) = C(\mathrm{d}/\mathrm{d}t)v_c(t)$. The behavior of the circuit is thus governed by the following differential equation:

$$RC\dot{v}_c(t) + v_c(t) = v(t). \quad (8)$$

A comparison with Eqs.(1)–(4) reveals that $g(t) = (RC)^{-1}\exp(t/RC)$ and that, therefore,

$$v_c(t) = (RC)^{-1}\exp(-t/RC)\int_{-\infty}^{t}\exp(t'/RC)\,v(t')\mathrm{d}t'. \quad (9)$$

Note that we have eliminated the constant of integration by choosing the lower limit of the integral in Eq.(9) as $t' = -\infty$. This choice reflects an underlying assumption that the capacitor has been empty at $t = -\infty$, which is equivalent to setting $v_c(-\infty) = 0$. Had we chosen the lower limit of the integral as $t' = 0$, the solution of Eq.(8) would have been written as follows:

$$v_c(t) = \left[v_c(0^+) + (RC)^{-1}\int_{0^+}^{t}\exp(t'/RC)\,v(t')\mathrm{d}t'\right]\exp(-t/RC); \quad t > 0. \quad (10)$$

In this version of the solution, the history of the circuit during $t < 0$ is encapsulated in $v_c(0^+)$, which proceeds to decay with a time-constant $\tau = RC$ as the capacitor empties into the closed circuit during $t > 0$. The history of $v(t)$ during $t < 0$ thus becomes irrelevant for the future evolution of the circuit, while a weighted integral of $v(t)$ during the time interval $(0, t)$ produces the remaining contribution to $v_c(t)$ in accordance with the second term in Eq.(10).

Returning to Eq.(9), it is easy to see that the application to the circuit of a δ-function as a voltage source, namely $v(t) = \delta(t)$, will produce a voltage $v_c(t)$ across the capacitor that is zero when $t < 0$ and is equal to $\exp(-t/RC)/(RC)$ when $t > 0$. The so-called *impulse response* of the circuit is thus given by the following function:

$$h(t) = \mathrm{step}(t)\exp(-t/RC)/(RC). \quad (11)$$

The general solution of Eq.(8) may now be written as a convolution between the input voltage $v(t)$ and the impulse response $h(t)$; that is, $v_c(t) = h(t) * v(t)$.

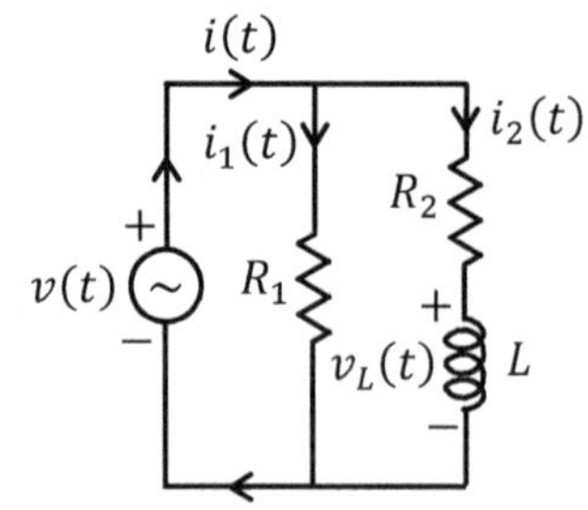

Example 4. A simple electric circuit consists of a resistor R_1 in parallel with the series combination of another resistor R_2 and an inductor L, as shown in the figure on the right. The differential equation governing this circuit is readily obtained upon applying Kirchhoff's current and voltage laws, namely,

$$v(t) = R_1 i_1(t) = R_2 i_2(t) + L\frac{\mathrm{d}}{\mathrm{d}t} i_2(t), \tag{12}$$

$$i(t) = i_1(t) + i_2(t) \quad \rightarrow \quad i_2(t) = i(t) - v(t)/R_1. \tag{13}$$

Substituting for $i_2(t)$ from Eq.(13) into Eq.(12), we find

$$LR_1 \frac{\mathrm{d}}{\mathrm{d}t} i(t) + R_1 R_2 i(t) = (R_1 + R_2)v(t) + L\frac{\mathrm{d}}{\mathrm{d}t} v(t). \tag{14}$$

The homogeneous solution of Eq.(14) is $i(t) = C\exp(-R_2 t/L)$, whose constant coefficient C will be determined upon enforcing the initial conditions. As for the particular solution of the equation, it is necessary to specify the applied voltage $v(t)$. In what follows, we shall solve the problem for three different excitation functions $v(t)$.

Case i) Let $v(t) = V_0 \text{step}(t)$, a step function that equals zero when $t < 0$ and V_0 when $t > 0$. Considering that the circuit has been dormant for $t < 0$, the current $i(t)$ must be zero during this period. As for $t > 0$, inspection of Eq.(14) reveals the constant current $i(t) = (R_1 + R_2)V_0/(R_1 R_2)$ to be an acceptable solution, which we thus adopt as the particular solution of the equation. The complete solution is, therefore, given by

$$i(t) = C\exp(-R_2 t/L) + (R_1 + R_2)V_0/(R_1 R_2). \tag{15}$$

Determination of the constant coefficient C requires that we pay close attention to the initial condition of the circuit at $t = 0$. The right-hand side of Eq.(14) contains a term $L\dot{v}(t)$, which, in the present case, is given by $LV_0\delta(t)$. The same δ-function must, therefore, appear on the left-hand side of the equation, which requires that the current $i(t)$, in going from $t = 0^-$ to $t = 0^+$, exhibit a discontinuous jump of V_0/R_1. The time derivative of $i(t)$ in the vicinity of $t = 0$ will then contain the term $(V_0/R_1)\delta(t)$, which, upon multiplication into LR_1, yields the desired $LV_0\delta(t)$ on the left-hand side of Eq.(14). All in all, we find that $i(t = 0^+) = V_0/R_1$. Substitution into Eq.(15) yields $C = -V_0/R_2$.

The physical explanation for the behavior of the current $i(t)$ is as follows: At $t = 0$, the applied voltage $v(t)$ suddenly jumps from 0 to V_0. The resistor R_1 immediately responds to this change by allowing the current $i_1(t) = V_0/R_1$ to flow through it. The current $i_2(t)$, however, cannot change instantaneously, because a jump in $i_2(t)$ would require an infinite voltage across the inductor L in accordance with $v_L(t) = L(\mathrm{d}/\mathrm{d}t)i_2(t)$, which is not feasible. We thus find $i(0^+) = i_1(0^+) + i_2(0^+) = V_0/R_1$.

The homogeneous solution $i(t) = -(V_0/R_2)\exp(-R_2 t/L)$ is a current that flows exclusively through the branch of the circuit consisting of the resistor R_2 and the inductor L. The corresponding voltage across this branch is then $R_2 i(t) + L(\mathrm{d}/\mathrm{d}t)i(t) = 0$, which is consistent with the fact that the resistor R_1 does not draw any fraction of the (homogeneous) current when the voltage source is short-circuited. As $t \rightarrow \infty$, the homogeneous current dies down, and the overall current reaches a constant value in the steady-state, where $(\mathrm{d}/\mathrm{d}t)i(t) = 0$. Thus, in the steady state, the voltage across the inductor vanishes, and the circuit reduces to a pair of parallel resistors R_1 and R_2. The equivalent resistor is then $R_1R_2/(R_1+R_2)$, which, in agreement with Eq.(15), produces the particular solution $i(t) = (R_1+R_2)V_0/(R_1R_2)$ in the steady state when $t \to \infty$.

Case ii) Here, we assume that the applied voltage is zero except during the time interval $0 \le t \le T$, when it has a constant value, say, $v(t) = V_0$. Thus, $i(t) = 0$ for $t < 0$, whereas between $t = 0$ and $t = T$ it is given by Eq.(15) with $C = -V_0/R_2$. We thus have

$$i(t) = (V_0/R_1) + (V_0/R_2)[1 - \exp(-R_2t/L)]; \qquad 0 \le t \le T. \tag{16}$$

To find the current $i(t)$ for the remaining period $t > T$, we note that, since the voltage source is turned off at $t = T$, the solution of Eq.(14) for $t > T$ consists solely of the homogeneous solution $i(t) = C\exp(-R_2t/L)$. The sudden jump in $v(t)$ at $t = T$ produces the δ-function $-LV_0\delta(t-T)$ on the right-hand side of Eq.(14). This requires a corresponding δ-function on the left-hand side of the equation and, accordingly, a sudden drop in the current by V_0/R_1. We thus find from Eq.(16) that, immediately after $t = T$, the current must drop to $i(T^+) = (V_0/R_2)[1 - \exp(-R_2T/L)]$. Consequently,

$$i(t) = (V_0/R_2)[\exp(R_2T/L) - 1]\exp(-R_2t/L); \qquad t > T. \tag{17}$$

In retrospect, it is easy to see that the currents $i_1(t)$ and $i_2(t)$ follow the voltage source independently of each other. The resistor R_1 draws the current $i_1(t) = v(t)/R_1$, irrespective of what happens in the other branch of the circuit. The second branch, which consists of the series combination of the resistor R_2 and the inductor L, responds to the voltage source with a certain sluggishness due to the fact that the inductor does not allow sudden jumps in the corresponding current $i_2(t)$.

Case iii) Suppose $v(t) = \text{step}(t)\sin(\omega t)$. Recalling that $\sin(\omega t) = \text{Imag}(e^{\mathrm{i}\omega t})$, we first solve Eq.(14) for $v(t) = \text{step}(t)\exp(\mathrm{i}\omega t)$, but in the end, retain only the imaginary part of the solution. For $t < 0$, the circuit is dormant and therefore $i(t) = 0$. For $t \ge 0$, the solution consists of a homogeneous part plus a particular part. The particular solution is obtained by setting $v(t) = \exp(\mathrm{i}\omega t)$ in Eq.(14), in which case the equation becomes

$$LR_1\frac{\mathrm{d}}{\mathrm{d}t}i(t) + R_1R_2i(t) = (R_1 + R_2 + \mathrm{i}L\omega)\exp(\mathrm{i}\omega t). \tag{18}$$

The particular solution of this equation is readily seen to be

$$i(t) = \left[\frac{R_1 + R_2 + \mathrm{i}L\omega}{R_1(R_2 + \mathrm{i}L\omega)}\right]\exp(\mathrm{i}\omega t) = I_0\exp[\mathrm{i}(\omega t - \varphi_0)], \tag{19a}$$

where

$$I_0 = (1/R_1)\sqrt{[(R_1+R_2)^2 + L^2\omega^2]/(R_2^2 + L^2\omega^2)}, \tag{19b}$$

$$\varphi_0 = \tan^{-1}\left(\frac{L\omega}{R_2}\right) - \tan^{-1}\left(\frac{L\omega}{R_1+R_2}\right) = \tan^{-1}\left[\frac{R_1L\omega}{R_2(R_1+R_2)+L^2\omega^2}\right]. \tag{19c}$$

The general solution of Eq.(14) is thus given by

$$i(t) = C\exp(-R_2t/L) + I_0\sin(\omega t - \varphi_0). \tag{20}$$

The constant coefficient C in Eq.(20) must be obtained by examining the initial condition of the circuit at $t = 0^+$. Considering that $v(t) = \text{step}(t)\sin(\omega t)$ is continuous at $t = 0$, the right-hand side of Eq.(14) does *not* have a δ-function and therefore neither should its left-hand side. The current $i(t)$ thus remains continuous at $t = 0$; that is, $i(0^+) = 0$ and, consequently, $C = I_0 \sin \varphi_0$.

Brain teaser. If you have a clock on which the hour and minute hands are identical, will there be any times that are ambiguous (along the lines of 1:21 and 4:07, say), and if so, how many?

8.3. Second-order ordinary differential equations with constant coefficients. In this section, we analyze the mass-and-spring system depicted below. Harmonic oscillators, of which the mass-and-spring system is a prominent example, are prevalent in numerous situations of practical interest in science and engineering. In our mass-and-spring system, the deviation $z(t)$ of the point mass from its equilibrium position at $z = 0$ is found by solving a second-order ordinary differential equation with constant coefficients. The overall force $\boldsymbol{F}(t)$ acting on the point particle of mass m is responsible for the particle's acceleration $\boldsymbol{a}(t)$ in accordance with Newton's second law of motion, $\boldsymbol{F}(t) = m\boldsymbol{a}(t)$. In the system depicted here, the total force is the sum of the applied external force $f(t)\hat{\boldsymbol{z}}$, the force $-\alpha z(t)\hat{\boldsymbol{z}}$ exerted by the spring, where $\alpha > 0$ is the spring constant, and the friction force $-\beta \dot{z}(t)\hat{\boldsymbol{z}}$, where $\beta > 0$ is the friction coefficient.

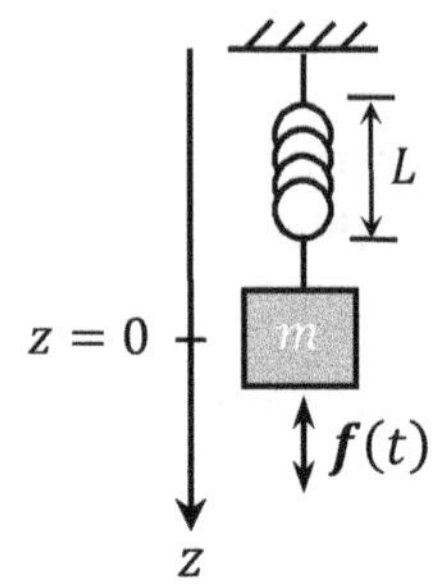

Note: If a constant external force, say, the force $mg\hat{\boldsymbol{z}}$ of gravity, happens to be present, it will be balanced by an equal force exerted by the spring in the opposite direction (i.e., the spring automatically elongates by $\Delta L = mg/\alpha$ in response to the gravitational force). In the system depicted here, we have already accounted for this (constant) lengthening of the spring by setting the equilibrium position of the particle's center of mass at $z = 0$. Any deviation of the particle from its equilibrium position will cause additional elongation (or compression) of the spring, which will subsequently exert the force $-\alpha z(t)\hat{\boldsymbol{z}}$ on the point particle. This exerted force is, of course, above and beyond what is needed to balance the pull of gravitation.

The equation of motion for our mass-and-spring system may thus be written as follows:

$$f(t) - \alpha z(t) - \beta \dot{z}(t) = m\ddot{z}(t). \tag{21}$$

Defining the constant parameters $\omega_0 = \sqrt{\alpha/m}$ and $\gamma = \beta/m$, one can streamline Eq.(21) and put it in the standard form for a generic harmonic oscillator. Given that the units of α and β are Newton/meter and $\text{Newton} \cdot \text{second/meter}$, respectively, both ω_0 and γ will have the units of $1/\text{sec}$. For reasons that will become clear during the following analysis, ω_0 is commonly referred to as the natural frequency, while γ is known as the damping coefficient of the system. With these new parameters, Eq.(21), after normalization and rearrangement, becomes

$$\frac{\mathrm{d}^2 z(t)}{\mathrm{d}t^2} + \gamma \frac{\mathrm{d}z(t)}{\mathrm{d}t} + \omega_0^2 z(t) = \frac{1}{m} f(t). \tag{22}$$

Upon Fourier transforming both sides of Eq.(22), we obtain the algebraic equation $(\mathrm{i}2\pi s)^2 Z(s) + \gamma(\mathrm{i}2\pi s)Z(s) + \omega_0^2 Z(s) = (1/m)F(s)$, which is readily solved to yield

$$Z(s) = \frac{(1/m)F(s)}{-(2\pi s)^2 + \gamma(\mathrm{i}2\pi s) + \omega_0^2}. \tag{23}$$

Let the externally applied force along the z-axis be $f(t) = F_0\, \text{step}(t) \exp(\mathrm{i}2\pi s_0 t)$, where s_0 is the real-valued frequency of the (one-sided) sinusoidal excitation $f(t)$, and F_0 is a complex constant. The actual excitation, of course, is $\text{Re}[f(t)]$, and the actual solution will be the real part of the function $z(t)$ obtained by solving Eq.(22). Upon Fourier transformation of $f(t)$, we will have

$$F(s) = \mathcal{F}\{f(t)\} = F_0\,\mathcal{F}\{\text{step}(t)\} * \mathcal{F}\{\exp(\mathrm{i}2\pi s_0 t)\} = F_0\left[\tfrac{1}{2}\delta(s) - \frac{\mathrm{i}}{2\pi s}\right] * \delta(s - s_0)$$

$$= \tfrac{1}{2}F_0\delta(s) * \delta(s - s_0) - \left(\frac{\mathrm{i}F_0}{2\pi s}\right) * \delta(s - s_0) = \tfrac{1}{2}F_0\delta(s - s_0) - \frac{\mathrm{i}F_0}{2\pi\,(s - s_0)}. \qquad (24)$$

Substitution into Eq.(23) yields

$$Z(s) = -\frac{F_0}{8\pi^2 m}\,\frac{\delta(s-s_0) - \mathrm{i}/[\pi(s-s_0)]}{s^2 - \mathrm{i}(\gamma/2\pi)s - (\omega_0/2\pi)^2}$$

$$= -\frac{F_0}{8\pi^2 m}\left\{\frac{\delta(s-s_0)}{s^2 - \left(\frac{\mathrm{i}\gamma}{2\pi}\right)s - \left(\frac{\omega_0}{2\pi}\right)^2} - \frac{(\mathrm{i}/\pi)}{(s-s_0)\left[s - \frac{\mathrm{i}\gamma}{4\pi} - \sqrt{\left(\frac{\omega_0}{2\pi}\right)^2 - \left(\frac{\gamma}{4\pi}\right)^2}\right]\left[s - \frac{\mathrm{i}\gamma}{4\pi} + \sqrt{\left(\frac{\omega_0}{2\pi}\right)^2 - \left(\frac{\gamma}{4\pi}\right)^2}\right]}\right\}. \qquad (25)$$

To arrive at Eq.(25), we have factorized the quadratic expression in the denominator of the second term, writing it as $s^2 - \mathrm{i}(\gamma/2\pi)s - (\omega_0/2\pi)^2 = (s - s_1)(s - s_2)$, where $s_{1,2} = \mathrm{i}(\gamma/4\pi) \pm \sqrt{(\omega_0/2\pi)^2 - (\gamma/4\pi)^2}$ are the roots of the quadratic expression.

The mass-and-spring system is said to be *underdamped* if $\gamma < 2\omega_0$, *critically damped* if $\gamma = 2\omega_0$, and *overdamped* if $\gamma > 2\omega_0$.

The inverse Fourier transform of $Z(s)$ may now be calculated using complex plane techniques, as follows:

Use sifting property of the δ-function.

$$z(t) = \mathcal{F}^{-1}\{Z(s)\} = \int_{-\infty}^{\infty} Z(s)\exp(\mathrm{i}2\pi st)\,\mathrm{d}s$$

$$= -\left(\frac{F_0}{8\pi^2 m}\right)\int_{-\infty}^{\infty}\frac{\delta(s-s_0)\exp(\mathrm{i}2\pi st)}{s^2 - \mathrm{i}(\gamma/2\pi)s - (\omega_0/2\pi)^2}\,\mathrm{d}s + \left(\frac{\mathrm{i}F_0}{8\pi^3 m}\right)\int_{-\infty}^{\infty}\frac{\exp(\mathrm{i}2\pi st)}{(s-s_0)(s-s_1)(s-s_2)}\,\mathrm{d}s$$

$$= -\left(\frac{F_0}{8\pi^2 m}\right)\frac{\exp(\mathrm{i}2\pi s_0 t)}{s_0^2 - \mathrm{i}(\gamma/2\pi)s_0 - (\omega_0/2\pi)^2} + \left(\frac{\mathrm{i}F_0}{8\pi^3 m}\right)\int_{-\infty}^{\infty}\frac{\exp(\mathrm{i}2\pi st)}{(s-s_0)(s-s_1)(s-s_2)}\,\mathrm{d}s. \qquad (26)$$

For the remaining integral in Eq.(26), the diagrams in figure 8.1 show the integration contours as well as the poles of the integrand in the complex ζ-plane. Note that s_0 is on the real axis, while s_1 and s_2 are in the upper-half of the complex plane. (Only the underdamped case corresponding to $\gamma < 2\omega_0$ is shown in the figure.)

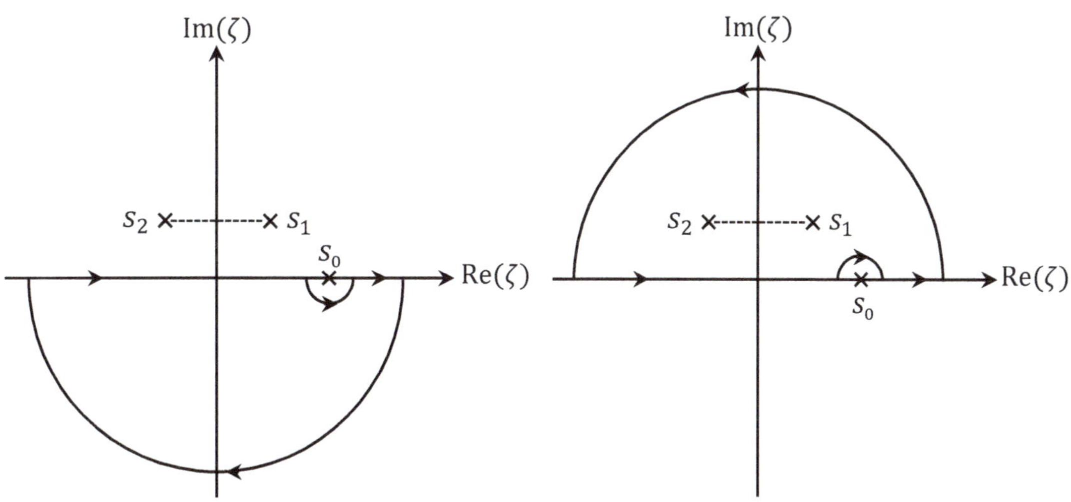

Figure 8.1. Complex plane diagrams showing integration contours and poles of the integrand for the inverse Fourier integral appearing on the right-hand side of Eq.(26).

For $t < 0$, the complex plane integral must be evaluated in the *lower* half plane, because $\exp(i2\pi\zeta t)$ on the large semicircle approaches zero only in the lower half plane. Invoking Jordan's lemma, the integral over the large semicircle vanishes, and the desired integral over the real axis becomes equal to $2\pi i$ multiplied by one-half the residue at $\zeta = s_0$; that is,

$$\int_{-\infty}^{\infty} \frac{\exp(i2\pi st)}{(s-s_0)(s-s_1)(s-s_2)} ds = -\frac{i\pi \exp(i2\pi s_0 t)}{(s_0-s_1)(s_0-s_2)} = -\frac{i\pi \exp(i2\pi s_0 t)}{s_0^2 - i(\gamma/2\pi)s_0 - (\omega_0/2\pi)^2}; \quad (t < 0). \quad (27)$$

The two terms in the expression for $z(t)$ in Eq.(26) now cancel out, yielding $z(t) = 0$ for $t < 0$. For $t \geq 0$, the complex plane integral in Eq.(26) must be evaluated in the upper half plane, where $\exp(i2\pi\zeta t)$ over the large semicircle approaches zero, thus allowing the application of Jordan's lemma. The desired integral over the real axis is then equal to $2\pi i$, multiplied by the sum of the residues at s_1 and s_2 plus one-half of the residue at s_0. We will have

$$z(t) = -\left(\frac{F_0}{8\pi^2 m}\right)\frac{\exp(i2\pi s_0 t)}{s_0^2 - i(\gamma/2\pi)s_0 - (\omega_0/2\pi)^2}$$

$$+2\pi i\left(\frac{iF_0}{8\pi^3 m}\right)\left[\frac{½\exp(i2\pi s_0 t)}{(s_0-s_1)(s_0-s_2)} + \frac{\exp(i2\pi s_1 t)}{(s_1-s_0)(s_1-s_2)} + \frac{\exp(i2\pi s_2 t)}{(s_2-s_0)(s_2-s_1)}\right]$$

$$= -\frac{F_0}{8\pi^2 m}\left[\frac{\exp(i2\pi s_0 t)}{s_0^2 - i(\gamma/2\pi)s_0 - (\omega_0/2\pi)^2} + \frac{\exp(i2\pi s_0 t)}{(s_0-s_1)(s_0-s_2)}\right]$$

$$-\frac{(F_0/m)}{4\pi^2(s_1-s_2)}\left[\frac{\exp(i2\pi s_1 t)}{s_1-s_0} - \frac{\exp(i2\pi s_2 t)}{s_2-s_0}\right]$$

$$= -\frac{(F_0/m)\exp(i2\pi s_0 t)}{4\pi^2 s_0^2 - i2\pi\gamma s_0 - \omega_0^2}$$

$$-\frac{(F_0/m)\exp(-\gamma t/2)}{2\,[\omega_0^2 - (\gamma/2)^2]^{½}}\left\{\frac{\exp\{i[\omega_0^2 - (\gamma/2)^2]^{½}\,t\}}{[\omega_0^2 - (\gamma/2)^2]^{½} - [2\pi s_0 - i(\gamma/2)]} + \frac{\exp\{-i[\omega_0^2 - (\gamma/2)^2]^{½}\,t\}}{[\omega_0^2 - (\gamma/2)^2]^{½} + [2\pi s_0 - i(\gamma/2)]}\right\}$$

$$= -\frac{(F_0/m)\exp(i2\pi s_0 t)}{4\pi^2 s_0^2 - i2\pi\gamma s_0 - \omega_0^2} - \frac{(F_0/m)\exp(-\gamma t/2)}{[\omega_0^2 - (\gamma/2)^2]^{½}}$$

$$\times \frac{[\omega_0^2 - (\gamma/2)^2]^{½}\cos\{[\omega_0^2 - (\gamma/2)^2]^{½}t\} + i[2\pi s_0 - i(\gamma/2)]\sin\{[\omega_0^2 - (\gamma/2)^2]^{½}\,t\}}{\omega_0^2 - (\gamma/2)^2 - [2\pi s_0 - i(\gamma/2)]^2}$$

$$= -\frac{(F_0/m)\exp(i2\pi s_0 t)}{4\pi^2 s_0^2 - i2\pi\gamma s_0 - \omega_0^2} + \frac{(F_0/m)\exp(-\gamma t/2)}{4\pi^2 s_0^2 - i2\pi\gamma s_0 - \omega_0^2}$$

$$\times\left[\cos\left(\sqrt{\omega_0^2 - (\gamma/2)^2}\,t\right) + \frac{(\gamma/2) + i2\pi s_0}{[\omega_0^2 - (\gamma/2)^2]^{½}}\sin\left(\sqrt{\omega_0^2 - (\gamma/2)^2}\,t\right)\right]. \quad (28)$$

It is easy to verify the continuity of $z(t)$ and its first derivative $\dot{z}(t)$ at $t = 0$; that is, $z(0^+) = 0$ and $\dot{z}(0^+) = 0$.

Similar solutions may be obtained for overdamped and critically damped systems. In the latter case, care must be taken to handle the second-order pole at $\zeta = s_1 = s_2$ properly. A straightforward method of arriving at the solution $z(t)$ for a critically damped system is to allow ω_0 in Eq.(28) to approach $\gamma/2$. We will have

$$z(t) = -\frac{(F_0/m)\exp(i2\pi s_0 t)}{[2\pi s_0 - i(\gamma/2)]^2} + \frac{(F_0/m)\exp(-\gamma t/2)}{[2\pi s_0 - i(\gamma/2)]^2} + \frac{(iF_0/m)\, t\exp(-\gamma t/2)}{2\pi s_0 - i(\gamma/2)}. \quad (29)$$

In the case of an overdamped system, where $\gamma > 2\omega_0$, the two poles s_1 and s_2 will be at $\zeta = s_{1,2} = i(\gamma/4\pi) \pm i\sqrt{(\gamma/4\pi)^2 - (\omega_0/2\pi)^2}$, which are located on the imaginary axis. Once again s_1 and s_2 are seen to be in the upper half of the complex ζ-plane.

8.4. Point particle subject to an external force in the presence of friction. Consider a particle of mass m moving along the x-axis within a medium that exerts a friction force on the particle. The particle is subject to an external force $\boldsymbol{f}(t) = f(t)\hat{\boldsymbol{x}}$, which is applied along the line of motion, namely the x-axis. The friction force is proportional to the particle velocity $v(t) = \dot{x}(t)$, with the proportionality constant being $-\beta$, where $\beta > 0$. Newton's second law yields the following equation of motion for the particle:

$$m\ddot{x}(t) + \beta\dot{x}(t) = f(t). \quad (30)$$

We assume that the particle has been at rest at $x = 0$ prior to the application of the external force at $t = 0$, that is, during $t < 0$, both its position and velocity have been equal to zero. In what follows, we shall first use the standard method of solving ODEs with constant coefficients to find $x(t)$ and $v(t)$ when the external force is given by $f(t) = f_0\text{step}(t)$. Subsequently, we will deduce the position and velocity of the particle in response to the external force $f(t) = \delta(t)$. The latter result will then be derived directly by applying the Fourier transform method to the equation of motion.

Let the homogeneous solution be in the form of $x(t) = x_0 \exp(\eta t)$. Substitution into Eq.(30) yields $m\eta^2 + \beta\eta = 0$. Consequently, $\eta_1 = 0$ and $\eta_2 = -\beta/m = -\gamma$, where γ is the so-called damping coefficient. For the external force $f(t) = f_0\text{step}(t)$, the particular solution of the equation (for $t \ge 0$) is readily found by inspection to be $x(t) = (f_0/\beta)t$. The complete solution thus turns out to be

$$x(t) = A_1 \exp(\eta_1 t) + A_2 \exp(\eta_2 t) + (f_0/\beta)t; \qquad t \ge 0. \quad (31)$$

At $t = 0^+$, the position and velocity of the particle are zero; that is, $x(0^+) = A_1 + A_2 = 0$, and $v(0^+) = \dot{x}(0^+) = \eta_1 A_1 + \eta_2 A_2 + (f_0/\beta) = 0$. Therefore, $A_1 = -A_2 = f_0/(\beta\eta_2)$. We will have

$$x(t) = -(f_0/\beta\gamma)[1 - \exp(-\gamma t) - \gamma t]\text{step}(t), \quad (32a)$$

$$v(t) = \dot{x}(t) = (f_0/\beta)[1 - \exp(-\gamma t)]\text{step}(t). \quad (32b)$$

It is seen that the particle moves with a velocity that is zero at the outset but reaches the steady state velocity of $v_{ss} = f_0/\beta$ as $t \to \infty$.

Next, recognizing that $\delta(t)$ is the derivative of $\text{step}(t)$, and that the equation of motion is linear, we obtain the position and velocity of the particle in response to $\delta(t)$ by differentiating (with respect to t) the response to $\text{step}(t)$ obtained above. We will have

$$x(t) = (1/\beta)[1 - \exp(-\gamma t)]\text{step}(t), \quad (33a)$$

$$v(t) = \dot{x}(t) = (1/m)\exp(-\gamma t)\,\text{step}(t). \quad (33b)$$

The particle is thus jolted at $t = 0$, acquiring an initial velocity $v(0^+) = 1/m$. Its velocity then decays exponentially (due to the action of the friction force) and approaches zero as $t \to \infty$. Here, a note about the units is in order. When the step force is represented by $f_0\text{step}(t)$, the units of f_0 are those of force (e.g., $Newton = kg \cdot m/s^2$ in SI) and the step function is dimensionless. However, when the step force is differentiated to yield the impulsive force $f_0\delta(t)$, the units of f_0 become those of force multiplied by time (e.g., $kg \cdot m/s$ in SI), and the δ-function acquires the units of inverse seconds $(1/s)$. Now, in deriving the expressions for $x(t)$ and $v(t)$ in Eq.(33), we have set $f_0 = 1$, which means that the numerator 1 of the coefficients $1/\beta$ and $1/m$ must be in units of $kg \cdot m/s$ (in SI). Needless to say, β has the units of mass divided by time (kg/s), while those of the damping coefficient γ are the inverse of time $(1/s)$.

Finally, applying the Fourier method, the equation of motion is Fourier transformed to yield

$$m(\mathrm{i}2\pi s)^2 X(s) + \beta(\mathrm{i}2\pi s)X(s) = F(s), \tag{34}$$

where

$$X(s) = \mathcal{F}\{x(t)\} = \int_{-\infty}^{\infty} x(t)\exp(-\mathrm{i}2\pi st)\,\mathrm{d}t, \tag{35a}$$

$$F(s) = \mathcal{F}\{f(t)\} = \int_{-\infty}^{\infty} f(t)\exp(-\mathrm{i}2\pi st)\,\mathrm{d}t. \tag{35b}$$

The Fourier transform $X(s)$ of the position function $x(t)$ is thus found (straightforwardly) to be

$$X(s) = -\frac{F(s)}{2\pi ms(2\pi s - \mathrm{i}\gamma)}. \tag{36}$$

Now, for the applied force function $f(t) = \delta(t)$, we have $F(s) = 1$, and the inverse Fourier transform $x(t) = \int_{-\infty}^{\infty} X(s)\exp(\mathrm{i}2\pi st)\,\mathrm{d}s$ may be evaluated by integration in the complex plane $s = s' + \mathrm{i}s''$. Note that $X(s)$ has two simple poles in the s-plane, one at $s_1 = 0$, and the other at $s_2 = \mathrm{i}\gamma/2\pi$. The integral must be evaluated in the lower half plane when $t < 0$, and in the upper half plane when $t > 0$, as shown in figure 8.2.

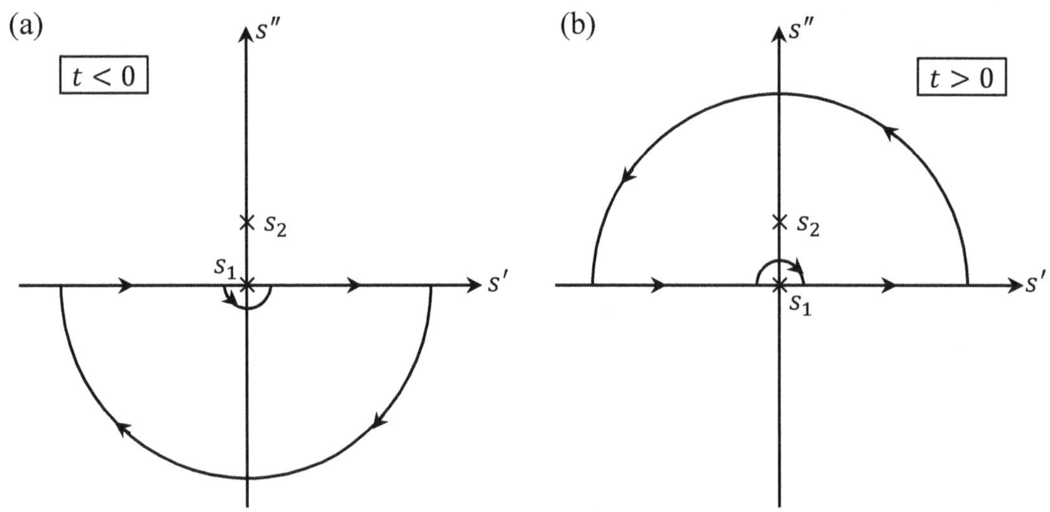

Figure 8.2. Complex plane diagrams showing integration contours and poles of the integrand for inverse Fourier transformation of the function $X(s)$ appearing in Eq.(36).

Since Jordan's lemma is applicable here, the large semicircles do not contribute to the integral. The residue of the integrand at s_1 is $1/(\mathrm{i}2\pi m\gamma)$, while that at s_2 is $-\exp(-\gamma t)/(\mathrm{i}2\pi m\gamma)$. We thus have

$$x(t) = \begin{cases} -1/(2m\gamma); & t < 0, \\ (1/m\gamma)[½ - \exp(-\gamma t)]; & t > 0. \end{cases} \tag{37}$$

Noting that $m\gamma = \beta$, one observes that, aside from a downward shift of $1/(2\beta)$, the position function $x(t)$ given by Eq.(37) is the same as that in Eq.(33a). The reason for this unusual behavior of the Fourier solution is that Eq.(30) is an equation for $\dot{x}(t)$, not $x(t)$. Consequently, a constant can always be added to $x(t)$ without changing the nature of the solution. That is how one should understand the difference $1/(2\beta)$ between the solution $x(t)$ obtained by the Fourier method and the actual solution given by Eq.(33a).

8.5. Solving a degenerate third-order differential equation. Consider the following homogeneous third-order ODE with arbitrary constant coefficients:

$$f'''(x) - 3f''(x) + 3f'(x) - f(x) = 0. \tag{38}$$

We guess the solution to be $f(x) = \exp(\lambda x)$. Substitution into the equation yields

$$\begin{aligned} f'''(x) - 3f''(x) + 3f'(x) - f(x) &= (\lambda^3 - 3\lambda^2 + 3\lambda - 1)\exp(\lambda x) \\ &= (\lambda - 1)^3 \exp(\lambda x) = 0. \end{aligned} \tag{39}$$

Since the only solution of Eq.(39) is $\lambda = 1$, the method yields the triply degenerate solution $f(x) = \exp(x)$.

In general, assuming $\exp(\lambda_1 x)$ and $\exp(\lambda_2 x)$ are two independent solutions of a linear differential equation (that is, $\lambda_1 \neq \lambda_2$), one can form a linear superposition of these two solutions in the form of $f(x) = A_0[\exp(\lambda_1 x) - \exp(\lambda_2 x)]/(\lambda_1 - \lambda_2)$, which would satisfy the differential equation. In what follows, we determine the limiting form of this superposition solution $f(x)$ when $\lambda_1 \to \lambda_2$ and confirm that the solution thus obtained does in fact satisfy our third-order differential equation. We write

$$\begin{aligned} f(x) &= A_0 \frac{\exp(\lambda_1 x) - \exp(\lambda_2 x)}{\lambda_1 - \lambda_2} = A_0 \exp(\lambda_2 x) \frac{\exp[(\lambda_1 - \lambda_2)x] - 1}{\lambda_1 - \lambda_2} \\ &= A_0 \exp(\lambda_2 x) \frac{1 + (\lambda_1 - \lambda_2)x + ½(\lambda_1 - \lambda_2)^2 x^2 + \cdots - 1}{\lambda_1 - \lambda_2} \\ &= A_0 \exp(\lambda_2 x)\,[x + ½(\lambda_1 - \lambda_2)x^2 + \cdots]. \end{aligned} \tag{40}$$

In the limit when $\lambda_1 \to \lambda_2 = \lambda$, this solution approaches $f(x) = A_0 x \exp(\lambda x)$. Setting $A_0 = 1$ (for convenience), we find

$$f(x) = x\exp(\lambda x) \quad \to \quad f'(x) = (\lambda x + 1)\exp(\lambda x)$$

$$\to \quad f''(x) = (\lambda^2 x + 2\lambda)\exp(\lambda x) \quad \to \quad f'''(x) = (\lambda^3 x + 3\lambda^2)\exp(\lambda x).$$

Substitution into Eq.(38) now yields

$$\begin{aligned} &f'''(x) - 3f''(x) + 3f'(x) - f(x) \\ &\quad = [(\lambda^3 x + 3\lambda^2) - 3(\lambda^2 x + 2\lambda) + 3(\lambda x + 1) - x]\exp(\lambda x) \\ &\quad = [(\lambda^3 - 3\lambda^2 + 3\lambda - 1)x + 3(\lambda^2 - 2\lambda + 1)]\exp(\lambda x) \\ &\quad = [(\lambda - 1)^3 x + 3(\lambda - 1)^2]\exp(\lambda x). \end{aligned} \tag{41}$$

Since our triply degenerate solution is $\lambda = 1$, the expression appearing on the right-hand side of Eq.(41) vanishes, confirming that $f(x) = x\exp(\lambda x)$ is indeed a solution of our original third-order differential equation.

Since a third-order linear ODE is expected to have three independent solutions, we are still short of one solution for Eq.(38). Let us begin by assuming that $\exp(\lambda_1 x)$, $\exp(\lambda_2 x)$, and $\exp(\lambda_3 x)$ are three independent solutions of a differential equation (that is, $\lambda_1 \neq \lambda_2$, $\lambda_2 \neq \lambda_3$, and $\lambda_1 \neq \lambda_3$). In the limit when $\lambda_1 \to \lambda_2 \to \lambda_3$, one can arrange the coefficients of the equation such that $\lambda_1 = \lambda_2 + \delta$ and $\lambda_3 = \lambda_2 - \delta$, with $\delta \to 0$. Subsequently, the following superposition of the three solutions would satisfy the differential equation:

$$f(x) = A_0[\exp(\lambda_1 x) - 2\exp(\lambda_2 x) + \exp(\lambda_3 x)]/\delta^2. \tag{42}$$

To determine the limiting form of this superposition solution when $\delta \to 0$, we write

$$\begin{aligned} f(x) &= A_0[\exp(\lambda_1 x) - 2\exp(\lambda_2 x) + \exp(\lambda_3 x)]/\delta^2 \\ &= A_0 \exp(\lambda_2 x)\{\exp[(\lambda_1 - \lambda_2)x] - 2 + \exp[(\lambda_3 - \lambda_2)x]\}/\delta^2 \\ &= A_0 \exp(\lambda_2 x)\,[\exp(\delta x) - 2 + \exp(-\delta x)]/\delta^2 \\ &= A_0 \exp(\lambda_2 x)\left[\left(1 + \delta x + \frac{\delta^2 x^2}{2!} + \frac{\delta^3 x^3}{3!} + \cdots\right) - 2\right. \\ &\qquad\qquad \left. + \left(1 - \delta x + \frac{\delta^2 x^2}{2!} - \frac{\delta^3 x^3}{3!} + \cdots\right)\right]/\delta^2 \\ &= A_0 \exp(\lambda_2 x)\,[x^2 + (\delta^2/12)x^4 + \cdots]. \end{aligned} \tag{43}$$

Thus, in the limit when $\delta \to 0$ (i.e., $\lambda_1, \lambda_3 \to \lambda_2 = \lambda$), the function $f(x) = A_0 x^2 \exp(\lambda x)$ becomes a solution of the third-order differential equation. To confirm that this solution does in fact satisfy the original third-order differential equation, we write

$$f(x) = x^2 \exp(\lambda x) \qquad \to f'(x) = (\lambda x^2 + 2x)\exp(\lambda x)$$

$$\to f''(x) = (\lambda^2 x^2 + 4\lambda x + 2)\exp(\lambda x) \quad \to f'''(x) = (\lambda^3 x^2 + 6\lambda^2 x + 6\lambda)\exp(\lambda x).$$

Substitution into the original differential equation, Eq.(38), now yields

$$\begin{aligned} &f'''(x) - 3f''(x) + 3f'(x) - f(x) \\ &\quad= [(\lambda^3 x^2 + 6\lambda^2 x + 6\lambda) - 3(\lambda^2 x^2 + 4\lambda x + 2) + 3(\lambda x^2 + 2x) - x^2]\exp(\lambda x) \\ &\quad= [(\lambda^3 - 3\lambda^2 + 3\lambda - 1)x^2 + 6(\lambda^2 - 2\lambda + 1)x + 6(\lambda - 1)]\exp(\lambda x) \\ &\quad= [(\lambda - 1)^3 x^2 + 6(\lambda - 1)^2 x + 6(\lambda - 1)]\exp(\lambda x). \end{aligned} \tag{44}$$

Since our triply degenerate solution is $\lambda = 1$, the right-hand side of Eq.(44) vanishes, confirming that $f(x) = x^2 \exp(\lambda x)$ is indeed a solution of our third-order ODE. The complete solution may thus be written as follows: $f(x) = (A_0 + A_1 x + A_2 x^2)\exp(\lambda x)$.

8.6. Bessel's equation and the method of Frobenius. The standard Bessel equation, a linear second-order ODE in the variable x, whose coefficients depend on x, is written as

$$x^2 \ddot{f}(x) + x\dot{f}(x) + (x^2 - \nu^2)f(x) = 0. \tag{45}$$

The Bessel equation's parameter ν can, in general, be an arbitrary complex constant, although we shall limit the following discussion to real-valued $\nu \geq 0$ and also discuss special cases where ν is a nonnegative integer. To arrive at a solution of Bessel's equation, we use the method of Frobenius, which starts by assuming the existence of a solution of the following form:

$$f(x) = x^s \sum_{n=0}^{\infty} a_n x^n. \tag{46}$$

In this expression, the parameter s (commonly referred to as the *index*) along with all the coefficients a_n must be determined by enforcing Eq.(45) for all values of x. If the index s turns out to be zero or a positive integer, then the form assumed for $f(x)$ in Eq.(46) is simply a Taylor series expansion of the sought-after function around the point $x = 0$. However, by introducing the factor x^s in Eq.(46), the method of Frobenius opens up the possibility of extending the solution of ODEs beyond the range of functions that are expandable in a Taylor series.

The various terms in the suggested expansion of $f(x)$ in accordance with Eq.(46) may be written as $a_n x^{n+s}$ and, on differentiation and substitution into Eq.(45), we find

$$x^2 \sum_{n=0}^{\infty} (n+s)(n+s-1)a_n x^{n+s-2} + x \sum_{n=0}^{\infty} (n+s) a_n x^{n+s-1}$$

$$+(x^2 - \nu^2) \sum_{n=0}^{\infty} a_n x^{n+s} = 0. \tag{47}$$

Upon rearrangement and simplification, the preceding equation can be written as follows:

$$\sum_{n=0}^{\infty} [(n+s)(n+s-1) + (n+s) - \nu^2] a_n x^{n+s} + \sum_{n=0}^{\infty} a_n x^{n+s+2} = 0. \tag{48}$$

The first sum on the left-hand side of Eq.(48) may be simplified, while the second sum is trivially modified by letting the summation dummy n range from 2 to ∞ (instead of 0 to ∞). Subsequently, Eq.(48) reduces to

$$\sum_{n=0}^{\infty} [(n+s)^2 - \nu^2] a_n x^{n+s} + \sum_{n=2}^{\infty} a_{n-2} x^{n+s} = 0. \tag{49}$$

We must now set to zero the coefficients of x^{n+s} for all values of n, as this would be the only way in which Eq.(49) can be satisfied for all x from $-\infty$ to ∞. From the first sum on the left-hand side of Eq.(49), we isolate the terms for $n = 0$ and $n = 1$. These will yield

$$(s^2 - \nu^2) a_0 = 0. \tag{50a}$$

$$[(1+s)^2 - \nu^2] a_1 = 0. \tag{50b}$$

The remaining terms in Eq.(49) correspond to $n \geq 2$, which may be combined to yield

$$[(n+s)^2 - \nu^2] a_n + a_{n-2} = 0; \qquad (n \geq 2). \tag{50c}$$

Equations 50a and 50b are known as the *indicial* equations—because they specify the values of the index s—whereas Eq.(50c) is commonly referred to as the *recursion* relation. From Eq.(50a), we learn that a_0 must be set to zero, unless the index s is chosen either as $s = \nu$ or as $s = -\nu$. Similarly, Eq.(50b) demands that a_1 be set to zero, unless the index s is chosen either as $s = \nu - 1$ or as $s = -(\nu + 1)$. Once the choices for s, a_0, and a_1 have been settled, the recursion relation can be used to find the remaining coefficients a_n for $n \geq 2$. In what follows, we examine the consequences of all the available options for selecting the values of s, a_0, and a_1.

Case i) Here, $s = \nu \geq 0$, a_0 is arbitrary, and a_1 must be set to zero. Since $a_1 = 0$, the recursion relation yields $a_3 = a_5 = a_7 = \cdots = 0$; that is, all a_n having odd values of n

must vanish. The recursion relation also specifies all a_n having even values of n in accordance with $a_n = a_{n-2}/[\nu^2 - (n+\nu)^2] = -\,a_{n-2}/[n(n+2\nu)]$. Given that n is now even, we may replace it with $2m$ to obtain $a_{2m} = -a_{2m-2}/[4m(\nu+m)]$. Thus,

$$a_2 = -\frac{a_0}{4(\nu+1)}, \tag{51a}$$

$$a_4 = -\frac{a_2}{4\times2\times(\nu+2)} = \frac{a_0}{4^2\times2\times(\nu+1)(\nu+2)}, \tag{51b}$$

$$a_6 = -\frac{a_4}{4\times3\times(\nu+3)} = -\frac{a_0}{4^3\times2\times3\times(\nu+1)(\nu+2)(\nu+3)}. \tag{51c}$$

The general trend is now easy to infer. Considering that Euler's gamma function (chapter 3, section 3) has the property $\Gamma(x+1) = x\Gamma(x)$, we write

$$a_{2m} = \frac{(-1)^m\Gamma(\nu+1)}{2^{2m}\, m!\, \Gamma(\nu+m+1)}a_0. \tag{52}$$

A solution of Eq.(45) in the general form of Eq.(46) is thus given by

$$f(x) = a_0\Gamma(\nu+1)x^\nu \sum_{m=0}^{\infty} \frac{(-1)^m}{m!\,\Gamma(\nu+m+1)}\left(\frac{x}{2}\right)^{2m}. \tag{53}$$

The constant a_0 is arbitrary and can be combined with the constant $\Gamma(\nu+1)$. A standard normalization of the solution given by Eq.(53) involves setting its arbitrary coefficient to $2^{-\nu}$. This solution of Eq.(45), generally known as a Bessel function of the first kind, order ν, is denoted by $J_\nu(x)$ and written as

$$J_\nu(x) = \sum_{m=0}^{\infty} \frac{(-1)^m(x/2)^{2m+\nu}}{m!\,\Gamma(\nu+m+1)}. \tag{54}$$

When ν is a nonnegative integer, the term $\Gamma(\nu+m+1)$, appearing in the denominator of Eq.(54), is written simply as $(\nu+m)!$. The first few Bessel functions of the first kind, integer order $\nu = 0, 1, 2, 3, \cdots$ are plotted in figure 8.3.

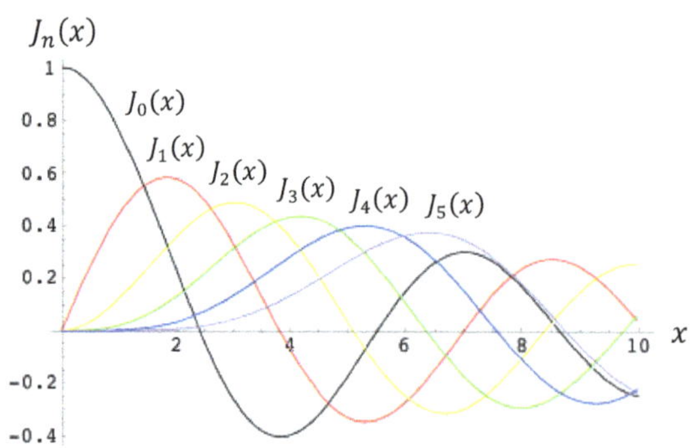

Figure 8.3. Plots of the first few Bessel functions of the first kind, integer order $\nu = 0, 1, 2, 3, \cdots$.

For $\nu = ½$, Eq.(52) can be written as

$$a_{2m} = \frac{(-1)^m}{2^{2m}m!\,(½+1)(½+2)\cdots(½+m)}a_0 = \frac{(-1)^m}{(2m+1)!}a_0. \tag{55}$$

Consequently,

$$f(x) = a_0 x^{½} \sum_{m=0}^{\infty} \frac{(-1)^m x^{2m}}{(2m+1)!} = a_0\frac{\sin x}{\sqrt{x}}. \tag{56}$$

Thus, aside from a constant coefficient, the general expression of $J_{½}(x)$ is $\sin x/\sqrt{x}$.

Case ii) For $\nu = ½$, the choice of $s = -\nu$ results in both indicial equations (50a) and (50b) being satisfied for arbitrary values of a_0 and a_1. The recursion relation, Eq.(50c), then yields $a_n = -a_{n-2}/[n(n-1)]$ for $n \geq 2$. We thus find

$$a_2 = -\frac{a_0}{2!}, \qquad a_4 = \frac{a_0}{4!}, \qquad a_6 = -\frac{a_0}{6!}, \qquad \cdots \qquad a_{2m} = \frac{(-1)^m a_0}{(2m)!}, \tag{57a}$$

$$a_3 = -\frac{a_1}{3!}, \qquad a_5 = \frac{a_0}{5!}, \qquad a_7 = -\frac{a_0}{7!}, \qquad \cdots \qquad a_{2m+1} = \frac{(-1)^m a_1}{(2m+1)!}. \tag{57b}$$

The general solution of the Bessel equation for $\nu = ½$ may now be written as follows:

$$f(x) = x^{-½}\left[a_0 \sum_{m=0}^{\infty} \frac{(-1)^m x^{2m}}{(2m)!} + a_1 \sum_{m=0}^{\infty} \frac{(-1)^m x^{2m+1}}{(2m+1)!}\right] = a_0\left(\frac{\cos x}{\sqrt{x}}\right) + a_1\left(\frac{\sin x}{\sqrt{x}}\right). \tag{58}$$

The solution given by Eq.(58) is a linear superposition of $J_{½}(x)$, the Bessel function of first kind, order ½, and $Y_{½}(x)$, the Bessel function of second kind, order ½. The latter function is proportional to $\cos x/\sqrt{x}$.

Case iii) For a positive integer ν, choosing $s = -\nu < 0$, allows a_0 to be arbitrary, but a_1 must be set to zero. The recursion relation of Eq.(50c) then becomes $n(n-2\nu)a_n + a_{n-2} = 0$, which must be satisfied for $n \geq 2$. This requires that $a_{2\nu-2} = a_{2\nu-4} = \cdots = a_0 = 0$. Also, given that $a_1 = 0$, the recursion relation yields $a_3 = a_5 = a_7 = \cdots = 0$. Therefore, the first nonzero coefficient in the expansion of $f(x)$ is $a_{2\nu}$. We now set $n = 2(\nu + m)$, where $m \geq 1$, and find $a_{2(\nu+m)} = -a_{2(\nu+m-1)}/[4m(\nu+m)]$. Thus,

$$a_{2(\nu+m)} = \frac{(-1)^m \Gamma(\nu+1)}{2^{2m}\, m!\, \Gamma(\nu+m+1)} a_{2\nu}; \qquad (m \geq 1). \tag{59}$$

A solution of Bessel's equation in the general form of Eq.(46) for integer values of $\nu > 0$ is thus given by

$$f(x) = a_{2\nu}\Gamma(\nu+1)2^{2\nu}x^{-\nu} \sum_{m=0}^{\infty} \frac{(-1)^m}{m!\, \Gamma(\nu+m+1)} (x/2)^{2(\nu+m)}. \tag{60}$$

Once this solution is properly normalized, it becomes identical to that given by Eq.(54). Thus, for integer values of $\nu \geq 0$, the two choices $s = \pm\nu$ of the index s yield exactly the same (normalized) solution $f(x) = J_\nu(x)$.

Case iv) For a positive half-integer ν (other than $\nu = ½$, which was already discussed), let $\nu = k + ½$, with $k \geq 1$ being a positive integer. Choosing $s = -\nu < 0$, allows a_0 to be arbitrary, but a_1 must be set to zero to satisfy the second indicial equation, Eq.(50b). The recursion relation of Eq.(50c) then becomes $n(n-2k-1)a_n + a_{n-2} = 0$, which must be satisfied for $n \geq 2$. It is seen that a_{2k+1} is no longer required to vanish, implying that only a finite number of odd coefficients, namely $a_1, a_3, \cdots, a_{2k-1}$, need be eliminated. For the remaining odd coefficients, setting $n = 2m + 1$ now yields $a_{2m+1} = -a_{2m-1}/[2(m-k)(2m+1)]$, with $m \geq k+1$; that is,

$$a_{2k+3} = -\frac{a_{2k+1}}{2(2k+3)}, \qquad a_{2k+5} = \frac{a_{2k+1}}{2\times4\times(2k+3)(2k+5)},$$

$$a_{2k+7} = -\frac{a_{2k+1}}{2\times4\times6\times(2k+3)(2k+5)(2k+7)}, \quad \cdots, a_{2\nu+2m} = \frac{(-1)^m}{2^{2m} m!\,(\nu+1)(\nu+2)\cdots(\nu+m)} a_{2\nu};\ m \geq 1.$$

As for the even coefficients, setting $n = 2m$ yields the recursion relation $a_{2m} = a_{2m-2}/[4m(\nu-m)]$, with $m \geq 1$; that is,

$$a_2 = \frac{a_0}{4(\nu-1)}, \qquad a_4 = \frac{a_0}{4^2 \times 2 \times (\nu-1)(\nu-2)}, \quad \cdots \quad a_{2m} = \frac{a_0}{2^{2m} m!\,(\nu-1)(\nu-2)\cdots(\nu-m)}.$$

The general solution of Bessel's equation for $\nu = k + ½$, with integer $k \geq 1$, obtained using the Frobenius method with $s = -\nu$, is thus seen to be

$$f(x) = a_0 \Gamma(1-\nu) x^{-\nu} \sum_{m=0}^{\infty} \frac{(-1)^m (x/2)^{2m}}{m!\,\Gamma(m+1-\nu)}$$

$$+ a_{2\nu} \Gamma(\nu+1) 2^{2\nu} x^{-\nu} \sum_{m=0}^{\infty} \frac{(-1)^m (x/2)^{2\nu+2m}}{m!\,\Gamma(\nu+m+1)}. \tag{61}$$

This is a linear combination of $J_{-\nu}(x)$ and $J_\nu(x)$, which are two independent solutions of Bessel's equation for non-integer values of ν. In general, the Bessel equation has two independent solutions, which are written as $J_\nu(x)$ and $Y_\nu(x)$, and referred to as Bessel functions of the first and second kinds, respectively. For non-integer ν, it can be shown that $Y_\nu(x)$ is a linear combination of $J_\nu(x)$ and $J_{-\nu}(x)$.

Case v) When the parameter $\nu > 0$ is neither an integer nor a half-integer, choosing $s = -\nu$ results in a_0 being arbitrary and $a_1 = 0$. The recursion relation then yields $a_3 = a_5 = a_7 = \cdots = 0$. Thus, all odd-indexed coefficients in the expression of $f(x)$ in Eq.(46) vanish. As for the even-indexed coefficients, we will have

$$a_n = \frac{a_{n-2}}{\nu^2 - (n+s)^2} = \frac{a_{n-2}}{n(2\nu - n)} \quad \rightarrow \quad a_{2m} = \frac{a_{2m-2}}{4m(\nu - m)}, \quad m \geq 1. \tag{62}$$

Consequently,

$$a_2 = \frac{a_0}{4(\nu-1)}, \qquad a_4 = \frac{a_0}{4^2 \times 2 \times (\nu-1)(\nu-2)}, \quad \cdots \quad a_{2m} = \frac{a_0}{2^{2m} m!\,(\nu-1)(\nu-2)\cdots(\nu-m)}. \tag{63}$$

The solution of Bessel's equation in this case is given by

$$f(x) = x^{-\nu}\left[a_0 + a_0 \sum_{m=1}^{\infty} \frac{(-1)^m (x/2)^{2m}}{m!\,(1-\nu)(2-\nu)\cdots(m-\nu)}\right] = a_0 \Gamma(1-\nu) \sum_{m=0}^{\infty} \frac{(-1)^m (x/2)^{2m-\nu}}{m!\,\Gamma(m+1-\nu)}. \tag{64}$$

Thus, for non-integer ν, the solution in Eq.(64), which is proportional to $J_{-\nu}(x)$, is independent of the other solution, $J_\nu(x)$, which we obtained for case i in Eq.(53). In general, the method of Frobenius yields both (linearly independent) solutions of the Bessel equation for non-integer values of the ν parameter.

Case vi) Here, $s = \nu - 1$, a_1 is arbitrary, and a_0 must be set to zero. Since $a_0 = 0$, the recursion relation yields $a_2 = a_4 = a_6 = \cdots = 0$. Thus, all even-indexed coefficients appearing in the expression of $f(x)$ in Eq.(46) vanish. As for odd-indexed coefficients, we will have

$$a_n = \frac{a_{n-2}}{\nu^2 - (n+s)^2} = -\frac{a_{n-2}}{(n-1)(2\nu+n-1)} \quad \rightarrow \quad a_{2m+1} = -\frac{a_{2m-1}}{4m(\nu+m)}, \quad m \geq 1. \tag{65}$$

Therefore,

$$a_3 = -\frac{a_1}{4 \times (\nu+1)}. \tag{66a}$$

$$a_5 = -\frac{a_3}{4 \times 2 \times (\nu+2)} = \frac{a_1}{4^2 \times 2 \times (\nu+1)(\nu+2)}. \tag{66b}$$

$$a_7 = -\frac{a_5}{4 \times 3 \times (\nu+3)} = -\frac{a_1}{4^3 \times 2 \times 3 \times (\nu+1)(\nu+2)(\nu+3)}. \tag{66c}$$

Once again, the general trend is easy to infer. We may write

$$a_{2m+1} = \frac{(-1)^m \Gamma(\nu+1)}{2^{2m}\, m!\, \Gamma(\nu+m+1)} a_1. \tag{67}$$

A solution of Bessel's equation in the general from of Eq.(46) is thus given by

$$f(x) = a_1 \Gamma(\nu + 1) x^{\nu-1} \sum_{m=0}^{\infty} \frac{(-1)^m}{2^{2m}\, m!\, \Gamma(\nu+m+1)} x^{2m+1}. \tag{68}$$

This solution is now seen to be the same as that given by Eq.(53). Consequently, the two indices $s = \nu$ and $s = \nu - 1$ yield the same solution, $J_\nu(x)$, for Bessel's equation.

In similar fashion, one can find solutions of the Bessel equation in the case when $s = -(\nu + 1)$. However, no new solutions will be found in this case—that is, solutions beyond linear combinations of $J_\nu(x)$ and $J_{-\nu}(x)$, which we have already identified. For non-integer values of ν, the two solutions thus obtained are linearly independent and, therefore, comprise the complete set of solutions of the Bessel equation. As for integer values of ν, one must resort to more advanced methods to arrive at the second independent solution, namely, Bessel functions of the second kind, order ν, denoted by $Y_\nu(x)$ and given by†

$$Y_n(x) = \frac{2}{\pi}[c + \ln(x/2)]\, J_n(x) - \frac{1}{\pi}(x/2)^{-n} \sum_{k=0}^{n-1} \frac{(n-k-1)!}{k!}(x/2)^{2k} - \frac{\left(\sum_{k=1}^{n} k^{-1}\right)(x/2)^n}{\pi(n!)}$$
$$- \frac{1}{\pi}(x/2)^n \sum_{k=1}^{\infty} \frac{(-1)^k (x/2)^{2k}}{k!\,(k+n)!}\left(\sum_{m=1}^{n+k} m^{-1} + \sum_{m=1}^{k} m^{-1}\right). \tag{69}$$

In this equation, $0!$ is defined as 1. Also, when the upper limit of a sum over k is below its lower limit, the term containing the sum must be set to zero; this occurs for $n = 0$ in the second and third terms on the right-hand side of Eq.(69), thus reducing the expression for $Y_0(x)$ to one containing the first and fourth terms only. Finally, the Euler constant c appearing in Eq.(69) is defined as

$$c = lim_{n\to\infty}\left(\sum_{k=1}^{n} k^{-1} - \ln n\right) = 0.577215\cdots. \tag{70}$$

Given below are expanded forms of $Y_0(x)$ and $Y_1(x)$; plots of these as well as some higher-order Bessel functions of the second kind appear in figure 8.4.

$$Y_0(x) = \frac{2}{\pi}[c + \ln(x/2)]\, J_0(x) + \frac{2}{\pi}\left[\frac{(x/2)^2}{(1!)^2} - (1 + ½)\frac{(x/2)^4}{(2!)^2} + (1 + ½ + ⅓)\frac{(x/2)^6}{(3!)^2} - \cdots\right]. \tag{71}$$

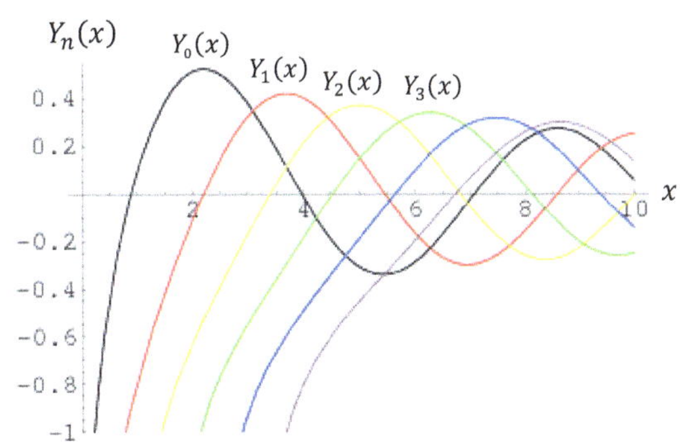

Figure 8.4. Plots of the first few Bessel functions of the second kind, integer order $\nu = 0, 1, 2, 3, \cdots$.

---

†See Eq.(**8.403**-2) in I. S. Gradshteyn and I. M. Ryzhik, *Table of Integrals, Series, and Products*, 7th edition, Academic Press, Burlington, MA (2007).

$$Y_1(x) = \frac{2}{\pi}[c + \ln(x/2)]\,J_1(x) - \frac{2}{\pi x} - \frac{x}{2\pi}\left[1 + \sum_{k=1}^{\infty} \frac{(-1)^k (x/2)^{2k}}{k!\,(k+1)!}\left(\sum_{m=1}^{k+1} m^{-1} + \sum_{m=1}^{k} m^{-1}\right)\right]. \quad (72)$$

For more information on Bessel functions of the second kind, integer order, the reader may consult the references listed at the end of the chapter.

8.7. Parametric oscillation. A harmonic oscillator in which the resonance frequency ω_0 and/or the damping coefficient γ are forced to oscillate at a frequency that is twice (or nearly twice) the resonance frequency of the oscillator could become unstable if the forcing is sufficiently strong. The differential equation governing the oscillations of such a parametric oscillator may be written

$$\ddot{x}(t) + \gamma[1 + \varepsilon_1 \cos(2\omega t)]\dot{x}(t) + \omega_0^2[1 + \varepsilon_2 \cos(2\omega t)]x(t) = 0. \quad (73)$$

In what follows, we shall assume that $|\varepsilon_1| \ll 1$ and $|\varepsilon_2| \ll 1$. Note that the forced oscillations of γ and ω_0^2, which are assumed to occur at the same frequency (2ω), are also in phase. Let us assume that the solution $x(t)$ of Eq.(73) is an exponentially growing/decaying oscillatory function of time, having a growth/decay coefficient η, frequency ω, and phase φ. (The reasoning behind our choice of the oscillation frequency ω of $x(t)$ at exactly one-half the frequency 2ω of the parametric oscillations will become clear shortly.) We thus write

$$\begin{aligned} x(t) &= x_0 \exp(\eta t)\cos(\omega t - \varphi) \\ &= ½x_0 \exp(-i\varphi)\exp[(\eta + i\omega)t] + ½x_0 \exp(i\varphi)\exp[(\eta - i\omega)t]. \end{aligned} \quad (74)$$

Substitution into Eq.(73) yields

$$\begin{aligned} &½x_0 \exp(-i\varphi)\{(\eta + i\omega)^2 + \gamma(\eta + i\omega) + ½\varepsilon_1\gamma\exp(+2i\varphi)(\eta - i\omega) + \omega_0^2 + ½\varepsilon_2\omega_0^2\exp(+2i\varphi)\} \\ &\qquad\qquad \times \exp[(\eta + i\omega)t] \\ &+½x_0 \exp(i\varphi)\{(\eta - i\omega)^2 + \gamma(\eta - i\omega) + ½\varepsilon_1\gamma\exp(-2i\varphi)(\eta + i\omega) + \omega_0^2 + ½\varepsilon_2\omega_0^2\exp(-2i\varphi)\} \\ &\qquad\qquad \times \exp[(\eta - i\omega)t] \\ &+¼x_0 \exp(-i\varphi)\,[(\eta + i\omega)\varepsilon_1\gamma + \varepsilon_2\omega_0^2]\exp[(\eta + 3i\omega)t] \\ &+¼x_0 \exp(+i\varphi)\,[(\eta - i\omega)\varepsilon_1\gamma + \varepsilon_2\omega_0^2]\exp[(\eta - 3i\omega)t] = 0. \end{aligned} \quad (75)$$

The last two terms on the left-hand side of Eq.(75), being of the same order of magnitude as ε_1 and ε_2, are small and may therefore be dropped. As for the remaining two terms, they are seen to be conjugates of each other; consequently, when one of their coefficients is set to zero, the other one will automatically vanish. It is thus only necessary that the following equation be satisfied:

$$[(\eta + i\omega)^2 + \gamma(\eta + i\omega) + \omega_0^2] + ½[\varepsilon_1\gamma(\eta - i\omega) + \varepsilon_2\omega_0^2]\exp(2i\varphi) = 0. \quad (76)$$

In the absence of parametric excitation, where $\varepsilon_1 = \varepsilon_2 = 0$, Eq.(76) yields the resonant solution

$$\eta_r + i\omega_r = -(\gamma/2) \pm i\sqrt{\omega_0^2 - (\gamma/2)^2}. \quad (77)$$

When a *weak* parametric excitation (i.e., $|\varepsilon_1| \ll 1$ and $|\varepsilon_2| \ll 1$) is present, the solution of Eq.(76) deviates slightly from the resonant solution to become $\eta + i\omega = (\eta_r + i\omega_r) + \delta$, where δ is a small complex frequency. Substitution into Eq.(76) and ignoring of the second-order terms δ^2 and $\varepsilon_1\delta^*$, now yields

$$2(\eta_r + \mathrm{i}\omega_r)\delta + \gamma\delta + ½[\varepsilon_1\gamma(\eta_r - \mathrm{i}\omega_r) + \varepsilon_2\omega_0^2]\exp(2\mathrm{i}\varphi) = 0. \tag{78}$$

This equation is readily solved for δ, yielding

$$\delta = -½\left[\frac{\varepsilon_2\omega_0^2 + \varepsilon_1\gamma(\eta_r - \mathrm{i}\omega_r)}{2(\eta_r + \mathrm{i}\omega_r) + \gamma}\right]\exp(2\mathrm{i}\varphi)$$

$$= ¼\mathrm{i}\left(\frac{\varepsilon_2\omega_0^2 - ½\varepsilon_1\gamma^2 - \mathrm{i}\varepsilon_1\gamma\omega_r}{\omega_r}\right)\exp(2\mathrm{i}\varphi)$$

$$= ¼\left[\varepsilon_1\gamma + \mathrm{i}(\varepsilon_2\omega_0^2 - ½\varepsilon_1\gamma^2)/\sqrt{\omega_0^2 - (\gamma/2)^2}\right]\exp(2\mathrm{i}\varphi). \tag{79}$$

We thus find the overall growth/decay coefficient η and the oscillation frequency ω as follows:

$$\eta = -½\gamma[1 - ½\varepsilon_1\cos(2\varphi)] - ¼(\varepsilon_2\omega_0^2 - ½\varepsilon_1\gamma^2)\sin(2\varphi)/\sqrt{\omega_0^2 - (\gamma/2)^2}, \tag{80}$$

$$\omega = \sqrt{\omega_0^2 - (\gamma/2)^2} + ¼\varepsilon_1\gamma\sin(2\varphi) + ¼(\varepsilon_2\omega_0^2 - ½\varepsilon_1\gamma^2)\cos(2\varphi)/\sqrt{\omega_0^2 - (\gamma/2)^2}. \tag{81}$$

In the special case of $\varepsilon_1 = 0$, if we set $\varphi = -45°$, we get $\omega = \omega_r = \sqrt{\omega_0^2 - (\beta_o/2)^2}$ (that is, no detuning from the resonance frequency) and

$$\eta = -½\gamma + ¼\varepsilon_2\omega_0^2/\sqrt{\omega_0^2 - (\gamma/2)^2}. \tag{82}$$

It is thus seen that the amplitude of $x(t)$ grows exponentially with time, provided that $\varepsilon_2 > 2(\gamma/\omega_o)\sqrt{1 - ¼(\gamma/\omega_o)^2} = 2\gamma\omega_r/\omega_0^2$. Given a set of parameters $\omega_o, \gamma, \varepsilon_1 = 0$, and ε_2, Eq.(80) reveals the threshold for the onset of gain to be $\sin(2\varphi) = 2\gamma\omega_r/(\varepsilon_2\omega_0^2)$, at which point the frequency detuning, according to Eq.(81), will be

$$\omega - \omega_r = \pm¼(\varepsilon_2\omega_0^2/\omega_r)\sqrt{1 - (2\gamma\omega_r/\varepsilon_2\omega_0^2)^2}. \tag{83}$$

Note that the frequency 2ω of the parametric oscillations will automatically set the oscillation frequency ω of $x(t)$. The system then adjusts the phase φ of $x(t)$ so that the detuning $\omega - \omega_r$ of the oscillation frequency would comply with Eq.(81). The overall amplification/damping coefficient η will then be given by Eq.(80).

Suggestions for further reading

1. G.B. Arfken and H.J. Weber, *Mathematical Methods for Physicists*, 6th edition, Academic Press, Cambridge, Massachusetts (2005).
2. F.B. Hildebrand, *Advanced Calculus for Applications*, 2nd edition, Prentice-Hall, Upper Saddle River, New Jersey (1976).
3. J. Mathews and R.L. Walker, *Mathematical Methods of Physics*, 2nd edition, Benjamin/Cummings Publishing, San Francisco, California (1970).
4. G. Stephenson and P.M. Radmore, *Advanced Mathematical Methods for Engineering and Science Students*, Cambridge University Press, Cambridge, United Kingdom (1990).
5. I.S. Gradshteyn and I.M. Ryzhik, *Table of Integrals, Series, and Products* (7th edition), Academic Press, Cambridge, Massachusetts (2007).

Problems

Problem 1) The output $f(x)$ of a system is related to its input $g(x)$ via the following n^{th} order linear differential equation with constant coefficients:

$$a_n \frac{\mathrm{d}^n f(x)}{\mathrm{d}x^n} + a_{n-1}\frac{\mathrm{d}^{n-1} f(x)}{\mathrm{d}x^{n-1}} + \cdots + a_1 \frac{\mathrm{d}f(x)}{\mathrm{d}x} + a_0 f(x) = g(x).$$

In the special case when $g(x) = \delta(x)$, the output $f(x)$ is referred to as the system's *impulse response* and denoted by $h(x)$. Use the Fourier transform method to show that, in general, the output of the system is given by $f(x) = g(x) * h(x)$, where $*$ denotes the convolution operation.

Problem 2) For the RC circuit shown, write a differential equation relating the capacitor's voltage $V_c(t)$ to the input voltage $V(t)$.

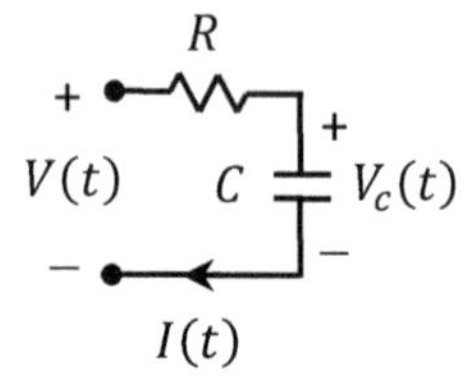

a) Use the Fourier transform method to determine $V_c(t)$ when $V(t) = V_0 \mathrm{step}(t)$; here, V_0 is a real-valued constant.

b) Use the Fourier transform method to determine $V_c(t)$ when $V(t) = \delta(t)$. Show that this result may also be arrived at by a simple differentiation of the result obtained in part (a).

c) Use the Fourier transform method to determine the capacitor voltage $V_c(t)$ when $V(t) = V_0 \exp(\mathrm{i}2\pi f t)\,\mathrm{step}(t)$; here, the amplitude V_0 is a complex constant, and the frequency f is a real-valued constant.

Problem 3) Consider the following ordinary second-order linear differential equation with constant coefficients:

$$\frac{\mathrm{d}^2\zeta(x)}{\mathrm{d}x^2} + a_1 \frac{\mathrm{d}\zeta(x)}{\mathrm{d}x} + a_0 \zeta(x) = 0.$$

Assuming that both a_0 and a_1 are real-valued positive constants, find the general solution to this homogeneous differential equation when (i) $a_1 > 2\sqrt{a_0}$, (ii) $a_1 < 2\sqrt{a_0}$, and (iii) $a_1 = 2\sqrt{a_0}$. Explore the conditions under which $\zeta(x)$ will be a real-valued function of x.

Problem 4) In the RC circuit shown on the right-hand side, the resistor R and/or the capacitor C are controlled by external signals such that the time-constant $\tau = RC$ of the circuit varies with time as $\tau(t) = \tau_0[1 + \alpha\cos(\omega t)]$. Here, $\tau_0 > 0$ is the average time-constant, α is the modulation amplitude ($|\alpha| < 1$), and $\omega > 0$ is the modulation frequency.

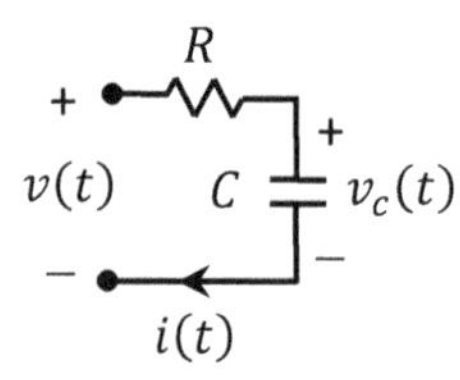

a) Invoking Kirchhoff's voltage law, write the first-order differential equation that relates the input voltage $v(t)$ to the output signal $v_c(t)$.

b) Find the general solution of the differential equation obtained in (a), and proceed to describe (qualitatively) the behavior of the solution $v_c(t)$.

Hint: To evaluate $\int_{-\infty}^{\omega t} \mathrm{d}\theta/(1 + \alpha\cos\theta)$, see Example 6 in chapter 3, section 1.

Problem 5) Use the method of Frobenius to find the solutions of the Legendre equation

$$(1 - x^2)f''(x) - 2xf'(x) + n(n+1)f(x) = 0.$$

Problem 6) An equi-dimensional, second-order, homogeneous ODE with coefficients a, b, and c (which are arbitrary constants) is written as follows:

$$ax^2 f''(x) + bxf'(x) + cf(x) = 0.$$

a) Find a solution $f(x)$ of the above equation.

b) Under what circumstances does the method that you employed to solve the equation yield only one solution instead of two independent solutions?

c) In general, when $s_1 \neq s_2$, the solution may be written as $f(x) = A_1 x^{s_1} + A_2 x^{s_2}$, where A_1, A_2 are arbitrary constants. Suppose that $A_1 = A_0/(s_1 - s_2)$ and $A_2 = -A_0/(s_1 - s_2)$. What would become of this particular solution, $f(x)$, in the limit when $s_1 \to s_2$? (**Hint**: $x^s = e^{s \ln x}$)

d) Confirm that the solution that you found in part (c) does in fact satisfy the differential equation in the special case when $b = a \pm 2\sqrt{ac}$.

Problem 7) a) Solve the first-order linear differential equation $f'(x) + \eta f(x) = \text{rect}(x)$ using the Fourier transform method. Here, η is a real-valued and positive constant.

b) Verify that your solution is, in fact, continuous at $x = \pm½$, where the excitation function $\text{rect}(x)$ is discontinuous.

c) Plot the function $f(x)$ over its entire domain $-\infty < x < \infty$.

Problem 8) The depicted mass-and-spring oscillator is governed by the second-order ordinary differential equation

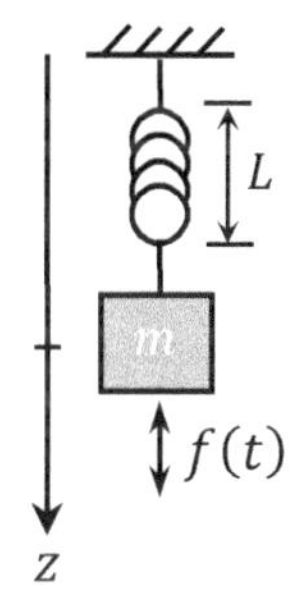

$$\frac{d^2 z(t)}{dt^2} + \gamma \frac{dz(t)}{dt} + \omega_0^2 z(t) = (1/m) f(t).$$

Here, m is the mass attached to the end of the spring, $\gamma = \beta/m$ is the damping coefficient (i.e., friction coefficient β normalized by m), $\omega_0 = \sqrt{\alpha/(Lm)}$ is the resonance frequency of the system associated with a spring of length L and spring constant α, and $f(t)$ is the external force applied along the z-axis. The function $z(t)$ is the deviation of the center of the mass m from its equilibrium position along the z-axis. This problem was solved in section 3 using the Fourier transform method. Here, we would like to arrive at the same solution using the alternative method in which the homogeneous and particular solutions of the differential equation are obtained separately.

a) Find the *homogeneous* solution of the differential equation for the following three cases: (i) $\gamma < 2\omega_0$, the so-called underdamped case; (ii) $\gamma = 2\omega_0$, the case of critical damping; and (iii) $\gamma > 2\omega_0$, the overdamped case.

b) Find the *particular* solution of the differential equation when the excitation function is given by $f(t) = F_0 \exp(i2\pi f_0 t) \,\text{step}(t)$. Here, the complex coefficient F_0 contains both the amplitude and the phase of the excitation function, while the real-valued f_0 is the excitation frequency.

c) Use the initial conditions, namely, $z(0) = z'(0) = 0$, to derive the *complete* solution to the problem (i.e., the superposition of the homogeneous and particular solutions that satisfies the initial conditions).

Problem 9) A critically damped mass-and-spring system is driven by a sinusoidal force of amplitude F_0 and frequency f_0; that is, $f(t) = F_0\text{step}(t)\sin(2\pi f_0 t)$; here, both F_0 and f_0 are real-valued. The second-order ODE governing the system is thus given by

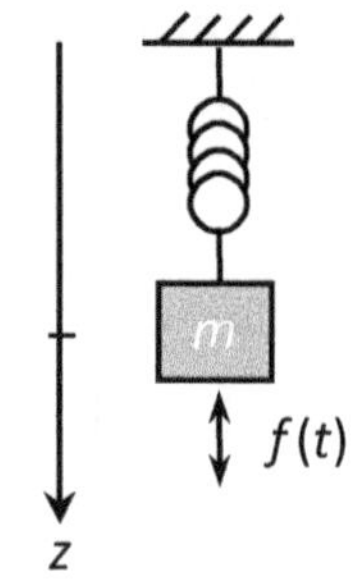

$$\ddot{z}(t) + 2\omega_0\dot{z}(t) + \omega_0^2 z(t) = (F_0/m)\text{step}(t)\sin(2\pi f_0 t).$$

As usual, m is the mass and ω_0 the natural frequency of the oscillator. (Note that, since the system is assumed to be critically damped, the damping coefficient γ is set equal to $2\omega_0$.)

a) Use the Fourier transform method to determine the displacement function $z(t)$ at all times t.

b) Describe the behavior of the system as the excitation frequency approaches the natural frequency of the system, that is, as $2\pi f_0 \to \omega_0$.

Problem 10) Use the method of Frobenius to find the solution to the following second-order ordinary differential equation:

$$x^2\frac{\mathrm{d}^2 f(x)}{\mathrm{d}x^2} + x\frac{\mathrm{d}f(x)}{\mathrm{d}x} + (x^2 - ¼)f(x) = 0.$$

Problem 11) Use the method of Frobenius to find the solutions of the Hermite equation

$$f''(x) - 2xf'(x) + 2nf(x) = 0.$$

Problem 12) Starting with the series expansion of the Bessel function of first kind, integer order n, that is, $J_n(x) = \sum_{m=0}^{\infty}\frac{(-1)^m(x/2)^{2m+n}}{m!\,(n+m)!}$, show that

a) $J_0'(x) = -J_1(x)$.

b) $J_n'(x) = ½[J_{n-1}(x) - J_{n+1}(x)]; \quad n \geq 1$.

c) $J_n(x) = (x/2n)[J_{n-1}(x) + J_{n+1}(x)]; \quad n \geq 1$.

Hint: In (b) and (c), you may begin by evaluating $\mathrm{d}[x^{\pm n}J_n(x)]/\mathrm{d}x$.

Problem 13) Show that the function $f(x) = \int_{-\pi}^{\pi}\exp(\mathrm{i}x\sin\theta)\,\mathrm{d}\theta$ satisfies the Bessel equation with order parameter $n = 0$. Considering that $J_0(0) = 1$, conclude that $J_0(x) = (2\pi)^{-1}\int_{-\pi}^{\pi}\exp(\mathrm{i}x\sin\theta)\,\mathrm{d}\theta$.

Problem 14) Show that $J_n(x) = (2\pi)^{-1}\int_{-\pi}^{\pi}\exp[\mathrm{i}(x\sin\theta - n\theta)]\,\mathrm{d}\theta$ satisfies Bessel's equation. Verify that this integral representation of $J_n(x)$ is also consistent with the identities (a)–(c) listed in problem 12.

Problem 15) Using a Taylor series expansion around the point $x = 0$ of the integral $(2\pi)^{-1}\int_{-\pi}^{\pi}\exp[\mathrm{i}(x\sin\theta - n\theta)]\,\mathrm{d}\theta$, show that the power series thus obtained is

$$J_n(x) = \sum_{m=0}^{\infty}\frac{(-1)^m(x/2)^{2m+n}}{m!\,(n+m)!}.$$

Problem 16) Bessel functions are solutions of the following linear, second-order, ordinary differential equation:

$$\left[x^2\frac{\mathrm{d}^2}{\mathrm{d}x^2} + x\frac{\mathrm{d}}{\mathrm{d}x} + (x^2 - \nu^2)\right] f_\nu(x) = 0.$$

The two independent solutions of the Bessel equation are typically written as $f_\nu^{(1)}(x) = J_\nu(x)$ and $f_\nu^{(2)}(x) = Y_\nu(x)$, where $J_\nu(x)$ is referred to as a Bessel function of the first kind, order ν, and $Y_\nu(x)$ is known as a Bessel function of the second kind, order ν. In general, ν is an arbitrary complex constant, but in many practical situations ν turns out to be a nonnegative integer $n = 0, 1, 2, \cdots$. An integral representation of Bessel functions of the first kind, integer order, is

$$J_n(x) = (2\pi)^{-1}\int_{-\pi}^{\pi} \exp[\mathrm{i}(x\sin\theta - n\theta)]\,\mathrm{d}\theta.$$

An important functional relationship involving the derivatives of $f_\nu(x)$ is

$$\frac{\mathrm{d}}{\mathrm{d}x} f_\nu(x) = ½[f_{\nu-1}(x) - f_{\nu+1}(x)].$$

Special cases of this relationship are $J_0'(x) = -J_1(x)$ and $Y_0'(x) = -Y_1(x)$. Using the above information, prove the following Fourier transform identities:

a) $\mathcal{F}\{J_0(2\pi x)\} = \dfrac{\mathrm{rect}(s/2)}{\pi\sqrt{1-s^2}}$.

b) $\mathcal{F}\{J_1(2\pi x)/(2x)\} = \sqrt{1-s^2}\,\mathrm{rect}(s/2)$.

c) $\mathcal{F}\{\mathrm{sign}(x)\,J_0(2\pi x)\} = \mathrm{i}\,\mathrm{sign}(s)\dfrac{\mathrm{rect}(s/2) - 1}{\pi\sqrt{s^2-1}}$.

Problem 17) The Bessel function of first kind, order n, has the following Taylor series expansion:

$$J_n(x) = \sum_{m=0}^{\infty} \frac{(-1)^m (x/2)^{2m+n}}{m!\,(n+m)!}.$$

Show by direct substitution into the following (Bessel) equation:

$$x^2\ddot{f}(x) + x\dot{f}(x) + (\alpha^2 x^2 - n^2)f(x) = 0,$$

that $J_n(\alpha x)$ satisfies the equation. Here, α is an arbitrary real-valued constant.

Problem 18) Use the method of Frobenius to find the general solution to the following linear, ordinary differential equation with constant coefficients:

$$\ddot{f}(x) + 2\dot{f}(x) + f(x) = 0.$$

Hint: The indicial equation has three solutions, $s_1 = 0$, $s_2 = 1$, and $s_3 = -1$. While s_1 and s_3 lead to the most general solution of the differential equation, it is easier to start with s_2 to obtain one of the two independent solutions. The other solution may then be found using either s_1 or s_3.

Problem 19) Use the method of Frobenius to find a solution to the following second-order ordinary differential equation:

$$x^2\ddot{f}(x) + x\dot{f}(x) - (x^2 + p^2)f(x) = 0.$$

The constant p is real valued and positive.

Note: The equation differs from Bessel's equation only in the sign of x^2 in the coefficient of $f(x)$.

Problem 20) Use the method of Frobenius to find the solutions of Airy's differential equation, namely,

$$f''(x) - xf(x) = 0.$$

Hint: There are four solutions to the indicial equations, but Airy's differential equation has only two independent solutions.

Problem 21) An integral representation of $J_0(x)$, the Bessel function of first kind, zeroth order, is

$$J_0(x) = (2\pi)^{-1} \int_0^{2\pi} \exp(ix \sin\theta)\, d\theta.$$

Use the method of stationary phase approximation to show that the asymptotic form of $J_0(x)$ for large values of x is $J_0(x) \sim \sqrt{2/(\pi x)} \cos\left(x - \frac{\pi}{4}\right)$.

Problem 22) Use the method of Fourier transformation to solve the following first-order linear ordinary differential equation with constant coefficients:

$$g'(x) + g(x) = \text{rect}(x) \cos(2\pi s_0 x).$$

The excitation function appearing on the right-hand side of the equation has a constant positive frequency s_0, and is limited to the interval $-½ \le x \le ½$. Your solution for $g(x)$ must cover the entire range of x from $-\infty$ to ∞.

Chapter 9

Partial Differential Equations

"I think that modern physics has definitely decided in favor of Plato. In fact the smallest units of matter are not physical objects in the ordinary sense; they are forms, ideas which can be expressed unambiguously only in mathematical language."

Werner Heisenberg (1901–1976)

9.1. Introduction. Partial differential equations (PDEs) are equations that are satisfied by functions of two or more variables in the context of one or more relations involving at least one of the derivatives of the functions under consideration. For example, if one desires to find a function of two variables, say, $f(x,y)$, which would adhere to the constraint $(\partial_x + \partial_y)f = 0$ over some subset of the (x,y) domain, then one has a PDE whose solution will be the sought-after function. In the case of this relatively simple equation, the solution is readily found to be $f(x,y) = g(x-y)$, where $g(\cdot)$ is any differentiable—but otherwise arbitrary—function of a single variable.

Another simple PDE is $\partial_{xy}f(x,y) = 0$. A first integration over x yields $\partial_y f = g(y)$, which is an arbitrary function of y. Then a second integration (this time over y) yields $f(x,y) = h(x) + \int g(y)\mathrm{d}y = h(x) + G(y)$, which is just the sum of two arbitrary single-variable functions. Clearly, the obtained solution embodies the restrictions imposed on it by the governing PDE, but there is a lack of specificity on the part of such solutions that can be lifted only after additional constraints, such as initial conditions or boundary conditions, have been imposed. A comparison might be helpful here to put into perspective the stark differences between PDEs and the ordinary differential equations (ODEs), which were the subject of the preceding chapter. While the solution of the elementary ODE, $f''(x) = 0$, is $f(x) = c_0 + c_1 x$, that of the corresponding PDE, namely $\partial_{xx}f(x,y) = 0$, is $f(x,y) = g_0(y) + x g_1(y)$. In the former case, additional information (e.g., initial conditions) is needed to fix the integration constants c_0 and c_1, whereas in the latter case, one is confronted with the task of identifying two arbitrary functions, $g_0(y)$ and $g_1(y)$, before arriving at a unique solution. This is the principal reason why PDEs are typically more difficult than ODEs to analyze and solve.

It is generally the case that the desired solution of a PDE is subject to additional constraints. Typically, a certain path, a contour, or a boundary within the domain of the function is specified, on which the sought-after solution is required to have certain definite properties. For example, the desired solution $f(x,y,z,t)$ of a PDE may have pre-specified values at $t = t_0$, say, $f(x,y,z,t_0) = f_0(x,y,z)$, or it could be the case that some derivative of the function at $t = t_0$ is specified in advance, such as $\partial_t f(x,y,z,t)|_{t=t_0} = f_1(x,y,z)$. It could happen that the solution $f(x,t)$ of a certain equation is required to remain fixed at, say, $x = x_0$, that is, $f(x_0,t) = c_0$, where c_0 is a constant, or there could be additional constraints on the same function, such as $\partial_x f(x,t)|_{x=x_1} = 0$. The solution may be pre-specified on a contour such as $\varphi(x,y,z) = c$, within the (x,y,z) domain of the desired function $f(x,y,z)$, or a combination of the function and some of its derivatives may likewise be specified in advance. In general, partial differential equations of interest in physical sciences are constrained by certain types of initial conditions, and may be subject to one or more boundary conditions as well. The reader will encounter several examples of such problems throughout the present chapter.

The *order* of a PDE is determined by the order of the highest derivative present in the equation. Thus, for example, the equation $a(x,y)\partial_x f + b(x,y)\partial_y f = c(x,y)$, where a, b, and c are known functions of the variables x and y, is a first-order equation, whereas $(\partial_{xx} - \partial_{yy})f(x,y) = g(x,y)$, where $g(x,y)$ is a known function, is a second-order equation in both variables x and y. A PDE might also contain mixed derivatives, such as ∂_{xy}, or various combinations of the sought-after function and its derivatives, as, for instance, in the equation $a(x,y)\partial_{xx}f + (\partial_{xy}f)^3 - f(x,y)(\partial_x f - \partial_{yy}f) + b(x,y) = 0$, where a and b are known functions of x and y. This second-order PDE is also an example of a *nonlinear* equation, in the sense that, if $f_1(x,y)$ and $f_2(x,y)$ are known to be solutions of the equation, there would be no guarantee that their linear combination $c_1 f_1 + c_2 f_2$ (where c_1 and c_2 are arbitrary constants) would also be a solution. A homogeneous equation such as $(a\partial_{xx} + b\partial_{yy} + c\partial_{tt} + d\partial_t)f(x,y,t) = 0$ is a good example of a linear PDE, irrespective of whether its a, b, c, d coefficients are constants or functions of the variables x, y, and t. This is because, if $f_1, f_2, \cdots, f_n$ were solutions of the equation, then any linear combination $\sum_{k=1}^{n} c_k f_k$ of these solutions (with c_k being constant coefficients), will also be a solution. An inhomogeneous version of the same equation, with the zero on the right-hand side being replaced by an arbitrary function $g(x,y,t)$, will also be linear in the sense that, if the "response" of the equation to different "excitations", say, g_1 and g_2, were denoted by f_1 and f_2, then the response to the linear combination $c_1 g_1 + c_2 g_2$ of these excitations would be the corresponding linear combination $c_1 f_1 + c_2 f_2$.

Partial differential equations are ubiquitous throughout the physical sciences, and the science of optics is no exception. Newton's laws of motion as applied to continua, Maxwell's equations of classical electrodynamics, the Schrödinger and Dirac equations of quantum mechanics, the fundamental equations of thermodynamics, etc., all give rise to PDEs with specific initial and/or boundary conditions in the context of analyzing the corresponding physical phenomena.[1-3†] In the field of optics and photonics, one frequently encounters partial differential equations when examining wave propagation in free space as well as in material media. Other examples include waveguides and resonators, lasers and photodetectors, semiconductor optical and electronic devices, laser-induced heat generation accompanied by thermal diffusion, nonlinear and quantum optical phenomena, opto-mechanics, and so on. The focus of the present chapter is on two important equations of mathematical physics that govern many of the aforementioned phenomena. These are the wave equation, which is a linear PDE of second order in time t and also second order in the spatial x, y, z coordinates, and the diffusion equation, which is a linear equation of first order in t and second order in the x, y, z coordinates. We will draw on examples from mechanical systems, such as vibrating strings and vibrating membranes, and also from distributed electrical systems, such as transmission lines, to illustrate the genesis of the wave equation. Once the basic equation is derived and the underlying physics understood, its method of solution will be quite general and applicable to similar equations that might arise in optics, electrodynamics, or numerous other contexts. Similarly, we will use examples from the field of thermal physics to derive the generic diffusion equation, then apply the methods of separation of variables and Fourier transformation to solve the equation. Once again, the solution techniques will be general and applicable to similar equations that arise in alternative physical settings. The advantage of our approach is that, by relying on familiar phenomena to arrive at the general structure of a particular partial differential equation, we minimize the requisite knowledge of physics that is expected from the student. Once the student grasps the significance of an equation in a specific physical context, it will be

[†]Superscript numbers throughout the chapter identify references listed at the end of the chapter.

easier for him or her to follow the subsequent logical and mathematical steps that must be taken to arrive at a solution of that equation.

9.2. Vibrating string and the wave equation in one spatial dimension. With reference to figure 9.1a, consider a uniform wire of length L and mass density ρ [kilogram/meter], stretched horizontally between two fixed points and subjected to a constant tensile stress that is represented by an internal force f [Newton]. Our goal, here, is to derive the equation of motion for the transverse vibrations of this wire, where any deviation from the equilibrium position along the x-axis is assumed to be along the vertical z-axis. The vibration amplitude, which is assumed to be small, is thus a function of the position x and time t, and might be expressed as $z(x,t)$. The boundary conditions for this problem are $z(x=0,t)=z(x=L,t)=0$. It is also necessary to specify the initial conditions of the wire, namely the initial position $z(x,t=0)$ and the initial velocity $\partial_t z(x,t)|_{t=0}$.

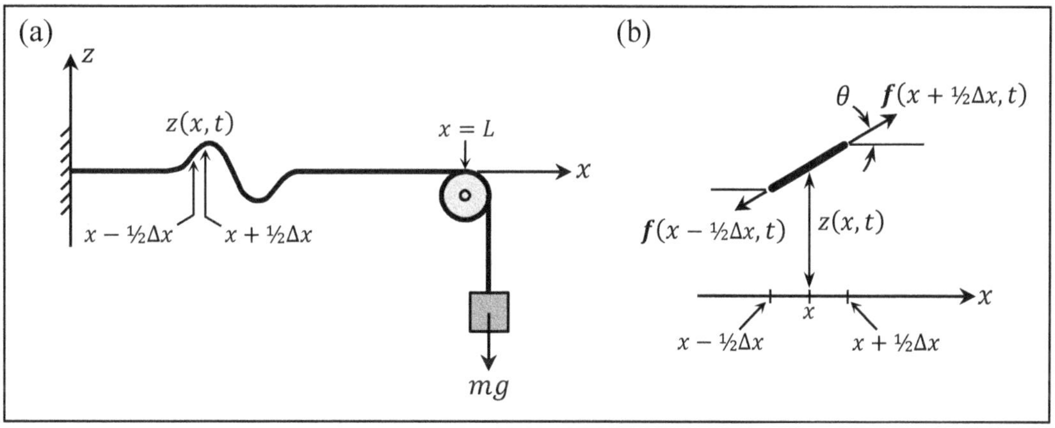

Figure 9.1. (a) A uniform wire of length L aligned with the x-axis is under a constant tensile stress $f=mg$. Small deviations of the wire away from its equilibrium position at $z=0$ are denoted by $z(x,t)$. (b) Magnified view of a short segment of the wire located between $x-½\Delta x$ and $x+½\Delta x$. The instantaneous slope $\partial_x z(x,t)$ of the wire at $x+½\Delta x$ is the tangent of the angle θ shown in the figure. For small vibration amplitudes $z(x,t)$, the magnitude f of the tensile force $\boldsymbol{f}$ that pulls the short wire segment on the right-hand side is the same as that on the left-hand side. However, the difference between $\theta(x+½\Delta x)$ and $\theta(x-½\Delta x)$ gives rise to a net force that acts on the wire in the vertical direction.

As shown in figure 9.1b, the vertical component of the tensile force at each end of a short segment of the wire is $f_z=f\sin\theta$. For small deviations from the wire's equilibrium position, the angle θ is fairly small and, consequently, its sine can be approximated with its tangent; that is, $\sin\theta\cong\tan\theta$. Given that $\tan\theta$ is the local slope of $z(x,t)$ when the latter is considered as a function of x at some fixed instant of time t, we will have $f_z\cong f\,\partial z(x,t)/\partial x$. The net force exerted in the vertical z direction by the wire's internal tensile stress on the short segment of the wire is therefore given by $f[\partial_x z(x,t)|_{x+½\Delta x}-\partial_x z(x,t)|_{x-½\Delta x}]$. Additionally, if any frictional forces act on the wire, their net contribution to the vertical force may be expressed as $-\beta\Delta x\,\partial z(x,t)/\partial t$, where β [Newton · sec/meter^2] is the coefficient of friction per unit length of the wire. Newton's law of motion then relates the total force acting on the short segment to its mass $\rho\Delta x$ and its acceleration $\partial^2 z(x,t)/\partial t^2$, as follows:

$$f[\partial_x z(x,t)|_{x+½\Delta x}-\partial_x z(x,t)|_{x-½\Delta x}]-\beta\Delta x\,\partial z(x,t)/\partial t=\rho\Delta x\,\partial^2 z(x,t)/\partial t^2. \quad (1)$$

We now divide both sides of Eq.(1) by Δx, then allow Δx to approach zero. Also, to simplify the notation, we define the new parameters $\gamma=\beta/\rho$ [sec^{-1}] and $v=\sqrt{f/\rho}$

[meter/sec], where γ is the damping coefficient and v is the velocity of propagation of the vibrational wave along the wire, as will become clear later on. The partial differential equation governing the vibration amplitude of our wire is thus found to be

$$v^2 \frac{\partial^2 z(x,t)}{\partial x^2} = \frac{\partial^2 z(x,t)}{\partial t^2} + \gamma \frac{\partial z(x,t)}{\partial t}. \qquad (2)$$

This PDE, which is second order in both the spatial coordinate x and the temporal coordinate t, is generally known as the (one-dimensional) wave equation. The goal of solving the PDE is to determine the function $z(x,t)$ for $0 \leq x \leq L$ and $t \geq 0$ such that it satisfies Eq.(2) while complying with the aforementioned boundary conditions as well as with the specified initial conditions.

9.3. Separation of variables. To tackle Eq.(2), we begin by searching for solutions that could be written as $z(x,t) = f(x)g(t)$. Substituting the suggested function into Eq.(2), we find

$$v^2 f''(x)g(t) = f(x)g''(t) + \gamma f(x)g'(t). \qquad (3)$$

Both sides of the above equation are now divided by $f(x)g(t)$, yielding

$$v^2 f''(x)/f(x) = [g''(t) + \gamma g'(t)]/g(t) \cdot \qquad (4)$$

Equation (4) has a function of x on the left, and a function of t on the right-hand side. The only way this equation can be satisfied is for both sides to be equal to some constant. Thus, the following equations must be simultaneously satisfied:

$$f''(x) = cf(x). \qquad (5)$$

$$g''(t) + \gamma g'(t) = cv^2 g(t). \qquad (6)$$

Now, if the separation constant c happens to be positive, the general solution of Eq.(5) will be $f(x) = A\sinh(\sqrt{c}x) + B\cosh(\sqrt{c}x)$. The boundary conditions, however, require that $f(0) = f(L) = 0$, which this solution cannot possibly satisfy. We conclude that the separation constant c must be negative, say, $c = -\eta^2$, in which case the general solution of Eq.(5) becomes $f(x) = A\sin(\eta x) + B\cos(\eta x)$. The boundary conditions at $x = 0$ and $x = L$ can now be satisfied by setting $B = 0$ and $\eta = n\pi/L$, where n can be any positive integer. The solution of Eq.(5) that satisfies the boundary conditions is thus $f(x) = A\sin(n\pi x/L)$, where A is an arbitrary constant, and $n = 1, 2, 3, \cdots$ specifies one among an infinite number of possible vibration modes for the string under consideration.

Substitution of $c = -\eta^2$ into Eq.(6) produces the standard harmonic oscillator differential equation $g''(t) + \gamma g'(t) + (\eta v)^2 g(t) = 0$. To solve this homogeneous ODE, we set $g(t) = \exp(\zeta t)$, then solve the resulting quadratic equation $\zeta^2 + \gamma\zeta + (\eta v)^2 = 0$ and obtain $\zeta_{1,2} = -(\gamma/2) \pm \sqrt{(\gamma/2)^2 - (n\pi v/L)^2}$. The behavior of $g(t)$ is, therefore, typical of the standard harmonic oscillator, which, depending on the values of n, v, L, and γ, could be overdamped, underdamped, or critically damped. In what follows, we shall assume that γ is sufficiently small, so that all the various vibration modes corresponding to different values of n are underdamped—even though, strictly speaking, this assumption is *not* necessary. The general solution of our PDE is, therefore, a superposition of all the separable solutions, which may be expressed as

$$z(x,t) = \exp(-\gamma t/2)\sum_{n=1}^{\infty} A_n \sin(n\pi x/L)\sin\left[\sqrt{(n\pi v/L)^2 - (\gamma/2)^2}\, t + \phi_n\right]. \qquad (7)$$

The unknown constants of this solution, that is, all amplitudes A_n and phase angles ϕ_n, must be determined by matching the initial conditions, namely the functional forms of $z(x, t = 0)$ and $\partial_t z(x, t)|_{t=0}$. Both these functions must be expanded in Fourier sine series, which would then yield, for each and every mode, two equations in the two unknowns A_n and ϕ_n.

Digression: The Fourier sine series for any function $Q(x)$ that is confined to the interval $0 \le x \le L$ is written as $Q(x) = \sum_{n=1}^{\infty} q_n \sin(n\pi x/L)$. The coefficients q_n are readily obtained by first extending $Q(x)$ to the interval $-L \le x \le 0$ in such a way as to create an odd function, as shown in figure 9.2. The odd function is denoted by $\tilde{Q}(x)$, has a width of $2L$, and has the property $\tilde{Q}(x) = Q(x)$ for $x \ge 0$ and $\tilde{Q}(x) = -Q(-x)$ for $x \le 0$. The Fourier series coefficients of $\tilde{Q}(x)$ are computed as follows:

$$q_n = (2/2L) \int_{-L}^{L} \tilde{Q}(x) \sin(2\pi n x/2L)\, dx = (2/L) \int_0^L Q(x) \sin(\pi n x/L)\, dx. \quad (8)$$

Needless to say, since, by construction, $\tilde{Q}(x)$ is an odd function of x, its Fourier series expansion does *not* contain a constant term, nor does it contain any cosine terms.

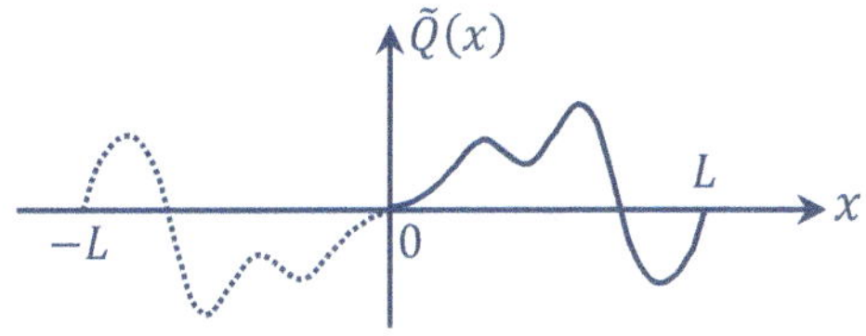

Figure 9.2. The function $Q(x)$, which is defined only over the interval $0 \le x \le L$, has been extended to the negative x-axis in the range $-L \le x \le 0$ in such a way as to create the odd function $\tilde{Q}(x)$. When $\tilde{Q}(x)$ is treated as a single period of a periodic function with period $2L$, its Fourier series will contain only sine functions of the form $\sin(n\pi x/L)$.

9.4. Wave equation in two spatial dimensions: The vibrating membrane. Figure 9.3 shows a rectangular drumhead covered with a tightly stretched membrane whose length and width along the x and y axes are L_x and L_y, respectively. The membrane is firmly attached to the rim of the drumhead, but is otherwise free to vibrate in the vertical direction (i.e., along the z-axis), with a vibration amplitude that will be denoted by $z(x, y, t)$. The tensile stress T [Newton/meter] within the membrane is assumed to be homogeneous and isotropic. Since the edges of the membrane are affixed to the drumhead, the boundary conditions may be written as $z(x = 0, y, t) = z(x = L_x, y, t) = z(x, y = 0, t) = z(x, y = L_y, t) = 0$.

Let us now examine a small rectangular segment of the membrane having dimensions $\Delta x \times \Delta y$ and imagine for a moment that a sharp knife has slit open one of the four edges of the little segment to a width of Δx or Δy. To hold the piece in place, one must now apply, in the direction perpendicular to the cut, a force (per unit length)

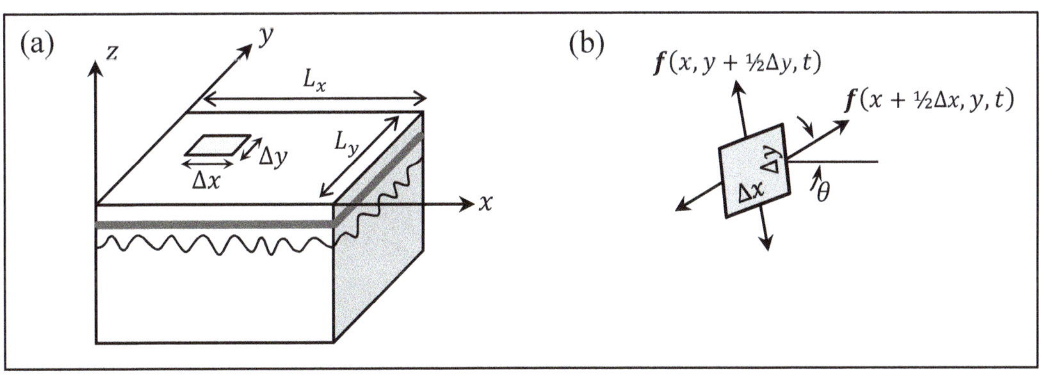

Figure 9.3. (a) A rectangular drumhead is covered with a tightly-stretched membrane in the xy-plane at $z = 0$. The edges of the membrane at $x = 0$, $x = L_x$, $y = 0$, and $y = L_y$ are firmly attached to the rim of the drumhead. (b) A small segment of the membrane, having length Δx and width Δy, is subject to tensile forces on its four sides. For small displacements $z(x, y, t)$, one may approximate $\sin\theta$ with $\tan\theta$, which is the local slope of the membrane at the midpoint of the corresponding edge. Thus, for instance, the vertical component of the force $\boldsymbol{f}(x + ½\Delta x, y, t)$ acting on the right-hand side of the small segment, may be written, to a good approximation, as $T\Delta y \partial_x z(x + ½\Delta x, y, t)$, where T is the tensile stress within the membrane.

that is equal to the tensile stress T within the membrane. We assume that the stress, in addition to being homogeneous and isotropic across the membrane, does *not* change very much when the membrane is vertically displaced, provided that the vibrational amplitude $z(x,y,t)$ is relatively small. Figure 9.3b shows that the small segment is subject to four tensile forces on its four sides. The vertical component of the tensile force acting on the right-hand side of the small segment is thus given by $T\Delta y \sin[\theta(x + ½\Delta x, y, t)]$. The assumption of small vibrational amplitudes now allows us to approximate $\sin\theta$ with $\tan\theta$, which is the local slope of the membrane at the midpoint of the corresponding edge. The vertical component of the tensile force acting on the right-hand side of the segment may now be expressed (to a good approximation) as $T\Delta y \partial_x z(x + ½\Delta x, y, t)$. Similar expressions, of course, can be written down for the forces acting on the remaining sides of the $\Delta x \times \Delta y$ segment.

In addition to the tensile force, the little segment of the membrane under consideration may by subject to a frictional force. Typically, the friction force is proportional to the membrane velocity $\partial_t z(x,y,t)$, with a proportionality constant of $-\beta\Delta x\Delta y$, where β [Newton · second/meter^3] is the friction coefficient per unit area. Denoting the areal mass-density of the membrane by ρ [kilogram/meter^2], Newton's law of motion for our tiny piece of the membrane can finally be written as follows:

$$\overbrace{\rho\Delta x\Delta y\partial_t^2 z(x,y,t)}^{\text{mass times acceleration}} = \overbrace{T\Delta y[\partial_x z(x+½\Delta x,y,t) - \partial_x z(x-½\Delta x,y,t)]}^{z\text{-component of tensile force initially pulling in the } x\text{-direction}}$$
$$+\underbrace{T\Delta x\big[\partial_y z(x,y+½\Delta y,t) - \partial_y z(x,y-½\Delta y,t)\big]}_{z\text{-component of tensile force initially pulling in the } y\text{-direction}} - \underbrace{\beta\Delta x\Delta y\partial_t z(x,y,t)}_{\text{friction force acting along } z}. \quad (9)$$

We now divide both sides of Eq.(9) by $\Delta x\Delta y$ and allow Δx as well as Δy to approach zero. We also introduce the damping coefficient $\gamma = \beta/\rho$ [1/sec], and the propagation velocity $v = \sqrt{T/\rho}$ [meter/sec], to arrive at the following two-dimensional wave equation that governs the vibration of the membrane along the z-axis:

$$v^2\left(\frac{\partial^2}{\partial x^2} + \frac{\partial^2}{\partial y^2}\right) z(x,y,t) = \left(\frac{\partial^2}{\partial t^2} + \gamma\frac{\partial}{\partial t}\right) z(x,y,t). \quad (10)$$

A comparison with Eq.(2) reveals that, in going from one to two spatial dimensions, it is only necessary to include the term $v^2\partial^2 z/\partial y^2$ on the left side of the wave equation.

The solution of Eq.(10) by the method of separation of variables is straightforward. Substituting $z(x,y,t) = f(x)g(y)h(t)$ into Eq.(10), followed by division of the resulting equation by $f(x)g(y)h(t)$, yields

$$v^2[f''(x)/f(x)] + v^2[g''(y)/g(y)] = [h''(t) + \gamma h'(t)]/h(t). \quad (11)$$

The only way for this sum of a function of x and a function of y to equal a function of t is for each term appearing in Eq.(11) to be constant; that is, $f''(x)/f(x) = c_1$, $g''(y)/g(y) = c_2$, and $[h''(t) + \gamma h'(t)]/h(t) = (c_1 + c_2)v^2$. The membrane's boundary conditions now demand that $f(0) = f(L_x) = 0$ and $g(0) = g(L_y) = 0$, which result in $f(x) = A\sin(m\pi x/L_x)$ and $g(y) = B\sin(n\pi y/L_y)$. Here, m and n are arbitrary positive integers that are fully independent of each other and, together, specify a single vibrational mode of the membrane. The arbitrary constant coefficients A and B, together with a similar coefficient for $h(t)$, will be combined further below to form a single coefficient A_{mn} associated with the mn^{th} vibrational mode of the membrane.

The acceptable solutions for $f(x)$ and $g(y)$ described above yield the values of the separation constants as $c_1 = -(m\pi/L_x)^2$ and $c_2 = -(n\pi/L_y)^2$. Consequently,

$$h''(t) + \gamma h'(t) + [(m\pi v/L_x)^2 + (n\pi v/L_y)^2]h(t) = 0. \qquad (12)$$

Once again, we see in Eq.(12) the general equation for the temporal evolution of a harmonic oscillator, which was analyzed at great length in the preceding chapter. Assuming, for the sake of simplicity, that the damping coefficient γ is small enough to render all the vibration modes underdamped, the general solution for the vibration amplitude of the membrane depicted in figure 9.3a can be written as follows:

$$z(x, y, t) = \exp(-\gamma t/2) \sum_{n=1}^{\infty} \sum_{m=1}^{\infty} A_{mn} \sin(m\pi x/L_x) \sin(n\pi y/L_y)$$

$$\times \sin\left[\sqrt{(m\pi v/L_x)^2 + (n\pi v/L_y)^2 - (\gamma/2)^2}\, t + \phi_{mn}\right]. \qquad (13)$$

The infinite set of the unknown parameters A_{mn} and ϕ_{mn} must now be determined by matching the initial conditions, namely the position $z(x, y, t = 0)$ and the velocity $\partial_t z(x, y, t)|_{t=0}$ of the membrane at time $t = 0$. Expansion of these initial position and velocity profiles into their respective Fourier sine series, then equating with the corresponding profiles obtained from Eq.(13), would yield, for each vibrational mode mn, two equations in two unknowns, which one must then solve to arrive at the desired values of A_{mn} and ϕ_{mn}.

9.5. Wave propagation on a transmission line. A typical transmission line consists of a parallel pair of copper cables designed to transmit electrical voltage and current from a generator (or source) to a receiver (or load). A discretized model of a transmission line is shown in figure 9.4. The line is characterized by its resistance per unit length, R [ohm/meter], inductance per unit length, L [henry/meter], and capacitance per unit length, C [farad/meter]. Each unit cell of the discretized model in figure 9.4 represents a short segment of length Δx and consists of a resistor $R\Delta x$ and an inductor $L\Delta x$ along the length of the cables, followed by a capacitor $C\Delta x$ placed between the two cables. The current and voltage along the line, being functions of the distance x from the source and the time t, are denoted by $i(x, t)$ and $v(x, t)$, respectively.

From the theory of lumped electrical circuits, it is well known that the passage of a current $I(t)$ through a resistor R produces a voltage drop that is given by $V(t) = RI(t)$. Similarly, for an inductor L the current-voltage relation is $V(t) = LI'(t)$, whereas for a capacitor C the relation is $I(t) = CV'(t)$.[4] Applying these rules along with Kirchhoff's current and voltage laws to a pair of adjacent loops in the circuit of figure 9.4 yields

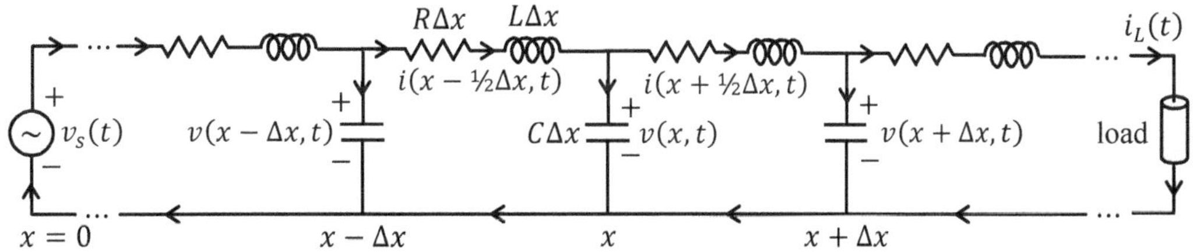

Figure 9.4. A transmission line, designed to transmit electrical voltage and current from a source to a load, is an example of distributed electric networks. The resistance, inductance, and capacitance per unit length of the line are denoted by R, L, and C, respectively. In this discretized model of a continuous transmission line, each segment of length Δx has resistance $R\Delta x$, inductance $L\Delta x$, and capacitance $C\Delta x$. The voltage $v(x, t)$ and current $i(x, t)$ in various segments are related to each other by the characteristic properties of the discrete elements and by the Kirchhoff laws.

$$v(x - \Delta x, t) = R\Delta x\, i(x - ½\Delta x, t) + L\Delta x\, \partial_t i(x - ½\Delta x, t) + v(x, t), \tag{14a}$$

$$v(x, t) = R\Delta x\, i(x + ½\Delta x, t) + L\Delta x\, \partial_t i(x + ½\Delta x, t) + v(x + \Delta x, t), \tag{14b}$$

$$i(x - ½\Delta x, t) - i(x + ½\Delta x, t) = C\Delta x\, \partial_t v(x, t). \tag{14c}$$

Subtracting Eq.(14b) from Eq.(14a), and substituting from Eq.(14c) into the resulting equation, we find

$$\frac{v(x-\Delta x,t) - v(x,t)}{\Delta x} = RC\Delta x\, \partial_t v(x, t) + LC\Delta x \partial_t^2 v(x, t) + \frac{v(x,t) - v(x+\Delta x,t)}{\Delta x}. \tag{15}$$

When $\Delta x \to 0$, the term on the left side of Eq.(15) approaches $-\partial_x v(x - ½\Delta x, t)$, whereas the analogous term on the right-hand side approaches $-\partial_x v(x + ½\Delta x, t)$. Upon normalizing by Δx, letting Δx to approach zero, and rearranging the various terms in Eq.(15), we arrive at

$$\frac{\partial^2}{\partial x^2} v(x, t) = \left(LC \frac{\partial^2}{\partial t^2} + RC \frac{\partial}{\partial t}\right) v(x, t). \tag{16}$$

The second-order PDE of Eq.(16), which governs the spatio-temporal evolution of the voltage $v(x, t)$ along our transmission line, is a wave equation similar to that encountered in Eq.(2) for the vibration of a string in a single spatial dimension.

Suppose now that the voltage source that drives the transmission line from the left-hand side in figure 9.4 provides the sinusoidal voltage $v_s(t) = v_0 \exp(i\omega t)$, where v_0 is the (constant) amplitude and ω the angular frequency of the oscillations. Because of the linearity of the circuit, the voltage and current everywhere along the transmission line will oscillate with the same frequency ω, so that the x and t variables can be separated, and the $v(x, t)$ of Eq.(16) replaced by $v(x) \exp(i\omega t)$. We will have

$$v''(x) + (LC\omega^2 - \mathrm{i}RC\omega) v(x) = 0. \tag{17}$$

The solution of this second-order, linear, ordinary differential equation is readily seen to be $v_\pm(x) = V_\pm \exp\left(\pm \mathrm{i}\sqrt{LC\omega^2 - \mathrm{i}RC\omega}\, x\right)$, where the plus and minus signs identify two independent solutions—in agreement with one's expectation from a second-order ODE. The corresponding constant amplitudes are denoted by $V_\pm$. In practice, transmission lines have a very small resistance R, so that the square root appearing in the above solution can be simplified (using the approximation $\sqrt{1 + \varepsilon} \cong 1 + ½\varepsilon$, when $|\varepsilon| \ll 1$), and the general solution of Eq.(16) may thus be written as follows:

$$\begin{aligned} v_\pm(x, t) &= V_\pm \exp\left[\pm \mathrm{i}\sqrt{LC}\omega\sqrt{1 - \mathrm{i}(R/L\omega)}\, x\right] \exp(\mathrm{i}\omega t) \\ &\cong V_\pm \exp\left\{\pm \mathrm{i}\sqrt{LC}\omega[1 - ½\mathrm{i}(R/L\omega)]x + \mathrm{i}\omega t\right\} \\ &= V_\pm \exp\left(\pm ½R\sqrt{C/L}\, x\right) \exp\left[\mathrm{i}\omega(t \pm \sqrt{LC}x)\right]. \end{aligned} \tag{18}$$

Of the two solutions that appear in Eq.(18), the one with the *minus* sign is a wave that propagates in the $+x$ direction (i.e., from the source toward the load); its propagation velocity is $1/\sqrt{LC}$, and its rate of (amplitude) attenuation along x is $½R\sqrt{C/L}$. Considering that the voltage at the source (located at $x = 0$) is $v_0 \exp(i\omega t)$, it is easy to infer from Eq.(18) that $V_- = v_0$.

The solution with the *plus* sign in Eq.(18) propagates backward (i.e., from the load toward the source), albeit with the same velocity and the same attenuation rate as the forward-propagating wave. The back-propagating wave is produced by reflection from the load—due to impedance mismatch between the load and the transmission line. Note that the attenuation rate in both directions is proportional to the resistance per unit length, R, of the transmission line, which is why the line must be designed to have the least possible resistance to reduce the so-called ohmic energy losses during transmission.

To determine the current $i(x,t)$, note that Eq.(14c) yields $\partial_x i(x,t) = -C\partial_t v(x,t)$. Consequently, for the forward and backward propagating voltages given by Eq.(18), the current is found to be

$$i_\pm(x,t) = \mp\left(C\omega/\sqrt{LC\omega^2 - \mathrm{i}RC\omega}\right)V_\pm \exp\left(\pm\mathrm{i}\sqrt{LC\omega^2 - \mathrm{i}RC\omega}\,x\right)\exp(\mathrm{i}\omega t)$$

$$\cong \mp\frac{1+½\mathrm{i}(R/L\omega)}{\sqrt{L/C}}V_\pm \exp\left(\pm½R\sqrt{C/L}\,x\right)\exp\left[\mathrm{i}\omega(t \pm \sqrt{LC}x)\right]$$

$1 + \mathrm{i}\varepsilon \cong e^{\mathrm{i}\varepsilon}$

$$\cong \mp\left(V_\pm/\sqrt{L/C}\right)\exp\left(\frac{\mathrm{i}R}{2L\omega}\right)\exp\left(\pm½R\sqrt{C/L}\,x\right)\exp\left[\mathrm{i}\omega(t \pm \sqrt{LC}x)\right]. \quad (19)$$

The parameter $\sqrt{L/C}$ [ohm] is known as the characteristic impedance of the transmission line. Thus, the amplitude of the current is obtained by dividing the amplitude of the voltage by the characteristic impedance of the line. Note also that the current is slightly phase-shifted (by $\varphi = R/2L\omega$) relative to the local voltage. To avoid impedance mismatch at the load (thereby eliminating the possibility of a reflected wave), the voltage-to-current ratio of the load should be as close to $\sqrt{L/C}\exp(-\mathrm{i}R/2L\omega)$ as is realistically possible. Another place where the characteristic impedance plays a role is in the attenuation coefficient of the line, $R/(2\sqrt{L/C})$, seen in Eqs.(18) and (19).

Brain teaser. In this infinite network of identical resistors, what is the resistance between A and B?

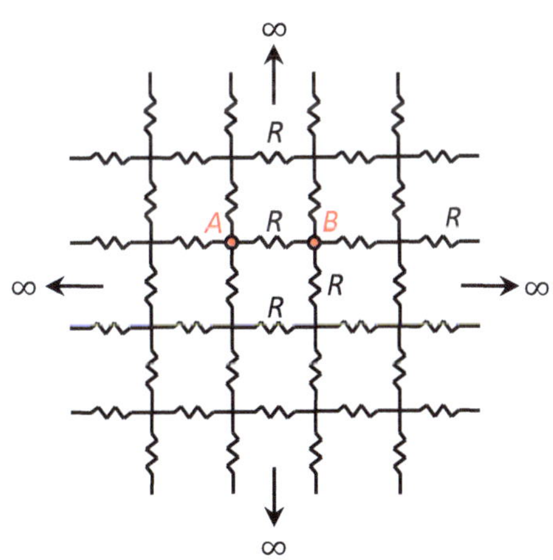

9.6. Solving the wave equation in cylindrical coordinates. While a partial differential equation, such as the wave equation, may represent precisely the same underlying physics and physical mechanisms, its mathematical expression might look quite different in different coordinate systems. The present section is devoted to an analysis of the wave equation for a circular membrane, whose natural geometry dictates the use of a cylindrical (as opposed to Cartesian) coordinate system. It will be seen that the equation of motion of the membrane (a wave equation), which continues to be a second-order PDE in spatial as well as temporal coordinates, not only acquires a different look than the one in Eq.(10), but also requires different kinds of functions for its solution.

Figure 9.5 shows a circular drumhead of radius R, with a tightly stretched membrane attached at the rim. As before, we shall assume that the tensile stress T within the membrane is homogeneous, isotropic, and, for small vibration amplitudes, independent of the vertical displacement $z(r, \varphi, t)$ of the membrane. Accounting for the tensile and frictional forces acting on a small, nearly trapezoidal segment of the membrane, as depicted in figure 9.5, one can immediately write down the equation of motion for the small trapezoid, as follows:

$$\overbrace{T\{[(r+½\Delta r)\partial_r z(r+½\Delta r,\varphi,t)-(r-½\Delta r)\partial_r z(r-½\Delta r,\varphi,t)]\Delta\varphi}^{z\text{-component of tensile force initially pulling in the } r\text{-direction}}$$

$$\underbrace{+r^{-1}\left[\partial_\varphi z(r,\varphi+½\Delta\varphi,t)-\partial_\varphi z(r,\varphi-½\Delta\varphi,t)\right]\Delta r\}}_{z\text{-component of tensile force initially pulling in the } \varphi\text{-direction}} - \underbrace{\beta r\Delta r\Delta\varphi\partial_t z(r,\varphi,t)}_{\text{friction force acting along } z}$$

$$= \underbrace{\rho r\Delta r\Delta\varphi\partial_t^2 z(r,\varphi,t)}_{\text{mass times acceleration}}. \tag{20}$$

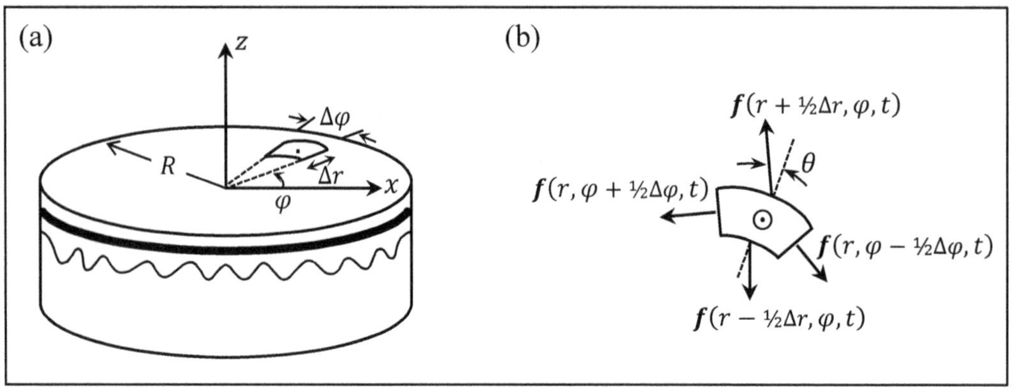

Figure 9.5. (a) A circular drumhead is covered with a tightly stretched membrane in the xy-plane at $z = 0$. The edge of the membrane at $r = R$ is firmly attached to the rim of the drumhead. (b) A small segment of the membrane in the shape of a (slightly curved) trapezoid is subject to tensile forces on its four sides. The dimensions of the trapezoid are $(r \pm ½\Delta r)\Delta\varphi$ at the bases and Δr on the slanted sides. For small vertical displacements $z(r, \varphi, t)$, one may approximate $\sin\theta$ with $\tan\theta$, which is the local slope of the membrane at the midpoint of the corresponding side. Thus, for instance, the vertical components of the force $\boldsymbol{f}(r \pm ½\Delta r, \varphi, t)$ acting on the bases of the small trapezoid may be written, to a good approximation, as the product of the tensile stress within the membrane, T, the width of each base, $(r \pm ½\Delta r)\Delta\varphi$, and the local slope of the membrane, $\partial_r z(r \pm ½\Delta r, \varphi, t)$.

Two differences between the above equation of motion for the trapezoidal segment and the corresponding equation for a rectangular segment given in Eq.(9) are noteworthy. First, since the trapezoid's bases have unequal lengths, the radial tensile forces acting on the opposite bases of the trapezoid have been multiplied by $(r \pm ½\Delta r)\Delta\varphi$. Second, the local azimuthal slopes of the membrane in the φ-direction are expressed as $r^{-1}\partial_\varphi z(r, \varphi \pm ½\Delta\varphi, t)$. This is because $\partial_\varphi z$ in and of itself is *not* a proper slope in the sense of the ratio of a change in elevation divided by the corresponding change in the distance at the base. Consequently, division by r is necessary to turn $\partial_\varphi z \cong \Delta z/\Delta\varphi$ into $r^{-1}\partial_\varphi z \cong \Delta z/(r\Delta\varphi)$, which is a proper ratio of two lengths. This is indeed what has been done in Eq.(20) with regard to the local φ-directed slope of the membrane.

Dividing both sides of Eq.(20) by $\rho r\Delta r\Delta\varphi$, then allowing Δr and $\Delta\varphi$ to approach zero, and, as usual, using normalized parameters $v = \sqrt{T/\rho}$ and $\gamma = \beta/\rho$, we will have

$$v^2\left[r^{-1}\partial_r(r\partial_r)+r^{-2}\partial_\varphi^2\right]z(r,\varphi,t)=(\partial_t^2+\gamma\partial_t)z(r,\varphi,t). \tag{21}$$

Streamlining this wave equation using the identity $\partial_r(r\partial_r)=r\partial_r^2+\partial_r$, we arrive at

$$v^2\left[\partial_r^2+r^{-1}\partial_r+r^{-2}\partial_\varphi^2\right]z(r,\varphi,t)=(\partial_t^2+\gamma\partial_t)z(r,\varphi,t). \tag{22}$$

The operator acting on $z(r,\varphi,t)$ on the left-hand sides of both Eqs.(21) and (22) is the Laplacian operator in the polar (r,φ) coordinates. In Cartesian coordinates, the same operator is written, in two-dimensional space, as $\nabla^2=\partial_x^2+\partial_y^2$, and in three dimensions as $\nabla^2=\partial_x^2+\partial_y^2+\partial_z^2$. Following the logic of Eq.(20), it is easy to see that the Laplacian operator in cylindrical coordinates (r,φ,z) must be $\nabla^2=\partial_r^2+r^{-1}\partial_r+r^{-2}\partial_\varphi^2+\partial_z^2$.

Our next task is to solve the equation of motion of the membrane, Eq.(22), subject to the boundary condition at the rim, namely $z(r=R,\varphi,t)=0$, and the initial conditions whereby $z(r,\varphi,t=0)$ and $\partial_t z(r,\varphi,t)|_{t=0}$ are pre-specified. Applying the method of separation of variables to Eq.(22) requires that we substitute $f(r)g(\varphi)h(t)$ for the function $z(r,\varphi,t)$, which results in the following normalized equation:

$$v^2\left[\frac{f''(r)}{f(r)}+r^{-1}\frac{f'(r)}{f(r)}+r^{-2}\frac{g''(\varphi)}{g(\varphi)}\right]=\frac{h''(t)+\gamma h'(t)}{h(t)}. \tag{23}$$

The above equation, consisting of a function of r and φ on the left, and a function of t on the right-hand side, can be satisfied only if both sides are equal to a constant. As it turns out, the separation constant must be a negative number, say, $-c^2$, or else the boundary condition $f(R)=0$ cannot be satisfied. (A positive separation constant will also result in an exponentially *growing* solution for $h(t)$, which, on physical grounds, is unacceptable.) We thus set $[h''(t)+\gamma h'(t)]/h(t)=-c^2$, and arrive at the standard harmonic oscillator ODE, namely,

$$h''(t)+\gamma h'(t)+c^2h(t)=0. \tag{24}$$

At this point, we are still in the dark as to the value of c^2, so we must leave Eq.(24) for the moment and return to Eq.(23) to determine $f(r)$ and $g(\varphi)$. It will be seen shortly that the boundary condition $f(R)=0$ imposed on the radial solution $f(r)$ will enable an evaluation of the separation constant $-c^2$. Equating the left side of Eq.(23) to $-c^2$ yields

$$\frac{r^2f''(r)+rf'(r)+(cr/v)^2f(r)}{f(r)}=-\frac{g''(\varphi)}{g(\varphi)}. \tag{25}$$

Once again, we have a situation in which a function of one variable, r, is equated with a function of a different variable, φ. We must, therefore, set the two sides of Eq.(25) equal to a constant. It is easy to see that the separation constant in this case must be m^2, where $m=0,1,2,\cdots$ is a nonnegative integer. The reason for this choice of the separation constant is that the solution of the resulting equation, $g''(\varphi)+m^2g(\varphi)=0$, will then be the periodic function $g(\varphi)=A\sin(m\varphi)+B\cos(m\varphi)$, which is precisely what one needs to ensure the repeatability of $g(\varphi)$ with every 2π increment of φ in our polar coordinate system. (The arbitrary coefficients A and B of $g(\varphi)$ will be eventually determined with the help of the initial conditions.)

Having determined that the separation constant for Eq.(25) is m^2, we now arrive at the following ordinary differential equation for $f(r)$:

$$r^2f''(r)+rf'(r)+[(cr/v)^2-m^2]f(r)=0. \tag{26}$$

This is the Bessel equation with an integer order parameter m, but it is not yet in the standard form because of the extra coefficient c/v (see chapter 8, section 6, Eq.(45)). A change of variable from r to $\tilde{r} = cr/v$ restores the standard form of the equation, enabling us to write the general solution of Eq.(26) as $f(r) = A'J_m(cr/v) + B'Y_m(cr/v)$, where $J_m(\cdot)$ and $Y_m(\cdot)$ are Bessel functions of the first and second kinds, order m, respectively, while A' and B' are arbitrary coefficients that must be determined with the aid of the initial conditions. In the present situation, however, the fact that $Y_m(\cdot)$ goes to infinity at the center of the membrane ($r = 0$) immediately informs us that B' must vanish. Also, the coefficient A' of $J_m(\cdot)$ can be absorbed into the coefficients A and B of $g(\varphi)$, so that the spatial part of the separable solution of Eq.(22) can be written as

$$f(r)g(\varphi) = J_m(cr/v)[A\sin(m\varphi) + B\cos(m\varphi)]. \tag{27}$$

Digression: Changing the variable in Eq.(26) from r to $\tilde{r} = cr/v$ requires the following substitutions:

$$\mathrm{d}f/\mathrm{d}r = (\mathrm{d}\tilde{r}/\mathrm{d}r)(\mathrm{d}f/\mathrm{d}\tilde{r}) = (c/v)\,\mathrm{d}f/\mathrm{d}\tilde{r}.$$

$$\mathrm{d}^2f/\mathrm{d}r^2 = \frac{\mathrm{d}}{\mathrm{d}r}(\mathrm{d}f/\mathrm{d}r) = \frac{\mathrm{d}}{\mathrm{d}r}[(c/v)\,\mathrm{d}f/\mathrm{d}\tilde{r}] = (c/v)\frac{\mathrm{d}\tilde{r}}{\mathrm{d}r}\times\frac{\mathrm{d}}{\mathrm{d}\tilde{r}}(\mathrm{d}f/\mathrm{d}\tilde{r}) = (c/v)^2\,\mathrm{d}^2f/\mathrm{d}\tilde{r}^2.$$

Equation (26) now assumes the standard form of Bessel's equation, namely,

$$\tilde{r}^2f''(\tilde{r}) + \tilde{r}f'(\tilde{r}) + (\tilde{r}^2 - m^2)f(\tilde{r}) = 0.$$

We have finally arrived at a point where the constant c can be specified. Since $f(r) = J_m(cr/v)$ must be zero at the boundary $r = R$ of the membrane, it is necessary to have $cR/v = r_{mn}$, where r_{mn} is the nth zero of the standard Bessel function $J_m(r)$. (With reference to chapter 8, figure 3, note that, for $m \neq 0$, the first zero of J_m is at $r = 0$, which does *not* yield a viable solution and must, therefore, be ignored.) The separation constant for the mn^{th} vibrational mode of the membrane is thus found to be $-c^2 = -(vr_{mn}/R)^2$. We may now solve the harmonic oscillator equation for $h(t)$, Eq.(24), and determine whether the oscillations of any given mode are overdamped, underdamped, or critically damped. For the sake of simplicity, let us assume that the damping coefficient γ is small enough to render *all* the modes underdamped. The general solution of Eq.(24) will then be $h(t) = \alpha\exp(-\gamma t/2)\sin(\omega_{mn}t + \phi)$, where $\omega_{mn} = \sqrt{(vr_{mn}/R)^2 - (\gamma/2)^2}$, while α and ϕ are arbitrary constants to be determined by matching the initial conditions.

Under the assumption that *all* the vibrational modes of the circular membrane are underdamped, the general solution of the wave equation can finally be written as

$$z(r,\varphi,t) = \exp(-\gamma t/2)\textstyle\sum_{m=0}^{\infty}\sum_{n=1}^{\infty}J_m(r_{mn}r/R)$$

$$\times\,[A_{mn}\sin(m\varphi)\sin(\omega_{mn}t + \phi_{mn}) + B_{mn}\cos(m\varphi)\sin(\omega_{mn}t + \psi_{mn})]. \tag{28}$$

When the initial conditions $z(r,\varphi,t = 0)$ and $\partial_t z(r,\varphi,t)|_{t=0}$ are expanded into their respective Fourier-Bessel series, a comparison with Eq.(28) will be all that is needed to determine the amplitudes (A_{mn}, B_{mn}) and phase angles (ϕ_{mn}, ψ_{mn}) that appear in the above solution of the wave equation.

9.7. Diffusion equation in one spatial dimension. An important equation in mathematical physics and applied sciences is the diffusion equation, which is a linear partial differential equation of the second order in spatial coordinates and first order in the time coordinate. In the present section, we derive the simplest form of the equation in the context of a heat diffusion problem in one-dimensional space and proceed to discuss its method of solution. Extensions to higher-dimensional spaces will be considered in subsequent sections.

In thermal physics, the specific heat of a material is defined as the amount of thermal energy needed to raise the temperature of a unit volume of the material by one degree Celsius (or Kelvin). The units of Joule/cubic centimeter/degree (abbreviated $\text{J}/\text{cm}^3/°\text{C}$) are commonly used for the specific heat, whose symbol is C. Another thermal parameter of material media is the thermal conductivity κ, which is defined as the proportionality constant between the local slope (i.e., gradient) of the temperature distribution, $\boldsymbol{\nabla} T(x,y,z,t)$, and the rate at which heat flows across a unit area internal surface that is perpendicular to the temperature gradient. The commonly used units of κ are $\text{Joule}/(\text{second} \cdot \text{centimeter} \cdot \text{degree})$, abbreviated as $\text{J}/(\text{cm} \cdot \text{sec} \cdot °\text{C})$. Thus, for example, the time rate of heat flow across a small $\Delta x \times \Delta y$ rectangular area that lies parallel to the xy-plane is $\kappa \Delta x \Delta y \partial_z T(x,y,z,t)$. A third parameter that is often used to characterize the thermal behavior of material media is the diffusivity D, which is simply defined as $D = \kappa/C$. As such, the dimensions (or units) of D are cm^2/sec.

Let us now consider a long, thin, straight, homogenous rod (or wire) that is aligned with the x-axis, as shown in figure 9.6. The rod has a small cross-sectional area A, specific heat C, thermal conductivity κ, and thermal diffusivity $D = \kappa/C$. If we assume that the rod's cross-sectional area A is small enough to render the temperature uniform within its thickness, then the temperature profile of the rod can be described by a function of two variables, say, $T(x,t)$. The wire is also assumed to have a pre-specified initial temperature distribution $T(x,t=0) = T_0(x)$.

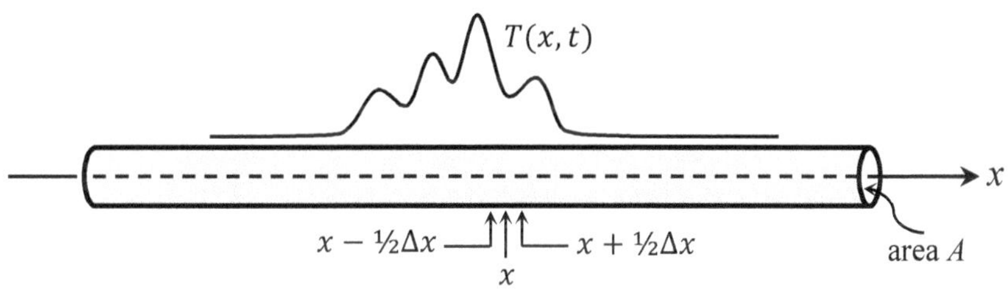

Figure 9.6. A long, thin, straight, homogeneous rod having cross-sectional area A, specific heat C, and thermal conductivity κ lies along the x-axis. The (above ambient) temperature of the rod is denoted by $T(x,t)$, indicating that the rod (or wire) is sufficiently thin to warrant the assumption of uniform temperature distribution within its cross-sectional area. The initial temperature profile $T(x,t=0) = T_0(x)$ evolves in time, in compliance with Eq.(30), the heat diffusion equation.

With reference to figure 9.6, a short segment of the wire, located between $x - ½\Delta x$ and $x + ½\Delta x$, receives thermal energy (i.e., heat) at its right-hand boundary at the rate of $\kappa A \partial_x T(x + ½\Delta x, t)$, while losing heat at the left-hand boundary at the rate of $\kappa A \partial_x T(x - ½\Delta x, t)$. The net heat gain (or less) must then cause a rise (or decline) in the temperature of the short segment at a rate of $CA\Delta x \partial_t T(x,t)$. Equating these two rates, we will have

$$\kappa A \partial_x T(x + ½\Delta x, t) - \kappa A \partial_x T(x - ½\Delta x, t) = CA\Delta x \partial_t T(x,t). \tag{29}$$

If we now divide both sides of this equation by $CA\Delta x$, allow Δx to approach zero, and recall that the ratio κ/C is defined as the diffusivity D, we will arrive at the following (one-dimensional) heat diffusion equation:

$$D \partial_x^2 T(x,t) = \partial_t T(x,t). \tag{30}$$

To solve this relatively simple diffusion equation by the method of separation of variables, we substitute $f(x)g(t)$ for $T(x,t)$ in Eq.(30) and find that $f''(x)/f(x) =$

$g'(t)/g(t) = c$, where c is the separation constant. Consequently, $g(t) = A\exp(ct)$, will be a properly decaying function of time if the constant c is chosen to be negative. We will then have $f(x) = A'\exp(\pm\mathrm{i}|c|^{1/2}x)$. Since the boundaries of the rod are placed at $x = \pm\infty$, the solution is not constrained by boundary conditions and, therefore, all negative values of c, the separation constant, are allowed. We must now expand the initial distribution $T_0(x)$ into a superposition of the eigen solutions $\exp(\pm\mathrm{i}|c|^{1/2}x)$, then multiply each eigen solution by $\exp(ct)$, to allow them to decay (exponentially) in time, independently of one another.

The expansion into the eigen solutions described in the preceding paragraph is, of course, nothing but a Fourier transformation. We can, therefore, reformulate the entire solution of Eq.(30) in terms of Fourier transform operations. To this end, let $\tilde{T}(s,t)$ be the one-dimensional Fourier transform of $T(x,t)$ over its x variable; that is,

$$\tilde{T}(s,t) = \int_{-\infty}^{\infty} T(x,t)\exp(-\mathrm{i}2\pi sx)\,\mathrm{d}x, \tag{31a}$$

$$T(x,t) = \int_{-\infty}^{\infty} \tilde{T}(s,t)\exp(\mathrm{i}2\pi sx)\,\mathrm{d}s. \tag{31b}$$

Substituting for $T(x,t)$ from Eq.(31b) into the diffusion equation, Eq.(30), yields

$$D\int_{-\infty}^{\infty}(\mathrm{i}2\pi s)^2\tilde{T}(s,t)\exp(\mathrm{i}2\pi sx)\,\mathrm{d}s = \int_{-\infty}^{\infty}\partial_t\tilde{T}(s,t)\exp(\mathrm{i}2\pi sx)\,\mathrm{d}s. \tag{32}$$

This equality of two Fourier transforms indicates the equality of the underlying functions being transformed; that is,

$$\partial_t\tilde{T}(s,t) = -4\pi^2 Ds^2\tilde{T}(s,t). \tag{33}$$

This first-order differential equation in t is readily integrated to yield

$$\tilde{T}(s,t) = \tilde{T}(s,t=0)\exp(-4\pi^2 Ds^2 t). \tag{34}$$

Note that the coefficient $\tilde{T}(s,t=0)$ appearing in Eq.(34) is just the Fourier transform $\tilde{T}_0(s)$ of the initial temperature distribution $T_0(x)$. What remains to be done is an inverse Fourier transformation of $\tilde{T}(s,t)$, which, being the product of two functions, ends up as the convolution between two individual inverse transforms. Now, the inverse transform of $\tilde{T}(s,t=0)$ is $T_0(x)$ and the inverse transform of $\exp(-4\pi^2Ds^2t)$, a Gaussian function of s, can be easily determined by recalling that the Fourier transform of $\exp(-\pi x^2)$ is $\exp(-\pi s^2)$. The scaling theorem of Fourier transformation asserts that $\exp[-\pi(x/\alpha)^2]$ transforms into $|\alpha|\exp(-\pi\alpha^2 s^2)$. Consequently, the inverse transform of $\exp(-4\pi^2Ds^2t)$ must be $(1/\sqrt{4\pi Dt})\exp(-x^2/4Dt)$. We thus arrive at

$$T(x,t) = T_0(x) * \left\{\exp[-(x/\sqrt{4Dt})^2]/\sqrt{4\pi Dt}\right\}. \tag{35}$$

This is the complete solution of the diffusion equation for the infinitely long, thin wire depicted in figure 9.6, subject to the initial condition that, at $t = 0$, the temperature along the length of the wire is given by $T_0(x)$. In words, Eq.(35) instructs us to convolve the initial distribution with the so-called "impulse response" of the system to arrive at the temperature distribution at time t. The system's transfer function $\exp(-4\pi^2Ds^2t)$, which, by definition, is the Fourier transform of the impulse response, equals 1 at $t = 0$. Therefore, the impulse response at $t = 0$ is the Dirac delta-function $\delta(x)$. We thus have $T(x,t=0) = T_0(x) * \delta(x)$, which, as expected, is $T_0(x)$. As time progresses, the transfer function narrows, the impulse response widens, and the convolution with $T_0(x)$ ends up smoothing out, broadening, and flattening the initial temperature distribution.

As a simple example, let the initial temperature profile be a Gaussian function of x, having width w_0 and normalized area 1.0; that is,

$$T_0(x) = (1/\sqrt{\pi}\, w_0) \exp[-(x/w_0)^2]. \tag{36}$$

Fourier transforming $T_0(x)$, multiplying the result by the transfer function of the rod, then carrying out the requisite inverse Fourier transformation, we find

$$\tilde{T}_0(s) = \exp(-\pi^2 w_0^2 s^2). \tag{37}$$

$$\tilde{T}(s,t) = \exp[-\pi^2 (w_0^2 + 4Dt) s^2]. \tag{38}$$

$$T(x,t) = \left[1/\sqrt{\pi(w_0^2 + 4Dt)}\right] \exp\left(-\frac{x^2}{w_0^2 + 4Dt}\right). \tag{39}$$

It is seen that, with increasing time, the width of the temperature profile, being w_0 at the outset, increases to become $w(t) = \sqrt{w_0^2 + 4Dt}$. This happens as the peak temperature declines in inverse proportion to the profile's width. Note that, in our system of units, the units of D are cm^2/sec, ensuring that w_0 and $\sqrt{4Dt}$ have the same units [cm].

In the limit where the width of the initial distribution approaches zero (i.e., $w_0 \to 0$), the starting profile, in accordance with Eq.(36), approaches a δ-function—otherwise known as an *impulse*. The impulse-like temperature profile then broadens in time, with a width that increases in proportion to the square root of time, that is, $w(t) = 2\sqrt{Dt}$, while the peak temperature drops as $1/\sqrt{4\pi Dt}$. Note that what we called the impulse response in Eq.(35) is the same function as in Eq.(39) in the limit when w_0 approaches zero.

9.8. Diffusion in two spatial dimensions: Cartesian coordinates. The physical phenomena of general interest in applied sciences are often modeled with the aid of partial differential equations that reside in the four-dimensional spacetime (x, y, z, t) of common experience. Depending on the nature of the problem under consideration, the space coordinates may be reduced by one or two, and, in the case of steady-state phenomena, the time dimension may be absent from the equations. In the preceding section, we saw how the diffusion equation could arise in the context of a temperature distribution along a thin, homogeneous wire. The equation governing the temporal evolution of the temperature along the length of the wire (which coincided with the x-axis) was seen to be a linear PDE of second order in x, and of first order in t. In the present section, we analyze a heat diffusion problem in two-dimensional space, where the temperature of a rectangular slab in the xy-plane will be denoted by $T(x, y, t)$. The governing equation will be seen to be, once again, a PDE of second order in the space coordinates x and y, and of first order in the time coordinate t. It is also possible to express the diffusion equation in polar coordinates, where, in two-dimensional space, the radial and azimuthal coordinates are denoted by r and φ. We will see in the following section that the PDE that governs the temperature profile $T(r, \varphi, t)$ of a circular object is of the second order in r, second order in φ, and first order in t. It is not difficult to see (as also elaborated in some of the problems at the end of the chapter) that, in three-dimensional space, the diffusion equation continues to be of second order in the spatial coordinates, and of first order in the time coordinate, irrespective of whether the geometry of the considered system dictates the adoption of a Cartesian (x, y, z), cylindrical (r, φ, z), or spherical (r, θ, φ) coordinate system.

Let a thin slab of uniform thickness τ, specific heat C, thermal conductivity κ, and diffusivity $D = \kappa/C$ sit in the xy-plane. We assume that the plate's thickness τ is sufficiently small, so that the temperature along the z-axis can be taken to be uniform. Let $T(x, y, t)$ denote the temperature of the slab at the point (x, y) at time t. The heat balance equation for a small $\Delta x \times \Delta y$ rectangular section may thus be written as follows:

$$\underbrace{\kappa\{[\partial_x T(x + ½\Delta x, y, t) - \partial_x T(x - ½\Delta x, y, t)]\tau\Delta y}_{\text{rate of heat flow into the small rectangle along the } x\text{-axis}}$$

$$+\underbrace{[\partial_y T(x, y + ½\Delta y, t) - \partial_y T(x, y - ½\Delta y, t)]\tau\Delta x\}}_{\text{rate of heat flow into the small rectangle along the } y\text{-axis}} = \underbrace{C\tau\Delta x\Delta y\partial_t T(x, y, t)}_{\text{heat capacity} \times \text{rate of change of } T}. \quad (40)$$

Dividing both sides of this equation by $C\tau\Delta x\Delta y$, then allowing Δx and Δy to approach zero, we will arrive at the following two-dimensional heat diffusion equation in Cartesian coordinates:

$$D(\partial_x^2 + \partial_y^2)T(x, y, t) = \partial_t T(x, y, t). \quad (41)$$

This diffusion equation can be solved (along the same lines as the wave equation in section 4), using the method of separation of variables, provided of course that the initial condition $T(x, y, t = 0)$ as well as appropriate boundary conditions are specified. In what follows, we shall use a reduced version of Eq.(41) to solve a slightly different problem.

With reference to figure 9.7, consider an $L_x \times L_y$ rectangular plate of negligible thickness, whose boundaries at $x = 0$, $x = L_x$, and $y = 0$ are kept at a fixed and uniform temperature T_0, whereas the boundary at $y = L_y$ is held at a constant but nonuniform temperature $T = T_0 + h(x)$, where $h(\cdot)$ is a pre-specified arbitrary function. Once the plate reaches a steady state, its temperature $T(x, y)$ will no longer vary with time. Therefore, the governing equation is obtained by setting $\partial_t T(x, y, t) = 0$ on the right-hand side of Eq.(41), which simplifies as

$$(\partial_x^2 + \partial_y^2)T(x, y) = 0. \quad (42)$$

Note that the diffusivity D has dropped out of Eq.(42), making the steady state temperature profile independent of the thermal constants of the material medium. In practice, after setting the temperature at the boundaries, the slab takes some time to reach the steady state. How fast or slow the steady state arrives depends, of course, on the value of D, but once the temperature stabilizes, its distribution across the slab will in no way

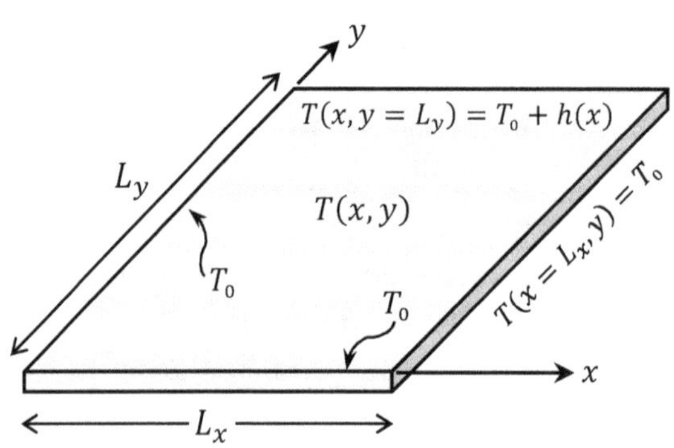

Figure 9.7. An $L_x \times L_y$ rectangular slab of negligible thickness τ sits in the xy-plane. The temperature of the boundaries at $x = 0$, $x = L_x$, and $y = 0$ is fixed at T_0, whereas the boundary at $y = L_y$ has a nonuniform (albeit time-independent) temperature profile $T = T_0 + h(x)$. Here, $h(x)$ is some arbitrary function of the x-coordinate. The temperature of the slab in the steady state is denoted by $T(x, y)$.

depend on D, but only on the boundary conditions. Another fact that needs to be mentioned at the outset is that a constant temperature can always be added to $T(x,y)$ without violating Eq.(42). Consequently, we reduce the boundary conditions initially to $T(0,y) = T(L_x,y) = T(x,0) = 0$ and $T(x,L_y) = h(x)$, then, after solving Eq.(42) subject to these reduced boundary conditions, we proceed to add the constant T_0 to our solution.

Invoking the method of separation of variables, we substitute $f(x)g(y)$ for $T(x,y)$ in Eq.(42), and arrive at $g''(y)/g(y) = c$ and $f''(x)/f(x) = -c$, where c is the separation constant. Considering that $f(x)$ vanishes at both $x = 0$ and $x = L_x$, the general solution $f(x) = A\sin(\sqrt{c}x) + B\cos(\sqrt{c}x)$ must have $B = 0$ and $\sqrt{c} = n\pi/L_x$, where $n = 1, 2, 3, \cdots$ is an arbitrary positive integer. Having determined the separation constant c, we now find the general solution of the remaining ODE, $g''(y) - cg(y) = 0$, as $g(y) = A'\sinh(\sqrt{c}y) + B'\cosh(\sqrt{c}y)$. Since the temperature at $y = 0$ is zero, we must have $B' = 0$. Consequently, the complete solution of the steady state problem is

$$T(x,y) = T_0 + \textstyle\sum_{n=1}^{\infty} A_n \sin(n\pi x/L_x)\sinh(n\pi y/L_x). \tag{43}$$

It remains to find the constant coefficients A_n by matching the boundary condition at $y = L_y$; that is, $h(x) = \sum_{n=1}^{\infty} A_n \sinh(n\pi L_y/L_x)\sin(n\pi x/L_x)$. This requires expanding $h(x)$ into a Fourier sine series over the interval $0 \le x \le L_x$, then equating the Fourier coefficients with $A_n\sinh(n\pi L_y/L_x)$, which would yield the values of A_n for all n. As discussed at the end of section 3, the function $h(x)$ has the Fourier sine series expansion $h(x) = \sum_{n=1}^{\infty} h_n \sin(n\pi x/L_x)$ where $h_n = (2/L_x)\int_0^{Lx} h(x)\sin(n\pi x/L_x)\,\mathrm{d}x$. It is seen, therefore, that $A_n = h_n/\sinh(n\pi L_y/L_x)$.

9.9. Diffusion in two spatial dimensions: polar coordinates. The figure below depicts an annular disk, having inner radius R_1, outer radius R_2, small thickness τ, specific heat C, thermal conductivity κ, and thermal diffusivity $D = \kappa/C$. The inner and outer rims of the annulus are kept at fixed but nonuniform temperatures $T(R_1,\varphi)$ and $T(R_2,\varphi)$. Our goal, here, is to determine the steady state temperature distribution $T(r,\varphi)$ within the annulus, where $R_1 \le r \le R_2$ and $0 \le \varphi < 2\pi$. The temperature profile through the thickness of the disk (i.e., in the z-direction) is assumed to be uniform.

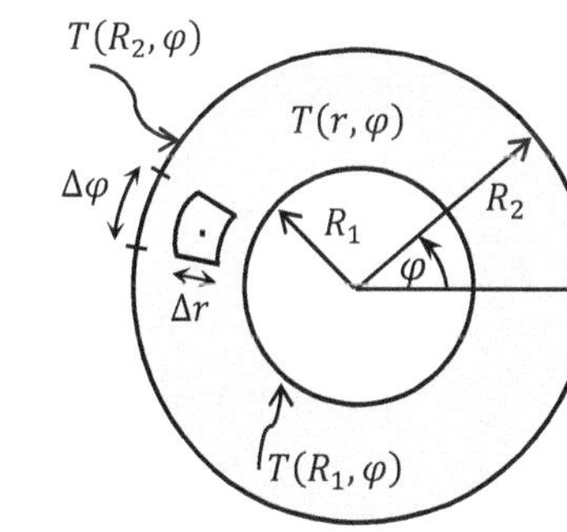

For a small, nearly trapezoidal segment of the disk, centered at an arbitrary point (r,φ) and having radial and azimuthal dimensions Δr and $\Delta\varphi$, the heat balance equation may be written as follows:

$$\underbrace{\kappa\{[(r + \tfrac{1}{2}\Delta r)\partial_r T(r + \tfrac{1}{2}\Delta r,\varphi,t) - (r - \tfrac{1}{2}\Delta r)\partial_r T(r - \tfrac{1}{2}\Delta r,\varphi,t)]\tau\Delta\varphi}_{\text{rate of heat flow into the trapezoid along the radial direction}}$$

$$\underbrace{+r^{-1}[\partial_\varphi T(r,\varphi + \tfrac{1}{2}\Delta\varphi,t) - \partial_\varphi T(r,\varphi - \tfrac{1}{2}\Delta\varphi,t)]\tau\Delta r\}}_{\text{rate of heat flow into the trapezoid along the azimuthal }\varphi\text{-direction}} = \underbrace{C\tau\, r\Delta\varphi\Delta r\partial_t T(r,\varphi,t)}_{\text{heat capacity}\times\text{rate of change of }T}. \tag{44}$$

Note in this equation that the azimuthal rate of heat flow is expressed as $r^{-1}\partial_\varphi T(r,\varphi \pm \tfrac{1}{2}\Delta\varphi,t)$, for the usual reason that the slope of the temperature profile in the φ direction is the infinitesimal limit of the rise-to-run ratio $\Delta T/(r\Delta\varphi)$ and not simply that of $\Delta T/\Delta\varphi$. Now, dividing both sides of Eq.(44) by $C\tau r\Delta\varphi\Delta r$, then allowing Δr and

$\Delta\varphi$ to approach zero, and using the identity $\partial_r(r\partial_r) = r\partial_r^2 + \partial_r$, we arrive at the two-dimensional heat diffusion equation in polar coordinates, namely,

$$D(\partial_r^2 + r^{-1}\partial_r + r^{-2}\partial_\varphi^2)T(r,\varphi,t) = \partial_t T(r,\varphi,t). \tag{45}$$

In the steady state, when the temperature has stabilized, it no longer varies with time and, therefore, its derivative with respect to t must vanish; that is; $\partial_t T = 0$. Thus, the PDE governing the steady state temperature distribution within the annular disk becomes

$$(r^2\partial_r^2 + r\partial_r + \partial_\varphi^2)T(r,\varphi) = 0. \tag{46}$$

Note that the diffusivity D has dropped out of the above equation, thus rendering the temperature profile $T(r,\varphi)$ independent of the material parameters κ and C. Equation (46) simply asserts that the rates at which the heat flows into and out of any small section of the disk precisely cancel each other out. Once the temperature at the boundaries $r = R_1, R_2$ is established (presumably by some external agent), the diffusivity D of the host material determines how long it takes before the system reaches a steady state. A large value of D hastens the arrival of the steady state, whereas small values of D make it longer to establish. In any case, once the temperature reaches a steady state, its distribution across the disk will be a function only of the boundary conditions and not of the thermal constants of the host material.

Invoking the method of separation of variables, we substitute $f(r)g(\varphi)$ for $T(r,\varphi)$ in Eq.(46), and find

$$\frac{r^2 f''(r) + rf'(r)}{f(r)} = -\frac{g''(\varphi)}{g(\varphi)}. \tag{47}$$

It is easy to see that the separation constant must be m^2, where $m = 0, 1, 2, \cdots$ is a nonnegative integer. The solution of the second-order ODE that governs the azimuthal temperature profile (i.e., $g''(\varphi) + m^2 g(\varphi) = 0$) is thus $g(\varphi) = A\sin(m\varphi) + B\cos(m\varphi)$. The constant coefficients A and B must be determined from the boundary conditions. On the left-hand side of Eq.(47), we now have

$$r^2 f''(r) + rf'(r) - m^2 f(r) = 0. \tag{48}$$

This is a so-called equi-dimensional ordinary differential equation, whose solutions are readily guessed to be in the form of $f(r) = r^s$, where s must be chosen to satisfy the equation. Substitution into the equation yields

$$s(s-1)r^s + sr^s - m^2 r^s = 0 \quad \rightarrow \quad s = \pm m. \tag{49}$$

The general solution of Eq.(48) is thus seen to be $f(r) = A'r^m + B'r^{-m}$, where A' and B' are arbitrary constant coefficients. This solution, however, is *not* adequate in the case of $m = 0$. Whereas for $m \neq 0$, the two solutions r^m and r^{-m} of Eq.(48) are independent of each other, for $m = 0$ these solutions coincide. One must, therefore, return to the original equation in search of a second independent solution. In the case of $m = 0$, Eq.(48) may be solved in the following way:

$$r^2 f''(r) + rf'(r) = 0 \quad \rightarrow \quad \frac{f''(r)}{f'(r)} = -\frac{1}{r} \quad \rightarrow \quad \frac{\mathrm{d}}{\mathrm{d}r}\ln[f'(r)] = -\frac{\mathrm{d}}{\mathrm{d}r}\ln r$$

$$\rightarrow \ln[f'(r)] = -\ln r + a_0 \xrightarrow{A_0 = \exp(a_0)} f'(r) = A_0/r \quad \rightarrow \quad f(r) = A_0 \ln r + B_0. \tag{50}$$

Finally, we are in a position to write the general solution of Eq.(46) as follows:

$$T(r,\varphi) = (A_0 \ln r + B_0) + \sum_{m=1}^{\infty} r^m [A_m \sin(m\varphi) + B_m \cos(m\varphi)]$$
$$+ \sum_{m=1}^{\infty} r^{-m} [C_m \sin(m\varphi) + D_m \cos(m\varphi)]. \qquad (51)$$

At the inner and outer rims $r = R_1, R_2$ of the annular disk, the temperature profile can now be written (in terms of all the unknown coefficients $A_0, B_0, A_m, B_m, C_m, D_m$) as

$$T(R_1, \varphi) = (A_0 \ln R_1 + B_0) + \sum_{m=1}^{\infty} [(A_m R_1^m + C_m R_1^{-m}) \sin(m\varphi) + (B_m R_1^m + D_m R_1^{-m}) \cos(m\varphi)],$$

$$T(R_2, \varphi) = (A_0 \ln R_2 + B_0) + \sum_{m=1}^{\infty} [(A_m R_2^m + C_m R_2^{-m}) \sin(m\varphi) + (B_m R_2^m + D_m R_2^{-m}) \cos(m\varphi)].$$

The unknown coefficients appearing in these equations will be uniquely determined once the pre-specified temperature distributions at the boundaries are expanded into their respective Fourier series. Note that each Fourier coefficient in the above equations has two unknowns, namely (A_0, B_0), (A_m, C_m), and (B_m, D_m). Matching the boundary conditions at $r = R_1$ and $r = R_2$ yields two linear equations for each pair of unknowns, the solutions of which will then uniquely identify all the unknown coefficients that appear in Eq.(51).

A solid disk (i.e., one *without* a hole at its center) is the limiting form of an annular disk with $R_1 = 0$. In the case of a solid disk, both functions $\ln r$ and r^{-m} diverge as $r \to 0$. Consequently, the coefficients A_0, C_m, and D_m in Eq.(51) must all be set to zero. The only boundary condition in this case, namely $T(R_2, \varphi) = T_0(\varphi)$, will then contain all the necessary information to determine the remaining coefficients B_0, A_m, and B_m. Note that the temperature at the center of the disk is now given by $T(r = 0, \varphi) = B_0$, which is the constant term in the Fourier series expansion of the temperature profile $T_0(\varphi)$ at the boundary $r = R_2$ of the solid disk. In other words, the temperature at the center equals the average temperature at the rim.

Another limiting case is the one where $R_1 \neq 0$ but $R_2 \to \infty$. Here, the $\ln r$ and r^m terms go to infinity as $r \to \infty$, which is unacceptable and, therefore, the coefficients A_0, A_m, and B_m of Eq.(51) must vanish. The remaining coefficient B_0, C_m, and D_m are then found by matching the boundary conditions at the inner rim, $r = R_1$. The temperature at the (far away) outer rim will be B_0, which is the average temperature of the inner rim. No amount of heat can enter or exit at the outer rim, where the radial gradient of the temperature, $\partial_r T(r,\varphi)|_{r\to\infty}$, vanishes. The heat entering the disk in some regions of the inner rim is balanced by the heat exiting the same rim at other (azimuthal) locations.

Digression: One may wonder why the Bessel functions do *not* appear in Eq.(51), which is, after all, a solution of the heat diffusion equation, Eq.(45). The answer is that, in the steady state, the term $\partial_t T$ on the right-hand side of the equation disappears, resulting in an equi-dimensional equation, Eq.(48), rather than the Bessel equation, for the radial solution $f(r)$. To better appreciate the ways in which steady state situations differ from dynamic ones—where the temperature evolves with time—let us consider a solid, homogeneous disk of radius R and negligible thickness τ, whose rim is kept at a fixed temperature, say, $T(r = R, \varphi, t) = T_0$, and whose initial temperature distribution is given as $T(r, \varphi, t = 0) = T_0 + T_1(r, \varphi)$. Application of the method of separation of variables to this problem then yields the following equations for the separated functions $f(r)$, $g(\varphi)$, $h(t)$, with m^2 and $-c^2$ being the usual separation constants:

$$r^2 f''(r) + r f'(r) + [(cr/\sqrt{D})^2 - m^2] f(r) = 0, \qquad (52)$$

$$g''(\varphi) + m^2 g(\varphi) = 0, \qquad (53)$$

$$h'(t) + c^2 h(t) = 0. \qquad (54)$$

The solutions of the Bessel equation, Eq.(52), are $J_m(cr/\sqrt{D})$ and $Y_m(cr/\sqrt{D})$. However, all Bessel functions of the second kind diverge at the center of the disk ($r = 0$), and are, therefore, unacceptable in the present case. At the boundary $r = R$, the radial solution must vanish, which requires that $cR/\sqrt{D} =$

r_{mn}, where r_{mn} is the nth zero of $J_m(\cdot)$, the standard Bessel function of the first kind, order m. Consequently, $c = \sqrt{D}r_{mn}/R$. The azimuthal equation, Eq.(53), yields $g(\varphi) = A\sin(m\varphi) + B\cos(m\varphi)$, where A and B are constant coefficients that must be determined by matching the initial condition at $t = 0$. The solution of the temporal equation, Eq.(54), is $h(t) = \exp(-c^2t) = \exp(-Dr_{mn}^2t/R^2)$, which indicates that all the various modes (associated with the pair of integers m and n) decay exponentially with time. (Recall that r_{mn} is dimensionless and that the units of D are cm^2/sec.) All in all, the general solution of the diffusion equation for the problem under consideration is

$$T(r, \varphi, t) = T_0 + \sum_{m=0}^{\infty} \sum_{n=1}^{\infty} J_m(r_{mn}r/R)[A_{mn}\sin(m\varphi) + B_{mn}\cos(m\varphi)]\exp(-Dr_{mn}^2t/R^2). \quad (55)$$

The coefficients A_{mn} and B_{mn} must be determined by expanding the initial temperature profile $T_1(r, \varphi)$ into a Fourier-Bessel series, then equating its coefficients with those given by Eq.(55) at $t = 0$.

Note that, according to Eq.(55), the decay rate of any given mode is proportional to the diffusion coefficient D, so that each and every mode would decay faster with time if the host medium happens to have a larger value of D. Also, higher-order modes have larger m and/or n values, which, in general, correspond to larger values of r_{mn}. The temporal decay of higher-order modes is, therefore, faster in time due to the presence of rapid variations in temperature across the disk surface and, consequently, larger thermal gradients along the r and/or φ directions.

9.10. The method of characteristics. This section describes a geometric approach to solving certain first-order partial differential equations in which the sought-after function depends on two variables. Not only does this method illustrate some fundamental facts concerning the nature of PDEs, but also it provides a powerful algorithmic approach for the numerical solution of certain types of partial differential equations.

Let us begin by considering the following first-order linear PDE, where the sought-after function is $f(\cdot)$ and its two variables are x and y:

$$a(x, y)\partial_x f(x, y) + b(x, y)\partial_y f(x, y) = c(x, y)f(x, y) + d(x, y). \quad (56)$$

The coefficients a, b, c, d of this equation are assumed to be smooth and well behaved functions of the variables x and y. As shown in figure 9.8a, a vector $\boldsymbol{v} = a\hat{\boldsymbol{x}} + b\hat{\boldsymbol{y}}$ may be associated with each and every point in the relevant domain within the xy-plane. One can then draw the *characteristic base curves* of the PDE under consideration, which are smooth and continuous curves that are everywhere tangent to the local $\boldsymbol{v}(x, y)$.

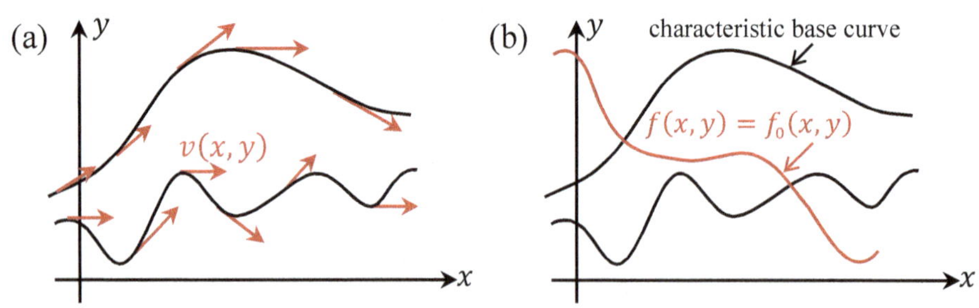

Figure 9.8. (a) The coefficients $a(x, y)$ and $b(x, y)$ of the PDE of Eq.(56) define a two-dimensional vector field $\boldsymbol{v}(x, y) = a\hat{\boldsymbol{x}} + b\hat{\boldsymbol{y}}$ in the xy-plane. Assuming that $a(x, y)$ and $b(x, y)$ are well behaved, one can draw, within the xy-plane, characteristic base curves that are tangent at every point to the local $\boldsymbol{v}(x, y)$. Here, two such curves are shown. (b) The initial condition of the PDE is specified on a predefined curve within the xy-plane, on which the sought-after function $f(x, y)$ is known to be equal to $f_0(x, y)$.

Each characteristic base curve can be parametrized by a real-valued variable, say, p, such that small increments $\mathrm{d}p$ along the curve would correspond to short distances traveled along the local vector $\boldsymbol{v}$. In other words, $a(x, y)\mathrm{d}p = \mathrm{d}x$ and $b(x, y)\mathrm{d}p = \mathrm{d}y$.

If we now represent each point of the xy-plane by its position vector $\boldsymbol{r} = x\hat{\boldsymbol{x}} + y\hat{\boldsymbol{y}}$, the left-hand side of Eq.(56) becomes $\boldsymbol{\nabla}f(\boldsymbol{r}) \cdot \mathrm{d}\boldsymbol{r}/\mathrm{d}p$, where $\boldsymbol{\nabla}f(\boldsymbol{r}) = (\partial_x f)\hat{\boldsymbol{x}} + (\partial_y f)\hat{\boldsymbol{y}}$

is the gradient of the desired function $f(\boldsymbol{r})$. Thus, as one travels a short distance $\mathrm{d}\boldsymbol{r}$ along any given characteristic base curve, the corresponding incremental change in the sought-after function will be $\mathrm{d}f(x,y) = \boldsymbol{\nabla} f(\boldsymbol{r}) \cdot \mathrm{d}\boldsymbol{r}$. It must now be clear that the PDE of Eq.(56) has become an ordinary differential equation, namely,

$$\mathrm{d}f(x,y)/\mathrm{d}p = c(x,y)f(x,y) + d(x,y). \quad (57)$$

In general, the initial condition of the PDE is specified on some curve within the xy-plane, such as that shown in figure 9.8b. On this so-called "initial curve," the desired $f(x,y)$ is known to be equal to some pre-specified function $f_0(x,y)$. It is now easy to describe a geometric (or numerical) method of solving the PDE of Eq.(56)—which is equivalent to the ODE of Eq.(57)—subject to the specified initial condition: Starting at each and every point on the initial curve, follow the local characteristic base curve and accumulate the increments $\mathrm{d}f$ of the desired function. This is done, in accordance with Eq.(57), by multiplying the local value of $c(x,y)f(x,y) + d(x,y)$ with the increments $\mathrm{d}p$ of the corresponding p, namely the parameter associated with the characteristic base curve that is being traversed.

For our next example, we pick a first-order *nonlinear* (or quasi-linear) PDE in which the sought-after function is $f(x,y)$, and the coefficients of the equation depend not only on the x, y variables, but also on the function $f(\cdot)$. The equation of interest is

$$a(x,y,f)\partial_x f(x,y) + b(x,y,f)\partial_y f(x,y) = c(x,y,f). \quad (58)$$

With reference to figure 9.9, note that any solution of Eq.(58) would be a continuous (and reasonably smooth) surface within the xyf-coordinate system. Such a surface is generally referred to as an *integral surface*. The initial condition would then be represented by a pre-specified curve that lies entirely on this integral surface. The initial curve, of course, indicates that, on a certain path within the xy-plane, the desired solution coincides with some known function $f_0(x,y)$. Also shown in figure 9.9 are some of the $\boldsymbol{v}$ vectors that are assigned to each point in the three-dimensional xyf-space in accordance with the formula $\boldsymbol{v}(x,y,f) = a\hat{\boldsymbol{x}} + b\hat{\boldsymbol{y}} + c\hat{\boldsymbol{f}}$, where $a, b,$ and c are the coefficients of Eq.(58).

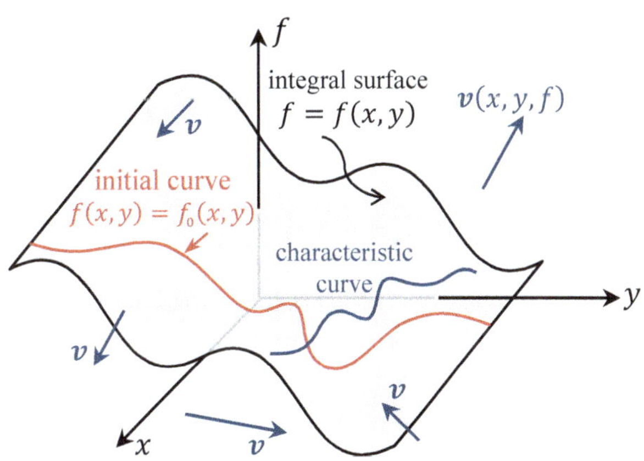

Figure 9.9. The solution of the PDE of Eq.(58) is a surface in the three-dimensional xyf coordinate system. The initial condition is typically a pre-specified curve lying entirely within the same surface. The vector field $\boldsymbol{v}(x,y,f) = a\hat{\boldsymbol{x}} + b\hat{\boldsymbol{y}} + c\hat{\boldsymbol{f}}$, where a, b, and c are the coefficients of the PDE, exists everywhere in the relevant domain of the xyf-space; here, only a few of the $\boldsymbol{v}$ vectors are shown. At all points located on the integral surface (i.e., the surface representing the sought-after solution), the $\boldsymbol{v}$ vectors are tangent to the local surface. Also shown is a single example of characteristic curves, which lie in the integral surface and are tangent to local v vectors.

If we now define the new function $F(\cdot)$ over the three-dimensional xyf-space as $F(x,y,f) = f(x,y) - f$, it goes without saying that the value of F everywhere on the integral surface will be zero. We also see from Eq.(58) that the gradient of $F(\cdot)$, namely $\boldsymbol{\nabla} F(x,y,f) = (\partial_x f)\hat{\boldsymbol{x}} + (\partial_y f)\hat{\boldsymbol{y}} - \hat{\boldsymbol{f}}$, satisfies the identity $\boldsymbol{v} \cdot \boldsymbol{\nabla} F = 0$ over the entire integral surface. Considering that $\boldsymbol{\nabla} F$ is perpendicular to the integral surface, the $\boldsymbol{v}$ vectors, being orthogonal to $\boldsymbol{\nabla} F$, must be the local tangents of the integral surface.

It is now easy to see how the geometric (or numerical) method of solution of the equation will proceed. One starts on the initial curve, then moves away along the local $\boldsymbol{v}$ vectors to construct the so-called *characteristic curves* of the equation. The collection of all such characteristic curves then constitutes the desired integral surface. In this construction, the integral surface $f = f(x,y)$ spreads outward and away from the initial curve $f = f_0(x,y)$ along innumerable characteristic curves, all of which adhere to the strict rule that their trajectories must always remain tangent to the local $\boldsymbol{v}$ vector.

9.11. Application of the method of characteristics to a second-order PDE. In certain special cases, it becomes possible to split a second-order PDE into a pair of first-order partial differentials, then apply the method of characteristics to the simpler equations that emerge. As an example, consider the one-dimensional wave equation along an infinitely long, thin, homogeneous wire in the absence of frictional losses, together with the relevant initial conditions, namely,

$$(v^2\partial_{xx} - \partial_{tt})f(x,t) = 0; \quad f(x,t=0) = f_0(x) \text{ and } \partial_t f(x,t)|_{t=0} = f_1(x). \tag{59}$$

Upon factorization, Eq.(59) becomes $(v\partial_x + \partial t)(v\partial_x - \partial t)f(x,t) = 0$, which, with the aid of an auxiliary function, say, $\psi(x,t)$, may be split into the following pair of coupled first-order PDEs:

$$(v\partial_x - \partial t)f(x,t) = \psi(x,t); \qquad f(x,t=0) = f_0(x). \tag{60a}$$

$$(v\partial_x + \partial t)\psi(x,t) = 0; \qquad \psi(x,t=0) = vf_0'(x) - f_1(x). \tag{60b}$$

Applying the method of characteristics to Eq.(60b), we find $a = v = \mathrm{d}x/\mathrm{d}p$ and $b = 1 = \mathrm{d}t/\mathrm{d}p$. Therefore, the characteristic base curves for this equation are the straight lines depicted in figure 9.10a, along which $\mathrm{d}\psi(x,t)/\mathrm{d}p = 0$ (i.e., ψ is constant). Each base curve satisfies the equation $x - vt = x_0$, with x_0 being its crossing point with the x-axis. At $(x,t) = (x_0, 0)$, the initial condition for $\psi(x,t)$ is given by Eq.(60b). Consequently,

$$\psi(x,t) = vf_0'(x - vt) - f_1(x - vt). \tag{61}$$

As for the remaining PDE, namely that given by Eq.(60a), we have $a = v = \mathrm{d}x/\mathrm{d}p$ and $b = -1 = \mathrm{d}t/\mathrm{d}p$. Consequently, the characteristic base curves for this equation are the straight lines parallel to the vector $v\hat{\boldsymbol{x}} - \hat{\boldsymbol{t}}$, as depicted in figure 9.10b. Along these base curves, $\mathrm{d}f/\mathrm{d}p = \psi$, or, given that $\mathrm{d}p = -\mathrm{d}t$, we have $\mathrm{d}f(x,t) = -\psi(x,t)\mathrm{d}t$. Each base curve in figure 9.10b obeys the equation $x + vt = x_0$, where x_0 is its crossing point with the x-axis. The ordinary differential equation along each base curve may now be written as

$$\mathrm{d}f(x,t)/\mathrm{d}t = -\psi(x_0 - vt, t) = -vf_0'(x_0 - 2vt) + f_1(x_0 - 2vt). \tag{62}$$

This equation should be integrated along each base curve depicted in figure 9.10b, starting at the initial point $(x,t) = (x_0, 0)$, which is located on the x-axis. Recognizing that the initial value of the sought-after function is $f(x_0, t=0) = f_0(x_0)$, and that, on these base curves, $x_0 = x + vt$, we find

$$f(x,t) = f_0(x_0) - v\int_0^t f_0'(x_0 - 2vt)\mathrm{d}t + \int_0^t f_1(x_0 - 2vt)\mathrm{d}t$$

$$= f_0(x_0) + ½\int_{x_0}^{x_0-2vt} f_0'(\zeta)\mathrm{d}\zeta - (1/2v)\int_{x_0}^{x_0-2vt} f_1(\zeta)\mathrm{d}\zeta$$

$$= f_0(x+vt) + ½[f_0(x-vt) - f_0(x+vt)] - (1/2v)\int_{x+vt}^{x-vt} f_1(\zeta)\mathrm{d}\zeta$$

$$= ½[f_0(x+vt) + f_0(x-vt)] + (1/2v)\int_{x-vt}^{x+vt} f_1(\zeta)\mathrm{d}\zeta. \quad (63)$$

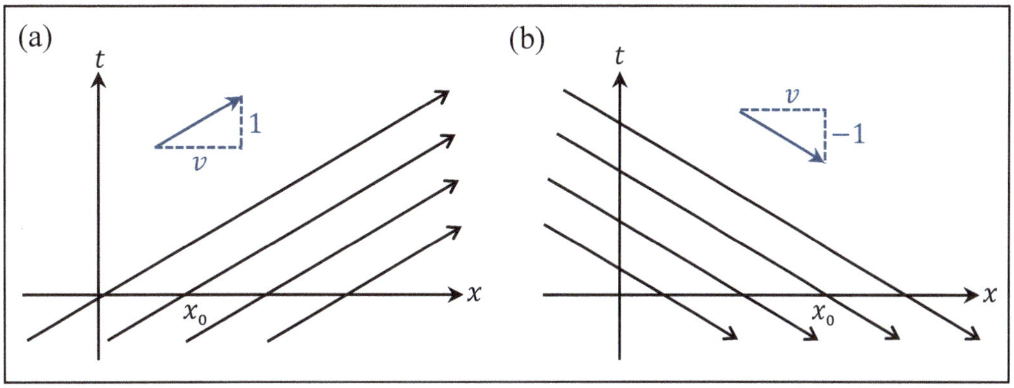

Figure 9.10. Characteristic base curves for two first-order partial differential equations obtained by splitting the second-order wave equation $(v^2\partial_x^2 - \partial_t^2)f(x,t) = 0$ into a pair of coupled equations. (a) The coefficients of $(v\partial_x + \partial_t)\psi(x,t) = 0$ specify the tangent vectors $v\hat{\boldsymbol{x}} + \hat{\boldsymbol{t}}$ of the corresponding base curves. (b) The coefficients of $(v\partial_x - \partial_t)f(x,t) = \psi(x,t)$ specify the tangent vectors $v\hat{\boldsymbol{x}} - \hat{\boldsymbol{t}}$ that define the corresponding base curves.

Equation (63) is the complete solution of the one-dimensional wave equation in the absence of friction, and also when the boundaries at $x = \pm\infty$ do not constrain the solution. This formula is credited to the French mathematician Jean-Baptiste le Rond d'Alembert (1717–1783). Problem 3 at the end of the chapter describes an alternative route to d'Alembert's formula.

9.12. A nonlinear partial differential equation. An electromagnetic wave-packet having a uniform cross-sectional profile in the xy-plane travels in free space along the z-axis. In the International System of Units (SI), where μ_0 and ε_0 represent the permeability and permittivity of free space,[4] the speed of light in vacuum is $c = 1/\sqrt{\mu_0\varepsilon_0}$, the impedance of free space is $Z_0 = \sqrt{\mu_0/\varepsilon_0}$, and the E- and H-fields of the (linearly polarized) wave-packet may be expressed as

$$\boldsymbol{E}(\boldsymbol{r},t) = E^{(\mathrm{i})}(t - z/c)\hat{\boldsymbol{x}}, \quad (64a)$$

$$\boldsymbol{H}(\boldsymbol{r},t) = E^{(\mathrm{i})}(t - z/c)\hat{\boldsymbol{y}}/Z_0. \quad (64b)$$

A typical plot of $E^{(\mathrm{i})}(\varphi)$ appears in figure 9.11a. At $z = 0$ the packet encounters a semi-infinite medium that is homogeneous, isotropic, and transparent, within which the displacement field $\boldsymbol{D}$ is some arbitrary function of the local and instantaneous E-field; that is, $\boldsymbol{D}(\boldsymbol{r},t) = \varepsilon_0\mathfrak{D}[E(\boldsymbol{r},t)]\hat{\boldsymbol{x}}$ (see figure 9.11b). At the interface between vacuum and the nonlinear medium, that is, at the xy-plane located at $z = 0$, a reflected wave is generated whose E- and H-fields may be written as follows:

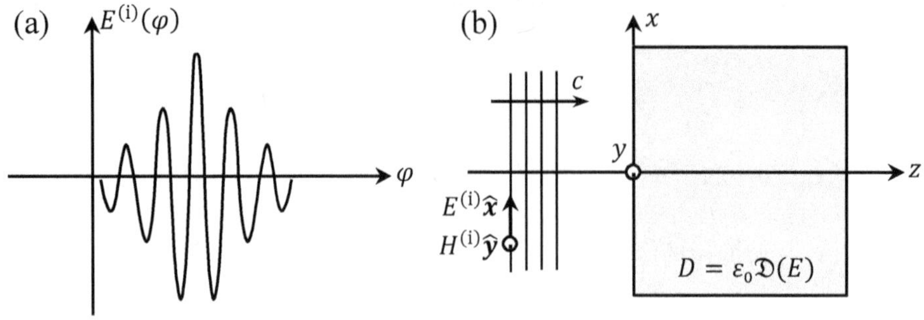

Figure 9.11. (a) Plot of a typical wave-packet's E-field that travels in free space in accordance with Eq.(64a). (b) The incident wave-packet arrives at the interface with a nonlinear medium that occupies the half-space $z \geq 0$.

$$\boldsymbol{E}(\boldsymbol{r},t) = E^{(\mathrm{r})}(t + z/c)\hat{\boldsymbol{x}}, \tag{65a}$$

$$\boldsymbol{H}(\boldsymbol{r},t) = -E^{(\mathrm{r})}(t + z/c)\hat{\boldsymbol{y}}/Z_0. \tag{65b}$$

We assume that $\mathfrak{D}(E)$ is a nonlinear function of the local and instantaneous E-field, such as that depicted in figure 9.12a. The medium is nonmagnetic, which implies that $\boldsymbol{B}(\boldsymbol{r},t) = \mu_0\boldsymbol{H}(\boldsymbol{r},t)$. Propagating inside the medium is a one-dimensional wave-packet whose E- and H-fields will be *assumed* here to have the following general structure:[5-7]

$$\boldsymbol{E}(\boldsymbol{r},t) = E^{(\mathrm{t})}[\varphi(z,t)]\hat{\boldsymbol{x}}\,, \tag{66a}$$

$$\boldsymbol{H}(\boldsymbol{r},t) = H^{(\mathrm{t})}[\varphi(z,t)]\hat{\boldsymbol{y}}. \tag{66b}$$

In the absence of free charges and free currents, all the aforementioned field profiles automatically satisfy Maxwell's first and fourth equations, namely $\boldsymbol{\nabla}\cdot\boldsymbol{D}(\boldsymbol{r},t) = 0$ and $\boldsymbol{\nabla}\cdot\boldsymbol{B}(\boldsymbol{r},t) = 0$.[4,7,8] Maxwell's second and third equations, $\boldsymbol{\nabla}\times\boldsymbol{H}(\boldsymbol{r},t) = \partial_t\boldsymbol{D}(\boldsymbol{r},t)$ and $\boldsymbol{\nabla}\times\boldsymbol{E}(\boldsymbol{r},t) = -\partial_t\boldsymbol{B}(\boldsymbol{r},t)$, become

$$\partial_z H_y(z,t) = -\partial_t D_x(z,t). \tag{67a}$$

$$\partial_z E_x(z,t) = -\mu_0\partial_t H_y(z,t). \tag{67b}$$

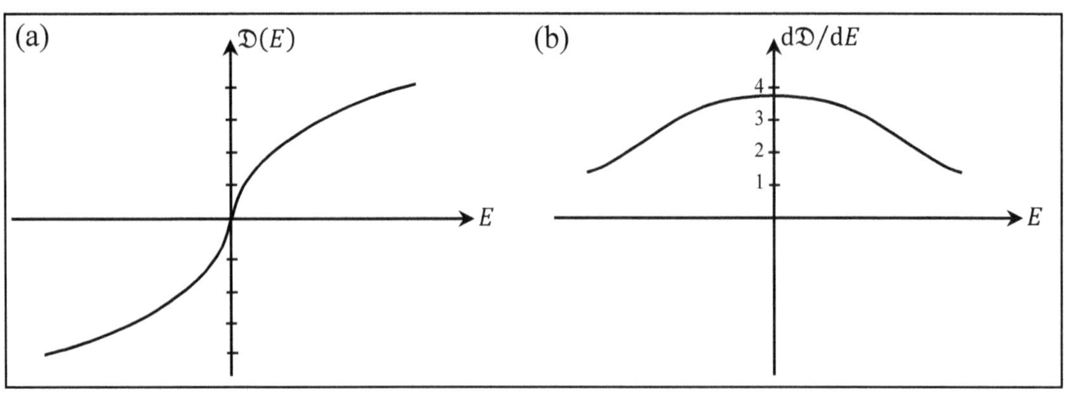

Figure 9.12. Inside the nonlinear medium, the displacement field $\boldsymbol{D}$ is assumed to be related to the local and instantaneous E-field via $\boldsymbol{D}(\boldsymbol{r},t) = \varepsilon_0\mathfrak{D}[E(\boldsymbol{r},t)]\hat{\boldsymbol{x}}$. (a) A typical plot of the function $\mathfrak{D}(E)$ versus the magnitude E of the local and instantaneous electric field. (b) The derivative of $\mathfrak{D}(E)$ with respect to E determines the "velocity" parameter v via Eq.(70), and the magnetic field inside the nonlinear medium, $H^{(\mathrm{t})}$, via Eq.(71).

It is now easy to verify that Eqs.(67) are satisfied by both the incident and reflected fields in the free space region given, respectively, by Eqs.(64) and (65). As for the fields inside the nonlinear medium, substitution of Eqs.(66a and b) into Eqs.(67a and b) yields

$$(\partial_z\varphi)\dot{H}^{(\mathrm{t})} = -\varepsilon_0[\mathrm{d}\mathfrak{D}(E)/\mathrm{d}E](\partial_t\varphi)\dot{E}^{(\mathrm{t})}. \tag{68a}$$

over-dot here indicates differentiation with respect to the argument φ of the function.

$$(\partial_z\varphi)\dot{E}^{(\mathrm{t})} = -\mu_0(\partial_t\varphi)\dot{H}^{(\mathrm{t})}. \tag{68b}$$

Simultaneous satisfaction of Eqs.(68a) and (68b) now demands that

$$\partial_z\varphi \pm \sqrt{\mu_0\varepsilon_0\,\mathrm{d}\mathfrak{D}(E)/\mathrm{d}E}\,\partial_t\varphi = 0. \tag{69a}$$

$$\dot{H}^{(\mathrm{t})} = \pm\sqrt{(\varepsilon_0/\mu_0)\mathrm{d}\mathfrak{D}(E)/\mathrm{d}E}\,\dot{E}^{(\mathrm{t})}. \tag{69b}$$

The $\pm$ signs appearing in Eqs.(69) correspond to wave-packets that propagate along the $\pm z$ directions. Given that the present concern is solely with propagation along the $+z$-direction, in what follows we shall omit the minus signs. Recalling that $c = 1/\sqrt{\mu_0\varepsilon_0}$ is the speed of light in vacuum, we now define the wave velocity v inside the nonlinear medium as a function of the local and instantaneous E-field, as follows:

$$v(E) = c/\sqrt{\mathrm{d}\mathfrak{D}(E)/\mathrm{d}E}. \tag{70}$$

Equation (69a) may thus be written as $[v(E)\partial_z + \partial_t]\varphi(z,t) = 0$, a first-order PDE for $\varphi(z,t)$, which can be analyzed via the method of characteristics. Considering that $[v(E)\hat{\boldsymbol{z}} + \hat{\boldsymbol{t}}]\cdot\boldsymbol{\nabla}\varphi(z,t) = 0$, the function $\varphi(z,t)$ must remain invariant along the vector $v(E)\hat{\boldsymbol{z}} + \hat{\boldsymbol{t}}$ in the (z,t) plane. Thus, starting anywhere on the t-axis at $z = 0$ and proceeding along $v(E)\hat{\boldsymbol{z}} + \hat{\boldsymbol{t}}$, one finds that $\varphi(z,t)$ remains constant, which, in accordance with Eqs.(66), results in the constancy of the E and H fields along the path.

Returning to Eq.(69b), the E- and H-field differentials are related via $Z_0\mathrm{d}H^{(\mathrm{t})} = \sqrt{\mathrm{d}\mathfrak{D}(E)/\mathrm{d}E}\,\mathrm{d}E^{(\mathrm{t})}$, where $Z_0 = \sqrt{\mu_0/\varepsilon_0}$ is the impedance of free space. Consequently,

$$Z_0H^{(\mathrm{t})} = \int_0^{E^{(\mathrm{t})}} \sqrt{\mathrm{d}\mathfrak{D}(E)/\mathrm{d}E}\,\mathrm{d}E. \tag{71}$$

Continuity of the tangential E- and H-fields at the boundary ($z = 0$) now demands that

$$E^{(\mathrm{i})} - E^{(\mathrm{r})} = \int_0^{E^{(\mathrm{i})}+E^{(\mathrm{r})}} \sqrt{\mathrm{d}\mathfrak{D}(E)/\mathrm{d}E}\,\mathrm{d}E. \tag{72}$$

This equation can now be solved numerically to arrive at the reflected field $E^{(\mathrm{r})}$ for each value of the incident field $E^{(\mathrm{i})}$. These values will in turn determine the transmitted field in accordance with $E^{(\mathrm{t})} = E^{(\mathrm{i})} + E^{(\mathrm{r})}$.

References

1. G. B. Arfken and H. J. Weber, *Mathematical Methods for Physicists*, 6th edition, Academic Press, Cambridge, Massachusetts (2005).
2. F. B. Hildebrand, *Advanced Calculus for Applications*, 2nd edition, Prentice-Hall, Upper Saddle River, New Jersey (1976).
3. J. Mathews and R. L. Walker, *Mathematical Methods of Physics*, 2nd edition, Benjamin/Cummings Publishing, California (1970).
4. J. D. Jackson, *Classical Electrodynamics*, 3rd edition, Wiley, New York (1999).
5. L. J. F. Broer, "Exact solution of the reflection problem in non-linear optics," *Physics Letters* **4**, 65 (1963).
6. N. Bloembergen, *Nonlinear Optics*, Addison-Wesley, Reading, Massachusetts (1965).
7. L. D. Landau, E. M. Lifshitz, and L. P. Pitaevskii, *Electrodynamics of Continuous Media*, Volume 8, 2nd edition, Butterworth-Heinemann, London, United Kingdom (1984).
8. M. Mansuripur, *Field, Force, Energy and Momentum in Classical Electrodynamics*, revised edition, Bentham Science Publishers, Sharjah, United Arab Emirates (2017).

Problems

Problem 1) With reference to figure a, a straight wire has a uniform mass density ρ [kg/m]. The wire also has a uniform spring constant α [Newton], such that if a segment of length L is elongated (or compressed) by ΔL, the longitudinal force needed to maintain this change of length will be $F = \pm\alpha\,\Delta L/L$. When a pressure wave travels along the length of the wire, a point originally located at x will be displaced from x by a distance $\zeta(x,t)$. The dynamic friction coefficient of the wire, denoted by β [Newton·sec/m^2], indicates that the braking force acting at time t on a segment of length Δx will be $F = -\beta\Delta x\,\partial\zeta(x,t)/\partial t$; here $\partial\zeta(x,t)/\partial t$ is the local velocity of the segment. In general, ρ, α and β are real, positive constants.

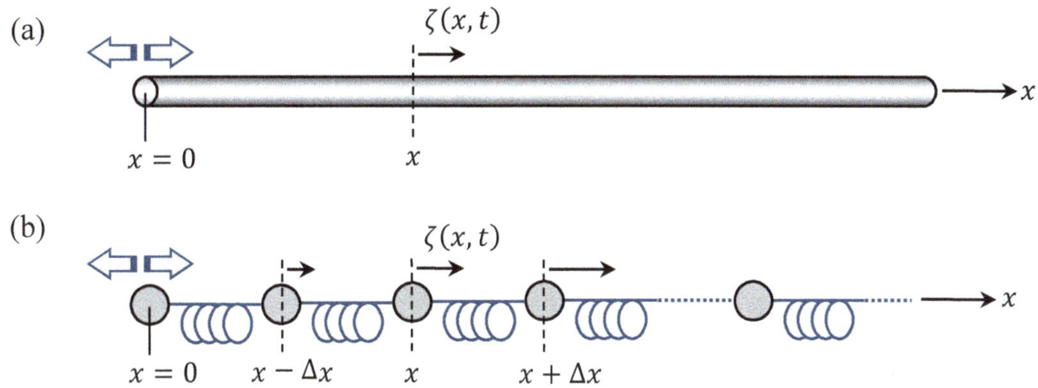

A good model for this wire is shown in figure b, where small balls of mass $\rho\Delta x$ are connected by short, massless springs having length Δx and spring constant α. The excitation source, located at the origin ($x = 0$), shakes the left-end of the wire with a predetermined motion such that $\zeta(x = 0, t) = \zeta_0(t)$.

a) Write the equation of motion for $\zeta(x,t)$.

b) Let the left-end excitation of the wire be given by $\zeta_0(t) = x_0 \exp(-\mathrm{i}2\pi f_0 t)$, where the complex constant x_0 represents the amplitude and phase of the oscillations, while the real-valued f_0 is the oscillation frequency. Assuming that the only acoustic wave traveling in the wire is the one that propagates from left to right, find an expression for the (complex) wave amplitude $\zeta(x,t)$.

c) In part (b), let β/ρ be sufficiently small (and/or f_0 sufficiently large) that the approximation $\sqrt{1 + \mathrm{i}\beta/(2\pi f_0\rho)} \cong 1 + \mathrm{i}\beta/(4\pi f_0\rho)$ would be justified. Under these circumstances, find the phase velocity and the attenuation coefficient of the acoustic wave propagating along the x-axis.

d) Let the Fourier transform of the excitation function $\zeta_0(t)$ contain only frequencies that satisfy the approximation used in part (c). Show that the corresponding displacement $\zeta(x,t)$ at an arbitrary point x along the wire has the same form as $\zeta_0(t)$, except for a time delay and an overall attenuation.

Problem 2) The Laplacian operator ∇^2 acting on a scalar function $\psi(x,y,z)$ of spatial coordinates in our three-dimensional space has the following physical significance. Take a small $\Delta x \times \Delta y \times \Delta z$ cubical volume centered at an arbitrary point (x,y,z). On each of the six surfaces of the cube, suppose the flux of the physical entity represented by ψ is proportional to the local slope of ψ along the corresponding surface normal. For instance, on the surface centered at $(x + ½\Delta x, y, z)$ the rate of outflow of ψ is given by the expression $\partial_x \psi(x + ½\Delta x, y, z)\Delta y \Delta z$. The net rate of flow of ψ out of the cube, normalized by its volume, will then be

$$\nabla^2 \psi = \left(\frac{\partial^2}{\partial x^2} + \frac{\partial^2}{\partial y^2} + \frac{\partial^2}{\partial z^2}\right)\psi.$$

a) Using a small volume cut out of a cylindrical shell, find an expression for $\nabla^2 \psi(\rho, \varphi, z)$ in the cylindrical coordinate system.

b) Using a small volume cut out of a spherical shell, find an expression for $\nabla^2 \psi(\rho, \theta, \varphi)$ in the spherical coordinate system.

Problem 3) Consider the one-dimensional wave equation with no damping, namely,

$$v^2 \frac{\partial^2 z(x,t)}{\partial x^2} = \frac{\partial^2 z(x,t)}{\partial t^2}.$$

The wave propagates in an infinitely long, thin wire aligned with the *x*-axis with a phase velocity v. The initial amplitude and vibration speed of the wave are specified as $z(x, t = 0) = f(x)$ and $\partial z(x,t)/\partial t|_{t=0} = g(x)$, respectively. Expressing the solution of the wave equation in terms of Fourier integrals, show that the general solution to this problem may be written as

$$z(x,t) = ½[f(x - vt) + f(x + vt)] + (1/2v) \int_{x-vt}^{x+vt} g(x')dx'.$$

This equation is known as the D'Alembert formula.

Problem 4) The membrane of a rectangular drumhead is under isotropic tension T [Newton/m] from all sides. The membrane is firmly affixed to the drumhead at three of its four boundaries, namely at $x = 0$, $y = 0$, and $x = L_x$, but is free to vibrate (in the z-direction) at the fourth boundary located at $y = L_y$. The membrane's mass density and friction coefficient (per unit area) are specified as ρ [kg/m^2] and β [kg/(m$^2 \cdot$ sec)], respectively.

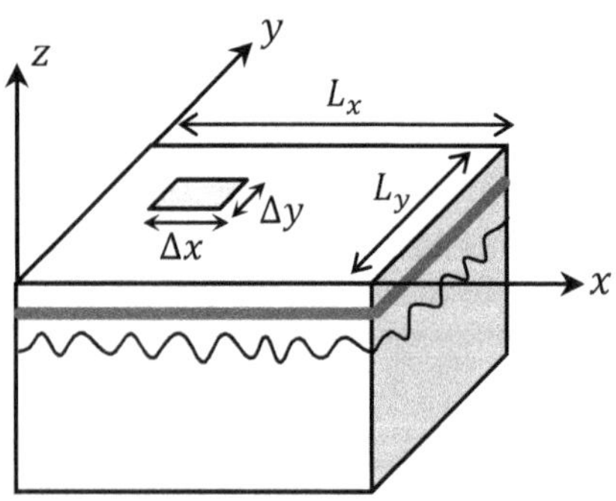

a) Making the usual assumption that the slope of the membrane's displacement $z(x, y, t)$ is small, write the governing equation of motion (in the absence of external forces) in terms of the wave velocity $v = \sqrt{T/\rho}$ and the damping coefficient $\gamma = \beta/\rho$.

b) Specify the relevant *boundary* conditions and *initial* conditions for the vibrating membrane.

c) Find all the admissible vibrational *modes* of the membrane by solving the partial differential equation obtained in part (a) subject to the boundary conditions specified in part (b).

d) Write a general expression for the vibration amplitude $z(x, y, t)$ as a *superposition* of all the various modes obtained in part (c). Explain (in words) how the unknown coefficients of this general expression can be determined.

Problem 5) A thin membrane of radius R vibrates with an amplitude $z(r,\phi,t)$. Ignoring friction losses and denoting by v the velocity of wave propagation in the membrane, the governing equation is found to be

$$v^2\left(\frac{\partial^2}{\partial r^2}+\frac{1}{r}\frac{\partial}{\partial r}+\frac{1}{r^2}\frac{\partial^2}{\partial \phi^2}\right)z(r,\phi,t)=\frac{\partial^2}{\partial t^2}z(r,\phi,t).$$

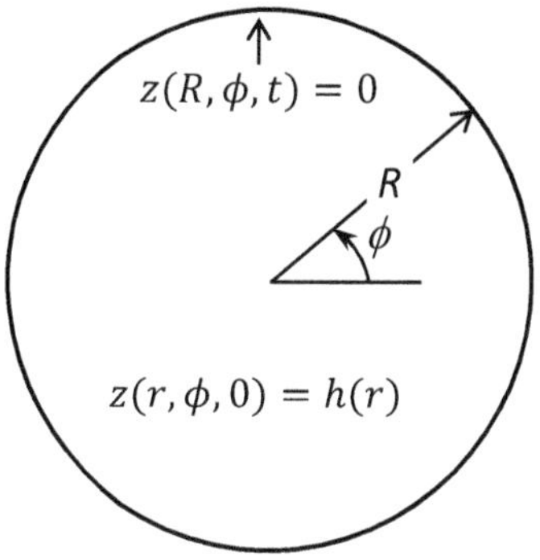

The membrane is firmly attached to a ring of radius R, so that $z(R,\phi,t)=0$. The initial amplitude of the vibrations is independent of the azimuthal coordinate ϕ and is given by $z(r,\phi,0)=h(r)$. The membrane's initial velocity is zero; that is, $\partial z(r,\phi,t)/\partial t|_{t=0}=0$. The vibration amplitude for all times t is thus given by

$$z(r,\phi,t)=\textstyle\sum_{n=1}^{\infty}c_n J_0(r_{0n}r/R)\cos(\omega_n t),$$

where r_{0n} is the nth zero of $J_0(r)$, the Bessel function of first kind, 0th order.

a) Verify that the initial velocity of the membrane at $t=0$ is in fact zero everywhere.

b) What is the value of ω_n in terms of r_{0n}, R and v?

c) Find the values of c_n in terms of $h(r)$ and other known parameters of the system.

Problem 6) An annular membrane of inner radius R_1, outer radius R_2 and negligible thickness, has a mass density ρ [kg/m^2] and friction coefficient β [kg/(m$^2\cdot$sec)]. Both edges of the membrane are firmly attached to a drumhead under a uniform and isotropic tensile stress T [Newton/m]. As usual, the wave equation governing the transverse vibrations of the membrane is

$$v^2\left(\frac{\partial^2}{\partial r^2}+\frac{\partial}{r\partial r}+\frac{\partial^2}{r^2\partial\varphi^2}\right)z(r,\varphi,t)=\left(\frac{\partial^2}{\partial t^2}+\gamma\frac{\partial}{\partial t}\right)z(r,\varphi,t),$$

where $\gamma=\beta/\rho\,[1/\text{sec}]$ is the damping coefficient and $v=\sqrt{T/\rho}\,[\text{m/sec}]$ is the wave propagation velocity within the membrane. The function $z(r,\varphi,t)$ represents the vibration amplitude at time t in a cylindrical coordinate system, where the membrane is located in the $z=0$ plane, is centered at the origin and is subject to the boundary conditions $z(R_1,\varphi,t)=z(R_2,\varphi,t)=0$. Describe the various modes of vibration of the membrane and their time dependence.

Problem 7) Consider a thin wire of length L and cross-sectional area A, having specific heat C (Joule/cm^3/°K), thermal conductivity κ (Joule/cm/sec/°K), and diffusivity $D=\kappa/C$ (cm^2/sec). The initial temperature distribution along the length of the wire is given by $T(x,t=0)$, where $0\le x\le L$. Explain how $T(x,t=0)$ should be expanded in a Fourier series to determine $T(x,t)$ for all times $t\ge 0$ under the following circumstances:

a) The end points of the wire are kept at zero temperature at all times; that is,

$$T(x=0,t)=0\quad\text{and}\quad T(x=L,t)=0.$$

b) The end points of the wire are isolated from the environment so that no heat can enter or exit the wire through these points; in other words,

$$\partial T(x=0,t)/\partial x=0\quad\text{and}\quad\partial T(x=L,t)/\partial x=0.$$

Problem 8) In the cylindrical coordinates system (r, φ, z), the (scalar) wave equation in three-dimensional space is written as follows:

$$v^2\nabla^2\psi(\boldsymbol{r}, t) = v^2\left(\frac{\partial^2\psi}{\partial r^2} + \frac{\partial\psi}{r\partial r} + \frac{\partial^2\psi}{r^2\partial\varphi^2} + \frac{\partial^2\psi}{\partial z^2}\right) = \frac{\partial^2\psi}{\partial t^2} + \gamma\frac{\partial\psi}{\partial t}.$$

Here, ψ represents the vibration amplitude at location $\boldsymbol{r} = (r, \varphi, z)$ in space and instant t in time, v is the velocity of wave propagation in space, and γ is the damping coefficient. In the depicted system, the wave is confined to the interior of a hollow cylindrical can of radius R and length L, where the boundary conditions are given by

$$\psi(r = R, \varphi, z, t) = 0; \qquad \psi(r, \varphi, z = 0, t) = 0; \qquad \psi(r, \varphi, z = L, t) = 0.$$

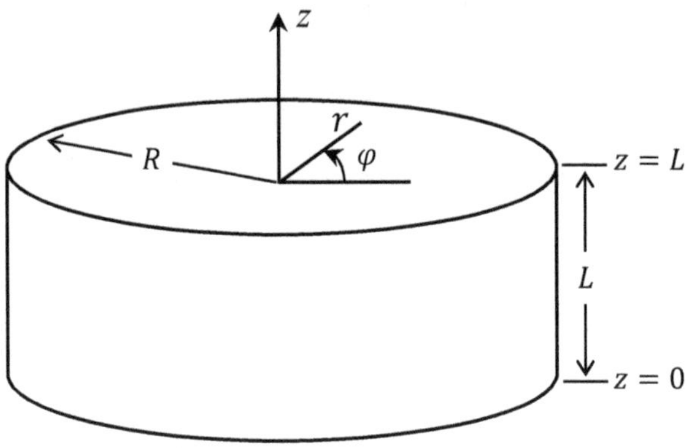

Use the method of separation of variables to determine the general form of the *real-valued* wave function $\psi(\boldsymbol{r}, t)$ inside the can. As usual, you may use ρ_{mn} to represent the nth zero of $J_m(\rho)$, a Bessel function of the first kind, order m. You may also use the positive integer ℓ to represent the mode index for the z-dependent part of the wave function. As for the time-dependent part, allow for the possibility of overdamping, critical damping, and underdamping.

Problem 9) A semi-infinite rod of uniform cross-sectional area A is centered on the positive x-axis, its surface insulated from the environment. The homogeneous material of the rod has thermal diffusivity D [cm^2/sec]. The initial temperature distribution is $T(x, t = 0) = f(x),\ 0 \le x < \infty$.

a) Assuming that the temperature at the end $x = 0$ is changed to zero and maintained at zero thereafter, that is, $T(x = 0, t) = 0,\ t \ge 0$, show that the temperature profile for $t > 0$ is given by

$$T(x, t) = \frac{1}{\sqrt{4\pi Dt}}\int_0^\infty f(x')\left\{\exp\left[-\frac{(x - x')^2}{4Dt}\right] - \exp\left[-\frac{(x + x')^2}{4Dt}\right]\right\}\mathrm{d}x'.$$

When the initial temperature profile is uniform (i.e., $T(x, t = 0) = f(x) = T_0,\ 0 < x < \infty$), show that $T(x, t) = T_0\,\mathrm{erf}(x/\sqrt{4Dt})$, where the so-called *error function* is defined as $\mathrm{erf}(x) = (2/\sqrt{\pi})\int_0^x \exp(-y^2)\,\mathrm{d}y$.

b) Assuming that the end $x = 0$ of the rod is insulated, that is, $\partial T(x, t)/\partial x|_{x=0} = 0,\ t \ge 0$, show that the temperature profile for $t > 0$ is given by

$$T(x, t) = \frac{1}{\sqrt{4\pi Dt}}\int_0^\infty f(x')\left\{\exp\left[-\frac{(x - x')^2}{4Dt}\right] + \exp\left[-\frac{(x + x')^2}{4Dt}\right]\right\}\mathrm{d}x'.$$

Problem 10) An infinitely long, thin, straight, uniform rod has an initial temperature distribution $T(x, t = 0) = T_0(x)$. In this problem, you are asked to use the method of separation of variables to determine the temperature profile $T(x, t)$ for all times $t \geq 0$ in order to confirm the solution to the same problem that was obtained in section 7 using Fourier transformation.

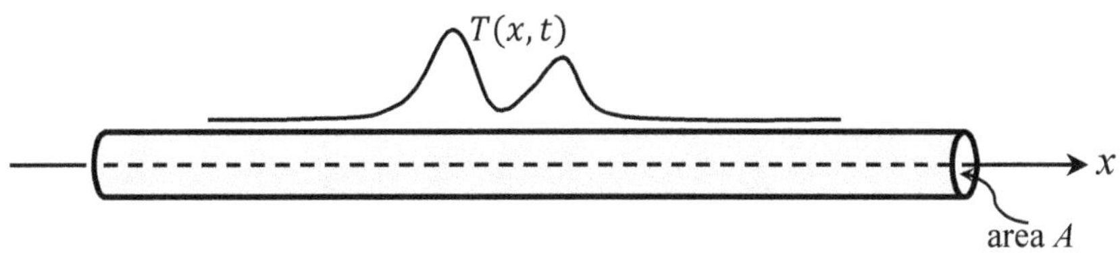

a) Write the one-dimensional heat diffusion equation for $T(x, t)$ assuming no heat loss to the environment. Identify the material parameter(s) that appear in the equation by their names and their units.

b) Use the method of separation of variables to solve the above diffusion equation and obtain the general form of its separable solutions. Explain why it is necessary for the separation constant to be negative.

c) Express the complete solution to the problem as a superposition of the various separable solutions obtained in part (b). Show that your solution is identical to the one obtained in section 7 by Fourier transformation.

Problem 11) Consider the following diffusion equation in one-dimensional space:

$$\alpha \frac{\partial^2}{\partial x^2} f(x,t) = \frac{\partial}{\partial t} f(x,t); \qquad \alpha > 0.$$

The initial condition is given as $f(x, t = 0) = \frac{\mathrm{d}}{\mathrm{d}x} \exp(-\pi x^2) = -2\pi x \exp(-\pi x^2)$. No boundary conditions are specified as the function $f(x, t)$ is allowed to spread along the x-axis with no restrictions other than $f(x, t) \to 0$ as $x \to \pm\infty$. In this problem, you are asked to solve the above partial differential equation and find $f(x, t)$ for $t \geq 0$.

a) Defining the Fourier transform $F(s,t) = \int_{-\infty}^{\infty} f(x,t) \exp(-\mathrm{i}2\pi s x)\, \mathrm{d}x$, write the corresponding partial differential equation for $F(s, t)$. (Note that the transformation is with respect to x, but *not* with respect to t.)

b) Find the transform of the initial condition, namely,

$$F(s, t = 0) = \int_{-\infty}^{\infty} f(x, t = 0) \exp(-\mathrm{i}2\pi s x)\, \mathrm{d}x.$$

c) Solve the partial differential equation obtained in part (a) to find $F(s, t)$ using the function $F(s, t = 0)$ as your initial condition.

d) Perform an inverse Fourier transformation to arrive at $f(x, t)$ for $t \geq 0$.

Problem 12) A homogeneous medium is specified by its specific heat C, thermal conductivity κ, and thermal diffusivity D. The temperature distribution in this medium in a cylindrical coordinate system is denoted by the function $T(r, \varphi, z, t)$.

a) Derive the general form of the differential equation that governs the temperature distribution in space and its evolution in time. Show all the steps, starting with a physical explanation of the procedure and ending with the complete diffusion equation in cylindrical coordinates.

b) Use the method of separation of variables to find the general solution for $T(r, \varphi, z, t)$.

c) Explain how the various unknown parameters can be obtained with the help of the boundary and initial conditions.

Problem 13) Let the temperature at five facets of a rectangular parallelepiped be maintained at T_0, as shown on the right. The sixth facet is maintained at the prescribed temperature distribution

$$T(x, y, z = L_Z) = T_0 + T_1(x, y)$$

until steady state conditions are attained. Determine the resultant temperature profile in the interior of the parallelepiped assuming that its constituent material is homogeneous.

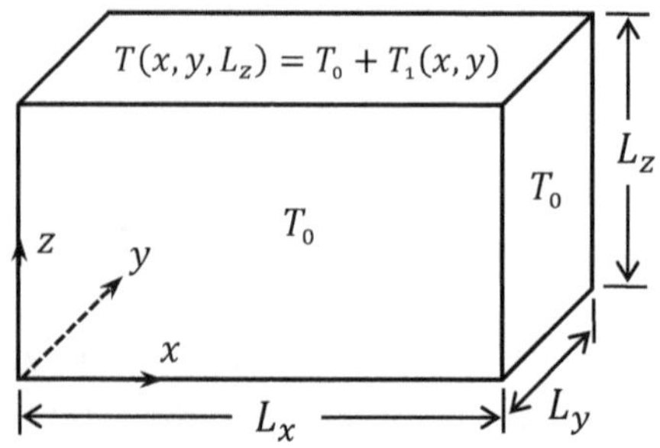

Problem 14) A thin strip of a homogeneous material is located in the xy-plane, in the region $0 \le x \le L$ and $0 \le y < \infty$. The slab is isolated from the environment on all sides, except at its lower edge along the x-axis, where the temperature is maintained at $T(x, y = 0) = T_0(x)$ over the entire interval $0 \le x \le L$. Considering that the steady state temperature distribution satisfies the $2D$ Laplace's equation

$$\left(\frac{\partial^2}{\partial x^2} + \frac{\partial^2}{\partial y^2}\right) T(x, y) = 0,$$

determine the temperature profile $T(x, y)$ of the slab.

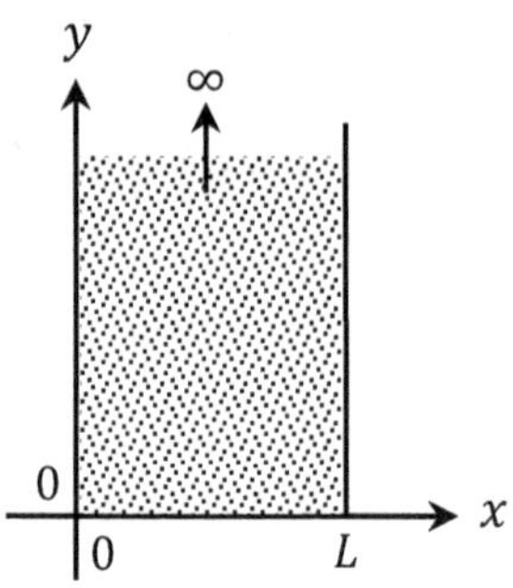

Problem 15) Consider a solid, homogeneous cylinder of radius R, length L, specific heat C, thermal conductivity κ, and diffusivity D. The cylinder is held at the ambient temperature T_0 on the sides, where $r = R$, and at the bottom, where $z = -L$. A thin layer on the upper surface of the cylinder is heated to above the ambient temperature, acquiring a circularly symmetric initial temperature profile $T(r, \phi, z, t = 0) = T_0 + f(r)\delta(z)$. The top facet is isolated from the environment so that no heat can enter or exit the cylinder from this surface. Describe the possible modes of temperature distribution within the cylinder and discuss their time dependence.

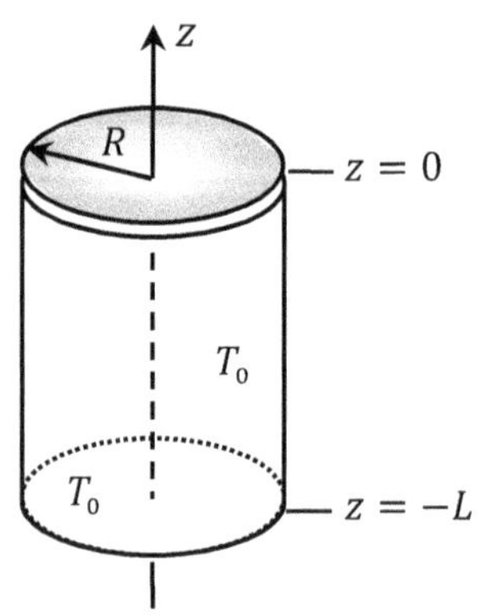

Problem 16) The rim of a thin, circular disk of radius R is kept at a constant (i.e., time-independent) temperature $T(r = R, \varphi) = f(\varphi)$.

a) Use separation of variables to express the steady state temperature distribution throughout the disk in terms of the Fourier series coefficients of $f(\varphi)$.

b) Show that the result obtained in part (a) can be simplified to yield the following remarkable integral (due to Poisson):

$$T(r, \varphi) = \frac{1}{2\pi} \int_0^{2\pi} \frac{R^2 - r^2}{R^2 - 2Rr\cos(\varphi - \theta) + r^2} f(\theta) \mathrm{d}\theta.$$

Problem 17) A thin, solid disk of radius R and thermal diffusivity D [cm^2/sec] has an initial temperature distribution $T(r,\phi,t=0)=T_0+f(r)\cos\phi$. Here, T_0 is the constant ambient temperature; $f(r)$, a function of the radial coordinate r, is specified in the interval $0\le r\le R$; and the azimuthal angle ϕ covers the entire available range from 0 to 2π. The boundary of the disk at $r=R$ is kept at the constant ambient temperature at all times $t\ge 0$, so that $T(r=R,\phi,t)=T_0$. (The disk is sufficiently thin, so that its temperature profile through the thickness may be assumed to be uniform.) Obtain the solution to the two-dimensional heat diffusion equation $D\,\nabla^2T(r,\phi,t)=\partial T(r,\phi,t)/\partial t$ for $t\ge 0$ using the method of separation of variables.

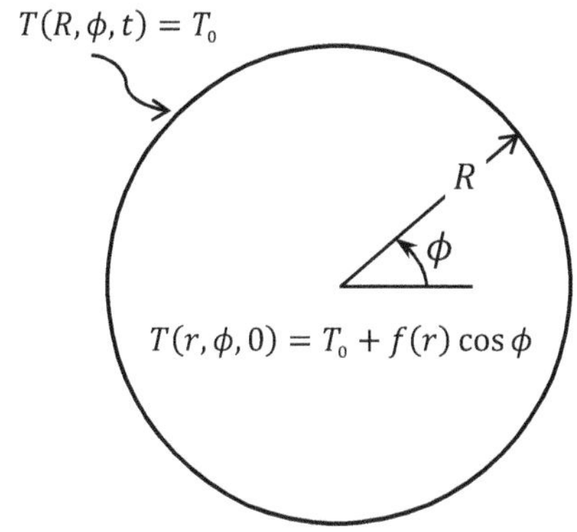

Hint: The Laplacian operator in cylindrical coordinates is $\nabla^2=\dfrac{\partial^2}{\partial r^2}+\dfrac{1}{r}\dfrac{\partial}{\partial r}+\dfrac{1}{r^2}\dfrac{\partial^2}{\partial\phi^2}+\dfrac{\partial^2}{\partial z^2}$.

Problem 18) A large, solid, homogeneous cylinder of height L and radius R sits atop the xy-plane and is centered on the z-axis. (For all practical purposes you may assume that R and L approach infinity.) The cylinder is isolated on all sides except on its bottom facet, where the temperature is maintained at $T(r,\varphi,z=0)=T_0(r)$. The steady state temperature distribution, being independent of the azimuthal angle φ, satisfies the two-dimensional Laplace equation in cylindrical coordinates, namely,

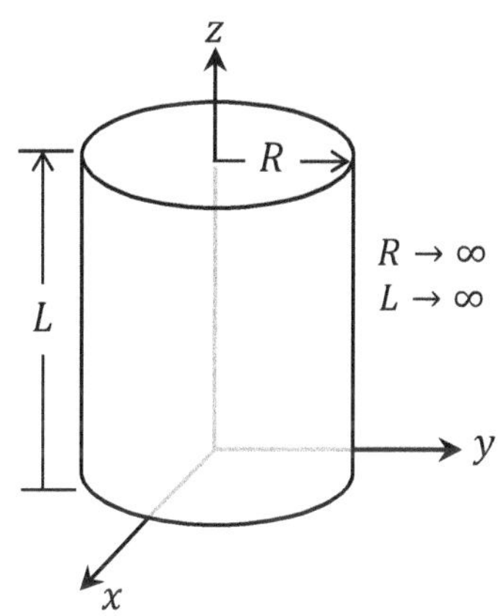

$$\left(\frac{\partial^2}{\partial r^2}+\frac{1}{r}\frac{\partial}{\partial r}+\frac{\partial^2}{\partial z^2}\right)T(r,z)=0.$$

Find the steady state temperature profile $T(r,z)$ of the cylinder.

Problem 19) Use the method of characteristics to solve the linear first-order partial differential equation $(\partial_x+t\partial_t)f(x,t)=0$ for $t>0$, subject to the initial condition $f(x,t=1)=f_1(x)$.

Problem 20) Use the method of characteristics to solve the first-order nonlinear partial differential equation $[f(x,t)\partial_x+\partial_t]f(x,t)=0$, subject to the initial condition that $f(x,t=0)=f_0(x)$. Explain how the solution fails when a pair of characteristic base curves cross each other, or, equivalently, when the local slope $\partial_x f(x,t)$ of the solution goes to infinity at some point x at some instant of time t. Examine the general solution thus obtained in the following special cases:

(i) $f_0(x)=-x$; (ii) $f_0(x)=\sin x$; (iii) $f_0(x)=1-x^2$.

Chapter 10

Special Functions

"Our job in physics is to see things simply, to understand a great many complicated phenomena in a unified way, in terms of a few simple principles."

Steven Weinberg (1933–)

"Discovery is the privilege of the child: the child who has no fear of being once again wrong, of looking like an idiot, of not being serious, of not doing things like everyone else."

Alexander Grothendieck (1928–2014)

10.1. Introduction. The present chapter is devoted to a discussion of the general properties of a number of special functions invented during the last several centuries by the likes of Leonhard Euler (1707–1783), Friedrich Wilhelm Bessel (1784–1846), George Biddell Airy (1801–1892), Carl Gustav Jacob Jacobi (1804–1851), Peter Gustav Lejeune Dirichlet (1805–1859), and Georg Friedrich Bernhard Riemann (1826–1866). We have already encountered some of these special functions in the preceding chapters. However, here we examine them from a broader perspective and explore their properties in greater detail.

The chapter begins in section 2 with a general description of the Sturm-Liouville theory, where ordinary differential equations belonging to a certain class are shown to have solutions that give rise to various sets of orthogonal functions. Arbitrary functions can thus be expanded in infinite series based on these orthogonal functions, in much the same way as sines and cosines (or, more generally, complex exponentials) are routinely used in the Fourier series expansion of general-purpose functions. In section 3 we introduce the Airy functions as solutions of Airy's differential equation and proceed to discuss their application in the study of caustics, a feature exhibited by electromagnetic fields under certain circumstances, which leads to stunning visual effects in optics. Euler's Gamma function and its analytic continuation to the entire complex plane are described in section 4, where some important characteristics of this magnificent function will be derived. The subject of section 5 is Jacobi's theta function, which has found applications in several branches of mathematical sciences.

Leonhard Euler

Finally, in section 6 we embark on a discussion of Riemann's zeta function $\zeta(s)$ and its analytic continuation to the entire complex plane, followed by an exploration of some of its intriguing properties. Riemann's zeta function is intimately connected with the distribution of prime numbers across the range of ordinary integers. A major unsolved problem in modern mathematics is the Riemann hypothesis, dating back to 1859, which speculates that all nontrivial zeros of $\zeta(s)$ lie along the straight line $\text{Re}(s) = ½$ in the complex s-plane. (The trivial zeros of the zeta function all lie on the real axis at $s = -2, -4, -6, \cdots$.) Riemann's hypothesis was included in the influential list of twenty three unsolved problems announced by the preeminent German mathematician David Hilbert at a Paris conference at the dawn of the twentieth century. Toward the end of his life, Hilbert declared: "If I were to awaken after having slept for a thousand years, my first question would be: Has the Riemann hypothesis been proven?" Although many of the problems on Hilbert's

Bernhard Riemann

original list have been solved in the intervening years, Riemann's hypothesis continues to challenge the best and the brightest mathematical minds of the world.

10.2. Orthogonal functions and the Sturm-Liouville theory. In this section, we consider the following second-order, linear, homogeneous, ordinary differential equation:

$$\frac{\mathrm{d}}{\mathrm{d}x}\left[p(x)\frac{\mathrm{d}}{\mathrm{d}x}f(x)\right] + [q(x) + \lambda r(x)]f(x) = 0. \tag{1}$$

We are interested in the behavior of the real-valued solutions $f(x)$ of this equation when $p(x)$, $q(x)$, and $r(x)$ are themselves real functions of x, defined over an interval $[a, b]$ on the x-axis, and also when λ is a real-valued parameter. The boundary conditions at $x = a$ and $x = b$ could be a mixture of the conditions listed below:

$$f(a) = 0; \qquad f(b) = 0; \qquad f'(a) = 0; \qquad f'(b) = 0;$$
$$f(a) + \gamma f'(a) = 0; \qquad f(b) + \gamma f'(b) = 0. \tag{2}$$

Depending on the behavior of $p(x)$ at and near the boundaries at $x = a$ and $x = b$, some of these boundary conditions may be relaxed. Many problems of interest in physical sciences can be formulated in terms of the above differential equation.

Example 1. $p(x) = 1$, $q(x) = 0$, $r(x) = 1$ leads to $f''(x) + \lambda f(x) = 0$, a second-order ordinary differential equation whose solutions are $f_1(x) = \sin(\sqrt{\lambda}x)$, $f_2(x) = \cos(\sqrt{\lambda}x)$.

Example 2. $p(x) = x$, $q(x) = -m^2/x$, $r(x) = x$ leads to the Bessel equation $xf''(x) + f'(x) + [\lambda x - (m^2/x)]f(x) = 0$, whose solutions are $J_m(\sqrt{\lambda}x)$ and $Y_m(\sqrt{\lambda}x)$, which are known, respectively, as Bessel functions of the first and second kinds, order m.

Example 3. $p(x) = 1 - x^2$, $q(x) = 0$, $r(x) = 1$, $\lambda = \nu(\nu + 1)$ leads to the Legendre equation, $(1 - x^2)f''(x) - 2xf'(x) + \nu(\nu + 1)f(x) = 0$, of which the Legendre polynomial of order ν, namely $P_\nu(x)$, is a solution when ν happens to be an integer.

The problem defined by the differential equation and its boundary conditions given in Eqs.(1) and (2) will have solutions $\phi_n(x)$ for discrete values of λ denoted by λ_n. Each such $\phi_n(x)$ satisfies the boundary conditions at $x = a$ and $x = b$. The discrete values of λ, arranged in ascending order and denoted by $\lambda_1, \lambda_2, \cdots, \lambda_n, \cdots$, are the characteristic values of the equation—also known as its eigenvalues. The corresponding solutions $\phi_1(x), \phi_2(x), \cdots, \phi_n(x), \cdots$ are known as the characteristic functions or eigenfunctions.

10.2.1. The Sturm-Liouville theorem. When $\lambda_m \neq \lambda_n$, the functions $\phi_m(x)$ and $\phi_n(x)$ will be mutually orthogonal, with the weighting function $r(x)$, over the $[a, b]$ interval. To prove the theorem, we write Eq.(1) for the two solutions $\phi_m(x)$ and $\phi_n(x)$, as follows:

$$\frac{\mathrm{d}}{\mathrm{d}x}\left[p(x)\frac{\mathrm{d}}{\mathrm{d}x}\phi_m(x)\right] + [q(x) + \lambda_m r(x)]\phi_m(x) = 0. \tag{3a}$$

$$\frac{\mathrm{d}}{\mathrm{d}x}\left[p(x)\frac{\mathrm{d}}{\mathrm{d}x}\phi_n(x)\right] + [q(x) + \lambda_n r(x)]\phi_n(x) = 0. \tag{3b}$$

Multiplying the first equation by $\phi_n(x)$ and the second one by $\phi_m(x)$, subtracting one equation from the other, then integrating from $x = a$ to $x = b$, we find

$$\int_a^b \{\phi_n(x)[p(x)\phi_m'(x)]' - \phi_m(x)[p(x)\phi_n'(x)]'\}\mathrm{d}x$$
$$+(\lambda_m - \lambda_n)\int_a^b r(x)\phi_m(x)\phi_n(x)\mathrm{d}x = 0. \tag{4}$$

Applying the method of integration by parts to the left-hand side of Eq.(4) now yields

$$p(x)[\phi_n(x)\phi_m'(x) - \phi_m(x)\phi_n'(x)]\Big|_a^b - \int_a^b \phi_n'(x)p(x)\phi_m'(x)\mathrm{d}x$$
$$+ \int_a^b \phi_m'(x)p(x)\phi_n'(x)\mathrm{d}x = (\lambda_n - \lambda_m)\int_a^b r(x)\phi_m(x)\phi_n(x)\mathrm{d}x. \quad (5)$$

The left-hand side of this equation vanishes under several circumstances, some of which are listed below:

i) $\phi(a) = \phi(b) = 0$.

ii) $\phi(a) = \phi'(b) = 0$.

iii) $p(a) = 0$, and $\phi(a)$ and $\phi'(a)$ finite, and $\phi(b) = 0$.

iv) $\phi'(a) = \phi'(b) = 0$.

v) Periodic boundary conditions: $\phi(a) = \phi(b)$ and $\phi'(a) = \phi'(b)$, and $p(a) = p(b)$.

Another interesting boundary condition is $\phi(a) + \gamma\phi'(a) = 0$ and $\phi(b) + \gamma\phi'(b) = 0$, where γ is an arbitrary parameter. In this case also, the left-hand side of Eq.(5) vanishes, as follows:

$$p(x)\{\phi_n(x)[\phi_m(x) + \gamma\phi_m'(x)] - \phi_m(x)[\phi_n(x) + \gamma\phi_n'(x)]\}\Big|_a^b = 0. \quad (6)$$

With the left-hand side of Eq.(5) going to zero under any one of the circumstances listed above, and with $\lambda_m \neq \lambda_n$, the integral appearing on the right-hand side of Eq.(5) must vanish; that is,

$$\int_a^b r(x)\phi_m(x)\phi_n(x)\mathrm{d}x = 0 \qquad (m \neq n). \quad (7)$$

Equation (7) states that the eigenfunctions $\phi_m(x)$ and $\phi_n(x)$ of the Sturm-Liouville equation are mutually orthogonal, with the weighting function $r(x)$, over the interval $[a, b]$. Note that, aside from $\lambda_m \neq \lambda_n$, the requirement for orthogonality is that the boundary conditions at $x = a$ and $x = b$ ensure that the left-hand side of Eq.(5) vanishes.

10.2.2. Can an eigenvalue be associated with more than one eigenfunction? We would like to show that, under many circumstances, a given eigenvalue λ_n can be associated with only one eigenfunction $\phi_n(x)$. Let us assume that $\phi_1(x)$ and $\phi_2(x)$ are different eigenfunctions associated with the same eigenvalue λ. We will have

$$\frac{\mathrm{d}}{\mathrm{d}x}\left[p(x)\frac{\mathrm{d}}{\mathrm{d}x}\phi_1(x)\right] + [q(x) + \lambda r(x)]\phi_1(x) = 0. \quad (8a)$$

$$\frac{\mathrm{d}}{\mathrm{d}x}\left[p(x)\frac{\mathrm{d}}{\mathrm{d}x}\phi_2(x)\right] + [q(x) + \lambda r(x)]\phi_2(x) = 0. \quad (8b)$$

If we now multiply the first of these equations by $\phi_2(x)$ and the second one by $\phi_1(x)$, then subtract one equation from the other, we will arrive at

$$\phi_2(x)[p(x)\phi_1'(x)]' - \phi_1(x)[p(x)\phi_2'(x)]' = 0. \quad (9)$$

Carrying out the differentiations in this equation now leads to

$$p'(x)[\phi_2(x)\phi_1'(x) - \phi_1(x)\phi_2'(x)] + p(x)[\phi_2(x)\phi_1''(x) - \phi_1(x)\phi_2''(x)]$$
$$= \{p(x)[\phi_2(x)\phi_1'(x) - \phi_1(x)\phi_2'(x)]\}' = 0.$$
$$\rightarrow \; p(x)[\phi_2(x)\phi_1'(x) - \phi_1(x)\phi_2'(x)] = \text{constant}. \quad (10)$$

Suppose the boundary conditions are such that, at least at one end of the $[a,b]$ interval, the function $p(x)[\phi_2(x)\phi_1'(x) - \phi_1(x)\phi_2'(x)]$ equals zero. The function must then be zero throughout the entire interval; that is,

$$\phi_2(x)\phi_1'(x) - \phi_1(x)\phi_2'(x) = 0 \quad \rightarrow \quad \phi_1'(x)/\phi_1(x) = \phi_2'(x)/\phi_2(x)$$

$$\rightarrow \ln \phi_1(x) = \ln \phi_2(x) + c \quad \rightarrow \quad \phi_1(x)/\phi_2(x) = \text{constant}. \tag{11}$$

Clearly, $\phi_1(x)$ and $\phi_2(x)$ are one and the same solution of the Sturm-Liouville equation.

10.2.3. Real-valued eigenvalues and eigenfunctions. In this section, we show that, under certain circumstances, the eigenvalues and eigenfunctions of the Sturm-Liouville equation are real. Suppose λ_n and its corresponding $\phi_n(x)$ are complex. Then, knowing that $p(x)$, $q(x)$, and $r(x)$ are real-valued, we can conjugate the Sturm-Liouville equation $[p(x)\phi_n'(x)]' + [q(x) + \lambda_n r(x)]\phi_n(x) = 0$, which would indicate that λ_n^* and $\phi_n^*(x)$ form another eigenvalue and eigenfunction pair. The procedure used to prove the Sturm-Liouville theorem in section 10.2.1 may now be repeated to yield

$$(\lambda_n - \lambda_n^*) \int_a^b r(x)\phi_n(x)\phi_n^*(x)\mathrm{d}x = 0. \tag{12}$$

Suppose now that $r(x) > 0$ over the (a,b) interval. Then, the integral appearing in Eq.(12) will be positive and, therefore, $\lambda_n = \lambda_n^*$, which indicates that λ_n is real-valued.

If we now allow the corresponding eigenfunction to be complex-valued; that is, $\phi_n(x) = u_n(x) + \mathrm{i}v_n(x)$, then, since λ_n is real, we will end up having both $u_n(x)$ and $v_n(x)$ as eigenfunctions associated with λ_n. However, as shown in section 10.2.2, each λ_n can have only one eigenfunction. Consequently, $u_n(x)$ must be the same function as $v_n(x)$, meaning that $\phi_n(x) = (1+\mathrm{i})u_n(x)$, with the coefficient $(1+\mathrm{i})$ simply being a complex constant. Thus, aside from multiplication by complex coefficients, when $r(x) > 0$ over the (a,b) interval, the eigenfunctions of the Sturm-Liouville problem are real-valued.

10.2.4. Positive eigenvalues. When $p(x) > 0$, $q(x) < 0$, and $r(x) > 0$, in the presence of certain other restrictions, the Sturm-Liouville eigenvalues will be positive. We know from the preceding section that the eigenvalues λ are real. If we now multiply both sides of the Sturm-Liouville equation by $\phi_n(x)$ and integrate over x from a to b, we obtain

$$\int_a^b \phi_n(x)[p(x)\phi_n'(x)]'\mathrm{d}x + \int_a^b [q(x) + \lambda_n r(x)]\phi_n^2(x)\mathrm{d}x = 0$$

$$\rightarrow \lambda_n \underbrace{\int_a^b r(x)\phi_n^2(x)\mathrm{d}x}_{\text{positive}} = \overbrace{p(a)\phi_n(a)\phi_n'(a) - p(b)\phi_n(b)\phi_n'(b)}^{\text{Usually zero, e.g., when } \phi = 0 \text{ or } \phi' = 0 \text{ at either boundary.}}$$

$$+ \underbrace{\int_a^b p(x)[\phi_n'(x)]^2\mathrm{d}x}_{\text{positive}} - \underbrace{\int_a^b q(x)\phi_n^2(x)\mathrm{d}x}_{\text{positive}}. \tag{13}$$

It is clear that, under many circumstances, the above equation will be satisfied if and only if the eigenvalue λ_n is positive. If the boundary conditions happen to be of the type $\phi(x) + \gamma\phi'(x) = 0$, there will be residual terms at the $x = a$ and $x = b$ boundaries, so that the positivity of the right-hand side of Eq.(13) can be guaranteed if $\gamma_b p(b)[\phi'(b)]^2$ turns out to be greater than or equal to $\gamma_a p(a)[\phi'(b)]^2$. It is seen that a variety of boundary conditions exist under which the positivity of λ_n can be assured.

10.2.5. Expanding an arbitrary function into an infinite series of $\phi_n(x)$. Consider an arbitrary function $f(x)$ defined over the interval $[a, b]$. Expressing this function as a sum over the eigenfunctions $\phi_n(x)$ of a Sturm-Liouville equation, and denoting the constant coefficients of the expansion by c_n, one may write

$$f(x) = \sum_{n=1}^{\infty} c_n \phi_n(x). \tag{14}$$

If we now multiply both sides of this equation by $r(x)\phi_m(x)$, with m being an arbitrary index, then integrate over x from a to b, we will have

$$\int_a^b f(x)r(x)\phi_m(x)\mathrm{d}x = \sum_{n=1}^{\infty} c_n \int_a^b r(x)\phi_m(x)\phi_n(x)\mathrm{d}x \underset{\uparrow}{=} c_m \int_a^b r(x)[\phi_m(x)]^2\mathrm{d}x.$$

$$\boxed{\text{orthogonality of } \phi_m(x) \text{ and } \phi_n(x) \text{ when } m \neq n} \tag{15}$$

The coefficient c_m of the expansion given in Eq.(14) is thus found to be

$$c_m = \int_a^b f(x)r(x)\phi_m(x)\mathrm{d}x \Big/ \int_a^b r(x)[\phi_m(x)]^2\mathrm{d}x. \tag{16}$$

Such expansions are generally found to converge, provided that $f(x)$ is sufficiently well behaved. Thus, the eigenfunctions $\phi_n(x)$ of the Sturm-Liouville equation form a complete set of orthogonal functions, with the weighting function $r(x)$, over the interval $[a, b]$. The weighted orthogonality is defined as $\int_a^b r(x)\phi_m(x)\phi_n(x)\mathrm{d}x = 0$ when $m \neq n$.

Example 1. We consider Bessel's equation of order m over the interval $[0, L]$. The Bessel function of second kind, order m, namely $Y_m(\sqrt{\lambda}x)$, is not a viable solution, because it diverges when $x \to 0$. The only acceptable solution is thus a Bessel function of the first kind, order m, namely $J_m(\sqrt{\lambda}x)$. Considering that the Strum-Liouville equation in this case has $p(x) = x$, $q(x) = -m^2/x$, and $r(x) = x$, we note that $p(0) = 0$ renders the boundary condition at $x = 0$ essentially irrelevant. The only requirement at $x = 0$ is for the solution $f(x)$ to remain finite, and also for its derivative at $x = 0$ to behave properly. By dropping the solution $Y_m(\sqrt{\lambda}x)$, we are assured of the proper behavior of the remaining solution $J_m(\sqrt{\lambda}x)$ at and around the boundary at $x = 0$.

At the opposite boundary, $x = L$, let the boundary condition be $f(x)|_{x=L} = 0$. Setting $J_m(\sqrt{\lambda}L) = 0$ yields $\sqrt{\lambda}L = x_{mn}$, where x_{mn} is the nth zero of $J_m(x)$. Consequently, $\lambda_n = (x_{mn}/L)^2$. The solution of the equation is, therefore, written as $J_m(x_{mn}x/L)$. We may now expand an arbitrary function $h(x)$ over the interval $[0, L]$ as follows:

$$h(x) = \sum_{n=1}^{\infty} c_n J_m(x_{mn}x/L), \qquad c_n = \frac{\int_0^L h(x)\overset{\swarrow r(x)}{x} J_m(x_{mn}x/L)\mathrm{d}x}{\int_0^L x J_m^2(x_{mn}x/L)\mathrm{d}x}. \tag{17}$$

A similar approach would work if the boundary condition at $x = L$ happened to be $f'(x)|_{x=L} = 0$, where x_{mn} would correspond to the successive peaks and valleys of $J_m(x)$.

Example 2. We examine the following inhomogeneous differential equation:

$$\frac{\mathrm{d}}{\mathrm{d}x}\left[p(x)\frac{\mathrm{d}}{\mathrm{d}x}f(x)\right] + [q(x) + \Lambda r(x)]f(x) = h(x). \tag{18}$$

For reasons that will soon become clear, one must assume that $\Lambda \neq \lambda_1, \lambda_2, \cdots, \lambda_n, \cdots$. Let $h(x)/r(x) = \sum_{n=1}^{\infty} a_n\phi_n(x)$, where $\phi_n(x)$ are the eigenfunctions associated with the eigenvalues λ_n of the corresponding homogeneous equation. Denoting a particular solution of Eq.(18) by $f(x) = \sum_{n=1}^{\infty} c_n\phi_n(x)$, substitution into the equation yields

$$\textstyle\sum_{n=1}^{\infty} c_n[p(x)\phi_n'(x)]' + \sum_{n=1}^{\infty} c_n[q(x) + \Lambda r(x)]\phi_n(x) = \sum_{n=1}^{\infty} a_n\phi_n(x)r(x). \quad (19)$$

If we now replace Λ with $(\Lambda - \lambda_n) + \lambda_n$, then rearrange the terms of Eq.(19), we find

$$\textstyle\sum_{n=1}^{\infty} c_n\{\overbrace{[p(x)\phi_n'(x)]' + [q(x) + \lambda_n r(x)]\phi_n(x)}^{\text{zero}}\}$$
$$\textstyle + \sum_{n=1}^{\infty} c_n(\Lambda - \lambda_n)r(x)\phi_n(x) = \sum_{n=1}^{\infty} a_n\phi_n(x)\, r(x). \quad (20)$$

Consequently, $c_n = a_n/(\Lambda - \lambda_n)$, with the caveat that $\Lambda \neq \lambda_1, \lambda_2, \cdots, \lambda_n, \cdots$. Note that the boundary conditions satisfied by the particular solution $f(x)$ are the same as those that were imposed on the Sturm-Liouville problem at the outset. These, of course, are the boundary conditions that have led to the determination of the eigenvalues λ_n and eigenfunctions $\phi_n(x)$.

As an example of situations where an inhomogeneous equation may arise in practice, consider the problem of a vibrating string discussed in chapter 9, section 2. Introducing an external z-directed force $f_z(x)\cos(\omega t)$, which is distributed along the length of the string, and in the absence of damping (i.e., when $\gamma = 0$), we will have

f_o is tensile force → $$f_o\partial_x^2 z(x,t) + f_z(x)\cos(\omega t) = \rho\partial_t^2 z(x,t). \quad (21)$$ ← ρ is mass density

Assuming that $z(x,t) = z(x)\cos(\omega t)$, this partial differential equation reduces to

$$(f_o/\rho)z''(x) + \omega^2 z(x) = -(1/\rho)f_z(x). \quad (22)$$

As expected, the presence of the external force makes the governing equation inhomogeneous. The role of Λ in Eq.(18) is now played by ω^2 in Eq.(22). Clearly, the external excitation frequency ω must differ from all the natural vibrational frequencies of the string, as the solution will blow up if ω approaches a natural frequency of the system.

10.3. Caustics and the Airy function. The Airy function, which is a solution of the second-order ordinary differential equation $f''(z) - zf(z) = 0$, is related to Bessel functions of order $\pm\tfrac{1}{3}$. To see this, consider the new function $g(z)$, which is related to $f(z)$ via the transformation $f(z) = z^{\frac{1}{2}}g(\tfrac{2}{3}\mathrm{i}z^{3/2})$. Since the derivative of the new variable $u(z) = \tfrac{2}{3}\mathrm{i}z^{3/2}$ is $u'(z) = \mathrm{i}z^{\frac{1}{2}}$, one could write

$$f(z) = z^{\frac{1}{2}}g(u), \quad (23a)$$

$$f'(z) = \tfrac{1}{2}z^{-\frac{1}{2}}g(u) + \mathrm{i}zg'(u), \quad (23b)$$

$$f''(z) = -\tfrac{1}{4}z^{-3/2}\, g(u) + \tfrac{1}{2}\mathrm{i}g'(u) + \mathrm{i}g'(u) - z^{3/2}g''(u). \quad (23c)$$

Consequently, recalling that $z^{3/2} = -3\mathrm{i}u(z)/2$, we will have

$$f''(z) - zf(z) = -\frac{\mathrm{i}}{6u}g(u) + \frac{3\mathrm{i}}{2}g'(u) + \frac{3\mathrm{i}}{2}ug''(u) + \frac{3\mathrm{i}}{2}ug(u)$$
$$= \frac{3\mathrm{i}}{2u}\left[u^2 g''(u) + ug'(u) + \left(u^2 - \frac{1}{3^2}\right)g(u)\right] = 0. \quad (24)$$

Ignoring the $3\mathrm{i}/2u$ pre-factor, Eq.(24) is readily recognized as the Bessel equation with the order parameter $\nu = \tfrac{1}{3}$. For $z > 0$, the solutions of the Airy equation are $f(z) = z^{\frac{1}{2}}J_{\pm\frac{1}{3}}(\tfrac{2}{3}\mathrm{i}z^{3/2})$, which may be expressed in terms of the modified Bessel functions $I_{\pm\frac{1}{3}}$. For $z < 0$, the solutions are $f(z) = \mathrm{i}|z|^{\frac{1}{2}}J_{\pm\frac{1}{3}}(\tfrac{2}{3}|z|^{3/2})$, which are ordinary Bessel functions of the first kind, order $\pm\tfrac{1}{3}$. The two solutions $\mathrm{Ai}(z)$ and $\mathrm{Bi}(z)$

of the Airy equation, commonly known as Airy functions of the first and second kind, are usually written as the sum and difference of the above Bessel functions, as follows:

$$\mathrm{Ai}(z) = \begin{cases} \sqrt{|z|/9}\,\left[J_{-1/3}\left(\tfrac{2}{3}|z|^{3/2}\right) + J_{1/3}\left(\tfrac{2}{3}|z|^{3/2}\right)\right], & z < 0 \\ \sqrt{z/9}\,\left[I_{-1/3}\left(\tfrac{2}{3}z^{3/2}\right) - I_{1/3}\left(\tfrac{2}{3}z^{3/2}\right)\right], & z > 0. \end{cases} \tag{25}$$

$$\mathrm{Bi}(z) = \begin{cases} \sqrt{|z|/3}\,\left[J_{-1/3}\left(\tfrac{2}{3}|z|^{3/2}\right) - J_{1/3}\left(\tfrac{2}{3}|z|^{3/2}\right)\right], & z < 0 \\ \sqrt{z/3}\,\left[I_{-1/3}\left(\tfrac{2}{3}z^{3/2}\right) + I_{1/3}\left(\tfrac{2}{3}z^{3/2}\right)\right], & z > 0. \end{cases} \tag{26}$$

Plots of the Airy functions $\mathrm{Ai}(x)$ and $\mathrm{Bi}(x)$ versus the real variable x are shown in figure 10.1. For a solution of Airy's equation using the method of Frobenius, see chapter 8, problem 20. A straightforward derivation of the Fourier transform of an Airy function of the first kind, namely, $F(s) = \mathcal{F}\{\mathrm{Ai}(x)\} = \exp[\tfrac{1}{3}\mathrm{i}(2\pi s)^3]$, is outlined in problem 2 at the end of the present chapter.

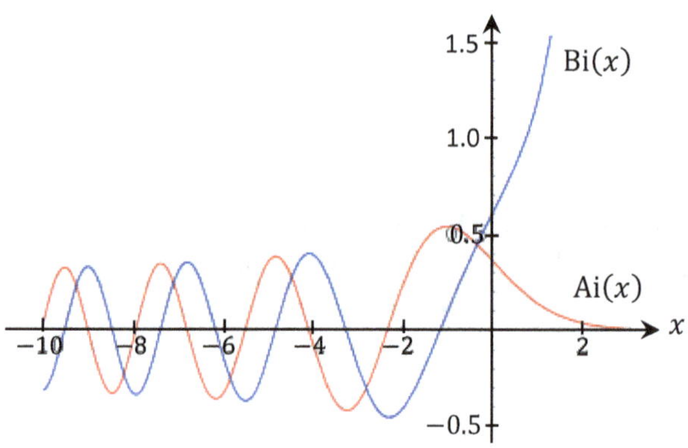

Figure 10.1. Plots of the Airy functions $\mathrm{Ai}(x)$ and $\mathrm{Bi}(x)$ versus the real variable x.

As a simple application of the Airy functions, we consider the appearance of caustics in a graded-index medium. (In geometric optics, a caustic is the envelope of rays that are reflected or refracted at a curved surface.) Recall that Maxwell's equations of classical electrodynamics in the presence of sources (i.e., $\rho_{\text{free}}, \boldsymbol{J}_{\text{free}}, \boldsymbol{P}, \boldsymbol{M}$) are written as

$$\boldsymbol{\nabla} \cdot \boldsymbol{D} = \rho_{\text{free}}, \tag{27}$$

$$\boldsymbol{\nabla} \times \boldsymbol{H} = \boldsymbol{J}_{\text{free}} + \partial_t \boldsymbol{D}, \tag{28}$$

$$\boldsymbol{\nabla} \times \boldsymbol{E} = -\partial_t \boldsymbol{B}, \tag{29}$$

$$\boldsymbol{\nabla} \cdot \boldsymbol{B} = 0. \tag{30}$$

In these equations, written in the international system of units (SI), $\boldsymbol{E}(\boldsymbol{r},t)$ and $\boldsymbol{H}(\boldsymbol{r},t)$ are the electric and magnetic fields, respectively, $\boldsymbol{D}(\boldsymbol{r},t) = \varepsilon_0 \boldsymbol{E}(\boldsymbol{r},t) + \boldsymbol{P}(\boldsymbol{r},t)$ is the displacement field, and $\boldsymbol{B}(\boldsymbol{r},t) = \mu_0 \boldsymbol{H}(\boldsymbol{r},t) + \boldsymbol{M}(\boldsymbol{r},t)$ is the magnetic induction field. The free charge density is $\rho_{\text{free}}(\boldsymbol{r},t)$, the free current density is $\boldsymbol{J}_{\text{free}}(\boldsymbol{r},t)$, the density of electric dipoles is the polarization $\boldsymbol{P}(\boldsymbol{r},t)$, and the density of magnetic dipoles is the magnetization $\boldsymbol{M}(\boldsymbol{r},t)$. Also appearing in these equations are the permittivity ε_0 and permeability μ_0 of free space. In terms of μ_0 and ε_0, the speed of light in vacuum is $c = (\mu_0\varepsilon_0)^{-1/2}$, and the impedance of free space is $Z_0 = (\mu_0/\varepsilon_0)^{1/2}$.

In linear, isotropic, nonmagnetic media, $\boldsymbol{D}(\boldsymbol{r},t)=\varepsilon_0\varepsilon(\boldsymbol{r})\boldsymbol{E}(\boldsymbol{r},t)$ and $\boldsymbol{B}(\boldsymbol{r},t)=\mu_0\boldsymbol{H}(\boldsymbol{r},t)$. Application of the curl operator to both sides of Eq.(29) yields

$$\boldsymbol{\nabla}\times(\boldsymbol{\nabla}\times\boldsymbol{E})=-\partial_t(\boldsymbol{\nabla}\times\boldsymbol{B})$$

$$\rightarrow\quad \boldsymbol{\nabla}(\boldsymbol{\nabla}\cdot\boldsymbol{E})-\boldsymbol{\nabla}^2\boldsymbol{E}=-\partial_t[\mu_0\boldsymbol{J}_{\text{free}}+\mu_0\varepsilon_0\varepsilon(\boldsymbol{r})\partial_t\boldsymbol{E}]. \tag{31}$$

In addition, from Eq.(27) we have

$$\varepsilon_0\boldsymbol{\nabla}\cdot[\varepsilon(\boldsymbol{r})\boldsymbol{E}]=\varepsilon_0\boldsymbol{E}\cdot\boldsymbol{\nabla}\varepsilon+\varepsilon_0\varepsilon(\boldsymbol{r})\boldsymbol{\nabla}\cdot\boldsymbol{E}=\rho_{\text{free}}. \tag{32}$$

Therefore, in a region of space where $\rho_{\text{free}}=0$, $\boldsymbol{J}_{\text{free}}=0$, and $\boldsymbol{E}(\boldsymbol{r},t)\perp\boldsymbol{\nabla}\varepsilon(\boldsymbol{r})$, we may write

$$\boldsymbol{\nabla}^2\boldsymbol{E}(\boldsymbol{r},t)-\mu_0\varepsilon_0\varepsilon(\boldsymbol{r})\partial_t^2\boldsymbol{E}(\boldsymbol{r},t)=0. \tag{33}$$

In a one-dimensional graded-index medium, where $\varepsilon(\boldsymbol{r})=\varepsilon(z)$, the condition $\boldsymbol{E}(\boldsymbol{r},t)\perp\boldsymbol{\nabla}\varepsilon(\boldsymbol{r})$ is satisfied for transverse electric (TE) waves (i.e., when $\boldsymbol{E}(\boldsymbol{r},t)=E_y(x,z,t)\hat{\boldsymbol{y}}$). This is a sufficient condition for the applicability of Eq.(33) to TE waves in such media. For a monochromatic wave of frequency ω, the fields are $\boldsymbol{E}(\boldsymbol{r})\exp(-i\omega t)$ and $\boldsymbol{H}(\boldsymbol{r})\exp(-i\omega t)$, and the wave equation becomes

$$\boldsymbol{\nabla}^2\boldsymbol{E}(\boldsymbol{r})+(\omega/c)^2\varepsilon(\boldsymbol{r})\boldsymbol{E}(\boldsymbol{r})=0. \tag{34}$$

Returning now to Maxwell's equations and applying the curl operator to Eq.(28), we find

$$\boldsymbol{\nabla}\times(\boldsymbol{\nabla}\times\boldsymbol{B})=\mu_0\boldsymbol{\nabla}\times\boldsymbol{J}_{\text{free}}+\mu_0\varepsilon_0\partial_t(\boldsymbol{\nabla}\times\varepsilon\boldsymbol{E})$$

$$\rightarrow\quad \boldsymbol{\nabla}(\boldsymbol{\nabla}\cdot\boldsymbol{B})-\boldsymbol{\nabla}^2\boldsymbol{B}=\mu_0\boldsymbol{\nabla}\times\boldsymbol{J}_{\text{free}}+\mu_0\varepsilon_0\partial_t(\boldsymbol{\nabla}\varepsilon\times\boldsymbol{E}+\varepsilon\boldsymbol{\nabla}\times\boldsymbol{E}). \tag{35}$$

Assuming $\boldsymbol{E}(\boldsymbol{r},t)\parallel\boldsymbol{\nabla}\varepsilon(\boldsymbol{r})$ and invoking Eqs.(29) and (30), the above equation is written

$$\boldsymbol{\nabla}^2\boldsymbol{B}-\mu_0\varepsilon_0\varepsilon(\boldsymbol{r})\partial_t^2\boldsymbol{B}=-\mu_0\boldsymbol{\nabla}\times\boldsymbol{J}_{\text{free}}. \tag{36}$$

For a monochromatic field in the absence of free currents, Eq.(36) reduces to the homogeneous Helmholtz equation, namely,

$$\boldsymbol{\nabla}^2\boldsymbol{B}(\boldsymbol{r})+(\omega/c)^2\varepsilon(\boldsymbol{r})\boldsymbol{B}(\boldsymbol{r})=0. \tag{37}$$

Note, however, that in a one-dimensional graded-index medium, where $\varepsilon(\boldsymbol{r})=\varepsilon(z)$, neither TE nor TM waves satisfy the condition $\boldsymbol{E}(\boldsymbol{r},t)\parallel\boldsymbol{\nabla}\varepsilon(\boldsymbol{r})$. Under such circumstances, the magnetic field would *not* be a solution of the Helmholtz equation.

Similar arguments apply to the vector and scalar potential fields $\boldsymbol{A}(\boldsymbol{r},t)$ and $\psi(\boldsymbol{r},t)$. From Maxwell's Eq.(30) we infer that $\boldsymbol{B}=\boldsymbol{\nabla}\times\boldsymbol{A}$, and from Eq.(29), $\boldsymbol{E}=-\boldsymbol{\nabla}\psi-\partial_t\boldsymbol{A}$. Substitution into Eq.(28) then yields

$$\boldsymbol{\nabla}\times(\boldsymbol{\nabla}\times\boldsymbol{A})=\mu_0\boldsymbol{J}_{\text{free}}+\mu_0\varepsilon_0\varepsilon(\boldsymbol{r})\partial_t(-\boldsymbol{\nabla}\psi-\partial_t\boldsymbol{A})$$

$$\rightarrow\quad \boldsymbol{\nabla}(\boldsymbol{\nabla}\cdot\boldsymbol{A})-\boldsymbol{\nabla}^2\boldsymbol{A}=\mu_0\boldsymbol{J}_{\text{free}}-\mu_0\varepsilon_0\varepsilon(\boldsymbol{r})\boldsymbol{\nabla}(\partial_t\psi)-\mu_0\varepsilon_0\varepsilon(\boldsymbol{r})\partial_t^2\boldsymbol{A}$$

$$\rightarrow\ \boldsymbol{\nabla}^2\boldsymbol{A}-\mu_0\varepsilon_0\varepsilon(\boldsymbol{r})\partial_t^2\boldsymbol{A}-\boldsymbol{\nabla}[\boldsymbol{\nabla}\cdot\boldsymbol{A}+\mu_0\varepsilon_0\varepsilon(\boldsymbol{r})\partial_t\psi]+\mu_0\varepsilon_0(\boldsymbol{\nabla}\varepsilon)\partial_t\psi=-\mu_0\boldsymbol{J}_{\text{free}}. \tag{38}$$

Working in the Lorenz gauge, we set $\boldsymbol{\nabla}\cdot\boldsymbol{A}+[\varepsilon(\boldsymbol{r})/c^2]\partial_t\psi=0$. However, on the left-hand side of Eq.(38), the term containing $\boldsymbol{\nabla}\varepsilon$ survives, thus spoiling the Helmholtz equation.

Similarly, substitution into Eq.(27) for the E-field in terms of the potentials $\boldsymbol{A}(\boldsymbol{r},t)$ and $\psi(\boldsymbol{r},t)$, followed by an invocation of the aforementioned Lorenz gauge, yields

$$\boldsymbol{\nabla} \cdot [\varepsilon_0 \varepsilon(\boldsymbol{r})(-\boldsymbol{\nabla}\psi - \partial_t \boldsymbol{A})] = \rho_{\text{free}}$$

$$\rightarrow \ \varepsilon(\boldsymbol{r}) \boldsymbol{\nabla} \cdot (\boldsymbol{\nabla}\psi + \partial_t \boldsymbol{A}) + (\boldsymbol{\nabla}\psi + \partial_t \boldsymbol{A}) \cdot \boldsymbol{\nabla}\varepsilon = -\rho_{\text{free}}/\varepsilon_0$$

$$\rightarrow \quad \boldsymbol{\nabla}^2\psi - \mu_0\varepsilon_0\varepsilon(\boldsymbol{r})\partial_t^2\psi + (\boldsymbol{\nabla}\psi + \partial_t \boldsymbol{A}) \cdot [(\boldsymbol{\nabla}\varepsilon)/\varepsilon(\boldsymbol{r})] = -\rho_{\text{free}}/[\varepsilon_0\varepsilon(\boldsymbol{r})]. \tag{39}$$

Once again, the term containing $\boldsymbol{\nabla}\varepsilon$ on the left-hand side of Eq.(39) spoils the Helmholtz equation, unless the *E*-field happens to be orthogonal to $\boldsymbol{\nabla}\varepsilon$, as would be the case for a TE-polarized wave in a one-dimensional graded-index medium. Note, however, that the scalar potential by itself is insufficient to determine the *E* and *B* fields. Nevertheless, ψ can be input to Eq.(38) to yield $\boldsymbol{A}(\boldsymbol{r}, t)$ as the solution to this inhomogeneous Helmholtz equation. All in all, resorting to electromagnetic potentials does not seem to provide any advantages in the context of graded-index media.

To solve Eq.(33) for a one-dimensional graded-index medium defined by $\varepsilon(\boldsymbol{r}) = (1 + \gamma z)n_0^2$, we substitute the TE-polarized wave $\boldsymbol{E}(\boldsymbol{r}, t) = E_y(z)\exp[\mathrm{i}(k_x x - \omega t)]\,\hat{\boldsymbol{y}}$ into the equation. We find

$$E_y'' + [(n_0\omega/c)^2(1 + \gamma z) - k_x^2]E_y = 0. \tag{40}$$

Defining $\mathbb{E}_y(k_z) = \int_{-\infty}^{\infty} E_y(z)\exp(-\mathrm{i}k_z z)\,\mathrm{d}z$ as the Fourier transform of $E_y(z)$, the above equation may be Fourier transformed, as follows:

$$[(n_0\omega/c)^2 - k_x^2 - k_z^2]\mathbb{E}_y + \mathrm{i}(n_0\omega/c)^2\gamma\mathbb{E}_y' = 0. \tag{41}$$

The solution of Eq.(41) is seen by inspection to be

$$\mathbb{E}_y(k_z) = \exp\{\mathrm{i}\gamma^{-1}[1 - (ck_x/n_0\omega)^2]k_z - \tfrac{1}{3}\mathrm{i}\gamma^{-1}(c/n_0\omega)^2 k_z^3\}. \tag{42}$$

Since the Fourier transform of the Airy function $\mathrm{Ai}(z)$ is known to be $\exp(\tfrac{1}{3}\mathrm{i}k^3)$ (see Problem 2 at the end of the chapter), the latter part of the function appearing in Eq.(42) is the Fourier transform of $\mathrm{Ai}[-\gamma^{1/3}(n_0\omega/c)^{2/3}z]$. The linear phase-factor that forms the first part of $\mathbb{E}_y(k_z)$ in Eq.(42) shifts the inverse transform to $z_0 = \gamma^{-1}[(ck_x/n_0\omega)^2 - 1]$. Consequently,

$$E_y(z) = \mathrm{Ai}[-\gamma^{1/3}(n_0\omega/c)^{2/3}(z - z_0)]. \tag{43}$$

Note that $k_x = (1 + \gamma z_0)^{1/2}(n_0\omega/c)$ indicates that the light rays at $z = z_0$ are parallel to the *x*-axis.

10.4. Euler's Gamma function. We encountered Euler's Gamma function for the first time in chapter 3, section 3. Here, we delve deeper, and discuss the intriguingly rich properties of this function in greater detail. The Gamma function is defined in the right half of the complex *z*-plane as follows:

$$\Gamma(z) = \int_0^{\infty} \exp(-t)\,t^{z-1}\mathrm{d}t, \qquad \mathrm{Re}(z) > 0. \tag{44}$$

Application of the method of integration by parts to Eq.(44) yields the fundamental functional relation $\Gamma(z + 1) = z\Gamma(z)$, which can subsequently be used to extend $\Gamma(z)$ to the left half of the *z*-plane. Within the strip $-1 < \mathrm{Re}(z) \leq 0$, for example, $\Gamma(z)$ is simply defined as $z^{-1}\Gamma(z + 1)$. By repeating this process of analytic continuation (one strip at a time), $\Gamma(z)$ can be defined over the entire *z*-plane. The function thus obtained is analytic everywhere except at $z = 0, -1, -2, -3, \cdots$, which constitute simple poles of the function. Figure 10.2 shows a hand-drawn graph of $|\Gamma(z)|$, the absolute value of the complex Gamma function, in the complex-plane $z = x + \mathrm{i}y$.

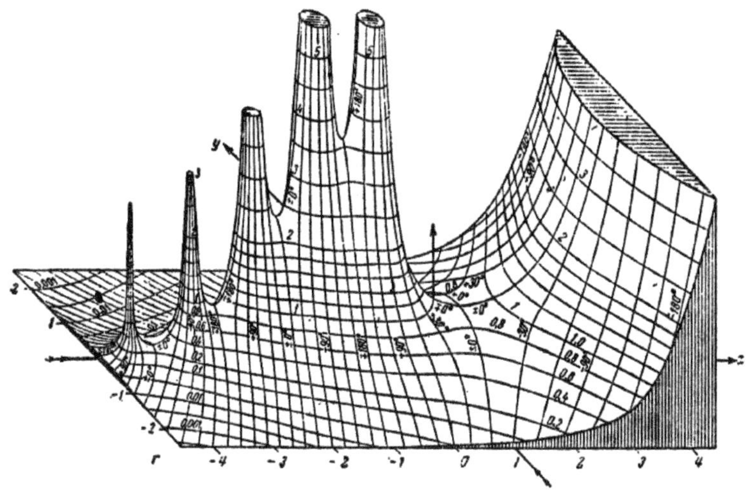

Figure 10.2. Graph of $|\Gamma(z)|$ in the complex-plane $z = x + \mathrm{i}y$. From E. Jahnke and F. Emde, *Tables of Higher Functions*, McGraw-Hill, New York, 1960 (1st edition published 1909).

An alternative (though equivalent) approach to analytic continuation involves the changing of the integration variable in Eq.(44) from t to $x = \ln t$, which results in

$$\Gamma(z) = \int_{-\infty}^{\infty} \exp[-\exp(x)] \exp(zx)\, \mathrm{d}x, \qquad \mathrm{Re}(z) > 0. \tag{45}$$

We now change the domain of integration to the complex ζ-plane, evaluating the integral on two straight lines parallel to the x-axis, namely the lines $x \pm \mathrm{i}\pi$, as shown in figure 10.3. We will have

$$\Gamma(z) = \exp(+\mathrm{i}\pi z) \int_{\text{lower line}} \exp[\exp(\zeta)] \exp(z\zeta)\, \mathrm{d}\zeta, \tag{46a}$$

$$\Gamma(z) = \exp(-\mathrm{i}\pi z) \int_{\text{upper line}} \exp[\exp(\zeta)] \exp(z\zeta)\, \mathrm{d}\zeta. \tag{46b}$$

When $\mathrm{Re}(z) > 0$, the integrand in Eq.(46) vanishes on the vertical line segment descending from $-\infty + \mathrm{i}\pi$ to $-\infty - \mathrm{i}\pi$ (see the left-hand side of figure 10.3). We may thus close the integration path on the left-hand side, then rearrange Eqs.(46) to arrive at

$$-2\mathrm{i} \sin(\pi z)\, \Gamma(z) = \int_{\Gamma} \exp[\exp(\zeta)] \exp(z\zeta)\, \mathrm{d}\zeta. \tag{47}$$

Equation (47) now yields the values of $\Gamma(z)$ over the entire z-plane (except, of course, at $z = 0, \pm 1, \pm 2, \cdots$, where $\sin(\pi z) = 0$). This is because, on the far left side of the contour in figure 10.3 (say, from ζ_1 to ζ_2), we will have $\exp[\exp(\zeta)] \cong 1$, yielding

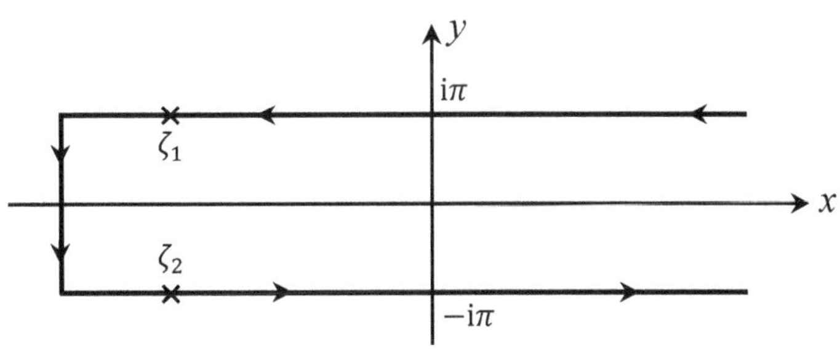

Figure 10.3. The integration contour in the complex plane $\zeta = x + \mathrm{i}y$ consists of two infinitely long straight lines, parallel to the x-axis located at $y = \pm\pi$, and one short, vertical segment at $x = -\infty$, running from $y = \pi$ to $-\pi$.

$$\int_{\zeta_1}^{\zeta_2} \exp[\exp(\zeta)] \exp(z\zeta)\, \mathrm{d}\zeta \cong z^{-1}[\exp(z\zeta_2) - \exp(z\zeta_1)], \tag{48}$$

which has a finite value, whether z resides in the right or left half of the complex plane. In Eq.(47) we now have a formula for evaluating $\Gamma(z)$ over the entire z-plane – aside from the ambiguities at $z = 0, \pm1, \pm2, \cdots$. This is the analytic extension to the entire z-plane of the function initially defined on the right half of the complex plane via Eq.(44).

10.4.1. The reflection formula. Euler's reflection formula, $\Gamma(z)\Gamma(1-z) = \pi/\sin(\pi z)$, is valid throughout the complex z-plane. We demonstrate its validity at first in the strip $0 < \mathrm{Re}(z) < 1$, where both $\Gamma(z)$ and $\Gamma(1-z)$ are expressed as Euler integrals; that is,

$$\begin{aligned}\Gamma(z)\Gamma(1-z) &= \left[\int_0^\infty \exp(-x)\, x^{z-1}\mathrm{d}x\right]\left[\int_0^\infty \exp(-y)\, y^{-z}\mathrm{d}y\right]\\ &= \iint_0^\infty \exp[-(x+y)]\, (x/y)^z x^{-1}\mathrm{d}x\mathrm{d}y.\end{aligned} \tag{49}$$

Let us switch to new independent variables $u = x + y$ and $v = x/y$, where u and v cover the first quadrant of the uv-plane whenever x and y cover the first quadrant of the xy-plane. Inverting these formulas, we find $x = uv/(1+v)$ and $y = u/(1+v)$. The Jacobian of this coordinate transformation is readily found to be

$$\begin{Vmatrix} \partial x/\partial u & \partial y/\partial u \\ \partial x/\partial v & \partial y/\partial v \end{Vmatrix} = \begin{Vmatrix} v/(1+v) & 1/(1+v) \\ u/(1+v)^2 & -u/(1+v)^2 \end{Vmatrix} = u/(1+v)^2. \tag{50}$$

Substitution into Eq.(49) now yields

G&R **3.194-4**

$$\Gamma(z)\Gamma(1-z) = \iint_0^\infty \frac{\exp(-u)v^z}{v(1+v)}\mathrm{d}u\mathrm{d}v = \int_0^\infty \exp(-u)\,\mathrm{d}u \int_0^\infty \frac{v^{z-1}}{1+v}\mathrm{d}v = \frac{\pi}{\sin(\pi z)}. \tag{51}$$

The range of validity of the last integral in Eq.(51) is $0 < \mathrm{Re}(z) < 1$, consistent with that of Eq.(49). The reflection formula can now be extended to the rest of the z-plane. Invoking the functional relations $\Gamma(z+1) = z\Gamma(z)$ and $\Gamma(1-z) = -z\Gamma(-z)$, together with the identity $\sin[\pi(z+1)] = -\sin(\pi z)$, Eq.(51) yields

$$\Gamma(z+1)\Gamma(-z) = \pi/\sin[\pi(z+1)]. \tag{52}$$

Euler's reflection formula is thus seen to be valid within the strip $-1 < \mathrm{Re}(z) < 2$. Repeated application of the functional relation ensures the validity of the formula throughout the z-plane.

10.4.2. The infinite product formula. For a positive integer N and an arbitrary complex parameter z, the analytic extension of the function $f(n) = n!$ may be written as follows:

$$\Gamma(z+1) = lim_{N\to\infty} \frac{1\cdot2\cdot3\cdots(N-1)\cdot N}{(1+z)(2+z)(3+z)\cdots(N-1+z)(N+z)}(N+1)^z. \tag{53}$$

This is because, for an arbitrary integer $n \geq 0$, this expression yields

$$\Gamma(n+1) = (n!)\, lim_{N\to\infty} \frac{(N+1)^n}{(N+1)(N+2)\cdots(N+n)} = n!. \tag{54}$$

We write the factor $(N+1)^z$ appearing in Eq.(53) as the following product of N terms:

$$(N+1)^z = \frac{2^z\cdot3^z\cdot4^z\cdots(N+1)^z}{1^z\cdot2^z\cdot3^z\cdots N^z}. \tag{55}$$

Substitution of Eq.(55) into Eq.(53) now enables one to write Eq.(53) as an infinite product, namely,

$$\Gamma(z+1) = \prod_{N=1}^{\infty} \left[\frac{N}{N+z} \cdot \frac{(N+1)^z}{N^z}\right]. \tag{56}$$

Straightforward algebraic manipulations finally yield

G&R **8.322** → $$\Gamma(z+1) = \prod_{N=1}^{\infty} \left(1+\frac{z}{N}\right)^{-1} \left(1+\frac{1}{N}\right)^{z}; \quad (\mathrm{Re}(z) > 0). \tag{57}$$

Note that the product function in Eq.(57) has poles at $z = -1, -2, -3, \cdots$, which are the well-known poles of $\Gamma(z+1)$. As an example of application of the product formula in Eq.(57), consider the product of $\Gamma(z+1)$ and $\Gamma(1-z)$, ignoring for the moment the possibility that the domains of convergence of the two functions may not overlap. We will have

$$\Gamma(z+1)\Gamma(1-z) = \prod_{N=1}^{\infty} \left(1+\frac{z}{N}\right)^{-1} \left(1+\frac{1}{N}\right)^{z} \left(1-\frac{z}{N}\right)^{-1} \left(1+\frac{1}{N}\right)^{-z}$$

$$= \prod_{N=1}^{\infty} [1-(z^2/N^2)]^{-1} = \frac{\pi z}{\sin(\pi z)}. \tag{58}$$

In arriving at Eq.(58), we have used the Euler expansion of $\sin(\pi z)/(\pi z)$; see chapter 4, section 8, Eq.(21). Using the functional relation $\Gamma(z+1) = z\Gamma(z)$, Eq.(58) may now be rewritten as

$$\Gamma(z)\Gamma(1-z)\sin(\pi z) = \pi. \tag{59}$$

This, of course, is Euler's reflection formula of Eq.(52).†

10.4.3. Legendre's duplication formula. In the present section, we prove Legendre's duplication formula $2\sqrt{\pi}\,\Gamma(2z) = 2^{2z}\Gamma(z)\Gamma(z+½)$, using another one of Euler's inventions, namely, the so-called beta function. Consider the product of $\Gamma(z)$ and $\Gamma(z')$, where $\mathrm{Re}(z) > 0$ and $\mathrm{Re}(z') > 0$; that is,

$$\Gamma(z)\Gamma(z') = \left[\int_0^{\infty} \exp(-x)\, x^{z-1} \mathrm{d}x\right]\left[\int_0^{\infty} \exp(-y)\, y^{z'-1} \mathrm{d}y\right]$$

$$= \iint_0^{\infty} \exp[-(x+y)]\, x^{z-1} y^{z'-1} \mathrm{d}x\mathrm{d}y. \tag{60}$$

This integral is simplified on changing the variables from (x, y) to (u, v), with $x = uv$ and $y = u(1-v)$, where $u \in [0, \infty)$ and $v \in [0,1]$. The new variables are thus $u = x + y$ and $v = x/(x+y)$. The Jacobian of the transformation is readily found to be

$$\left\|\begin{matrix} \partial x/\partial u & \partial y/\partial u \\ \partial x/\partial v & \partial y/\partial v \end{matrix}\right\| = \left\|\begin{matrix} v & 1-v \\ u & -u \end{matrix}\right\| = u. \tag{61}$$

We will have

$$\Gamma(z)\Gamma(z') = \left[\int_0^{\infty} \exp(-u)\, u^{z+z'-1} \mathrm{d}u\right]\left[\int_0^{1} v^{z-1}(1-v)^{z'-1} \mathrm{d}v\right]$$

$$= \Gamma(z+z')B(z, z'). \tag{62}$$ ← Euler's beta function; $\mathrm{Re}(z) > 0$, $\mathrm{Re}(z') > 0$

---

† Also listed as identity **8.334**-3 in I. S. Gradshteyn and I. M. Ryzhik, *Table of Integrals, Series, and Products*, seventh edition, Academic Press, Burlington, Massachusetts (2007).

Now, the beta function has the following properties:

$$B(z,z) = \int_0^1 [v(1-v)]^{z-1} \mathrm{d}v = \frac{1}{2}\int_{-1}^{1} \left[\left(\frac{1+x}{2}\right)\left(\frac{1-x}{2}\right)\right]^{z-1} \mathrm{d}x = 2^{2(1-z)} \int_0^1 (1-x^2)^{z-1}\,\mathrm{d}x. \quad (63)$$

$$B(z,½) = \int_0^1 v^{z-1}(1-v)^{-½}\mathrm{d}v = 2\int_0^1 (1-x^2)^{z-1}\mathrm{d}x. \quad (64)$$

Consequently, $B(z,z) = 2^{1-2z}B(z,½)$. Returning to Eq.(62), we now write

$$\frac{\Gamma(z)\Gamma(z)}{\Gamma(2z)} = B(z,z) = 2^{1-2z}B(z,½) = 2^{1-2z}\frac{\Gamma(z)\Gamma(½)}{\Gamma(z+½)}$$

$$\rightarrow\ 2\sqrt{\pi}\,\Gamma(2z) = 2^{2z}\Gamma(z)\Gamma(z+½). \leftarrow \boxed{\Gamma(½) = \sqrt{\pi}} \quad (65)$$

This duplication formula is readily extended to the left half of the complex plane by repeated application of the functional relation $\Gamma(z+1) = z\Gamma(z)$. Multiplying both sides of Eq.(65) with $4z(z+½)$ yields

$$2\sqrt{\pi}(2z+1)(2z)\Gamma(2z) = 2^{2(z+1)}z\Gamma(z)(z+½)\Gamma(z+½)$$

$$\rightarrow\quad 2\sqrt{\pi}\,\Gamma[2(z+1)] = 2^{2(z+1)}\Gamma(z+1)\Gamma\left(z+\frac{3}{2}\right). \quad (66)$$

It is thus seen that Legendre's duplication formula is valid for $-1 < \mathrm{Re}(z) < 0$, since multiplication with $4z(z+½)$ has replaced z with $z+1$, which resides in the right half of the plane. Repeating this procedure confirms the validity of the duplication formula across the entire z-plane.

10.5. Jacobi's theta function. Jacobi's theta function is defined as follows:

$$\Theta(t) = \sum_{n=-\infty}^{\infty} \exp(-\pi n^2 t);\qquad t > 0. \quad (67)$$

An important property of the theta function is

$$\Theta(1/t) = \sqrt{t}\,\Theta(t). \quad (68)$$

To prove this identity, note that the Fourier transform of $f(x) = \exp(-\pi x^2)$ is $F(s) = \exp(-\pi s^2)$. Therefore, in accordance with the scaling theorem of Fourier transform theory, which states that the Fourier transform of $f(\alpha x)$ is equal to $|\alpha|^{-1}F(s/\alpha)$, we will have

$$\mathcal{F}\{\exp(-\pi t x^2)\} = \frac{1}{\sqrt{t}}\exp\left(-\frac{\pi s^2}{t}\right). \quad (69)$$

In this equation, $t > 0$ is an arbitrary real-valued constant. Now, multiplying a function $f(x)$ by the comb function, namely, $\mathrm{comb}(x) = \sum_{n=-\infty}^{\infty}\delta(x-n)$, results in the convolution of the Fourier transform $F(s)$ of $f(x)$ with the Fourier transform $\mathrm{comb}(s)$ of $\mathrm{comb}(x)$. Consequently,

$$\mathcal{F}\{\exp(-\pi t x^2)\mathrm{comb}(x)\} = \frac{1}{\sqrt{t}}\exp\left(-\frac{\pi s^2}{t}\right) * \mathrm{comb}(s). \quad (70)$$

Considering that the area under a function $f(x)$ is equal to the value of its Fourier transform $F(s)$ at $s = 0$, Eq.(70) yields

$$\sum_{n=-\infty}^{\infty}\exp(-\pi t n^2) = \frac{1}{\sqrt{t}}\sum_{n=-\infty}^{\infty}\exp\left(-\frac{\pi n^2}{t}\right). \quad (71)$$

The proof is thus complete.

The following important identity relating $\Theta(t)$, $\Theta'(t)$ and $\Theta'(1/t)$ is obtained upon differentiating Eq.(68):

$$\Theta(t) + 2t\Theta'(t) + 2t^{-3/2}\Theta'(1/t) = 0. \tag{72}$$

In particular, at $t = 1$, we find $\Theta(1) + 4\Theta'(1) = 0$.

10.6. Riemann's zeta function. Riemann's zeta function $\zeta(s) = \sum_{m=1}^{\infty} m^{-s}$ can be readily shown to exist when $\operatorname{Re}(s) > 1$. Analytic extension of the function to the rest of the complex s-plane begins by observing that

$$\begin{aligned}\int_0^\infty \exp(-mt)\, t^{n-1}\mathrm{d}t &= \overset{\nearrow 0}{-\frac{\exp(-mt)}{m} t^{n-1}\Big|_0^\infty} + \left(\frac{n-1}{m}\right)\int_0^\infty \exp(-mt)t^{n-2}\mathrm{d}t \\ &= \frac{1\cdot 2 \cdots (n-2)\cdot(n-1)}{m^{n-1}}\int_0^\infty \exp(-mt)\,\mathrm{d}t = \frac{(n-1)!}{m^n}; \qquad n \geq 1. \end{aligned} \tag{73}$$

Summing this equation over all positive integers m yields

$$\begin{aligned}\sum_{m=1}^{\infty}\int_0^\infty \exp(-mt)\, t^{n-1}\mathrm{d}t &= \int_0^\infty \Big[\sum_{m=1}^{\infty} \exp(-mt)\Big]t^{n-1}\mathrm{d}t \\ &= \int_0^\infty \frac{\exp(-t)}{1-\exp(-t)} t^{n-1}\mathrm{d}t = \int_0^\infty \frac{t^{n-1}}{\exp(t)-1}\mathrm{d}t \\ &= \sum_{m=1}^{\infty}(n-1)!/m^n = (n-1)!\,\zeta(n); \qquad n > 1. \end{aligned} \tag{74}$$

At this point, the zeta function $\zeta(s)$ is defined only for integers $s = n > 1$. However, the integral from $t = 0$ to ∞ of $t^{s-1}/[\exp(t) - 1]$ appearing in Eq.(74) exists for any complex-valued s whose real part is greater than 1.0. Similarly, $(n-1)!$ can be extended to the complex s-plane and replaced with $\Gamma(s)$. We thus have

Substitution: $t = \exp(x)$

$$\Gamma(s)\zeta(s) = \int_0^\infty \frac{t^{s-1}}{\exp(t)-1}\mathrm{d}t = \int_{-\infty}^{\infty} \frac{\exp(sx)}{\exp[\exp(x)]-1}\mathrm{d}x; \qquad \operatorname{Re}(s) > 1. \tag{75}$$

Note: When $x \to -\infty$, the integrand on the right-hand side of Eq.(75) may be written as follows:

$$\frac{\exp(sx)}{\exp[\exp(x)]-1} = \frac{\exp(sx)}{\cancel{1}+\exp(x)+\frac{1}{2!}\exp(2x)+\cdots-\cancel{1}} = \frac{\exp[(s-1)x]}{1+\frac{1}{2!}\exp(x)+\frac{1}{3!}\exp(2x)+\cdots} \to \exp[(s-1)x]. \tag{76}$$

The resulting expression approaches zero and is integrable, so long as $\operatorname{Re}(s) > 1$. At the other extreme, when $x \to \infty$, the denominator grows much faster than the numerator, irrespective of the value of s, making the function integrable again.

The preceding argument is deficient in that $\zeta(s)$ represented by the integral in Eq.(75) is proven to coincide with $\sum_{m=1}^{\infty} m^{-s}$ only at integer values of s greater than 1. Under the circumstances, the uniqueness of analytic continuation cannot be invoked in conjunction with $\zeta(s)$ of Eq.(75), as the overlap between its domain and that of $\sum_{m=1}^{\infty} m^{-s}$ is restricted to the *discrete* set of points $s = 2, 3, 4, \cdots$ in the complex s-plane. A more powerful argument would use a change of variable from mt to t', yielding $\int_0^\infty \exp(-mt)\, t^{s-1}\mathrm{d}t = m^{-s}\int_0^\infty \exp(-t')\, t'^{s-1}\mathrm{d}t' = m^{-s}\Gamma(s)$, which leads directly to Eq.(75), thus confirming that $\zeta(s)$ represented by the integral in Eq.(75) coincides with $\sum_{m=1}^{\infty} m^{-s}$ over the entire half plane where $\operatorname{Re}(s) > 1$.

Next, we endeavor to extend the range of s to the entire complex plane. To this end, consider, within the complex plane $z = x + \mathrm{i}y$, two straight lines parallel to the x-axis, having $y = \pm\pi$, as shown in figure 10.4. On each line, the integral from $-\infty \pm \mathrm{i}\pi$ to $\infty \pm \mathrm{i}\pi$ of the complex-valued function $\exp(sz)/\{\exp[-\exp(z)] - 1\}$ may be written as

$$\int_{-\infty\pm\mathrm{i}\pi}^{\infty\pm\mathrm{i}\pi} \frac{\exp(sz)}{\exp[-\exp(z)]-1}\mathrm{d}z = \int_{-\infty}^{\infty} \frac{\exp[s(x\pm\mathrm{i}\pi)]}{\exp[-\exp(x\pm\mathrm{i}\pi)]-1}\mathrm{d}x = \exp(\pm\mathrm{i}\pi s)\int_{-\infty}^{\infty} \frac{\exp(sx)}{\exp[\exp(x)]-1}\mathrm{d}x$$
$$= \exp(\pm\mathrm{i}\pi s)\,\Gamma(s)\zeta(s). \qquad (77)$$

In light of the preceding arguments, the integrand on the left-hand side of Eq.(77) is seen to be integrable when $z \to \infty \pm \mathrm{i}\pi$, irrespective of the value of $s = s' + \mathrm{i}s''$. At the other extreme, when $z \to -\infty \pm \mathrm{i}\pi$, the integrand is integrable, provided that $s' = \mathrm{Re}(s) > 1$. Connecting the two infinitely long horizontal lines via a short segment parallel to the y-axis between $-\infty \pm \mathrm{i}\pi$ will not affect the integral, as the integrand, namely $\exp(sz)/\{\exp[-\exp(z)] - 1\}$, is also vanishingly small on the vertical segment. We may thus write the (counterclockwise) integral over the three-legged contour depicted in figure 10.4 as follows:

$$\int_{\sqsubset\,\text{contour}} \frac{\exp(sz)}{\exp[-\exp(z)]-1}\mathrm{d}z = \exp(-\mathrm{i}\pi s)\,\Gamma(s)\zeta(s) - \exp(\mathrm{i}\pi s)\,\Gamma(s)\zeta(s)$$
$$= -2\mathrm{i}\sin(\pi s)\,\Gamma(s)\zeta(s); \qquad \mathrm{Re}(s) > 1. \qquad (78)$$

It is now interesting to note that, on closing the integration path at $x = -\infty$, it is no longer necessary to restrict $\mathrm{Re}(s)$ to values greater than 1.0. Given that $\sin(\pi s)$ and $\Gamma(s)$ are already defined over the entire s-plane (except at $s = 0, -1, -2, \cdots$, which will be addressed shortly), the function $\zeta(s)$ *everywhere* in the complex s-plane (aside from the singularity at $s = 1$) may now be defined by Eq.(78). Note that, on the extreme left-hand side of the contour near the vertical segment, the integrand is well approximated by the function $-\exp[(s-1)z]$, whose integral from $z_1 = x_o + \mathrm{i}\pi$ to $z_2 = x_o - \mathrm{i}\pi$, where x_o is a sufficiently large negative number, is given by

$$\frac{\exp[(s-1)(x_0+\mathrm{i}\pi)]-\exp[(s-1)(x_0-\mathrm{i}\pi)]}{s-1} = \frac{2\mathrm{i}\exp[(s-1)x_0]\sin[(s-1)\pi]}{s-1}. \qquad (79)$$

Thus, the divergent behavior, when $\mathrm{Re}(s) \le 1$, of each integral on the left-hand side of its corresponding horizontal path is brought under control when the two integration paths are conjoined via the short vertical segment. The ambiguity of $\zeta(s)$ at $s = 2, 3, 4, \cdots$

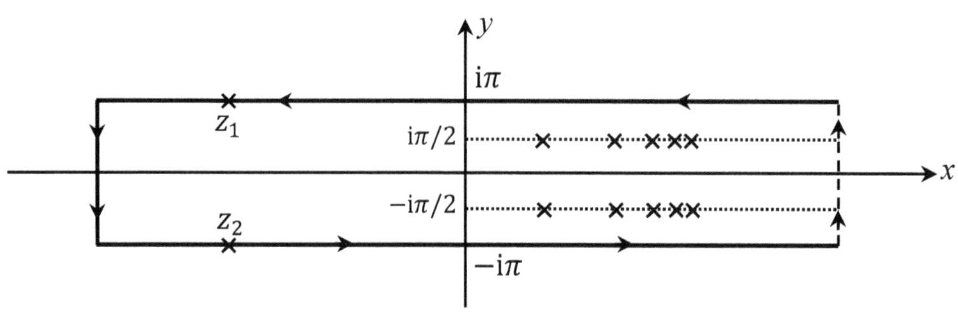

Figure 10.4. The integration contour in the complex plane $z = x + \mathrm{i}y$ consists of two infinitely long straight lines parallel to the x-axis, located at $y = \pm\pi$, and one short, vertical segment at $x = -\infty$, running from $y = \pi$ to $-\pi$. When $\mathrm{Re}(s) < 0$, it becomes possible to close the contour on the extreme right-hand side via a second short vertical segment at $x = \infty$, running from $y = -\pi$ to π. Also shown are the first few poles of the integrand appearing in Eq.(80), namely $z_n = \ln(2\pi|n|) \pm \mathrm{i}\pi/2$.

is readily resolved if we recognize that, for these values of s, Eq.(78) is identical to Eq.(75), which is unambiguous. To resolve the ambiguity of $\zeta(s)$ at $s = 0, -1, -2, \cdots$, we invoke the identity $\sin(\pi s)\,\Gamma(s)\Gamma(1-s) = \pi$, given in Eq.(59), to rewrite Eq.(78) as

$$\zeta(s) = -\frac{\Gamma(1-s)}{2\pi \mathrm{i}} \int_{\sqsubset \text{contour}} \frac{\exp(sz)}{\exp[-\exp(z)]-1} \mathrm{d}z; \qquad (s \neq 1, 2, 3, \cdots). \tag{80}$$

The poles of the integrand in Eq.(80) are solutions of the equation $\exp(z_n) = \mathrm{i}2\pi n$, where $n \neq 0$ is a positive or negative integer. The poles are thus located at $z_n = \ln(2\pi|n|) \pm \mathrm{i}\pi/2$, as shown in figure 10.4. (Similar rows of poles, obtained by adding and subtracting integer multiples of $2\pi\mathrm{i}$ to z_n, also exist elsewhere in the z-plane.)

When $\mathrm{Re}(s) < 0$, the integrand in Eq.(80) approaches zero on the vertical segment, shown as a dashed line on the right-hand side of the contour in figure 10.4. Under this circumstance, the integration path can be closed. The closed-loop integral is then evaluated by adding up the residues of the integrand at its poles z_n, followed by multiplication by $2\pi\mathrm{i}$. Considering that the derivative of the integrand's denominator, namely, $-\exp(z)\exp[-\exp(z)]$, equals $-\exp(z_n)$ at the poles, the residues of the integrand will be $-\exp[(s-1)z_n]$, yielding

$$\begin{aligned}\zeta(s) &= \Gamma(1-s)\textstyle\sum_n \exp[(s-1)z_n] \\ &= \Gamma(1-s)\{\exp[\tfrac{1}{2}\mathrm{i}(s-1)\pi] + \exp[-\tfrac{1}{2}\mathrm{i}(s-1)\pi]\}\textstyle\sum_{n=1}^{\infty}(2\pi n)^{s-1} \\ &= 2\Gamma(1-s)\sin(\tfrac{1}{2}\pi s)(2\pi)^{s-1}\zeta(1-s) \quad \leftarrow \boxed{\zeta(s) = 0 \text{ when } s = -2, -4, -6, \cdots} \\ &= \pi^{\left(s-\frac{3}{2}\right)}\Gamma\left(\frac{1-s}{2}\right)\Gamma\left(1-\frac{s}{2}\right)\sin(\tfrac{1}{2}\pi s)\,\zeta(1-s) \quad \leftarrow \boxed{\text{G\&R } \mathbf{8.335}\text{-1}} \\ &= \pi^{(s-\frac{1}{2})}\Gamma\left(\frac{1-s}{2}\right)\left[\Gamma\left(\frac{s}{2}\right)\right]^{-1}\zeta(1-s); \qquad \mathrm{Re}(s) < 0. \quad \leftarrow \boxed{\text{G\&R } \mathbf{8.334}\text{-3}}\end{aligned} \tag{81}$$

Consequently,

$$(1/\sqrt{\pi})^s\,\Gamma\left(\frac{s}{2}\right)\zeta(s) = (1/\sqrt{\pi})^{1-s}\,\Gamma\left(\frac{1-s}{2}\right)\zeta(1-s); \qquad \mathrm{Re}(s) < 0. \tag{82}$$

Note: On the vertical leg on the right-hand side of the integration contour (shown as a dashed line in figure 10.4), the behavior of the integrand in Eq.(80) needs elaboration. As $x \to \infty$, the integrand's denominator,

$$\exp[-\exp(x+\mathrm{i}y)] - 1 = \exp[-\exp(x)\cos(y)]\exp[-\mathrm{i}\exp(x)\sin(y)] - 1, \tag{83}$$

behaves differently, depending on whether $|y| < \tfrac{1}{2}\pi$, $|y| = \tfrac{1}{2}\pi$, or $|y| > \tfrac{1}{2}\pi$. When $\tfrac{1}{2}\pi < |y| \leq \pi$ and $x \to \infty$, the denominator approaches ∞, and when $-\tfrac{1}{2}\pi < y < \tfrac{1}{2}\pi$, the denominator approaches -1. The denominator is highly oscillatory (and might vanish) in the immediate neighborhood of $y = \pm\tfrac{1}{2}\pi$. As for the integrand's numerator, $\exp(sz) = \exp[(s' + \mathrm{i}s'')(x+\mathrm{i}y)] = \exp(s'x - s''y)\exp[\mathrm{i}(s''x + s'y)] \to 0$ on the entire vertical leg when $s' < 0$ and $x \to \infty$. Overall, the contribution to the integral of the vertical leg on the right-hand side of the contour becomes negligible as $x \to \infty$, provided that $\mathrm{Re}(s) < 0$.

So far, we have demonstrated the symmetry of the function $(1/\sqrt{\pi})^s\Gamma(s/2)\zeta(s)$ with respect to the central point $s = \tfrac{1}{2}$ between the two semi-infinite domains $s < 0$ and $s > 1$. The fact that this function is obtained by analytic continuation over the entire s-plane suffices to conclude that the same symmetry must extend to the regions $0 \leq s \leq \tfrac{1}{2}$ and $\tfrac{1}{2} \leq s \leq 1$.

It is straightforward to show that $\zeta(s^*) = \zeta^*(s)$. This is obviously true of the function $\zeta(s) = \sum_{m=1}^{\infty} m^{-s}$ within its domain of convergence (i.e., $\mathrm{Re}(s) > 1$). By virtue of analytic continuation, the same property must extend to the entire s-plane. Needless to say, the inversion symmetry of $(1/\sqrt{\pi})^s\Gamma(s/2)\zeta(s)$ with respect to $s = ½$ also implies that the identity $\zeta(s^*) = \zeta^*(s)$, once demonstrated for $\text{Re}(s) > 1$, automatically extends to the region $\text{Re}(s) < 0$. Alternatively, one may prove the identity by conjugating both sides of Eq.(80) and invoking the reflection symmetry (with respect to the x-axis) of the integrand over the contour of integration.

10.6.1. Connecting Riemann's zeta function to prime numbers. Invoking the geometric series $\sum_{n=0}^{\infty} x^n = 1/(1-x)$ where $|x| < 1$, it is straightforward to assert the following identities involving prime numbers $2, 3, 5, \cdots, p, \cdots$ for $\text{Re}(s) > 1$:

$$1 + \left(\frac{1}{2^s}\right) + \left(\frac{1}{2^s}\right)^2 + \left(\frac{1}{2^s}\right)^3 + \cdots = \frac{1}{1-2^{-s}}, \tag{84a}$$

$$1 + \left(\frac{1}{3^s}\right) + \left(\frac{1}{3^s}\right)^2 + \left(\frac{1}{3^s}\right)^3 + \cdots = \frac{1}{1-3^{-s}}, \tag{84b}$$

$$1 + \left(\frac{1}{5^s}\right) + \left(\frac{1}{5^s}\right)^2 + \left(\frac{1}{5^s}\right)^3 + \cdots = \frac{1}{1-5^{-s}}, \tag{84c}$$

$$\vdots$$

$$1 + \left(\frac{1}{p^s}\right) + \left(\frac{1}{p^s}\right)^2 + \left(\frac{1}{p^s}\right)^3 + \cdots = \frac{1}{1-p^{-s}}, \tag{84d}$$

$$\vdots$$

Euler observed that, if the left-hand sides of these equations are multiplied together, the individual terms within the product will be of the form $1/(p_1^{k_1} p_2^{k_2} \cdots p_n^{k_n})^s$, which would then uniquely represent individual elements $1/m^s$ of the zeta function. Consequently,

$$\zeta(s) = \textstyle\sum_{m=1}^{\infty} m^{-s} = \prod_p \left(\frac{1}{1-p^{-s}}\right) = \prod_p \left(\frac{p^s}{p^s - 1}\right); \qquad \text{Re}(s) > 1. \tag{85}$$

This identity may be used to demonstrate that none of the zeros of $\zeta(s)$ resides in the region $\text{Re}(s) > 1$. To see this, observe that $\ln \zeta(s) = -\sum_p \ln(1 - p^{-s})$. Since $p \geq 2$ and $\text{Re}(s) = s' > 1$, the absolute values of the real parts of the individual terms in the above series are less than $(2 \ln 2)/p^{s'}$. Therefore, the real part of the series is convergent, indicating that $\ln|\zeta(s)|$ has a finite value, and that $\zeta(s)$ cannot become zero in the region $\text{Re}(s) > 1$.

10.6.2. Dirichlet's eta function. The eta function $\eta(s) = \sum_{m=1}^{\infty}(-1)^{m-1} m^{-s}$ is similar to Riemann's zeta function, except that the signs of its even terms are reversed. When the steps in Eq.(74) are repeated for the eta function, we find

$$\Gamma(s)\eta(s) = \int_0^{\infty} \frac{t^{s-1}}{\exp(t) + 1} \text{d}t\,; \qquad \text{Re}(s) > 0. \tag{86}$$

Note that the integral converges for values of s whose real part is greater than zero. Thus, the initial domain of convergence of $\eta(s)$ is larger than that of $\zeta(s)$. A simple identity relates the two functions, as shown below.

$$\begin{aligned}\zeta(s) - \eta(s) &= \left(1 + \frac{1}{2^s} + \frac{1}{3^s} + \frac{1}{4^s} + \frac{1}{5^s} \cdots\right) - \left(1 - \frac{1}{2^s} + \frac{1}{3^s} - \frac{1}{4^s} + \frac{1}{5^s} - \cdots\right) \\ &= \frac{2}{2^s}\left(1 + \frac{1}{2^s} + \frac{1}{3^s} + \frac{1}{4^s} + \frac{1}{5^s} \cdots\right) = 2^{1-s}\zeta(s) \quad \rightarrow \quad \zeta(s) = \frac{\eta(s)}{1 - 2^{1-s}}.\end{aligned} \tag{87}$$

Considering that the domain of convergence of $\eta(s)$ is the entire upper half of the plane, where $\text{Re}(s) > 0$, and that analytic continuation of $\zeta(s)$ is unique, Eq.(87) provides an alternative expression for the continuation of $\zeta(s)$ to the critical strip

$0 < \mathrm{Re}(s) < 1$. Subsequently, the *functional equation* given by Eq.(82) extends the zeta function to the left half of the complex s-plane.

10.6.3. Zeta function expressed as an integral involving Jacobi's theta function. Riemann derived a second integral representation for $\zeta(s)$, valid for all complex values of s, by resorting to Jacobi's theta function $\Theta(x) = \sum_{m=-\infty}^{\infty} \exp(-\pi m^2 x)$. Riemann began by evaluating the following integral using a change of variable from $\pi m^2 x$ to t:

$$\int_0^{\infty} \exp(-\pi m^2 x)\, x^{(s-2)/2} \mathrm{d}x = \frac{1}{(\pi m^2)^{s/2}} \int_0^{\infty} \exp(-t)\, t^{(s-2)/2} \mathrm{d}t$$

$$= \left(1/\sqrt{\pi}\right)^s \Gamma(s/2) m^{-s}, \qquad \mathrm{Re}(s) > 0. \tag{88}$$

Summing both sides of Eq.(88) over m from 1 to ∞ yields

$$\left(1/\sqrt{\pi}\right)^s \Gamma(s/2)\zeta(s) = \int_0^{\infty} [\,\textstyle\sum_{m=1}^{\infty} \exp(-\pi m^2 x)] x^{(s-2)/2} \mathrm{d}x$$

$$= \int_0^{\infty} ½[\Theta(x) - 1] x^{(s-2)/2} \mathrm{d}x, \qquad \mathrm{Re}(s) > 1. \tag{89}$$

This integral may be split into two parts, one ranging from $x = 0$ to 1, the other from $x = 1$ to ∞. The first integral is then evaluated using the theta function identity $\Theta(1/t) = \sqrt{t}\Theta(t)$, given in Eq.(68), as follows:

$$½ \int_0^1 [\Theta(x) - 1] x^{(s-2)/2} \mathrm{d}x = -\frac{1}{s} + \frac{1}{2}\int_1^{\infty} \Theta(1/t) t^{-(s+2)/2} \mathrm{d}t$$

$$= -\frac{1}{s} + \frac{1}{2}\int_1^{\infty} \Theta(t) t^{-(s+1)/2} \mathrm{d}t$$

$$\boxed{\mathrm{Re}(S) > 1} \rightarrow \; = -\frac{1}{s} + \frac{1}{s-1} + \frac{1}{2}\int_1^{\infty} [\Theta(t) - 1] t^{-(s+1)/2} \mathrm{d}t. \tag{90}$$

Combining Eqs.(89) and (90), we find

$$\left(1/\sqrt{\pi}\right)^s \Gamma(s/2)\zeta(s) = \frac{1}{s(s-1)} + ½ \int_1^{\infty} [\Theta(x) - 1]\left[x^{(s-2)/2} + x^{-(s+1)/2}\right] \mathrm{d}x$$

$$= \frac{1}{s(s-1)} + \int_1^{\infty} [\Theta(x) - 1] x^{-¾} \cosh[½(s - ½) \ln x]\, \mathrm{d}x. \tag{91}$$

The preceding integral may be further simplified via integration by parts (twice) to yield

$$\int_1^{\infty} [\Theta(x) - 1] x^{-¾} \cosh[½(s - ½) \ln x]\, \mathrm{d}x$$

$$= -\frac{2}{s-½} \int_1^{\infty} \{¼[\Theta(x) - 1] x^{-¾} + \Theta'(x) x^{¼}\} \sinh[½(s - ½) \ln x]\, \mathrm{d}x$$

$$= \frac{4}{(s-½)^2} \{¼[\Theta(1) - 1] + \Theta'(1)\}$$

$$+ \frac{4}{(s-½)^2} \int_1^{\infty} \left\{\frac{1}{16}[\Theta(x) - 1] x^{-¾} + \Theta''(x) x^{5/4} + \frac{3}{2}\Theta'(x) x^{¼}\right\} \cosh[½(s - ½) \ln x]\, \mathrm{d}x$$

$$= \frac{4\Theta'(1) + \Theta(1) - 1}{(s-½)^2} + \frac{1}{4(s-½)^2} \int_1^{\infty} [\Theta(x) - 1] x^{-¾} \cosh[½(s - ½) \ln x]\, \mathrm{d}x$$

$$+ \frac{4}{(s-½)^2} \int_1^{\infty} \left[\Theta''(x) x^{3/2} + \frac{3}{2}\Theta'(x) x^{½}\right] x^{-¼} \cosh[½(s - ½) \ln x]\, \mathrm{d}x. \tag{92}$$

Rearranging this equation, and taking into account the theta function identity $4\Theta'(1) + \Theta(1) = 0$ proven at the end of section 5, we will have

$$\int_1^\infty [\Theta(x) - 1]x^{-3/4} \cosh[\tfrac{1}{2}(s - \tfrac{1}{2}) \ln x]\, dx = -\frac{1}{s(s-1)}$$

$$+ \frac{4}{s(s-1)} \int_1^\infty \frac{d[x^{3/2}\Theta'(x)]}{dx} x^{-1/4} \cosh[\tfrac{1}{2}(s - \tfrac{1}{2}) \ln x]\, dx. \qquad (93)$$

Substituting Eq.(93) into Eq.(91) now yields

$$\left(1/\sqrt{\pi}\right)^s \Gamma(s/2)\zeta(s) = \frac{4}{s(s-1)} \int_1^\infty \frac{d[x^{3/2}\Theta'(x)]}{dx} x^{-1/4} \cosh[\tfrac{1}{2}(s - \tfrac{1}{2}) \ln x]\, dx. \qquad (94)$$

Note that the integral appearing in Eq.(94) exists for all points in the complex s-plane. The symmetry with respect to $s = \tfrac{1}{2}$ of the function on the left-hand side of Eq.(94) should now be obvious, as the substitution of $(1 - s)$ for s on the right-hand side does *not* alter the function.

References

1. Stephen Hawking, *God Created the Integers: The Mathematical Breakthroughs That Changed History*, 2nd edition, Running Press, Philadelphia (2007).
2. G.B. Arfken and H.J. Weber, *Mathematical Methods for Physicists*, 6th edition, Academic Press, Burlington, Massachusetts (2005).
3. F.B. Hildebrand, *Advanced Calculus for Applications*, 2nd edition, Prentice-Hall, Upper Saddle River, New Jersey (1976).
4. J. Mathews and R.L. Walker, *Mathematical Methods of Physics*, 2nd edition, Benjamin/Cummings Publishing, San Francisco, California (1970).
5. G. Stephenson and P.M. Radmore, *Advanced Mathematical Methods for Engineering and Science Students*, Cambridge University Press, Cambridge, United Kingdom (1990).
6. G.H. Hardy and E.M. Wright, "The Zeta Function," in *An Introduction to the Theory of Numbers*, 5th edition, Clarendon Press, Oxford, United Kingdom, pp. 245-247 and 255 (1979).
7. S.J. Patterson, *An Introduction to the Theory of the Riemann Zeta-Function*, Cambridge University Press, New York (1988).
8. I.S. Gradshteyn and I.M. Ryzhik, *Table of Integrals, Series, and Products*, seventh edition, Academic Press, Burlington, Massachusetts (2007).

Problems

Problem 1) Consider the second-order linear differential equation

$$a_0(x)\frac{\mathrm{d}^2 f(x)}{\mathrm{d}x^2} + a_1(x)\frac{\mathrm{d}f(x)}{\mathrm{d}x} + [a_2(x) + \lambda a_3(x)]f(x) = 0.$$

a) Multiply the equation with an arbitrary function $g(x)$, then determine $g(x)$ such that the equation may be written in the standard Sturm-Liouville form, namely,

$$\frac{\mathrm{d}}{\mathrm{d}x}\left[p(x)\frac{\mathrm{d}f(x)}{\mathrm{d}x}\right] + [q(x) + \lambda r(x)]f(x) = 0.$$

In terms of the original coefficients $a_0(x)$, $a_1(x)$, $a_2(x)$, and $a_3(x)$, derive explicit expressions for $p(x)$, $q(x)$, and $r(x)$.

b) Reduce each of the following equations to the standard Sturm-Liouville form:

$$\frac{\mathrm{d}^2 f(x)}{\mathrm{d}x^2} + \cot(x)\frac{\mathrm{d}f(x)}{\mathrm{d}x} + \lambda f(x) = 0.$$

$$x\frac{\mathrm{d}^2 f(x)}{\mathrm{d}x^2} + (1-x)\frac{\mathrm{d}f(x)}{\mathrm{d}x} + (x^2 + \lambda)f(x) = 0.$$

Problem 2) The solutions of Airy's differential equation, $f''(x) - xf(x) = 0$, denoted by $\mathrm{Ai}(x)$ and $\mathrm{Bi}(x)$, are commonly known as Airy functions of the first and second kind, respectively. Since $\mathrm{Bi}(x)$ diverges rapidly as $x \to \infty$, its Fourier transform does not exist. The Fourier transform of $\mathrm{Ai}(x)$, however, is readily determined by applying the differentiation theorem of Fourier transforms to Airy's differential equation (see chapter 7, section 6.6). Show that the Fourier transform of $\mathrm{Ai}(x)$ is $F(s) = \exp[⅓\mathrm{i}(2\pi s)^3]$.

Problem 3) Starting with Euler's definition of the Gamma function, namely,

$$\Gamma(z) = \int_0^\infty \exp(-t)\, t^{z-1}\mathrm{d}t, \qquad \mathrm{Re}(z) > 0,$$

use the method integration by parts to prove the following identity:

$$\Gamma(z) = \int_0^\infty \exp(-t)\,(t-z)t^{z-1}\ln t\,\mathrm{d}t, \qquad \mathrm{Re}(z) > 0.$$

Problem 4) Euler's Gamma function $\Gamma(z)$ is analytic everywhere in the complex z-plane except at the (simple) poles of the function $z = 0, -1, -2, -3, \cdots$. Use Euler's reflection formula $\Gamma(z)\Gamma(1-z) = \pi/\sin(\pi z)$ to determine the residues of $\Gamma(z)$ at its poles.

Problem 5) Assuming that the parameter y is real-valued, prove the following identities:

a) $$|\Gamma(\mathrm{i}y)|^2 = \frac{\pi}{y\sinh(\pi y)}.$$

b) $$|\Gamma(½ + \mathrm{i}y)|^2 = \frac{\pi}{\cosh(\pi y)}.$$

c) $$\Gamma(1 + \mathrm{i}y)\Gamma(1 - \mathrm{i}y) = \frac{\pi y}{\sinh(\pi y)}.$$

Problem 6) Assuming that x and y are real valued, prove the following identity:

$$\Gamma(1 + x + \mathrm{i}y)\Gamma(1 - x + \mathrm{i}y)\Gamma(1 + x - \mathrm{i}y)\Gamma(1 - x - \mathrm{i}y) = \frac{2\pi^2(x^2+y^2)}{\cosh(2\pi y) - \cos(2\pi x)}.$$

Problem 7) The Gauss multiplication formula is a generalization of the Legendre duplication formula (described in section 10.4.3), as follows:

$$\prod_{k=0}^{n-1} \Gamma\left(z+\frac{k}{n}\right) = (\sqrt{2\pi}\,)^{n-1}(n^{½\,-\,nz})\Gamma(nz).$$

Use the infinite product formula given in section 10.4.2, Eq.(53), and Stirling's asymptotic expression for the factorial of N, namely $N! \sim \sqrt{2\pi N}(N/e)^N$, to prove the above identity.

Hint: It suffices to prove the formula in the special case of $n = 3$; that is,

$$\Gamma(z)\Gamma(z+⅓)\Gamma(z+⅔) = 2\pi(3^{½-3z})\Gamma(3z).$$

The proof in the general case will then be a straightforward extension of the proof for the case of $n = 3$.

Chapter 11

Elementary Matrix Algebra

"Since the beginning of physics, symmetry considerations have provided us with an extremely powerful and useful tool in our effort to understand nature. Gradually they have become the backbone of our theoretical formulation of physical laws."

Tsung-Dao Lee (1926–)

11.1. Introduction. The rudiments of linear algebra were initially developed for solving systems of linear equations. Gottfried Wilhelm Leibniz (1646–1716) considered determinants as early as 1693, and Gabriel Cramer (1704–1752) employed them to provide explicit solutions for linear systems. In the early 1800s, Gauss developed the method now known as Gaussian elimination, which proved to be a simple, powerful, and elegant approach to solving systems of linear equations.

The study of matrix algebra first emerged in the United Kingdom in the mid-1800s. The term matrix (Latin for womb) was introduced by James Sylvester in 1848. Arthur Cayley (1821–1895), who defined matrix multiplication and matrix inverses in the context of compositions of linear transformations, appears to have understood the intimate connection between matrices and determinants. The modern definition of a vector space was introduced in 1888 by the Italian mathematician Giuseppe Peano (1858–1932). At the dawn of the twentieth century, a theory of linear transformations of finite-dimensional vector spaces had finally emerged.

The use of matrices in quantum mechanics, special relativity, and statistics helped spread the subject of linear algebra beyond pure mathematics. The advent of electronic computers during the second half of the twentieth century accelerated the pace of efforts to develop efficient algorithms for Gaussian elimination and matrix decompositions. In modern science and its widespread technological applications, linear algebra has become an indispensable tool for modeling and for numerical simulations. In the fields of optics and photonics, the use of matrices is ubiquitous in geometric-optical ray tracing, lens design, beam propagation algorithms, multilayer stack design and analysis, studies of resonant cavities, design and evaluation of image-forming systems, and quantum optics.[1†]

The present chapter begins by defining some of the basic properties of matrices and their standard arithmetic operations such as addition, subtraction, and multiplication. In sections 2 and 3, we introduce the norms of $1 \times N$ and $N \times 1$ dimensional matrices (also referred to as row and column vectors) and discuss some elementary properties of such matrices. Inversion of square matrices and an introduction to the concept of determinants form the subject of section 4. We then briefly describe elementary row and column operations (as well as their associated matrices) in section 5. These elementary operations are invoked in section 6 to establish the important property that the determinant of the product of two matrices equals the product of their determinants.

Section 7 describes the Gram-Schmidt process, which is an algorithm for converting a set of equidimensional and linearly independent row (or column) vectors into an orthonormal set. We then turn to the important topic of eigenvalues and eigenvectors in section 8, where we introduce the characteristic equation, whose roots are the eigenvalues of a square matrix, and establish the concepts of right and left eigenvectors. The discussion of eigenvalues and eigenvectors leads naturally to the process of diagonalization of square matrices. Certain operations that involve diagonalized matrices

---

[†]Superscript numbers throughout the chapter identify references listed at the end of the chapter.

(e.g., raising a matrix to a power, or computing the exponential of a matrix) are subsequently elaborated in section 9. In section 10, we show how matrix operations (such as inversion, diagonalization, and exponentiation) can be employed in the service of solving systems of first-order linear ordinary differential equations with constant coefficients.

Following a brief discussion of the Cayley-Hamilton theorem in section 11, we turn our attention to some of the basic properties of unitary matrices in section 12, and to those of Hermitian matrices in section 13. Finally, the Schur decomposition of square matrices will be discussed in section 14, where, among other things, we prove that unitary matrices as well as Hermitian matrices can always be diagonalized with the aid of unitary matrices. Throughout the chapter, several examples are used to illustrate the various matrix operations and also to help clarify the newly introduced concepts and notions.

11.2. Definitions and basic matrix operations. A matrix is a collection of numbers (real or complex) that are arranged in a specific way. An N-dimensional matrix A with elements a_n, could be arranged either as a "row vector" or as a "column vector," as shown below:

$$A_1 = (a_1, a_2, \cdots, a_N), \qquad A_2 = \begin{pmatrix} a_1 \\ a_2 \\ \vdots \\ a_N \end{pmatrix}. \tag{1}$$

The row vector A_1 is said to be a $1 \times N$ matrix, whereas the column vector A_2 is an $N \times 1$ matrix. An $M \times N$ dimensional matrix will look like this:

$$A = \begin{pmatrix} a_{11} & a_{12} & \cdots & a_{1N} \\ a_{21} & a_{22} & \cdots & a_{2N} \\ \vdots & \vdots & \vdots & \vdots \\ a_{M1} & a_{M2} & \cdots & a_{MN} \end{pmatrix}. \tag{2}$$

It is often easier to write a matrix as $[a_{mn}]$, or simply as a_{mn}, recognizing that the first index identifies the row, while the second index identifies the column in which the element is located.

The matrices A and B are said to be equal if they have the same number of rows, M, and the same number of columns, N, and also the same elements in corresponding locations; that is, $a_{mn} = b_{mn}$.

Multiplying a matrix by a constant c entails multiplying each and every element of the matrix by c. Thus, $cA = Ac = [ca_{mn}]$. The multiplication by zero of any $M \times N$ matrix yields an all zeros $M \times N$ matrix O.

Two matrices A and B can be added together or subtracted from each other provided that they have the same dimensions. Thus, if both A and B are $M \times N$ matrices, we can write $A + B = [a_{mn} + b_{mn}]$. In other words, the corresponding elements of the two matrices are added together, one by one, to form a new $M \times N$ matrix, say, C, where $c_{mn} = a_{mn} + b_{mn}$.

The *transpose* of the $M \times N$ matrix A whose elements are specified as a_{mn}, is an $N \times M$ matrix denoted by A^T, whose elements are a_{nm}. Thus, the mth row of A, when rotated clockwise through 90°, becomes the mth column of A^T. Alternatively, one could say that rotating the nth column of A counterclockwise through 90° turns it into the nth row of A^T.[2,3]

The *conjugate transpose* of a complex matrix is obtained by conjugating each and every element of the matrix before (or after) its transposition; that is, the element mn of A^{*T} is a_{nm}^*. (Recall that the conjugate of $c = c' + \mathrm{i}c''$ is $c^* = c' - \mathrm{i}c''$.) A *Hermitian* matrix is an $N \times N$ matrix that happens to be its own conjugate transpose; in other words, A is Hermitian if $A^{*T} = A$. An anti-Hermitian (also known as skew-Hermitian) matrix is then defined as a square matrix that satisfies $A^{*T} = -A$.

Two matrices A and B can be multiplied to create the new matrix $C = AB$, provided that the number of columns of A equals the number of rows of B. Thus, if A is an $M \times N$ matrix and B an $N \times P$ matrix, then their product C will be an $M \times P$ matrix whose elements, by definition, are $c_{mp} = \sum_{n=1}^{N} a_{mn} b_{np}$ (see figure 11.1). The triple matrix product $A_1 A_2 A_3$ can be evaluated either as $A_1(A_2 A_3)$ or as $(A_1 A_2) A_3$ (see problem 1 at the end of the chapter). Given the definition of the product of two matrices, it is not difficult to see that $(AB)^T = B^T A^T$. It is often convenient to drop the summation symbol Σ and simply write $c_{mp} = a_{mn} b_{np}$, the sum over the repeated index being implicitly assumed. This is called the *Einstein notation*.[4,5]

$$\begin{pmatrix} a_{11} & a_{12} & \cdots & a_{1N} \\ a_{21} & a_{22} & \cdots & a_{2N} \\ \vdots & \vdots & \ddots & \vdots \\ a_{M1} & a_{M2} & \cdots & a_{MN} \end{pmatrix} \begin{pmatrix} b_{11} & b_{12} & \cdots & b_{1P} \\ b_{21} & b_{22} & \cdots & b_{2P} \\ \vdots & \vdots & \ddots & \vdots \\ b_{N1} & b_{N2} & \cdots & b_{NP} \end{pmatrix} = \begin{pmatrix} c_{11} & c_{12} & \cdots & c_{1P} \\ c_{21} & c_{22} & \cdots & c_{2P} \\ \vdots & \vdots & \ddots & \vdots \\ c_{M1} & c_{M2} & \cdots & c_{MP} \end{pmatrix}$$

Figure 11.1. To multiply the $M \times N$ matrix A into the $N \times P$ matrix B, take each row of A, say, row m, and multiply it into each column of B, say, column n. This operation entails multiplying the corresponding elements a_{mn} and b_{np} of the selected row and column, followed by adding up the products. The resulting sum becomes the element c_{mp} of the product matrix C; that is, $c_{mp} = \sum_{n=1}^{N} a_{mn} b_{np}$.

By definition, the norm (or magnitude) of a row vector, such as A_1 of Eq.(1), is $\|A_1\| = \sqrt{A_1 A_1^T} = (\sum_{n=1}^{N} a_n^2)^{½}$. Similarly, the norm (or magnitude) of a column vector, say, A_2 of Eq.(1), is defined as $\|A_2\| = \sqrt{A_2^T A_2} = (\sum_{n=1}^{N} a_n^2)^{½}$. In the case of complex vectors, it is often desirable to conjugate either the vector or its transpose so that the norm will end up being $(\sum_{n=1}^{N} |a_n|^2)^{½}$. The inner product of the $1 \times N$ row vector A and the $N \times 1$ column vector B would then be defined as $AB^* = \sum_{n=1}^{N} a_n b_n^*$. The squared magnitude of the inner product is thus given by $|AB^*|^2 = (AB^*)(A^*B)$. Given that AB^* is a scalar entity, its transpose equals itself; that is, $AB^* = (AB^*)^T = B^{*T} A^T$. Note, however, that AB^* is the complex conjugate of A^*B.

A *square* matrix is one with an equal number of rows and columns; that is, the $M \times N$ matrix A is square if $M = N$. A square matrix A is said to be diagonal if all the elements a_{mn} having $m \neq n$ are zeros. Each diagonal element a_{mm} of the matrix may then be zero or nonzero. A special diagonal matrix is the identity matrix I, whose diagonal elements are all equal to 1; that is,

$$I = \begin{pmatrix} 1 & 0 & \cdots & 0 \\ 0 & 1 & \cdots & 0 \\ 0 & 0 & \cdots & 0 \\ \vdots & \vdots & \ddots & \vdots \\ 0 & 0 & \cdots & 1 \end{pmatrix}. \tag{3}$$

The product of any $M \times N$ matrix A and the $N \times N$ identity matrix I is simply $AI = A$. Similarly, the product of the $M \times M$ identity matrix I and any $M \times N$ matrix A is $IA = A$.

Two square matrices A and B can be multiplied together either as AB or as BA. In general, matrix multiplication is a noncommutative operation; that is, $AB \neq BA$. If A and B are $N \times N$ matrices, their *commutator* $C = [A, B]$ is also an $N \times N$ matrix defined as $[A, B] = AB - BA$. A square matrix that commutes with its own conjugate transpose is known as a *normal* matrix; that is, A is normal if $A^{*T}A = AA^{*T}$.

11.3. The Schwarz inequality. The inner product AB^{*T} of the (generally complex-valued) $1 \times N$ row vectors A and B satisfies the Schwarz inequality

$$|AB^{*T}|^2 \leq \|A\|^2 \times \|B\|^2. \tag{4}$$

The proof of this inequality is similar to that given for a pair of arbitrary functions $f(x)$ and $g(x)$ in chapter 1, section 9. Here, we use the somewhat simpler method of proof that was introduced in chapter 1, problem 8. Consider the (row) vector $A + \lambda B$, where λ is an arbitrary complex constant. In general, one can write

$$0 \leq \|A + \lambda B\|^2 = (A + \lambda B)(A + \lambda B)^{*T} = \|A\|^2 + \lambda BA^{*T} + \lambda^* AB^{*T} + \lambda\lambda^* \|B\|^2. \tag{5}$$

To minimize the right-hand side of the expression, we treat λ and λ^* as two independent variables, then minimize the expression with respect to λ by setting its partial derivatives with respect to both λ and λ^* equal to zero. We find

$$\begin{pmatrix} \partial_\lambda \\ \partial_{\lambda^*} \end{pmatrix} \|A + \lambda B\|^2 = \begin{pmatrix} BA^{*T} + \lambda^* \|B\|^2 \\ AB^{*T} + \lambda \|B\|^2 \end{pmatrix} = \begin{pmatrix} 0 \\ 0 \end{pmatrix} \quad \rightarrow \quad \begin{pmatrix} \lambda^* \\ \lambda \end{pmatrix} = -\begin{pmatrix} BA^{*T} \\ AB^{*T} \end{pmatrix} / \|B\|^2. \tag{6}$$

Substitution for λ and λ^* into Eq.(5) now yields the desired Schwarz inequality; that is,

$$(AB^{*T})(BA^{*T}) = (AB^{*T})(A^* B^T) = |AB^{*T}|^2 \leq \|A\|^2 \times \|B\|^2. \tag{7}$$

Digression: The justification for λ and λ^* in this derivation of the Schwarz inequality being treated as independent variables is as follows: Let $\lambda = \lambda' + \mathrm{i}\lambda''$, which leads to $\lambda^* = \lambda' - \mathrm{i}\lambda''$. To minimize a function $f(\cdot)$ of the two real-valued variables λ' and λ'', we must set $\partial f/\partial\lambda' = 0$ and $\partial f/\partial\lambda'' = 0$. However, if the function happens to be in the form of $f(\lambda, \lambda^*)$, we would have

$$\partial f/\partial\lambda' = (\partial f/\partial\lambda)(\partial\lambda/\partial\lambda') + (\partial f/\partial\lambda^*)(\partial\lambda^*/\partial\lambda') = (\partial f/\partial\lambda) + (\partial f/\partial\lambda^*) = 0,$$

$$\partial f/\partial\lambda'' = (\partial f/\partial\lambda)(\partial\lambda/\partial\lambda'') + (\partial f/\partial\lambda^*)(\partial\lambda^*/\partial\lambda'') = \mathrm{i}(\partial f/\partial\lambda) - \mathrm{i}(\partial f/\partial\lambda^*) = 0.$$

It is seen that the necessary and sufficient conditions for the partial derivatives of f with respect to λ' and λ'' to vanish are that $\partial f/\partial\lambda = 0$ and $\partial f/\partial\lambda^* = 0$.

Occasionally, it is helpful to draw analogies between ordinary vectors in three-dimensional Euclidean space and N-dimensional row and/or column vectors of linear algebra. In vector calculus, the inner product (or dot-product) of two vectors $\boldsymbol{U}$ and $\boldsymbol{V}$ is given by $UV \cos\theta$, where U and V are the lengths of $\boldsymbol{U}$ and $\boldsymbol{V}$, while θ is the angle between the two vectors. One may thus invoke the Schwarz inequality for the pair of real-valued N-dimensional vectors A and B to define the cosine of the angle θ between A and B as the inner product of A and B, normalized by the product $\|A\| \times \|B\|$ of their norms.

11.4. Matrix inversion and the concept of determinant. The square $N \times N$ matrix A is said to be nonsingular if and only if it is invertible. The inverse of A, which is also an $N \times N$ matrix denoted by A^{-1}, will then have the property $AA^{-1} = A^{-1}A = I$. (Note that

A^{-1} commutes with A.) A square matrix is singular if and only if its determinant is zero. The determinant of a 2×2 matrix is defined as follows:

$$\text{Determinant of } A = |A| = \begin{vmatrix} a_{11} & a_{12} \\ a_{21} & a_{22} \end{vmatrix} = a_{11}a_{22} - a_{12}a_{21}. \quad (8)$$

The rule for computing the determinant of higher-dimensional $N \times N$ matrices is as follows:

i) Pick a single row (or a single column).

ii) Compute the *cofactor* for each of the N elements in that row (or column).

iii) Multiply each element of the chosen row (or column) with its own cofactor.

iv) Add up the N products found in step (iii) to arrive at the determinant of the matrix.

To compute the cofactor c_{mn} associated with a_{mn}, remove the corresponding row (row m) and the corresponding column (column n) from the original matrix, compute the determinant of the remaining $(N-1) \times (N-1)$ matrix, then multiply the result by $(-1)^{m+n}$. The entire process is best illustrated by computing the determinant of a 3×3 matrix, as follows:

$$\begin{vmatrix} a_{11} & a_{12} & a_{13} \\ a_{21} & a_{22} & a_{23} \\ a_{31} & a_{32} & a_{33} \end{vmatrix} = a_{11}(a_{22}a_{33} - a_{23}a_{32}) - a_{12}(a_{21}a_{33} - a_{23}a_{31}) + a_{13}(a_{21}a_{32} - a_{22}a_{31}). \quad (9)$$

Here, we have chosen the first row of the matrix for computing its determinant. Of course, one could choose any other row, or any column for that matter, and the result would have been the same.

The inverse A^{-1} of the nonsingular $N \times N$ matrix A is found as follows:

i) Form an $N \times N$ matrix C whose elements c_{mn} are the cofactors of the corresponding elements a_{mn} of the matrix A.

ii) Divide C by the (nonzero) determinant of A; that is, replace each element c_{mn} of C with $c_{mn}/|A|$.

iii) The inverse of A is the transpose of the matrix found in step (ii); that is, $A^{-1} = C^T/|A|$.

Note: In the trivial case of $N = 1$, the determinant of $A = (a_{11})$ is $|A| = a_{11}$, its matrix of cofactors is $C = (1)$ and its inverse is $A^{-1} = (1/a_{11})$.

To prove the validity of the procedures listed above for computing the determinant $|A|$ and the inverse A^{-1} of A, we rely to a large extent on the method of proof by induction. It is straightforward to demonstrate the following properties for all 2×2 matrices. We shall then assume their validity for arbitrary $N \times N$ matrices and proceed to prove them for arbitrary $(N+1) \times (N+1)$ matrices.

Property 1) For row m of the $N \times N$ matrix A, the sum $\sum_{n=1}^{N} a_{mn}c_{mn}$ is the same as that for any other row. This sum also equals $\sum_{m=1}^{N} a_{mn}c_{mn}$ for any column n of the matrix.

Property 2) For any pair of rows of A, say, rows m_1 and m_2, where $m_1 \neq m_2$, the sum $\sum_{n=1}^{N} a_{m_1,n}c_{m_2,n} = 0$. Similarly, for any pair of columns such as n_1 and n_2, where $n_1 \neq n_2$, the sum $\sum_{m=1}^{N} a_{m,n_1}c_{m,n_2} = 0$.

Property 3) The determinant of A^T is the same as that of A. Consequently, the matrix of cofactors of A^T is C^T, namely the transpose of the matrix C of cofactors of A.

Property 3 follows directly from property 1 via the method of induction. It also indicates that the inverse of A^T is $(A^{-1})^T$, although one could deduce the latter result more directly by noting that the transpose of the identity $AA^{-1} = A^{-1}A = I$ is $(A^{-1})^T A^T = A^T (A^{-1})^T = I^T = I$.

To prove property 1, pick a pair of rows m_1 and m_2. The sum for row m_1 is $\sum_{n=1}^{N} a_{m_1,n} c_{m_1,n}$, where the cofactor $c_{m_1,n}$ is $(-1)^{m_1+n}$ times the determinant of an $(N-1) \times (N-1)$ matrix. By assumption, the determinant of this smaller matrix can be evaluated over any row, so pick row m_2 and proceed to evaluate the sum of $a_{m_2,n}$ times the corresponding cofactor for the $(N-1) \times (N-1)$ matrix, which, aside from a ± 1 coefficient, is the determinant of a still smaller $(N-2) \times (N-2)$ matrix. This reduced matrix is what is left of the original matrix A after removing two of its rows, m_1 and m_2, and also two of its columns, n_1 and n_2 (see figure 11.2). The determinant of the reduced matrix is now multiplied by $\pm(-1)^{m_1+n_1+m_2+n_2} a_{m_1,n_1} a_{m_2,n_2}$, and the result is summed over all pairs n_1, n_2, where $n_1 \neq n_2$. (The leading $\pm$ sign in this expression accounts for the fact that the indices m_1, n_1, m_2, n_2 of the matrix A might differ by one unit from those of the smaller $(N-1) \times (N-1)$ submatrices. This $\pm$ sign, however, remains the same when the roles of m_1, n_1 and m_2, n_2 are reversed.) It is thus clear that one could switch the roles of m_1 and m_2 without changing the final result. Consequently, $|A|$ could be evaluated by the sum $\sum_{n=1}^{N} a_{mn} c_{mn}$ for any row m of the matrix A. Similar arguments can be advanced for a pair of columns or for a pair consisting of one row and one column.

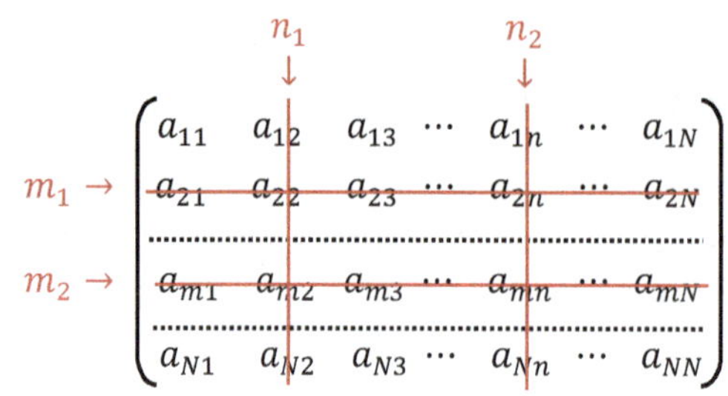

Figure 11.2. Eliminating rows m_1 and m_2, as well as columns n_1 and n_2, reduces the original $N \times N$ matrix to a smaller $(N-2) \times (N-2)$ matrix. The proof of properties 1 and 2 of the cofactors c_{mn} involves the determinant of the reduced $(N-2) \times (N-2)$ matrix together with certain combinations of the elements $a_{m_1,n_1}, a_{m_1,n_2}, a_{m_2,n_1}, a_{m_2,n_2}$ of the original matrix.

To prove property 2, note that the cofactor $c_{m_2,n}$ is $(-1)^{m_2+n}$ times the determinant of the $(N-1) \times (N-1)$ matrix obtained by removing row m_2 and column n from matrix A. To find this determinant, choose row m_1, then add up the products of a_{m_1,n_1} and its corresponding cofactor. The $(N-2) \times (N-2)$ matrix, whose determinant is now being evaluated, is obtained by removing rows m_1 and m_2 and also columns n and n_1 from the original matrix A (see figure 11.2). However, a similar term is obtained if the roles of n and n_1 are switched, except for a minus sign that now appears before the product of $a_{m_1,n_1} a_{m_1,n}$ and the determinant of the reduced $(N-2) \times (N-2)$ matrix. The two terms thus cancel out, and the sum over an entire row ends up being zero.

11.5. Elementary row and column operations. In addition to the method described in the preceding section, it is possible to invert matrices, and also to find their determinants, using a sequence of elementary operations on the rows and/or columns of the matrix. In numerical computations involving large matrices, such elementary row/column operations are much easier to carry out than the alternative method based on the computation of cofactors. The concepts surrounding these elementary row/column operations also prove useful in conceptualizing various problems and proving certain theorems of linear algebra.

The present section is devoted to a brief discussion of elementary row/column operations. We will show how these operations enable one to put any matrix into an upper (or lower) triangular form. If the matrix happens to be square and nonsingular, further elementary operations can eventually convert it to a diagonal matrix, or, upon normalization, an identity matrix. The inverse of a nonsingular matrix is then found by repeating the same sequence of elementary row/column operations on an identity matrix. We begin by defining three types of elementary operations, whose effects on the determinant of the matrix are readily inferred from the methods of the preceding section.

i) Swapping two rows, say, m_1 and m_2 (or swapping two columns, say, n_1 and n_2), will result in multiplying the determinant of the matrix by -1.

ii) Multiplying a row (or a column) by a constant c results in multiplying the determinant of the entire matrix by c.

iii) Adding a multiple of one row to another row, say, adding c times row m_1 to row m_2, does not change the determinant of the matrix. (The same is true for adding a multiple of one column to another column.)

These operations on, say, a 5×5 matrix A can be formally represented by multiplying A on the left-hand side with slightly modified identity matrices, such as those shown below. The product matrix $E_{\text{i}}A$ is similar to A, except for its third and fifth rows being swapped. The product $E_{\text{ii}}A$ is the matrix A with its second row multiplied by c. The matrix $E_{\text{iii}}A$ is similar to A except that its fourth row is modified by the addition of c times the second row of A. Note that $|E_{\text{i}}| = -1$, $|E_{\text{ii}}| = c$, and $|E_{\text{iii}}| = 1$.

$$E_{\text{i}} = \begin{pmatrix} 1 & 0 & 0 & 0 & 0 \\ 0 & 1 & 0 & 0 & 0 \\ 0 & 0 & 0 & 0 & 1 \\ 0 & 0 & 0 & 1 & 0 \\ 0 & 0 & 1 & 0 & 0 \end{pmatrix}; \quad E_{\text{ii}} = \begin{pmatrix} 1 & 0 & 0 & 0 & 0 \\ 0 & c & 0 & 0 & 0 \\ 0 & 0 & 1 & 0 & 0 \\ 0 & 0 & 0 & 1 & 0 \\ 0 & 0 & 0 & 0 & 1 \end{pmatrix}; \quad E_{\text{iii}} = \begin{pmatrix} 1 & 0 & 0 & 0 & 0 \\ 0 & 1 & 0 & 0 & 0 \\ 0 & 0 & 1 & 0 & 0 \\ 0 & c & 0 & 1 & 0 \\ 0 & 0 & 0 & 0 & 1 \end{pmatrix}.$$

All the elementary matrices introduced above are invertible. The inverse of E_{i} is exactly the same matrix (i.e., two successive swaps of the same pair of rows result in no swap at all). The inverse of E_{ii} has the same structure as E_{ii}, except for the constant c being replaced by $1/c$ (simply because multiplying a row by c, followed by multiplying the same row by $1/c$, does not change the matrix). The inverse of E_{iii} also retains the structure of E_{iii}, except for the substitution of the constant c by $-c$ (that is, the addition of a row, followed by subtraction of the same row with the same coefficient, leaves the matrix intact).

The elementary matrices $E_{\text{i}}, E_{\text{ii}}, E_{\text{iii}}$, if multiplied on the right-hand side of a matrix A, will perform their corresponding operations on the various *columns* of A. Thus, for a 5×5 matrix A, the product AE_{i} will have the third and fifth columns of A swapped, the matrix AE_{ii} has the second column of A multipled by c, and the matrix AE_{iii} has c times the fourth column of A added to the second column of A.

When a matrix is in upper triangular form (i.e., all the elements below the diagonal are zeros), or in the lower triangular form (i.e., all the elements above the diagonal are zeros), its determinant will be the product of its diagonal elements. This is easy to demonstrate by calculating the cofactors and determinants of the matrix and its various submatrices by always picking a row (or a column) that has a single nonzero element.

A matrix can be transformed into an upper triangular matrix (also called the row echelon form) using the process of *Gauss elimination*, as follows. Start with column 1 of the matrix A. If it so happens that the entire column 1 consists of zeros, leave this column alone, move on to column 2, and treat the matrix as though its first column did not exist. (In what follows we shall assume that the first column has at least one nonzero element.) If the first element of the first row (i.e., a_{11}) is zero, swap the first row with some other row whose first element is not zero (elementary operation of type i). Once the first element of the first row becomes nonzero, proceed to divide (or normalize) all elements of that first row by the first element of the row (elementary operation of type ii). At this point the leading element of the first row equals 1. In the next step, subtract a multiple of the (normalized) first row from any other row whose first element is nonzero, in such a way as to force the first element of the targeted row to vanish (elementary operation of type iii, repeatedly applied to as many rows as needed). In this way, the entire column 1, except for its top element, becomes zero. You must now ignore row 1 and column 1 and repeat the same procedure for the remaining $(N-1) \times (N-1)$ matrix. When all the rows are thus processed, you will end up with an upper triangular matrix. If any element on the diagonal of the triangular matrix happens to be zero, then the matrix is singular—because its determinant, which is proportional to the product of the diagonal elements of the upper triangular matrix, is going to be zero. In this case, the process comes to a halt since the matrix is found to be noninvertible. The Gaussian elimination process has thus revealed that $|A| = 0$.

If, however, all the diagonal elements of the upper triangular matrix are found to be equal to 1, we can proceed to eliminate the elements that are above the diagonal. To this end, start with the row at the very bottom and apply elementary operations of type iii to force every element on the last column (aside from the bottom element, of course) to vanish. You then ignore the last row and last column and repeat the process for the remaining $(N-1) \times (N-1)$ matrix. (The literature sometimes refers to this method of removing the nonzero elements above the diagonal of an upper triangular matrix as the *Gauss-Jordan elimination.*) All in all, it must be clear that, by repeated application of elementary operations of type i, ii, and iii, one can transform a nonsingular square matrix into an identity matrix, as follows:

$$E_k \cdots E_j E_i A = I. \tag{10}$$

Each elementary operation that appears in Eq.(10) either multiplies the determinant of A by -1 (type i), or multiplies it by a constant c (type ii), or does not change the determinant at all (type iii). In the end, recalling that A is assumed to be nonsingular, it will be transformed into an identity matrix whose determinant is 1. It is thus seen that $|A|$ is the inverse of the product of all the constants (including -1s) associated with $E_i, E_j, \cdots E_k$. Since these elementary matrices are invertible, Eq.(10) yields

$$A = E_i^{-1} E_j^{-1} \cdots E_k^{-1} I. \tag{11}$$

Considering that the determinant of an inverted elementary matrix is the inverse of its determinant, one must conclude that $|A|$ is equal to the product of the determinants of $E_i^{-1}, E_j^{-1}, \cdots, E_k^{-1}$.

Digression: An alternative approach to transforming an upper triangular matrix (whose diagonal elements are all equal to 1) into an identity matrix is via elementary column operations of type iii. Here, the first column must be used to set all the elements of the first row to zero—aside from the leading element, of course. The first row and the first column are subsequently ignored, and the procedure is repeated for the remaining $(N-1) \times (N-1)$ matrix. The process continues until all the columns are exhausted, at which point we have arrived at the desired identity matrix. In this way, a sequence of elementary operations, first acting on the rows, then on the columns, succeeds in reducing the nonsingular matrix A, first to an upper triangular matrix whose diagonal elements are all 1s, and subsequently to an identity matrix. In terms of the elementary matrices $E_{\text{i}}, E_{\text{ii}}, E_{\text{iii}}$, the sequence of operations can be expressed as follows:

$$E_k \cdots E_j E_i \, A \, E_l E_m \cdots E_n = I. \tag{12}$$

Multiplying both side of Eq.(12) by the inverses of the elementary matrices in correct order, we find

$$A = E_i^{-1} E_j^{-1} \cdots E_k^{-1} I E_n^{-1} \cdots E_m^{-1} E_l^{-1}. \tag{13}$$

The identity matrix in the middle of this expression makes no contribution to the product and can be eliminated. We thus see the nonsingular matrix A written as the product of a number of elementary matrices. Once again, $|A|$ is the product of the determinants of $E_i^{-1}, E_j^{-1}, \cdots, E_k^{-1}, E_l^{-1}, E_m^{-1}, \cdots E_n^{-1}$.

11.6. Determinant of the product is the product of the determinants. Our goal in the present section is to prove an important property of determinants, namely that $|AB| = |A||B|$, where A and B are arbitrary $N \times N$ matrices. We have already proven this property in cases where either A or B is an elementary matrix, $E_{\text{i}}, E_{\text{ii}}, E_{\text{iii}}$, or a product of such matrices. Extension to arbitrary matrices requires only two trivial observations.

If the matrix A happens to be singular; that is, if $|A| = 0$, then the product matrix AB must be singular as well. To prove this statement, suppose that AB is *not* singular. Then the matrix $C = (AB)^{-1}$ exists, which would have the property $ABC = I$. This would imply that the matrix BC is the inverse of A. However, A cannot have an inverse, since its determinant is assumed to be zero. Consequently, the determinant of the product matrix AB must also be zero. It thus remains to prove the theorem in the case of $|A| \neq 0$.

It was shown in the preceding section that, when $|A| \neq 0$, one can express A as a product of elementary matrices, say, $A = E_1 E_2 \cdots E_k$, with $|A| = |E_1||E_2| \cdots |E_k|$. Consequently, $AB = E_1 E_2 \cdots E_k B$. The sequence of elementary operations $E_k, \cdots, E_2, E_1$ acting on B causes the determinant of B to change either by multiplication by -1 (elementary type i operation), or multiplication by a constant c (type ii operation), or the determinant does not change at all (type iii operation). This is tantamount to multiplying the determinant of B by the product of the determinants of $E_1, E_2, \cdots, E_k$, which is the same as the determinant of A. The inevitable conclusion is that $|AB| = |A||B|$.

11.7. The Gram-Schmidt process. It is often useful to construct an orthonormal set of vectors out of a set of equidimensional and linearly independent column vectors $V_1, V_2, \cdots, V_N$. The procedure used for such constructions is known as the Gram-Schmidt process. (The same procedure is applicable to equidimensional and linearly independent row vectors $U_1, U_2, \cdots, U_N$.) By definition, the equidimensional vectors $V_1, V_2, \cdots, V_N$ are linearly independent if the identity $c_1 V_1 + c_2 V_2 + \cdots + c_N V_N = 0$ is satisfied only when the (generally complex-valued) coefficients $c_1, c_2, \cdots, c_N$ are all equal to zero. Given a set of such vectors, the Gram-Schmidt algorithm proceeds along the following lines:

i) Compute the norm of V_1 in accordance with $\|V_1\| = \sqrt{V_1^{*T} V_1}$, then normalize V_1 by its own norm to obtain the unit vector $W_1 = V_1 / \|V_1\|$.

ii) Subtract from V_2 its projection onto W_1, then normalize the resulting vector to obtain a second unit vector W_2, namely $W_2 = [V_2 - (W_1^{*T} V_2) W_1] / \|V_2 - (W_1^{*T} V_2) W_1\|$.

iii) Subtract from V_3 its projections onto W_1 and W_2, then normalize the resulting vector to obtain $W_3 = [V_3 - (W_1^{*T} V_3) W_1 - (W_2^{*T} V_3) W_2] / \|V_3 - (W_1^{*T} V_3) W_1 - (W_2^{*T} V_3) W_2\|$.

iv) Continue the process by subtracting from each V_n all of its projections onto the preceding orthonormalized vectors $W_1, W_2, \cdots, W_{n-1}$, then normalize the resulting vector to obtain W_n.

It is now easy to verify that the unit vectors $W_1, W_2, \cdots, W_N$ thus obtained are pair-wise orthogonal; that is, $W_m^{*T} W_n = 0$ for $m \neq n$. The process as described applies to real as well as complex vectors. (Note that one could also use an alternative definition for the inner product of two complex vectors in which neither vector is conjugated. The Gram-Schmidt process remains applicable to complex vectors in cases where the definition of the inner product of a pair of vectors requires the conjugation neither of the first nor of the second vector.) The square matrix A whose rows (or columns) consist of orthonormal vectors can be inverted simply by conjugate transposition; that is, $A^{-1} = A^{*T}$. Such matrices are generally referred to as unitary matrices. Some of the interesting properties of unitary matrices will be explored in section 11.

11.8. Eigenvalues and eigenvectors. For the $N \times N$ square matrix A, the right eigen-vector V is an $N \times 1$ (column) vector with the property $AV = \lambda V$, where the constant λ is referred to as the eigenvalue of the matrix A associated with its eigenvector V. The defining equation, when expressed as $(A - \lambda I)V = 0$, shows that $V = (A - \lambda I)^{-1}(0)$ is identically zero unless the matrix $(A - \lambda I)$ happens to be singular. Consequently, for a nontrivial eigenvector V to exist, the determinant of $(A - \lambda I)$ must vanish. The resulting N^{th} order polynomial equation in λ, namely $|A - \lambda I| = 0$, can then be solved for its N roots $\lambda_1, \lambda_2, \cdots, \lambda_N$, which constitute the N eigenvalues of the matrix A.

Each eigenvalue λ_n may be used to solve the equation $(A - \lambda_n I)V_n = 0$ for the corresponding eigenvector V_n. Considering that the coefficient matrix $(A - \lambda_n I)$ of the equation is singular, the solution V_n is not expected to be unique. However, if the roots $\lambda_1, \lambda_2, \cdots, \lambda_N$ of the so-called characteristic equation $|A - \lambda I| = 0$ happen to be distinct from one another (i.e., no degenerate roots), then each V_n turns out to be unique to within a scale factor. This arbitrariness (to within a scale factor) of the eigenvectors of nondegenerate matrices should not be surprising, since the equation $(A - \lambda_n I)V_n = 0$ clearly indicates that V_n will continue to satisfy this (homogeneous) equation after being multiplied by any constant coefficient. The freedom to scale the eigenvectors is generally used to normalize them by adjusting the scale-factor such that $V^{*T} V = 1$.

The left eigenvectors and their corresponding eigenvalues for the $N \times N$ matrix A are similarly defined as $1 \times N$ (row) vectors U, with the property that $UA = \lambda U$. The equation may now be expressed as $U(A - \lambda I) = 0$, which reveals the characteristic equation $|A - \lambda I| = 0$ for left eigenvalues to be the same as that for right eigenvalues. Thus, in general, every $N \times N$ matrix has N eigenvalues (although some may be degenerate, i.e., overlapping), and, for every eigenvalue λ_n, there exists at least one right eigenvector V_n as well as one left eigenvector U_n. The set of left eigenvectors U_n is related to the set of right eigenvectors V_n, as will be seen shortly.

The triple matrix product $U_n A V_n$ equals $\lambda_n U_n V_n$ unambiguously. However the triple product $U_m A V_n$, where $m \neq n$, could be written as $\lambda_m U_m V_n$ or as $\lambda_n U_m V_n$, depending on which pair of matrices are first multiplied together. Obviously, both answers cannot be correct if $\lambda_m \neq \lambda_n$, unless $U_m V_n = 0$. The right and left eigenvectors associated with differing eigenvalues λ_m and λ_n of a given matrix are thus seen to have a zero inner product—inner product means that the row vector U_m precedes the column vector V_n. Such vectors are said to be mutually orthogonal.

Let us now construct three new $N \times N$ matrices, $\widetilde{U}, \widetilde{V}, \Lambda$, by arranging the left eigenvectors $U_1, U_2, \cdots, U_N$ as successive rows of $\widetilde{U}$, the right eigenvectors $V_1, V_2, \cdots, V_N$ as successive columns of $\widetilde{V}$, and the eigenvalues $\lambda_1, \lambda_2, \cdots, \lambda_N$ as successive elements of the diagonal matrix Λ, as follows:

$$\widetilde{U} = \begin{pmatrix} U_1 \\ U_2 \\ \vdots \\ U_N \end{pmatrix}, \qquad \widetilde{V} = (V_1 \mid V_2 \mid \cdots \mid V_N), \qquad \Lambda = \begin{pmatrix} \lambda_1 & 0 & \cdots & 0 \\ 0 & \lambda_2 & \cdots & 0 \\ \vdots & \vdots & \ddots & \vdots \\ 0 & 0 & \cdots & \lambda_N \end{pmatrix}. \quad (14)$$

We may now write $\widetilde{U}A = \Lambda\widetilde{U}$ and $A\widetilde{V} = \widetilde{V}\Lambda$. If the eigenvalues of A turn out to be nondegenerate (i.e., no repeated eigenvalues), it can be shown (see problems 7 and 8 at the end of the chapter) that the stacked eigenvector matrices $\widetilde{U}$ and $\widetilde{V}$ will be nonsingular (i.e., invertible), in which case one can diagonalize the matrix A by multiplying it on the left and right either by $\widetilde{U}$ and $\widetilde{U}^{-1}$, or by $\widetilde{V}^{-1}$ and $\widetilde{V}$. In other words, $\widetilde{U}A\widetilde{U}^{-1} = \Lambda$, and $\widetilde{V}^{-1}A\widetilde{V} = \Lambda$. When the eigenvectors are properly scaled, it can be shown that $\widetilde{U}$ and $\widetilde{V}$ are the inverses of each other.

Example. Consider the case of the simple 2×2 matrix $A = \begin{pmatrix} 2 & 1 \\ 6 & -3 \end{pmatrix}$, whose characteristic equation is written

$$|A - \lambda I| = \begin{vmatrix} 2-\lambda & 1 \\ 6 & -3-\lambda \end{vmatrix} = (2-\lambda)(-3-\lambda) - 1 \times 6 = \lambda^2 + \lambda - 12 = 0. \quad (15)$$

The solutions $\lambda_1 = 3$ and $\lambda_2 = -4$ of the characteristic equation correspond to the right eigenvectors V_1 and V_2, which may be found as follows:

$$(A - \lambda_1 I)V_1 = 0 \quad \rightarrow \quad \begin{pmatrix} -1 & 1 \\ 6 & -6 \end{pmatrix}\begin{pmatrix} v_{11} \\ v_{12} \end{pmatrix} = 0 \quad \rightarrow \quad v_{11} = v_{12}. \quad (16)$$

$$(A - \lambda_2 I)V_2 = 0 \quad \rightarrow \quad \begin{pmatrix} 6 & 1 \\ 6 & 1 \end{pmatrix}\begin{pmatrix} v_{21} \\ v_{22} \end{pmatrix} = 0 \quad \rightarrow \quad v_{22} = -6v_{21}. \quad (17)$$

Here, $v_{12} \neq 0$ and $v_{21} \neq 0$ are arbitrary constants. The matrix $\widetilde{V}$, formed by stacking these right eigenvectors as columns, and the inverse $\widetilde{V}^{-1}$ of $\widetilde{V}$, are readily found to be

$$\widetilde{V} = \begin{pmatrix} v_{12} & v_{21} \\ v_{12} & -6v_{21} \end{pmatrix}; \qquad \widetilde{V}^{-1} = \frac{1}{7}\begin{pmatrix} 6/v_{12} & 1/v_{12} \\ 1/v_{21} & -1/v_{21} \end{pmatrix}. \quad (18)$$

The matrix A may thus be decomposed as follows:

$$A = \widetilde{V}\Lambda\widetilde{V}^{-1} = \frac{1}{7}\begin{pmatrix} v_{12} & v_{21} \\ v_{12} & -6v_{21} \end{pmatrix}\begin{pmatrix} 3 & 0 \\ 0 & -4 \end{pmatrix}\begin{pmatrix} 6/v_{12} & 1/v_{12} \\ 1/v_{21} & -1/v_{21} \end{pmatrix}. \quad (19)$$

A similar set of calculations for the left eigenvectors yields

$$U_1(A - \lambda_1 I) = 0 \quad \rightarrow \quad (u_{11} \quad u_{12})\begin{pmatrix} -1 & 1 \\ 6 & -6 \end{pmatrix} = 0 \quad \rightarrow \quad u_{11} = 6u_{12}. \quad (20)$$

$$U_2(A - \lambda_2 I) = 0 \quad \rightarrow \quad (u_{21} \quad u_{22})\begin{pmatrix} 6 & 1 \\ 6 & 1 \end{pmatrix} = 0 \quad \rightarrow \quad u_{21} = -u_{22}. \quad (21)$$

Once again, admitting the arbitrary constants $u_{12} \neq 0$ and $u_{21} \neq 0$, we find

$$\tilde{U} = \begin{pmatrix} 6u_{12} & u_{12} \\ u_{21} & -u_{21} \end{pmatrix}; \qquad \tilde{U}^{-1} = \frac{1}{7}\begin{pmatrix} 1/u_{12} & 1/u_{21} \\ 1/u_{12} & -6/u_{21} \end{pmatrix}. \tag{22}$$

$$A = \tilde{U}^{-1}\Lambda\tilde{U} = \frac{1}{7}\begin{pmatrix} 1/u_{12} & 1/u_{21} \\ 1/u_{12} & -6/u_{21} \end{pmatrix}\begin{pmatrix} 3 & 0 \\ 0 & -4 \end{pmatrix}\begin{pmatrix} 6u_{12} & u_{12} \\ u_{21} & -u_{21} \end{pmatrix}. \tag{23}$$

Clearly, Eqs.(19) and (23) are equivalent decompositions of A, provided that the arbitrary coefficients are chosen such that $u_{12} = 1/v_{12}$ and $u_{21} = 1/v_{21}$. Note also that the right eigenvector V_1 associated with λ_1 is orthogonal to the left eigenvector U_2 associated with λ_2. Similarly, the right eigenvector V_2 associated with λ_2 is orthogonal to the left eigenvector U_1 associated with λ_1. In other words,

$$U_1V_2 = (6u_{12} \quad u_{12})\begin{pmatrix} v_{21} \\ -6v_{21} \end{pmatrix} = 0, \tag{24}$$

$$U_2V_1 = (u_{21} \quad -u_{21})\begin{pmatrix} v_{11} \\ v_{11} \end{pmatrix} = 0. \tag{25}$$

In general, a nondegenerate $N \times N$ matrix A can always be diagonalized with the aid of the matrix $\tilde{V}$ whose columns are the right eigenvectors of A, as $\tilde{V}^{-1}A\tilde{V} = \Lambda$, and also with the aid of the matrix $\tilde{U}$, whose rows are the left eigenvectors of A, as $\tilde{U}A\tilde{U}^{-1} = \Lambda$. Moreover, assigning the arbitrary scale factors of U_n and V_n such that $U_nV_n = 1$ for all $1 \leq n \leq N$, it is easy to verify that $\tilde{U}$ and $\tilde{V}$ end up being the inverses of each other.

11.9. Matrix operations aided by eigenvalues and eigenvectors. When the square matrix A is diagonalizable, it can be written as $A = \tilde{V}\Lambda\tilde{V}^{-1}$. Here, the diagonal matrix Λ contains the eigenvalues $\lambda_1, \lambda_2, \cdots, \lambda_N$ of A on its diagonal, and the square matrix $\tilde{V}$ has the corresponding eigenvectors $V_1, V_2, \cdots, V_N$ of A as its columns. It is readily seen that $|A| = |\tilde{V}||\Lambda||\tilde{V}^{-1}|$ reduces to $|\Lambda|$, which in turn reduces to the product $\lambda_1\lambda_2 \cdots \lambda_N$ of the eigenvalues of A. Also, given that the inverse of the triple matrix product XYZ is $Z^{-1}Y^{-1}X^{-1}$, one sees immediately that $A^{-1} = \tilde{V}\Lambda^{-1}\tilde{V}^{-1}$; in other words, while the inverse of A shares the eigenvectors $V_1, V_2, \cdots, V_n$ with A, the corresponding eigenvalues of A^{-1} are $\lambda_1^{-1}, \lambda_2^{-1}, \cdots, \lambda_N^{-1}$. This property is directly verified by noting that $AV_n = \lambda_n V_n$, when multiplied by $\lambda_n^{-1}A^{-1}$ on both sides, yields $\lambda_n^{-1}V_n = A^{-1}V_n$.

Digression: The determinant of a matrix always equals the product of its eigenvalues, whether or not the matrix can be diagonalized. This is due to the fact that the characteristic equation $|A - \lambda I| = \eta_N\lambda^N + \eta_{N-1}\lambda^{N-1} + \cdots + \eta_1\lambda + \eta_0 = 0$ has two important properties: (i) The coefficient η_N of the highest-order term is $(-1)^N$, and (ii) the lowest-order term, η_0, equals $|A|$. To prove the first of these assertions, express the determinant in terms of the matrix elements and their cofactors, then observe that the highest-order term of the characteristic equation arises from the product $(a_{11} - \lambda)(a_{22} - \lambda)\cdots(a_{NN} - \lambda)$. To prove the second property, simply set $\lambda = 0$ in the characteristic equation to find that its left-hand side becomes $|A|$, while its right-hand side reduces to η_0. If we now rewrite the characteristic equation in terms of its roots as $(\lambda_1 - \lambda)(\lambda_2 - \lambda)\cdots(\lambda_N - \lambda) = 0$, it is clear that the coefficient of λ^N will be $(-1)^N$, and that the lowest-order term (obtained by setting $\lambda = 0$) will be the product $\lambda_1\lambda_2\cdots\lambda_N$ of the roots of the characteristic equation. We have thus proved that, under any and all circumstances, the determinant of a square matrix equals the product of its eigenvalues.

There exists a host of algebraic operations on diagonalizable matrices, which reduce to operations on the eigenvalues of the matrix under consideration. For instance, the square of the diagonalizable matrix A can be expressed as

$$A^2 = AA = (\tilde{V}\Lambda\tilde{V}^{-1})(\tilde{V}\Lambda\tilde{V}^{-1}) = \tilde{V}\Lambda\cancelto{I}{(\tilde{V}^{-1}\tilde{V})}\Lambda\tilde{V}^{-1} = \tilde{V}\Lambda^2\tilde{V}^{-1}. \tag{26}$$

Thus, if the matrix A, whose eigenvalues are $\lambda_1, \lambda_2, \cdots, \lambda_N$, is diagonalizable, then its square A^2, while sharing the eigenvectors $V_1, V_2, \cdots, V_N$ with A, will have the eigenvalues $\lambda_1^2, \lambda_2^2, \cdots, \lambda_N^2$. An alternative way of arriving at the same conclusion is by multiplying both sides of $AV_n = \lambda_n V_n$ with A, which leads to $A^2 V_n = \lambda_n A V_n = \lambda_n^2 V_n$. (The latter result, of course, is valid whether or not A is diagonalizable.) Needless to say, similar arguments can be made for the left eigenvectors $U_1, U_2, \cdots, U_N$ of A. It is also obvious that A^k, where k is any positive integer, has the same eigenvectors as A, while its eigenvalues are $\lambda_1^k, \lambda_2^k, \cdots, \lambda_N^k$. In the case of A^{-k}, where k is some positive integer, if one assumes that the matrix A is invertible (i.e., none of its eigenvalues vanish), then repeated multiplication of A^{-1} into the identity $AV_n = \lambda_n V_n$ (or the identity $U_n A = \lambda_n U_n$) reveals that A^{-k} shares its eigenvectors with A, but has eigenvalues $\lambda_1^{-k}, \lambda_2^{-k}, \cdots, \lambda_N^{-k}$.

The exponential of a square matrix A is defined by analogy with the Taylor series expansion of e^x, where x is a real or complex number; that is, $\exp(x) = \sum_{n=0}^{\infty}(x^n/n!)$. Thus, by definition,

$$\exp(A) = I + A + \tfrac{1}{2!}A^2 + \tfrac{1}{3!}A^3 + \cdots + \tfrac{1}{n!}A^n + \cdots. \tag{27}$$

For a diagonalizable matrix A, we have $A^n = \tilde{V}\Lambda^n\tilde{V}^{-1}$ and, therefore,

$$\exp(A) = \tilde{V}\left(I + \Lambda + \tfrac{1}{2!}\Lambda^2 + \tfrac{1}{3!}\Lambda^3 + \cdots + \tfrac{1}{n!}\Lambda^n + \cdots\right)\tilde{V}^{-1} = \tilde{V}[\exp(\Lambda)]\tilde{V}^{-1}. \tag{28}$$

Note that $\exp(\Lambda)$ is a simple diagonal matrix, with the elements on the diagonal being $\exp(\lambda_n)$, where λ_n is an eigenvalue of A.

Example. Consider the 2×2 matrix $\begin{pmatrix} A & B \\ C & D \end{pmatrix}$ whose determinant is equal to 1; that is, $AD - BC = 1$. In what follows, we shall find a general expression for the nth power of this so-called $ABCD$ matrix, where n is an arbitrary positive integer. The eigenvalues λ_1 and λ_2 of the matrix are readily evaluated, as follows:

$$\begin{pmatrix} A & B \\ C & D \end{pmatrix}\begin{pmatrix} v_1 \\ v_2 \end{pmatrix} = \lambda\begin{pmatrix} v_1 \\ v_2 \end{pmatrix} \;\rightarrow\; \begin{vmatrix} A-\lambda & B \\ C & D-\lambda \end{vmatrix} = 0 \;\rightarrow\; (A-\lambda)(D-\lambda) - BC = 0$$

$$\rightarrow\; \lambda^2 - (A+D)\lambda + \cancelto{1}{(AD - BC)} = 0 \;\rightarrow\; \lambda_{1,2} = ½(A+D) \pm \sqrt{¼(A+D)^2 - 1}.$$

Introducing the parameter θ, where $\cos\theta = ½(A+D)$, reduces the characteristic equation to $\lambda^2 - 2\lambda\cos\theta + 1 = 0$, which yields $\lambda_{1,2} = \cos\theta \pm \mathrm{i}\sin\theta = \exp(\pm\mathrm{i}\theta)$. Next, we determine the right eigenvectors $V = (v_1 \quad v_2)^T$ of the matrix, as follows:

$$Av_1 + Bv_2 = \lambda v_1 \;\rightarrow\; v_2 = \left(\frac{\lambda - A}{B}\right)v_1 \;\rightarrow\; V_{1,2} = \begin{pmatrix} 1 \\ [\exp(\pm\mathrm{i}\theta) - A]/B \end{pmatrix}v_1.$$

The parameter v_1 is arbitrary and can be set to 1. The 2×2 matrix $\tilde{V}$, whose columns are the eigenvectors of the $ABCD$ matrix, along with its inverse $\tilde{V}^{-1}$, are thus given by

$$\tilde{V} = \begin{pmatrix} 1 & 1 \\ [\exp(\mathrm{i}\theta) - A]/B & [\exp(-\mathrm{i}\theta) - A]/B \end{pmatrix},$$

$$\tilde{V}^{-1} = \frac{\mathrm{i}B}{2\sin\theta} \begin{pmatrix} [\exp(-\mathrm{i}\theta) - A]/B & -1 \\ -[\exp(\mathrm{i}\theta) - A]/B & 1 \end{pmatrix}.$$

The nth power of the $ABCD$ matrix may now be computed as follows:

$$\begin{aligned}
\begin{pmatrix} A & B \\ C & D \end{pmatrix}^n &= \tilde{V}\Lambda^n\tilde{V}^{-1} \\
&= \frac{\mathrm{i}}{2B\sin\theta} \begin{pmatrix} B & B \\ \exp(\mathrm{i}\theta) - A & \exp(-\mathrm{i}\theta) - A \end{pmatrix} \begin{pmatrix} \exp(\mathrm{i}n\theta) & 0 \\ 0 & \exp(-\mathrm{i}n\theta) \end{pmatrix} \begin{pmatrix} \exp(-\mathrm{i}\theta) - A & -B \\ A - \exp(\mathrm{i}\theta) & B \end{pmatrix} \\
&= \frac{\mathrm{i}}{2B\sin\theta} \begin{pmatrix} B & B \\ \exp(\mathrm{i}\theta) - A & \exp(-\mathrm{i}\theta) - A \end{pmatrix} \begin{pmatrix} [\exp(-\mathrm{i}\theta) - A]\exp(\mathrm{i}n\theta) & -B\exp(\mathrm{i}n\theta) \\ [A - \exp(\mathrm{i}\theta)]\exp(-\mathrm{i}n\theta) & B\exp(-\mathrm{i}n\theta) \end{pmatrix} \\
&= \frac{\mathrm{i}}{2B\sin\theta} \begin{pmatrix} 2\mathrm{i}B\sin[(n-1)\theta] - 2\mathrm{i}AB\sin(n\theta) & -2\mathrm{i}B^2\sin(n\theta) \\ 2\mathrm{i}[1 + A^2 - 2A\cos\theta]\sin(n\theta) & -2\mathrm{i}B\sin[(n+1)\theta] + 2\mathrm{i}AB\sin(n\theta) \end{pmatrix} \\
&= \frac{1}{\sin\theta} \begin{pmatrix} A\sin(n\theta) - \sin[(n-1)\theta] & B\sin(n\theta) \\ C\sin(n\theta) & \sin[(n+1)\theta] - A\sin(n\theta) \end{pmatrix}.
\end{aligned}$$

Upon further simplification, the term appearing on the lower right-hand side of the matrix becomes

$$\begin{aligned}
\sin[(n+1)\theta] - A\sin(n\theta) &= \sin(n\theta)\cos\theta + \cos(n\theta)\sin\theta - A\sin(n\theta) \\
&= 2\sin(n\theta)\cos\theta - \sin(n\theta)\cos\theta + \cos(n\theta)\sin\theta - A\sin(n\theta) \\
&= D\sin(n\theta) - \sin[(n-1)\theta].
\end{aligned}$$

Substitution into the preceding expression for the nth power of the $ABCD$ matrix finally yields

$$\begin{pmatrix} A & B \\ C & D \end{pmatrix}^n = \frac{1}{\sin\theta} \begin{bmatrix} A\sin(n\theta) - \sin(n-1)\theta & B\sin(n\theta) \\ C\sin(n\theta) & D\sin(n\theta) - \sin(n-1)\theta \end{bmatrix}. \quad (29)$$

This formula, occasionally cited with a reference to Sylvester's theorem,[6] finds application in optical physics and, in particular, the analysis of resonant cavities.

11.10. Systems of first-order linear ordinary differential equations. The solution of a set of N first-order linear ODEs with constant coefficients in N unknown functions can be obtained with the aid of matrix algebra in ways that closely parallel the solution of a single first-order ODE in only one unknown function. Given a first-order linear ODE with constant coefficients, say, $af'(x) + bf(x) = c$, the solution is the sum of a particular solution, c/b, and the homogeneous solution, $h_0\exp(-bx/a)$, with the arbitrary coefficient h_0 determined after satisfying the initial condition, say, $f(x)|_{x=x_0} = f_0$. Let us now examine the following set of N equations in the N unknown functions $f_1(x)$, $f_2(x), \cdots, f_N(x)$, where the unknowns are arranged in the $N \times 1$ column vector $F(x)$:

$$AF'(x) + BF(x) = C. \tag{30}$$

In this equation, A and B are $N \times N$ constant matrices, whereas C is a constant $N \times 1$ column vector. Assuming that A and B are invertible, we find the particular solution of Eq.(30) as $B^{-1}C$ and the homogeneous solution as $[\exp(-A^{-1}Bx)]H_0$, with H_0 being an arbitrary $N \times 1$ column vector to be determined by satisfying the initial conditions, say, $F(x = x_0) = F_0$. The general solution of Eq.(30) is thus given by

$$F(x) = (B^{-1}C) + [\exp(-A^{-1}Bx)]H_0. \tag{31}$$

Denoting the eigenvalues of the product matrix $A^{-1}B$ by $\lambda_1, \lambda_2, \cdots, \lambda_N$, it is easy to see that each solution function $f_n(x)$ can be written as a linear combination of the exponential functions $\exp(-\lambda_1 x), \exp(-\lambda_2 x), \cdots, \exp(-\lambda_N x)$, plus an additive constant term derived from the vector $B^{-1}C$.

11.11. The Cayley-Hamilton theorem. A square matrix A always satisfies its own characteristic equation; that is, if one substitutes the $N \times N$ matrix A for λ in the characteristic polynomial $|A - \lambda I| = \sum_{k=0}^{N} \eta_k \lambda^k$, one will find that $\sum_{k=0}^{N} \eta_k A^k = O$, where O is the $N \times N$ null matrix (i.e., the matrix whose elements are all zeros). The proof of the theorem is quite easy for diagonalizable matrices, because replacing A with $\tilde{V}\Lambda\tilde{V}^{-1}$ in the matrix polynomial yields

$$\sum_{k=0}^{N} \eta_k A^k = \tilde{V}(\sum_{k=0}^{N} \eta_k \Lambda^k)\tilde{V}^{-1}. \tag{32}$$

The matrix $\sum_{k=0}^{N} \eta_k \Lambda^k$ appearing in Eq.(32) is a diagonal matrix, its diagonal elements being $\sum_{k=0}^{N} \eta_k \lambda_n^k$, where λ_n is the nth eigenvalue of A. Since all λ_ns satisfy the characteristic equation $\sum_{k=0}^{N} \eta_k \lambda^k = 0$, we see that $\sum_{k=0}^{N} \eta_k \Lambda^k$ is a null matrix and, therefore, $\sum_{k=0}^{N} \eta_k A^k = O$. The proof of the Cayley-Hamilton theorem in the more general case of a non-diagonalizable A is somewhat more complicated and will not be pursued here; the interested reader may consult the literature for the general proof.[7-9]

11.12. Unitary matrices. A unitary matrix A is one that preserves the norm of any (row or column) vector which it multiplies. Thus, if $V_2 = AV_1$, then $\|V_2\|^2 = V_2^{*T}V_2 = V_1^{*T}A^{*T}AV_1$ will equal $\|V_1\|^2$, provided that $A^{*T}A = I$. Similarly, if $U_2 = U_1A$, then $\|U_2\|^2 = U_2U_2^{*T} = U_1AA^{*T}U_1^{*T}$ will equal $\|U_1\|^2$, provided that $AA^{*T} = I$.

By definition, a (square) matrix A is unitary if its conjugate transpose A^{*T} is also its inverse A^{-1}. Considering that the right and left eigenvectors of A satisfy $AV = \lambda V$ and $UA = \lambda U$, respectively, the preservation of $\|U\|$ and $\|V\|$ demands that $\lambda\lambda^* = 1$, or, equivalently, that each eigenvalue λ of a unitary matrix be in the form of $\exp(\mathrm{i}\varphi)$, where φ is some real-valued phase angle. Thus, the eigenvalues of unitary matrices reside on the unit circle in the complex plane.

Given that the conjugate transpose of $AV = \lambda V$ is $V^{*T}A^{*T} = \lambda^* V^{*T}$, and that, for a unitary matrix, $A^{*T} = A^{-1}$, it is seen that the right eigenvector V of A, when conjugate transposed, becomes the left eigenvector of A^{-1}, with a corresponding eigenvalue of $\lambda^* = \exp(-\mathrm{i}\varphi)$. Since, as argued in the preceding section, a matrix A and its inverse A^{-1} share their set of left eigenvectors (in addition to their set of right eigenvectors), we conclude that V^{*T} is the left eigenvector of A corresponding to its right eigenvector V; that is, V and V^{*T} have the same eigenvalue $\lambda = \exp(\mathrm{i}\varphi)$. Suppose now that two right eigenvectors, say, V_m and V_n, have differing eigenvalues, λ_m and λ_n. Then the left eigenvector V_m^{*T} must be orthogonal to the right eigenvector V_n; that is, $V_m^{*T}V_n = 0$.

If we proceed to normalize all the eigenvectors, so that $V_n^{*T}V_n = 1$, and assume that none of the eigenvalues $\lambda_1, \lambda_2, \cdots, \lambda_N$ of A coincide with each other, then the $N \times N$ matrix $\tilde{V}$, whose columns are the (orthonormal) right eigenvectors $V_1, V_2, \cdots, V_N$ of A, will have $\tilde{V}^{*T}$ as its inverse. The matrix $\tilde{V}$ formed by the (normalized) eigenvectors of a unitary matrix A is thus seen to be a unitary matrix in its own right.

That a unitary matrix A can be diagonalized by means of another unitary matrix $\tilde{V}$ is a general property that does not require the non-degeneracy of the eigenvalues of A invoked in the preceding paragraph. Section 14 provides a general proof that all normal matrices (of which unitary matrices form a subset) are diagonalizable by unitary matrices.

11.13. Hermitian matrices. A Hermitian matrix is one that equals its own conjugate transpose; that is, $A^{*T} = A$. This requires that all the diagonal elements a_{nn} of A be real, and that the off-diagonal elements satisfy the relation $a_{mn}^* = a_{nm}$. (It is customary to denote the conjugate transpose of a matrix by a † superscript, in which case the Hermitian character of A is expressed by the identity $A^\dagger = A$.) As it turns out, all eigenvalues of a Hermitian matrix A are real valued, and the eigenvectors that correspond to differing eigenvalues of A are mutually orthogonal.

To prove the first of the above statements, multiply (on the left) both sides of the equation $AV = \lambda V$ by V^{*T} to arrive at $V^{*T}AV = \lambda V^{*T}V$. The conjugate transpose of the latter equation is $V^{*T}A^{*T}V = \lambda^* V^{*T}V$. However, since $A^{*T} = A$, the equation and its conjugate transpose must be identical, which is possible only if $\lambda = \lambda^*$ (i.e., if λ is real).

To prove the second statement, consider the conjugate transpose of $AV = \lambda V$; that is, $V^{*T}A^{*T} = V^{*T}A = \lambda V^{*T}$, which identifies V^{*T} as a left eigenvector of A that shares the eigenvalue λ with the right eigenvector V. When a pair of eigenvalues differ from each other, the left eigenvector associated with one will be orthogonal to the right eigenvector associated with the other. Consequently, when none of the eigenvalues $\lambda_1, \lambda_2, \cdots, \lambda_N$ of A overlap each other, the corresponding right eigenvectors, after normalization, will form an orthonormal set. (The same is true of the left eigenvectors, of course.) The matrix $\tilde{V}$, whose columns consist of the (normalized) right eigenvectors of the Hermitian matrix A, is thus seen to be a unitary matrix. The inverse of $\tilde{V}$ is its conjugate transpose, having the left eigenvectors of A as its rows.

That a Hermitian matrix A can be diagonalized by means of a unitary matrix $\tilde{V}$ is quite general and does not require the non-degeneracy of the eigenvalues of A invoked in the preceding paragraph. In section 14, we provide a general proof that all normal matrices (of which Hermitian matrices form a subset) can be diagonalized via unitary matrices.

In similar fashion, one can readily prove that an anti-Hermitian (or skew-Hermitian) matrix has purely imaginary eigenvalues, and that the eigenvectors corresponding to differing eigenvalues of an anti-Hermitian matrix are orthogonal to one another. (Recall that, by definition, the square matrix A is anti-Hermitian if $A^{*T} = -A$; that is, if all its diagonal elements are imaginary, while its off-diagonal elements satisfy the identity $a_{mn}^* = -a_{nm}$.) Given that anti-Hermitian matrices are *normal*, the arguments of the following section demonstrate that they too can be diagonalized via unitary matrices, irrespective of eigenvalue degeneracies. An interesting property of the anti-Hermitian matrix A is that $\exp(A)$ is unitary, simply because $\exp(A)\exp(A^{*T}) = \exp(O) = I$.

11.14. The Schur decomposition. An important theorem of linear algebra, due to Issai Schur (1875–1941), asserts that any $N \times N$ matrix A can be decomposed as $A = STS^{*T}$, where S is a unitary $N \times N$ matrix, and T an upper (or lower) triangular $N \times N$ matrix. In the special case of a *normal* matrix A (i.e., when $A^{*T}A = AA^{*T}$), the matrix T of the Schur decomposition turns out to be diagonal, its diagonal elements being the eigenvalues $\lambda_1, \lambda_2, \cdots, \lambda_N$ of A. Thus, the Schur decomposition of normal matrices (which include all Hermitian, skew-Hermitian, unitary, and symmetric real matrices) ensures that such matrices, irrespective of their eigenvalue degeneracies, can always be diagonalized by means of a unitary matrix S.

To demonstrate the feasibility of the Schur decomposition, we note that any square matrix A has at least one eigenvalue λ_1 and a corresponding eigenvector V_1. We begin by assuming that V_1 is normalized; that is, $V_1^{*T}V_1 = 1$, then select $N-1$ orthonormal vectors $\mathbb{V}_2, \mathbb{V}_3, \cdots, \mathbb{V}_N$ from the available space of N-dimensional column vectors in such a way that the $N \times N$ matrix $S_1 = [V_1 \mid \mathbb{V}_2 \mid \mathbb{V}_3 \mid \cdots \mid \mathbb{V}_N]$ is unitary. (To this end, one may rely on the Gram-Schmidt process to construct the desired set of N orthonormal vectors.) It is now easy to see that $S_1^{*T}AS_1$ has the following structure

$$\begin{bmatrix} V_1^{*T} \\ \mathbb{V}_2^{*T} \\ \mathbb{V}_3^{*T} \\ \vdots \\ \mathbb{V}_N^{*T} \end{bmatrix} \begin{bmatrix} a_{11} & a_{12} & \cdots & a_{1N} \\ a_{21} & a_{22} & \cdots & a_{2N} \\ \vdots & \vdots & \ddots & \vdots \\ a_{N1} & a_{N2} & \cdots & a_{NN} \end{bmatrix} [V_1 \ \mathbb{V}_2 \ \mathbb{V}_3 \ \cdots \ \mathbb{V}_N] = \begin{bmatrix} \lambda_1 & \beta_1 & \cdots & \beta_{N-1} \\ 0 & b_{11} & \cdots & b_{1,N-1} \\ \vdots & \vdots & \ddots & \vdots \\ 0 & b_{N-1,1} & \cdots & b_{N-1,N-1} \end{bmatrix}. \quad (33)$$

(In Eq. (33), the $b_{m,n}$ block is marked B.)

The square matrix on the right-hand side of Eq.(33) has the eigenvalue λ_1 of A in its upper-left corner, has zeros elsewhere in its first column, has elements $\beta_1, \beta_2, \cdots, \beta_{N-1}$ in its first row, and has an embedded $(N-1) \times (N-1)$ square matrix B, whose elements are denoted by $b_{m,n}$ in the equation.

Let us now apply to matrix B the same procedure as was applied to matrix A. In other words, since B must have at least one eigenvalue λ_2 and a corresponding eigenvector, we can construct a unitary matrix R so that $R^{*T}BR$ has the same structure as the matrix appearing on the right-hand side of Eq.(33), albeit with the reduced dimensions of $(N-1) \times (N-1)$, and with λ_2 residing at its upper-left corner. We will then have

$$\begin{bmatrix} 1 & 0 & \cdots & 0 \\ 0 & r_{11}^* & \cdots & r_{N-1,1}^* \\ \vdots & \vdots & \ddots & \vdots \\ 0 & r_{1,N-1}^* & \cdots & r_{N-1,N-1}^* \end{bmatrix} \begin{bmatrix} \lambda_1 & \beta_1 & \cdots & \beta_{N-1} \\ 0 & b_{11} & \cdots & b_{1,N-1} \\ \vdots & \vdots & \ddots & \vdots \\ 0 & b_{N-1,1} & \cdots & b_{N-1,N-1} \end{bmatrix} \begin{bmatrix} 1 & 0 & \cdots & 0 \\ 0 & r_{11} & \cdots & r_{1,N-1} \\ \vdots & \vdots & \ddots & \vdots \\ 0 & r_{N-1,1} & \cdots & r_{N-1,N-1} \end{bmatrix}$$

$$= \begin{bmatrix} \lambda_1 & \gamma_1 & \gamma_2 & \cdots & \gamma_{N-1} \\ 0 & \lambda_2 & \delta_1 & \cdots & \delta_{N-2} \\ 0 & 0 & c_{11} & \cdots & c_{1,N-2} \\ \vdots & \vdots & \vdots & \ddots & \vdots \\ 0 & 0 & c_{N-2,1} & \cdots & c_{N-2,N-2} \end{bmatrix}. \quad (34)$$

(In Eq. (34), the blocks are marked R^{*T}, B, R and C, respectively.)

It is not difficult to verify that the $N \times N$ matrix $S_2 = \begin{bmatrix} 1 & 0 \\ 0 & R \end{bmatrix}$ is unitary; that is, its columns are orthonormal. The left-hand side of Eq.(34) is now seen to be $S_2^{*T}S_1^{*T}AS_1S_2$.

Also, S_1S_2 is unitary, because $(S_1S_2)^{-1} = S_2^{-1}S_1^{-1} = S_2^{*T}S_1^{*T} = (S_1S_2)^{*T}$. We have thus constructed a unitary matrix S_1S_2, which transforms A into the matrix shown on the right-hand side of Eq.(34). Repeated application of the same procedure to matrix C and its descendants, in the manner of Eq.(34), would nullify all the elements below the diagonal of the transformed matrix, resulting in a final (transformed) matrix that is upper triangular. This completes our proof of Schur's decomposition algorithm.

The eigenvalues of A are the same as the eigenvalues of the upper-triangular matrix T, which turn out to be the diagonal elements λ_n of the T matrix. This is because the characteristic equations of A and T are intimately related, as shown below:

$$|A - \lambda I| = |STS^{*T} - \lambda SS^{*T}| = |S||T - \lambda I||S^{*T}| = |T - \lambda I| = \prod_{n=1}^{N}(\lambda - \lambda_n) = 0. \quad (35)$$

In the special case of a normal matrix A, the Schur decomposition yields, as always, a unitary matrix S and an upper diagonal matrix T so that $A = STS^{*T}$. We thus have

$$\begin{aligned} TT^{*T} &= (S^{*T}AS)(S^{*T}AS)^{*T} = S^{*T}ASS^{*T}A^{*T}S = S^{*T}AA^{*T}S \\ &= S^{*T}A^{*T}AS = S^{*T}A^{*T}SS^{*T}AS = (S^{*T}AS)^{*T}(S^{*T}AS) = T^{*T}T. \end{aligned} \quad (36)$$

Considering that T is an upper triangular-matrix, we may write TT^{*T} as follows:

$$TT^{*T} = \begin{bmatrix} \lambda_1 & t_{1,2} & t_{1,3} & \cdots & t_{1,N} \\ 0 & \lambda_2 & t_{2,3} & \cdots & t_{2,N} \\ 0 & 0 & \vdots & \cdots & \vdots \\ 0 & 0 & \vdots & \cdots & \vdots \\ 0 & 0 & 0 & \lambda_{N-1} & t_{N-1,N} \\ 0 & 0 & 0 & 0 & \lambda_N \end{bmatrix} \times \begin{bmatrix} \lambda_1^* & 0 & 0 & 0 & 0 \\ t_{1,2}^* & \lambda_2^* & 0 & 0 & 0 \\ t_{1,3}^* & t_{2,3}^* & \cdots & \cdots & 0 \\ \vdots & \vdots & \cdots & \cdots & 0 \\ \vdots & \vdots & 0 & \lambda_{N-1}^* & 0 \\ t_{1,N}^* & t_{2,N}^* & 0 & t_{N-1,N}^* & \lambda_N^* \end{bmatrix}. \quad (37)$$

The first element of the product matrix TT^{*T} is $|\lambda_1|^2 + |t_{1,2}|^2 + |t_{1,3}|^2 + \cdots + |t_{1,N}|^2$. In contrast, the product $T^{*T}T$ is written as

$$T^{*T}T = \begin{bmatrix} \lambda_1^* & 0 & 0 & 0 & 0 \\ t_{1,2}^* & \lambda_2^* & 0 & 0 & 0 \\ t_{1,3}^* & t_{2,3}^* & \cdots & \cdots & 0 \\ \vdots & \vdots & \cdots & \cdots & 0 \\ \vdots & \vdots & 0 & \lambda_{N-1}^* & 0 \\ t_{1,N}^* & t_{2,N}^* & 0 & t_{N-1,N}^* & \lambda_N^* \end{bmatrix} \times \begin{bmatrix} \lambda_1 & t_{1,2} & t_{1,3} & \cdots & t_{1,N} \\ 0 & \lambda_2 & t_{2,3} & \cdots & t_{2,N} \\ 0 & 0 & \vdots & \cdots & \vdots \\ 0 & 0 & \vdots & \cdots & \vdots \\ 0 & 0 & 0 & \lambda_{N-1} & t_{N-1,N} \\ 0 & 0 & 0 & 0 & \lambda_N \end{bmatrix}. \quad (38)$$

The first element of the product matrix $T^{*T}T$ is just $|\lambda_1|^2$. Since, according to Eq.(36), T is a normal matrix, that is, $TT^{*T} = T^{*T}T$, and since $|t_{1,2}|^2 \geq 0$, $|t_{1,3}|^2 \geq 0$, and so on, we conclude that $t_{1,2} = t_{1,3} = \cdots = t_{1,N} = 0$. We now turn to the second row and second column of the matrices that are multiplied together in Eqs.(37) and (38) and use a similar argument to conclude that $t_{2,3} = t_{2,4} = \cdots = t_{2,N} = 0$. Repeated application of the argument then ensures that all the upper-diagonal elements of T are zeros and, consequently, T is a diagonal matrix. This completes the proof that all normal matrices are diagonalizable via a unitary matrix.

References

1. The introductory remarks in this section have been adapted from Wikipedia's article on *Linear Algebra* (https://en.wikipedia.org/wiki/Linear_algebra).
2. L. Mirsky, *An Introduction to Linear Algebra*, Clarendon Press, Oxford, United Kingdom (1955).
3. W. Nef, *Linear Algebra*, McGraw-Hill, New York (1967).
4. A. G. Hamilton, *Linear Algebra*, Cambridge University Press, Cambridge, United Kingdom (1989).
5. T. A. Whitelaw, *Introduction to Linear Algebra*, Chapman & Hall, New York (1991).
6. P. W. Milonni and J. H. Eberly, *Laser Physics*, Wiley, Hoboken, New Jersey (2010).
7. R. A. Horn and C. R. Johnson, *Topics in Matrix Analysis*, Cambridge University Press, Cambridge, United Kingdom (1991).
8. R. Bhatia, *Matrix Analysis*, Graduate Texts in Mathematics 169, Springer, New York (1997).
9. F. Zhang, *Matrix Theory*, Springer, New York (2011).

Problems

Problem 1) The matrices A, B, and C have dimensions $K \times L$, $L \times M$, and $M \times N$, respectively. Show that, when computing the product matrix ABC, it is immaterial whether A is multiplied into B at first, then their product multiplied into C, or if B is first multiplied into C, followed by multiplication of A into the product of B and C. In other words, show that $(AB)C = A(BC)$.

Problem 2) Using direct calculation of the cofactors and the determinant for an arbitrary 3×3 matrix, show that the matrix does in fact possess the general properties 1, 2, and 3 listed in section 4.

Problem 3) Show that the inverse of the matrix product AB, where A and B are nonsingular $N \times N$ matrices, is the matrix $B^{-1}A^{-1}$.

Problem 4) Given the $N \times N$ square matrix A, show that, if an $N \times N$ matrix B can be found, which satisfies the equation $AB = I$, then it is always true that $BA = I$.

Problem 5) Use the method of Gauss elimination to solve the following sets of three linear equations in three unknowns.

$$\text{(a)} \begin{cases} 2x_1 + 5x_2 - 6x_3 = 3, \\ 7x_1 - 4x_2 + 9x_3 = 4, \\ 5x_1 - 6x_2 - 8x_3 = 7. \end{cases} \quad \text{(b)} \begin{cases} -x + 4y - 5z = -3, \\ 3x - 12y + 13z = 8, \\ 2x - 8y + 17z = 12. \end{cases} \quad \text{(c)} \begin{cases} 2\alpha - 4\beta + 6\gamma = 1, \\ \alpha + \beta + 4\gamma = 3, \\ 2\alpha + 2\beta + 8\gamma = 6. \end{cases}$$

Hint: Gaussian elimination reduces the 3×3 coefficients matrix to an upper-triangular form. The reduced equations are subsequently solved by starting at the bottom row and finding the unknowns while moving up toward the upper row.

Problem 6) Use the Gram-Schmidt process to convert the columns of the 3×3 matrix shown into a set of orthonormal column vectors. Show that the inverse of the 3×3 matrix formed by the orthonormalized vectors equals its own conjugate transpose.

$$\begin{pmatrix} 1+\mathrm{i} & 6 & 3\mathrm{i} \\ 2-3\mathrm{i} & 9-\mathrm{i} & 5-8\mathrm{i} \\ 7 & 5+4\mathrm{i} & 2-\mathrm{i} \end{pmatrix}.$$

Problem 7) Prove that an $N \times N$ matrix whose N eigenvalues $\lambda_1, \lambda_2, \cdots, \lambda_N$ are distinct (i.e., no eigenvalue degeneracies) will have N linearly independent right eigenvectors $V_1, V_2, \cdots, V_N$, and also N linearly independent left eigenvectors $U_1, U_2, \cdots, U_N$. (The proof for the left eigenvectors is the duplicate of that for the right eigenvectors.)

Hint: Start with two of the eigenvectors, say, V_1 and V_2, and show that any linear combination of the two, namely, $c_1V_1 + c_2V_2$, cannot be an eigenvector for eigenvalues other than λ_1 and λ_2. Consequently, the eigenvector V_3 associated with λ_3 cannot be a linear combination of V_1 and V_2. A similar argument can then be used for $c_1V_1 + c_2V_2 + c_3V_3$, and so on.

Problem 8) The $N \times N$ identity matrix I has N identical eigenvalues ($\lambda_1 = \lambda_2 = \cdots = \lambda_N = 1$), but it also has N linearly independent eigenvectors. Use other examples (preferably drawn from simple 2×2 and 3×3 matrices) to show that, although non-degeneracy of the eigenvalues of an $N \times N$ matrix is sufficient for N linearly independent eigenvectors to exist, it is by no means necessary. In other words, show that in some instances of degenerate eigenvalues, the eigenvectors are linearly independent, whereas in some other cases it is impossible to find N linearly independent eigenvectors.

Problem 9) Explore the conditions under which a 2×2 matrix fails to be diagonalized.

Problem 10) Show that a square $N \times N$ matrix which has a zero eigenvalue is singular; in other words, if $\lambda = 0$ is an eigenvalue of A, then A is noninvertible.

Problem 11) The Cartesian coordinate system $x'y'$ is obtained from the xy system by a rotation through the angle θ. Show that the coordinates (x_0, y_0) of an arbitrary point in the xy-plane are related to the coordinates (x'_0, y'_0) of the same point represented in the $x'y'$-plane via a 2×2 rotation matrix, as follows:

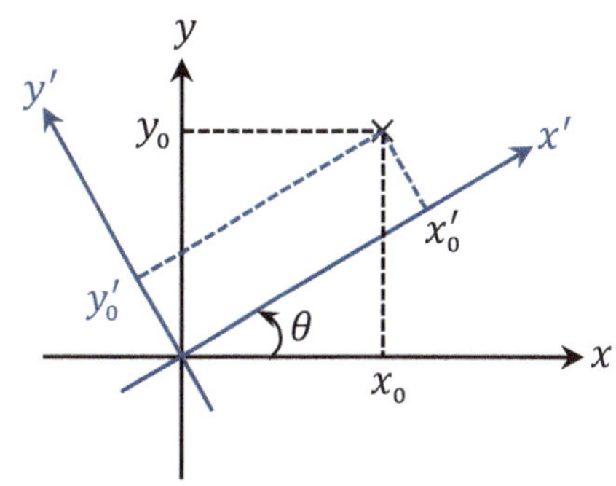

$$\begin{pmatrix} x'_0 \\ y'_0 \end{pmatrix} = \begin{pmatrix} \cos\theta & \sin\theta \\ -\sin\theta & \cos\theta \end{pmatrix} \begin{pmatrix} x_0 \\ y_0 \end{pmatrix}.$$

Diagonalize the rotation matrix with the aid of its eigenvalues and eigenvectors.

Problem 12) Prove that $\exp(A)\exp(B) = \exp(A + B)$, where both A and B are square $N \times N$ matrices.

Problem 13) In the example at the end of section 9, we found a general expression for the nth power of the $ABCD$ matrix, where n was an arbitrary positive integer. Repeat the same calculation *without* invoking the assumption that the determinant $AD - BC$ of the 2×2 $ABCD$ matrix equals 1.

Problem 14) Let A be a diagonalizable $N \times N$ matrix. Define the square root of A and use simple examples drawn from 2×2 and 3×3 matrices to confirm that $\sqrt{A}$, when multiplied by itself, turns out to be A. How many different square roots can A have?

Problem 15) Extend the argument of problem 14 to define the power of the diagonalizable $N \times N$ matrix A for any rational power m/n. Extend the concept still further by defining A^{α}, where α is any real or complex number.

Problem 16) The following second-order linear ODE with constant coefficients governs the response of a harmonic oscillator to a step function excitation:

$$f''(t) + \gamma f'(t) + \omega_0^2 f(t) = \text{step}(t).$$

As usual, the constant parameters ω_0 and γ appearing in the equation represent the natural frequency and the damping coefficient of the oscillator, respectively.

a) Defining a new (as yet unknown) function $g(t) = f'(t)$, convert the above second-order ODE into a set of two first-order ODEs, then use the method described in section 10 to solve the equations for the initial conditions $f(0) = f_0$ and $g(0) = f'(0) = g_0$.

b) Analyze the cases of $\gamma < 2\omega_0$ (underdamped), $\gamma = 2\omega_0$ (critically damped), and $\gamma > 2\omega_0$ (overdamped).

Problem 17) Find all unitary 2×2 matrices (other than I, the identity matrix, and $-I$), which have a single (i.e., degenerate) eigenvalue.

Problem 18) Prove that a square matrix whose rows are orthonormal will automatically have orthonormal columns as well.

Chapter 12

Vector Calculus

"The first principle is that you must not fool yourself and you are the easiest person to fool."

Richard P. Feynman (1918–1988)

12.1. Introduction. In its simplest form, a vector is an arrow with a specific length and a specified direction in three-dimensional Euclidean space. There is a rule for multiplying a vector and a scalar constant; there are also rules for adding, subtracting, and multiplying two or more vectors. One typically uses boldface letters and symbols to identify vector entities, although the use of an arrow above the designated symbol is also commonplace. Thus, $\boldsymbol{V}$ and $\vec{V}$ both represent the same vector, whose length (or magnitude) is sometimes written as $|\boldsymbol{V}|$ or $|\vec{V}|$, or, quite often, simply as V. To specify the direction of a vector, one must first identify a coordinate system, such as the Cartesian system (x, y, z), the cylindrical system (ρ, ϕ, z), or the spherical system (r, θ, ϕ) (see figure 12.1). In the Cartesian system, the unit vectors along the x, y, z axes are denoted by $\hat{\boldsymbol{x}}$, $\hat{\boldsymbol{y}}$, and $\hat{\boldsymbol{z}}$, respectively. Given the projections (or components) V_x, V_y, V_z of the vector $\boldsymbol{V}$ along the coordinate axes, one may proceed to specify the vector as $\boldsymbol{V} = V_x\hat{\boldsymbol{x}} + V_y\hat{\boldsymbol{y}} + V_z\hat{\boldsymbol{z}}$. Note that the location of the vector relative to the origin of coordinates is immaterial in this case, since any parallel transportation of the vector in three-dimensional Euclidean space does not affect its projections onto the x, y, z axes.

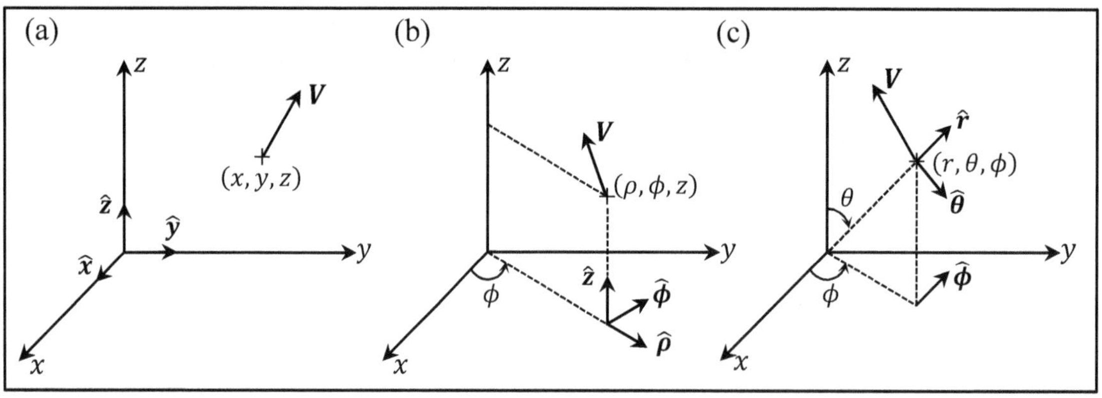

Figure 12.1. (a) In a Cartesian coordinate system, a vector $\boldsymbol{V}$, located at the point (x, y, z), is specified by its projections along the unit vectors $\hat{\boldsymbol{x}}, \hat{\boldsymbol{y}}, \hat{\boldsymbol{z}}$. (b) In a cylindrical coordinate system, the unit vector $\hat{\boldsymbol{z}}$ is always aligned with the z-axis, but the directions of $\hat{\boldsymbol{\rho}}$ and $\hat{\boldsymbol{\phi}}$ depend on the point (ρ, ϕ, z) at which the vector $\boldsymbol{V}$ is located. (c) In a spherical coordinate system, the mutually orthogonal unit vectors $\hat{\boldsymbol{r}}, \hat{\boldsymbol{\theta}}, \hat{\boldsymbol{\phi}}$ are defined at the location (r, θ, ϕ) of the vector $\boldsymbol{V}$.

In contrast with the Cartesian system, the location of the vector in a cylindrical or a spherical coordinate system dictates the use of the local unit vectors $(\hat{\boldsymbol{\rho}}, \hat{\boldsymbol{\phi}}, \hat{\boldsymbol{z}})$ in the former case, and $(\hat{\boldsymbol{r}}, \hat{\boldsymbol{\theta}}, \hat{\boldsymbol{\phi}})$ in the latter case. Once the orientation of the (mutually orthogonal) unit vectors is specified, the projections of the vector $\boldsymbol{V}$ on these unit vectors can be determined and, subsequently, the vector is written as $\boldsymbol{V} = V_\rho\hat{\boldsymbol{\rho}} + V_\phi\hat{\boldsymbol{\phi}} + V_z\hat{\boldsymbol{z}}$ in cylindrical coordinates, or as $\boldsymbol{V} = V_r\hat{\boldsymbol{r}} + V_\theta\hat{\boldsymbol{\theta}} + V_\phi\hat{\boldsymbol{\phi}}$ in spherical coordinates.

A vector can be a function of time t, in the sense that its magnitude and/or its orientation can vary with time. Thus, the wind velocity at a given point in space can be specified as $\boldsymbol{V}(t)$, which represents the time dependence of both the speed and the direction of the wind at the given location. The instantaneous velocity of a point particle can be represented by a vector, as can its instantaneous linear momentum, angular momentum, or acceleration. The force acting on a point particle at time t is represented by a vector, $\boldsymbol{f}(t)$, which is related to the instantaneous momentum $\boldsymbol{p}(t)$ of the particle via Newton's second law of motion, namely $\boldsymbol{f}(t) = \mathrm{d}\boldsymbol{p}(t)/\mathrm{d}t$.

Algebraic operations on vectors are defined such that they depend on the magnitudes and orientations of the individual vectors involved, independently of the coordinate system used to describe the vectors. Figure 12.2 shows that multiplication of a vector by a constant, addition and subtraction of two vectors, and inner and outer multiplication of two vectors (also known as the dot- and cross-products), should be understood as operations that do *not* involve any references to specific coordinate systems.

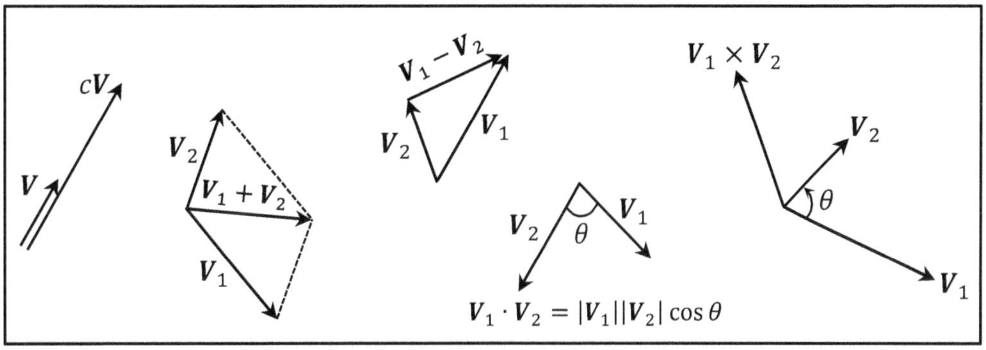

Figure 12.2. Algebraic operations on vectors are independent of the particular coordinate system used to specify them. When a vector $\boldsymbol{V}$ is multiplied by a constant c, the result is a vector aligned with $\boldsymbol{V}$, whose magnitude is $c|\boldsymbol{V}|$. (If c happens to be negative, $c\boldsymbol{V}$ will be antiparallel to $\boldsymbol{V}$.) To add the vectors $\boldsymbol{V}_1$ and $\boldsymbol{V}_2$, transport $\boldsymbol{V}_2$ to the end of $\boldsymbol{V}_1$, then draw an arrow from the beginning of $\boldsymbol{V}_1$ to the end point of the displaced $\boldsymbol{V}_2$. The difference $\boldsymbol{V}_1 - \boldsymbol{V}_2$ between two vectors $\boldsymbol{V}_1$ and $\boldsymbol{V}_2$ is a vector drawn from the end point of $\boldsymbol{V}_2$ to the end point of $\boldsymbol{V}_1$. (Note that the addition of the vector $\boldsymbol{V}_1 - \boldsymbol{V}_2$ to $\boldsymbol{V}_2$ will result in $\boldsymbol{V}_1$.) The inner product (or dot-product) of $\boldsymbol{V}_1$ and $\boldsymbol{V}_2$ is a scalar entity given by the product of the length $|\boldsymbol{V}_1|$ of $\boldsymbol{V}_1$, the length $|\boldsymbol{V}_2|$ of $\boldsymbol{V}_2$, and the cosine of the angle θ between $\boldsymbol{V}_1$ and $\boldsymbol{V}_2$. The outer product (or cross-product) of $\boldsymbol{V}_1$ and $\boldsymbol{V}_2$ is a vector that is perpendicular to the plane of $\boldsymbol{V}_1$ and $\boldsymbol{V}_2$ and has magnitude $|\boldsymbol{V}_1||\boldsymbol{V}_2|\sin\theta$. Along the perpendicular to the $\boldsymbol{V}_1\boldsymbol{V}_2$-plane, the orientation of $\boldsymbol{V}_1 \times \boldsymbol{V}_2$ is determined by the right-hand rule, namely the direction pointed by the thumb when the remaining fingers close along the path from $\boldsymbol{V}_1$ to $\boldsymbol{V}_2$. The cross-product rule is noncommutative; that is, $\boldsymbol{V}_1 \times \boldsymbol{V}_2 = -\boldsymbol{V}_2 \times \boldsymbol{V}_1$.

Despite the fact that the algebraic rules of vector manipulations are independent of the coordinate system in which the vectors are specified, it is possible to formulate these rules in terms of the various components of the vectors along the specified coordinate axes. For example, in Cartesian coordinates, $\boldsymbol{V}_1 \pm \boldsymbol{V}_2 = (V_{x1} \pm V_{x2})\hat{\boldsymbol{x}} + (V_{y1} \pm V_{y2})\hat{\boldsymbol{y}} + (V_{z1} \pm V_{z2})\hat{\boldsymbol{z}}$. Similarly, in spherical coordinates, $\boldsymbol{V}_1 \cdot \boldsymbol{V}_2 = V_{r1}V_{r2} + V_{\theta 1}V_{\theta 2} + V_{\phi 1}V_{\phi 2}$, with the implicit understanding that both $\boldsymbol{V}_1$ and $\boldsymbol{V}_2$ are located at the same point (r, θ, ϕ) in space, and that, therefore, they are projected onto the same unit vectors $\hat{\boldsymbol{r}}$, $\hat{\boldsymbol{\theta}}$, and $\hat{\boldsymbol{\phi}}$. The mutual orthogonality of the unit vectors leads to the following cross-products:

$$\text{Cartesian (universal):}\quad \hat{\boldsymbol{x}} \times \hat{\boldsymbol{y}} = \hat{\boldsymbol{z}}, \qquad \hat{\boldsymbol{y}} \times \hat{\boldsymbol{z}} = \hat{\boldsymbol{x}}, \qquad \hat{\boldsymbol{z}} \times \hat{\boldsymbol{x}} = \hat{\boldsymbol{y}}. \tag{1}$$

$$\text{Cylindrical (local):}\quad \hat{\boldsymbol{\rho}} \times \hat{\boldsymbol{\phi}} = \hat{\boldsymbol{z}}, \qquad \hat{\boldsymbol{\phi}} \times \hat{\boldsymbol{z}} = \hat{\boldsymbol{\rho}}, \qquad \hat{\boldsymbol{z}} \times \hat{\boldsymbol{\rho}} = \hat{\boldsymbol{\phi}}. \tag{2}$$

$$\text{Spherical (local):}\quad \hat{\boldsymbol{r}} \times \hat{\boldsymbol{\theta}} = \hat{\boldsymbol{\phi}}, \qquad \hat{\boldsymbol{\theta}} \times \hat{\boldsymbol{\phi}} = \hat{\boldsymbol{r}}, \qquad \hat{\boldsymbol{\phi}} \times \hat{\boldsymbol{r}} = \hat{\boldsymbol{\theta}}. \tag{3}$$

Thus, for instance, in cylindrical coordinates, one could write

$$\begin{aligned}\boldsymbol{V}_1 \times \boldsymbol{V}_2 &= (V_{\rho 1}\hat{\boldsymbol{\rho}} + V_{\phi 1}\hat{\boldsymbol{\phi}} + V_{z1}\hat{\boldsymbol{z}}) \times (V_{\rho 2}\hat{\boldsymbol{\rho}} + V_{\phi 2}\hat{\boldsymbol{\phi}} + V_{z2}\hat{\boldsymbol{z}})\\ &= (V_{\phi 1}V_{z2} - V_{z1}V_{\phi 2})\hat{\boldsymbol{\rho}} + (V_{z1}V_{\rho 2} - V_{\rho 1}V_{z2})\hat{\boldsymbol{\phi}} + (V_{\rho 1}V_{\phi 2} - V_{\phi 1}V_{\rho 2})\hat{\boldsymbol{z}}.\end{aligned} \tag{4}$$

Similar formulas can be readily obtained for dot- and cross-products in various coordinate systems by expressing each vector in terms of its components along the coordinate axes, multiplying term by term, then adding the results of individual products.

Ordinarily, one thinks of the projections (or components) of a vector along the coordinate axes as a collection of three real numbers. However, there is no good reason to limit the concept of a vector to an arrow whose projections on the coordinate axes are real numbers. Considering that the rules of vector algebra can be expressed in terms of addition, subtraction, and multiplication of the three components associated with each and every vector, and given that such rules exist not only for real numbers but also for complex numbers, one can readily extend the concept by considering a generalized vector to be an object having three complex-valued components which, as a whole, obeys the rules of vector algebra. Of course, the unit vectors associated with specific coordinate systems must remain unchanged and as depicted in figure 12.1. In the simplest possible example, a complex vector $\boldsymbol{V} = (V_x' + \mathrm{i}V_x'')\hat{\boldsymbol{x}} + (V_y' + \mathrm{i}V_y'')\hat{\boldsymbol{y}} + (V_z' + \mathrm{i}V_z'')\hat{\boldsymbol{z}}$ multiplied by a complex constant $c = c' + \mathrm{i}c''$ yields the new vector $\boldsymbol{W} = c\boldsymbol{V}$, as follows:

$$\boldsymbol{W} = c\boldsymbol{V} = (c' + \mathrm{i}c'')(\boldsymbol{V}' + \mathrm{i}\boldsymbol{V}'') = (c'\boldsymbol{V}' - c''\boldsymbol{V}'') + \mathrm{i}(c'\boldsymbol{V}'' + c''\boldsymbol{V}') = \boldsymbol{W}' + \mathrm{i}\boldsymbol{W}''. \quad (5)$$

All other operations on complex vectors can be similarly carried out using straightforward applications of the rules of complex algebra and vector algebra, regardless of the coordinate system under consideration. For example, the cross-product of a pair of arbitrary complex vectors $\boldsymbol{U} = \boldsymbol{U}' + \mathrm{i}\boldsymbol{U}''$ and $\boldsymbol{V} = \boldsymbol{V}' + \mathrm{i}\boldsymbol{V}''$ may be computed as follows:

$$\begin{aligned}\boldsymbol{U} \times \boldsymbol{V} &= (\boldsymbol{U}' + \mathrm{i}\boldsymbol{U}'') \times (\boldsymbol{V}' + \mathrm{i}\boldsymbol{V}'')\\ &= (\boldsymbol{U}' \times \boldsymbol{V}' - \boldsymbol{U}'' \times \boldsymbol{V}'') + \mathrm{i}(\boldsymbol{U}' \times \boldsymbol{V}'' + \boldsymbol{U}'' \times \boldsymbol{V}') = \boldsymbol{W}' + \mathrm{i}\boldsymbol{W}'' = \boldsymbol{W}. \quad (6)\end{aligned}$$

Finally, it is possible to have not only individual vectors at fixed points in three-dimensional space, but also a *vector field* where each point $\boldsymbol{r}$ of the space has its own (real or complex) vector $\boldsymbol{V}(\boldsymbol{r})$. Moreover, a vector field does not have to be static, in the sense that each vector assigned to a given point in space would remain the same forever. Rather, the vectors at each and every point $\boldsymbol{r}$ can change their magnitudes and directions as functions of time t. Such dynamic vector fields are generally denoted by $\boldsymbol{V}(\boldsymbol{r}, t)$. In the remainder of this chapter, we explore the rules for differentiation and integration of such fields—a subject that is commonly known as vector calculus.

12.2. Scalar and vector fields. When a number, generally complex valued, is associated with each point or event $(\boldsymbol{r}, t)$ in space-time, we have a complex function $f(\boldsymbol{r}, t)$, generally referred to as a *scalar field.* The temperature T, specified at each point $\boldsymbol{r}$ in a room and at each instant of time t is a good example of a real-valued scalar field $T(\boldsymbol{r}, t)$.

Similarly, when a vector $\boldsymbol{V}$ is associated with each point $(\boldsymbol{r}, t)$ in space-time, we have a vector field $\boldsymbol{V}(\boldsymbol{r}, t)$. In 3D space, vectors are generally specified by their three components along specific directions. Thus, in a Cartesian coordinate system we have, at each point $(\boldsymbol{r}, t)$, three numbers (V_x, V_y, V_z) that specify the vector field. Similarly, in a cylindrical coordinate system, the vector field is identified by (V_ρ, V_ϕ, V_z), while in spherical coordinates the components of the field are $(V_\rho, V_\theta, V_\phi)$. As mentioned earlier, there are no good reasons to restrict the components of vectors to real numbers. Thus, in general, a vector field assigns to each point $(\boldsymbol{r}, t)$ a complex vector (i.e., a vector whose three components in Euclidean space are complex valued). One way to visualize a complex vector field is to imagine a pair of ordinary, real-valued vectors $(\boldsymbol{V}', \boldsymbol{V}'')$ attached to each point in space, while the magnitudes and directions of both $\boldsymbol{V}'$ and $\boldsymbol{V}''$ at each and every location $\boldsymbol{r}$ change arbitrarily with time. The complex vector field is then identified as $\boldsymbol{V}(\boldsymbol{r},t) = \boldsymbol{V}'(\boldsymbol{r},t) + \mathrm{i}\boldsymbol{V}''(\boldsymbol{r},t)$, where, in Cartesian coordinates,

$$V_x(\boldsymbol{r},t) = V_x'(\boldsymbol{r},t) + \mathrm{i}V_x''(\boldsymbol{r},t), \tag{7a}$$

$$V_y(\boldsymbol{r},t) = V_y'(\boldsymbol{r},t) + \mathrm{i}V_y''(\boldsymbol{r},t), \tag{7b}$$

$$V_z(\boldsymbol{r},t) = V_z'(\boldsymbol{r},t) + \mathrm{i}V_z''(\boldsymbol{r},t). \tag{7c}$$

Similar expressions may be written for the components of a complex vector field in other coordinate systems as well.

Of course, what distinguishes a vector field (real or complex) from a mere collection of three scalar fields is *vector algebra*, namely, the rules of addition, subtraction, dot-multiplication, and cross-multiplication of vectors. For concreteness, we shall describe these rules in Cartesian coordinates only, although the same applies in any orthonormal coordinate system as well. For addition and subtraction of vector fields we have

$$\boldsymbol{V}_1(\boldsymbol{r},t) \pm \boldsymbol{V}_2(\boldsymbol{r},t) = [V_{x1}(\boldsymbol{r},t) \pm V_{x2}(\boldsymbol{r},t)]\hat{\boldsymbol{x}} + [V_{y1}(\boldsymbol{r},t) \pm V_{y2}(\boldsymbol{r},t)]\hat{\boldsymbol{y}} + [V_{z1}(\boldsymbol{r},t) \pm V_{z2}(\boldsymbol{r},t)]\hat{\boldsymbol{z}}. \tag{8}$$

Note that the corresponding components of the two fields are simply added together or subtracted from each other. The addition and subtraction rules for complex numbers being well known, the rule for addition and subtraction of vector fields clearly applies to real as well as complex vector fields.

For dot- and cross-multiplication of two vector fields such as $\boldsymbol{V}_1(\boldsymbol{r},t)$ and $\boldsymbol{V}_2(\boldsymbol{r},t)$, one must express each field as the sum of its three components, $V_x\hat{\boldsymbol{x}} + V_y\hat{\boldsymbol{y}} + V_z\hat{\boldsymbol{z}}$, then proceed to multiply all the components of $\boldsymbol{V}_1$ into all the components of $\boldsymbol{V}_2$ using the following rules:

Dot: $\hat{\boldsymbol{x}}\cdot\hat{\boldsymbol{x}} = \hat{\boldsymbol{y}}\cdot\hat{\boldsymbol{y}} = \hat{\boldsymbol{z}}\cdot\hat{\boldsymbol{z}} = 1;\quad \hat{\boldsymbol{x}}\cdot\hat{\boldsymbol{y}} = \hat{\boldsymbol{y}}\cdot\hat{\boldsymbol{x}} = \hat{\boldsymbol{x}}\cdot\hat{\boldsymbol{z}} = \hat{\boldsymbol{z}}\cdot\hat{\boldsymbol{x}} = \hat{\boldsymbol{y}}\cdot\hat{\boldsymbol{z}} = \hat{\boldsymbol{z}}\cdot\hat{\boldsymbol{y}} = 0.$ (9)

Cross:

$$\hat{\boldsymbol{x}}\times\hat{\boldsymbol{x}} = \hat{\boldsymbol{y}}\times\hat{\boldsymbol{y}} = \hat{\boldsymbol{z}}\times\hat{\boldsymbol{z}} = 0;$$

$$\hat{\boldsymbol{x}}\times\hat{\boldsymbol{y}} = -\hat{\boldsymbol{y}}\times\hat{\boldsymbol{x}} = \hat{\boldsymbol{z}};\ \ \hat{\boldsymbol{y}}\times\hat{\boldsymbol{z}} = -\hat{\boldsymbol{z}}\times\hat{\boldsymbol{y}} = \hat{\boldsymbol{x}};\ \ \hat{\boldsymbol{z}}\times\hat{\boldsymbol{x}} = -\hat{\boldsymbol{x}}\times\hat{\boldsymbol{z}} = \hat{\boldsymbol{y}}. \tag{10}$$

Once again, the multiplication rule for complex numbers being well known, it makes no difference whether the components of the vector fields being multiplied together are real or complex valued.

A good example of a real-valued vector field is the wind velocity within a wind tunnel, described as a function of time t for each point $\boldsymbol{r}$ in 3D space. To each point $\boldsymbol{r}$ is thus assigned a vector $\boldsymbol{V}(t)$, whose magnitude and orientation describe the speed of the wind and its direction at time t. As time progresses and the wind changes in speed and direction, the individual vectors remain affixed to their location in space but get shorter or longer, while turning to point in different directions.

In the classical theory of electrodynamics, the electric field $\boldsymbol{E}(\boldsymbol{r},t)$ and the magnetic field $\boldsymbol{H}(\boldsymbol{r},t)$ are vector fields whose relations to each other and to their various sources are governed by Maxwell's equations. The sources of the electromagnetic fields are free charge density $\rho_{\text{free}}(\boldsymbol{r},t)$, which is a scalar field, and free current density $\boldsymbol{J}_{\text{free}}(\boldsymbol{r},t)$, polarization $\boldsymbol{P}(\boldsymbol{r},t)$, and magnetization $\boldsymbol{M}(\boldsymbol{r},t)$, which are vector fields. Other important fields in the Maxwell-Lorentz theory of electrodynamics are the Poynting vector $\boldsymbol{S}(\boldsymbol{r},t) = \boldsymbol{E}(\boldsymbol{r},t)\times\boldsymbol{H}(\boldsymbol{r},t)$, the scalar potential $\psi(\boldsymbol{r},t)$, and the vector potential $\boldsymbol{A}(\boldsymbol{r},t)$.

The rules of vector algebra may be applied to three or more vectors in various combinations and arrangements, provided that all the vectors are located at a fixed point $(\boldsymbol{r},t)$ in space-time. One can then prove identities such as $(\boldsymbol{A}+\boldsymbol{B})\cdot\boldsymbol{C} = \boldsymbol{A}\cdot\boldsymbol{C} + \boldsymbol{B}\cdot\boldsymbol{C}$ or

$\boldsymbol{A} \times (\boldsymbol{B} \times \boldsymbol{C}) = (\boldsymbol{A} \cdot \boldsymbol{C})\boldsymbol{B} - (\boldsymbol{A} \cdot \boldsymbol{B})\boldsymbol{C}$ or $(\boldsymbol{A} \times \boldsymbol{B}) \cdot \boldsymbol{C} = (\boldsymbol{B} \times \boldsymbol{C}) \cdot \boldsymbol{A} = (\boldsymbol{C} \times \boldsymbol{A}) \cdot \boldsymbol{B}$. Table 1 lists some of the important identities of vector algebra. The vectors $\boldsymbol{A}$, $\boldsymbol{B}$, $\boldsymbol{C}$, and $\boldsymbol{D}$ appearing in this table could be real or complex valued. For a proof of the listed identities, see the problems at the end of the chapter.

$$\boldsymbol{A} \cdot \boldsymbol{B} = \boldsymbol{B} \cdot \boldsymbol{A}$$
$$\boldsymbol{A} \cdot (\boldsymbol{B} + \boldsymbol{C}) = \boldsymbol{A} \cdot \boldsymbol{B} + \boldsymbol{A} \cdot \boldsymbol{C}$$
$$\boldsymbol{A} \times \boldsymbol{B} = -\boldsymbol{B} \times \boldsymbol{A}$$
$$\boldsymbol{A} \times (\boldsymbol{B} + \boldsymbol{C}) = \boldsymbol{A} \times \boldsymbol{B} + \boldsymbol{A} \times \boldsymbol{C}$$
$$(\boldsymbol{A} + \boldsymbol{B}) \times \boldsymbol{C} = \boldsymbol{A} \times \boldsymbol{C} + \boldsymbol{B} \times \boldsymbol{C}$$
$$\boldsymbol{A} \cdot (\boldsymbol{B} \times \boldsymbol{C}) = \boldsymbol{B} \cdot (\boldsymbol{C} \times \boldsymbol{A}) = \boldsymbol{C} \cdot (\boldsymbol{A} \times \boldsymbol{B})$$
$$\boldsymbol{A} \times (\boldsymbol{B} \times \boldsymbol{C}) = (\boldsymbol{A} \cdot \boldsymbol{C})\boldsymbol{B} - (\boldsymbol{A} \cdot \boldsymbol{B})\boldsymbol{C}$$
$$(\boldsymbol{A} \times \boldsymbol{B}) \cdot (\boldsymbol{C} \times \boldsymbol{D}) = (\boldsymbol{A} \cdot \boldsymbol{C})(\boldsymbol{B} \cdot \boldsymbol{D}) - (\boldsymbol{A} \cdot \boldsymbol{D})(\boldsymbol{B} \cdot \boldsymbol{C})$$

Table 12.1. Important vector algebraic identities.

12.3. Gradient of a scalar field. It is possible to construct a vector field from a (generally complex-valued) scalar field $f(\boldsymbol{r},t) = f'(\boldsymbol{r},t) + \mathrm{i}f''(\boldsymbol{r},t)$ using the procedure outlined below. First, freeze the time at $t = t_0$. Then pick an arbitrary point in space, say $\boldsymbol{r} = \boldsymbol{r}_0$, and draw an infinitesimally small sphere of radius Δr centered at $\boldsymbol{r}_0$. Among all the points $\boldsymbol{r}$ on the surface of this sphere, find the point $\boldsymbol{r}_{\max}$ at which $\Delta f' = f'(\boldsymbol{r},t_0) - f'(\boldsymbol{r}_0,t_0)$ is a maximum. Define the vector $\boldsymbol{\nabla} f'$ in the direction $(\boldsymbol{r}_{\max} - \boldsymbol{r}_0)/\Delta r$ and give it the magnitude $\Delta f'/\Delta r$. Next, determine $\boldsymbol{\nabla} f''$ by repeating the same procedure for $f''(\boldsymbol{r},t)$. The gradient of $f(\boldsymbol{r},t)$ at $(\boldsymbol{r}_0,t_0)$ is then given by $\boldsymbol{\nabla} f = \boldsymbol{\nabla} f' + \mathrm{i}\boldsymbol{\nabla} f''$.

The gradient of the scalar field $f(\boldsymbol{r},t)$, evaluated at all points $(\boldsymbol{r},t)$ in space-time, is thus a (generally complex valued) vector field. Note that the procedure for determining the gradient of $f(\boldsymbol{r},t)$ is totally independent of any coordinate system, which may be used to locate the points $\boldsymbol{r}$ in space. Therefore, $\boldsymbol{\nabla} f(\boldsymbol{r},t)$ is a property of the scalar field $f(\boldsymbol{r},t)$ and its spatial distribution at any fixed time t.

To determine the gradient of the scalar field $f(\boldsymbol{r},t)$ by analytic means, one must work within a coordinate system (e.g., a Cartesian system) and use the first-order terms of the Taylor series expansion of $f(\boldsymbol{r},t)$ as follows:

$$\begin{aligned}\Delta f(\boldsymbol{r},t) &= f(\boldsymbol{r},t_0) - f(\boldsymbol{r}_0,t_0) = (\partial f/\partial x)\Delta x + (\partial f/\partial y)\Delta y + (\partial f/\partial z)\Delta z \\ &= [(\partial f/\partial x)\hat{\boldsymbol{x}} + (\partial f/\partial y)\hat{\boldsymbol{y}} + (\partial f/\partial z)\hat{\boldsymbol{z}}] \cdot [(\Delta x)\hat{\boldsymbol{x}} + (\Delta y)\hat{\boldsymbol{y}} + (\Delta z)\hat{\boldsymbol{z}}] \\ &= [(\partial f/\partial x)\hat{\boldsymbol{x}} + (\partial f/\partial y)\hat{\boldsymbol{y}} + (\partial f/\partial z)\hat{\boldsymbol{z}}] \cdot \Delta\boldsymbol{r}. \qquad (11)\end{aligned}$$

In this equation, the first term, $(\partial f/\partial x)\hat{\boldsymbol{x}} + (\partial f/\partial y)\hat{\boldsymbol{y}} + (\partial f/\partial z)\hat{\boldsymbol{z}}$, is a fixed complex vector derived from $f(\boldsymbol{r},t_0)$ by evaluating its partial derivatives at $\boldsymbol{r}_0$. The second term, $\Delta\boldsymbol{r}$, however, although fixed in magnitude (i.e., Δr = radius of infinitesimal sphere), can assume any arbitrary orientation in 3D space. Now the dot-product of two vectors $\boldsymbol{V}_1$ and $\boldsymbol{V}_2$ is $\boldsymbol{V}_1 \cdot \boldsymbol{V}_2 = |\boldsymbol{V}_1||\boldsymbol{V}_2|\cos\theta$, where θ is the angle between $\boldsymbol{V}_1$ and $\boldsymbol{V}_2$. For given lengths $|\boldsymbol{V}_1|$ and $|\boldsymbol{V}_2|$ of the two vectors, therefore, the dot-product is a maximum when the two vectors overlap; that is, when $\theta = 0$. Applying this logic to Eq.(11), we see immediately that

$$\boldsymbol{\nabla} f(\boldsymbol{r},t) = (\partial f/\partial x)\hat{\boldsymbol{x}} + (\partial f/\partial y)\hat{\boldsymbol{y}} + (\partial f/\partial z)\hat{\boldsymbol{z}}. \qquad (12)$$

The above procedure may be used to obtain analytic expressions for the gradient operator in other coordinate systems as well. For example, in cylindrical coordinates where $\boldsymbol{r} = (\rho, \phi, z)$, we find

$$\begin{aligned}\Delta f(\boldsymbol{r}, t) &= f(\boldsymbol{r}, t_0) - f(\boldsymbol{r}_0, t_0) = (\partial f/\partial\rho)\Delta\rho + (\partial f/\partial\phi)\Delta\phi + (\partial f/\partial z)\Delta z \\ &= [(\partial f/\partial\rho)\hat{\boldsymbol{\rho}} + (\partial f/\partial\phi)\hat{\boldsymbol{\phi}} + (\partial f/\partial z)\hat{\boldsymbol{z}}] \cdot [(\Delta\rho)\hat{\boldsymbol{\rho}} + (\Delta\phi)\hat{\boldsymbol{\phi}} + (\Delta z)\hat{\boldsymbol{z}}] \\ &= [(\partial f/\partial\rho)\hat{\boldsymbol{\rho}} + \rho^{-1}(\partial f/\partial\phi)\hat{\boldsymbol{\phi}} + (\partial f/\partial z)\hat{\boldsymbol{z}}] \cdot [(\Delta\rho)\hat{\boldsymbol{\rho}} + (\rho\Delta\phi)\hat{\boldsymbol{\phi}} + (\Delta z)\hat{\boldsymbol{z}}] \\ &= [(\partial f/\partial\rho)\hat{\boldsymbol{\rho}} + \rho^{-1}(\partial f/\partial\phi)\hat{\boldsymbol{\phi}} + (\partial f/\partial z)\hat{\boldsymbol{z}}] \cdot \Delta\boldsymbol{r}. \end{aligned} \tag{13}$$

Note that, in going from the second to the third line in the above derivation, to obtain the correct ϕ-component for $\Delta\boldsymbol{r}$, we have replaced $\Delta\phi$ by $\rho\Delta\phi$, and, as compensation for this modification, we have had to multiply $\partial f/\partial\phi$ by ρ^{-1}. Thus, the gradient operator in cylindrical coordinates is

$$\boldsymbol{\nabla} f(\boldsymbol{r}, t) = (\partial f/\partial\rho)\hat{\boldsymbol{\rho}} + \rho^{-1}(\partial f/\partial\phi)\hat{\boldsymbol{\phi}} + (\partial f/\partial z)\hat{\boldsymbol{z}}. \tag{14}$$

By the same token, the gradient operator in spherical coordinates may be written as follows:

$$\boldsymbol{\nabla} f(\boldsymbol{r}, t) = (\partial f/\partial\rho)\hat{\boldsymbol{\rho}} + \rho^{-1}(\partial f/\partial\theta)\hat{\boldsymbol{\theta}} + (\rho\sin\theta)^{-1}(\partial f/\partial\phi)\hat{\boldsymbol{\phi}}. \tag{15}$$

Example 1. Consider the scalar function $f(\boldsymbol{r}, t) = a_0 \exp[\mathrm{i}(\boldsymbol{k} \cdot \boldsymbol{r} - \omega t)]$, where $a_0 = a_0' + \mathrm{i}a_0''$ and $\omega = \omega' + \mathrm{i}\omega''$ are two complex constants, while $\boldsymbol{k} = \boldsymbol{k}' + \mathrm{i}\boldsymbol{k}''$ is a constant complex vector. Thus, associated with each point $(\boldsymbol{r}, t)$ in space-time is the complex number $f(\boldsymbol{r}, t)$. At a fixed point $\boldsymbol{r}$, the time dependence of $f(\boldsymbol{r}, t)$ may be written $\exp(\omega'' t)\,[\cos(\omega' t) - \mathrm{i}\sin(\omega' t)]$; that is, depending on the sign of ω'', the magnitude of the oscillating function either increases or decreases exponentially with time.

At fixed time, the $\boldsymbol{r}$-dependence of $f(\boldsymbol{r}, t)$ is $\exp(-\boldsymbol{k}'' \cdot \boldsymbol{r})\,[\cos(\boldsymbol{k}' \cdot \boldsymbol{r}) + \mathrm{i}\sin(\boldsymbol{k}' \cdot \boldsymbol{r})]$. Thus, the arrow of $\boldsymbol{k}''$ specifies the direction in 3D space along which the amplitude of the function decays exponentially with distance; the rapidity of such exponential decay is determined by the magnitude of $\boldsymbol{k}''$. Similarly, $\boldsymbol{k}'$ specifies both the direction and magnitude of the change of phase in 3D space. Along the vector $\boldsymbol{k}'$, the phase velocity V_ϕ may be found as follows:

$$\boldsymbol{k}' \cdot \boldsymbol{r} - \omega' t = \text{constant} \quad \rightarrow \quad k'\Delta r - \omega'\Delta t = 0 \quad \rightarrow \quad V_\phi = \Delta r/\Delta t = \omega'/k'. \tag{16}$$

The gradient of $f(\boldsymbol{r}, t)$ is readily obtained from Eq.(12) by straightforward differentiation, namely,

$$\begin{aligned}\boldsymbol{\nabla}\{a_0 \exp[\mathrm{i}(\boldsymbol{k} \cdot \boldsymbol{r} - \omega t)]\} &= \mathrm{i}(a_0 k_x \hat{\boldsymbol{x}} + a_0 k_y \hat{\boldsymbol{y}} + a_0 k_z \hat{\boldsymbol{z}}) \exp[\mathrm{i}(\boldsymbol{k} \cdot \boldsymbol{r} - \omega t)] \\ &= \mathrm{i}a_0 \boldsymbol{k} \exp[\mathrm{i}(\boldsymbol{k} \cdot \boldsymbol{r} - \omega t)]. \end{aligned} \tag{17}$$

The gradient operator $\boldsymbol{\nabla}$ acting on the scalar "plane wave" $a_0 \exp[\mathrm{i}(\boldsymbol{k} \cdot \boldsymbol{r} - \omega t)]$ thus yields a vector field that is also a plane wave having the same frequency ω and wave vector (or k-vector) $\boldsymbol{k}$, but its amplitude is obtained by multiplying a_0 with the complex vector $\mathrm{i}\boldsymbol{k}$.

A plane wave vector field $\boldsymbol{A}_0 \exp[\mathrm{i}(\boldsymbol{k} \cdot \boldsymbol{r} - \omega t)]$ whose amplitude $\boldsymbol{A}_0$ is proportional (or "parallel") to its k-vector is known as a *longitudinal* vector field. In general, when $\boldsymbol{k}$ and $\boldsymbol{A}_0$ are complex valued, it does not make sense to speak of their parallelism, as each complex vector in reality is represented by two real-valued vectors, namely $(\boldsymbol{k}', \boldsymbol{k}'')$ and $(\boldsymbol{A}_0', \boldsymbol{A}_0'')$, neither of which may be parallel to any other. However, if $\boldsymbol{A}_0$ happens to be proportional to $\boldsymbol{k}$ (with, in general, a complex proportionality constant), one may define the vector field as being longitudinal.

Another useful operation performed on the scalar plane wave $a_0 \exp[\mathrm{i}(\boldsymbol{k}\cdot\boldsymbol{r} - \omega t)]$ is time differentiation, which yields

$$\partial f(\boldsymbol{r},t)/\partial t = (\partial/\partial t)a_0 \exp[\mathrm{i}(\boldsymbol{k}\cdot\boldsymbol{r} - \omega t)] = -\mathrm{i}\omega a_0 \exp[\mathrm{i}(\boldsymbol{k}\cdot\boldsymbol{r} - \omega t)]. \quad (18)$$

This is yet another scalar plane wave with the same $(\boldsymbol{k},\omega)$ as the original plane wave, but with a different amplitude.

Brain teaser. Stand in front of a large, flat mirror, and see your image staring back at you. The image represents your appearance in just about every respect, except for a few subtle differences, such as your right hand becoming the image's left hand, and vice versa. Of course, there is nothing special about the left-right symmetry of the mirror that differs from its up-down symmetry. Why is it then that your head and feet do not switch places in the mirror?

12.4. Integration of fields over time and/or space. The scalar field $f(\boldsymbol{r},t)$ may be integrated in several different ways. For example, integrating at fixed $\boldsymbol{r}$ over a time interval (t_1,t_2) yields $\int_{t_1}^{t_2} f(\boldsymbol{r},t)\mathrm{d}t$. This operation is useful, for example, when the time-averaged field at a given location in space is desired.

Another useful integral is $\int_{\text{volume}} f(\boldsymbol{r},t)\mathrm{d}v$, where t is kept constant, while the function is summed over a volume of space. A good example is volume integration of electric charge density $\rho(\boldsymbol{r},t)$, which yields the total charge within the volume of interest as a function of time.

The vector field $\boldsymbol{V}(\boldsymbol{r},t)$ may also be integrated in a variety of ways. For example, integrating at fixed $\boldsymbol{r}$ over a time interval (t_1,t_2) yields $\int_{t_1}^{t_2} \boldsymbol{V}(\boldsymbol{r},t)\mathrm{d}t$. Again, this operation is useful when the time-averaged field at a given location in space is desired.

The line integral $\int_{\boldsymbol{r}_1}^{\boldsymbol{r}_2} \boldsymbol{V}(\boldsymbol{r},t_0)\cdot \mathrm{d}\boldsymbol{\ell}$ is obtained at a fixed instant of time t_0, by dot-multiplying the vector field with the short linear segment $\Delta\boldsymbol{\ell}$ of a continuous curve that connects a fixed point $\boldsymbol{r}_1$ to another fixed point $\boldsymbol{r}_2$ (see figure 12.3). Summing up the resultant complex numbers would yield the result of this integration as a complex number whose value generally depends not only on t_0, $\boldsymbol{r}_1$, and $\boldsymbol{r}_2$, but also on the integration path.

With reference to figure 12.3, the surface integral $\int_{\text{surface}} \boldsymbol{V}(\boldsymbol{r},t_0)\cdot \mathrm{d}\boldsymbol{s}$ is obtained at a fixed instant of time t_0, by dot-multiplying the vector field with a small, oriented surface element $\Delta\boldsymbol{s}$ of a continuous surface, which surface may or may not have a boundary. Summing up the resultant complex numbers would yield the result of the surface integral as a complex scalar whose value depends on t_0 as well as on the chosen surface of integration.

When the integration surface has no boundaries, it is called a closed surface. By convention, the vectors $\Delta\boldsymbol{s}$ assigned to small elements of a closed surface always point *outward*, away from the enclosed volume. In the case of an open surface, care must be exercised to define the *direction* of the surface normals of various elements in such a way as to avoid discontinuous changes in the orientation direction of the adjacent vectors $\Delta\boldsymbol{s}$, although occasionally such discontinuities may be unavoidable (e.g., when the surface of interest is that of a Möbius strip).

Another useful integral is $\int_{\text{volume}} \boldsymbol{V}(\boldsymbol{r},t)\mathrm{d}v$, where t is kept constant, while the function is vectorially summed over within a volume of space. Note that Δv is treated as a scalar, since no direction can be assigned to a volume element. A good example from the field of magneto-statics is $\int_{\text{volume}} |\boldsymbol{r}-\boldsymbol{r}'|^{-1}\boldsymbol{J}(\boldsymbol{r}')\mathrm{d}v'$, where the electric current density $\boldsymbol{J}(\boldsymbol{r}')$ divided by the distance $|\boldsymbol{r}-\boldsymbol{r}'|$ to the observation point $\boldsymbol{r}$, then integrated over the volume of current-carrying material, is found to be proportional to the vector potential $\boldsymbol{A}(\boldsymbol{r})$ at the observation point.

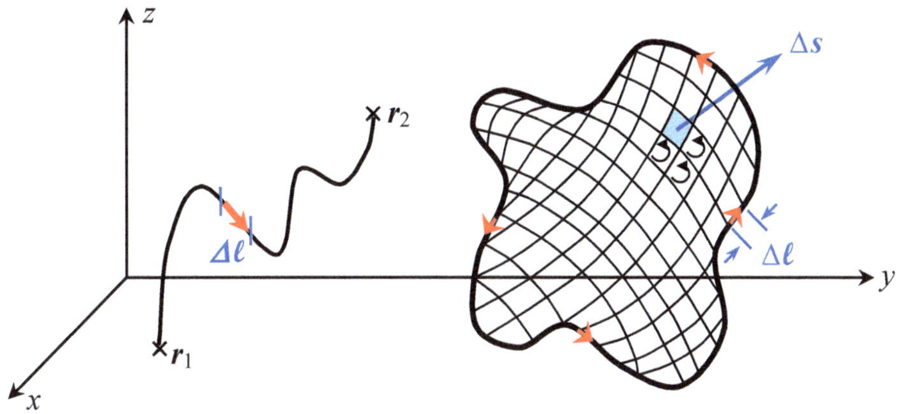

Figure 12.3. Line and surface integrals of a vector field. The line integral requires an integration path, such as the open curve from $\boldsymbol{r}_1$ to $\boldsymbol{r}_2$ shown on the left, or the closed loop that marks the boundary of the surface on the right-hand side. The line element $\Delta\boldsymbol{\ell}$ is an infinitesimal vector whose length is the length of the line segment, while its orientation is that of the line segment at its mid-point. The direction of travel (e.g., from $\boldsymbol{r}_1$ to $\boldsymbol{r}_2$, or counterclockwise around the loop) is also conveyed by the vector $\Delta\boldsymbol{\ell}$. Surface integrals require the integration surface to be divided into infinitesimally small, flat surface elements Δs. The vector $\Delta\boldsymbol{s}$, associated with a given surface element, has a length equal to the surface area Δs and is oriented perpendicular to the surface element while directed in accordance with the right-hand rule in conjunction with the chosen direction of travel around the boundary of the surface element.

12.5. Divergence of a vector field. The divergence operator acts on a vector field $\boldsymbol{V}(\boldsymbol{r},t)$ to produce a scalar field $\boldsymbol{\nabla}\cdot\boldsymbol{V}(\boldsymbol{r},t)$. To determine the divergence of $\boldsymbol{V}(\boldsymbol{r},t)$ at a given point $\boldsymbol{r}_0$ in space, first freeze the time everywhere at $t=t_0$. Choose an infinitesimal closed surface of volume Δv surrounding $\boldsymbol{r}_0$—the shape of the closed surface being irrelevant so long as all the points on this surface are sufficiently close to $\boldsymbol{r}_0$. Divide the closed surface into small surface elements Δs, with the surface-normal pointing outward everywhere. Evaluate the integral of $\boldsymbol{V}(\boldsymbol{r},t_0)$ over the closed surface, then normalize the result by Δv to obtain the divergence of $\boldsymbol{V}(\boldsymbol{r},t)$ at $(\boldsymbol{r}_0,t_0)$; that is,

$$\boldsymbol{\nabla}\cdot\boldsymbol{V}(\boldsymbol{r}_0,t_0)=lim_{\Delta v\to 0}\frac{\oint \boldsymbol{V}(\boldsymbol{r},t_0)\cdot \mathrm{d}\boldsymbol{s}}{\Delta v}. \tag{19}$$

Note that normalization by Δv is absolutely necessary, otherwise the surface integral goes to zero as the volume shrinks around $\boldsymbol{r}_0$. The definition of divergence as given by Eq.(19) is independent of the chosen shape of the closed surface. When this operation is repeated everywhere in space-time, one obtains the scalar field $\boldsymbol{\nabla}\cdot\boldsymbol{V}(\boldsymbol{r},t)$.

In Cartesian coordinates, it is convenient to choose, for the surface surrounding $\boldsymbol{r}_0$, a small cube whose edges run parallel to the coordinate axes. Using the first-order terms in the Taylor series expansion of $\boldsymbol{V}(\boldsymbol{r},t)$ around $\boldsymbol{r}_0$, one can readily show that

$$\boldsymbol{\nabla}\cdot\boldsymbol{V}(\boldsymbol{r},t)\;=(\partial V_x/\partial x)+(\partial V_y/\partial y)\;+(\partial V_z/\partial z). \tag{20}$$

Similar expressions for the divergence of a vector field in cylindrical and spherical coordinates may be readily obtained if one chooses, in the vicinity of $\boldsymbol{r}_0$, not a cube but a small section of a cylindrical or spherical shell as the surface of integration in Eq.(19).

12.5.1. Divergence operator in cylindrical coordinates. In cylindrical coordinates, one considers an infinitesimally small volume element centered at an arbitrary point $\boldsymbol{r} = (\rho, \phi, z)$, as shown in figure 12.4. The volume element is cut from a hollow cylinder centered on the z-axis and having inner and outer radii $\rho \pm ½\Delta\rho$. Subsequently, the vector field $\boldsymbol{V}(\boldsymbol{r}) = V_\rho(\rho,\phi,z)\hat{\boldsymbol{\rho}} + V_\phi(\rho,\phi,z)\hat{\boldsymbol{\phi}} + V_z(\rho,\phi,z)\hat{\boldsymbol{z}}$ is integrated over the surface of the volume element and the resulting integral is normalized by the volume $\rho\Delta\rho\Delta\phi\Delta z$ of the element, as follows:

$$\begin{aligned}\boldsymbol{\nabla}\cdot\boldsymbol{V}(\rho,\phi,z) = lim_{\Delta\rho,\Delta\phi,\Delta z\to 0}(\rho\Delta\phi\Delta\rho\Delta z)^{-1}\big[&V_\rho(\rho+½\Delta\rho,\phi,z)(\rho+½\Delta\rho)\Delta\phi\Delta z\\ &-V_\rho(\rho-½\Delta\rho,\phi,z)(\rho-½\Delta\rho)\Delta\phi\Delta z\\ &+V_\phi(\rho,\phi+½\Delta\phi,z)\Delta\rho\Delta z - V_\phi(\rho,\phi-½\Delta\phi,z)\Delta\rho\Delta z\\ &+V_z(\rho,\phi,z+½\Delta z)\rho\Delta\phi\Delta\rho - V_z(\rho,\phi,z-½\Delta z)\rho\Delta\phi\Delta\rho\big]\\ = \rho^{-1}\partial_\rho(\rho V_\rho) + \rho^{-1}\partial_\phi V_\phi + \partial_z V_z. \qquad (21)\end{aligned}$$

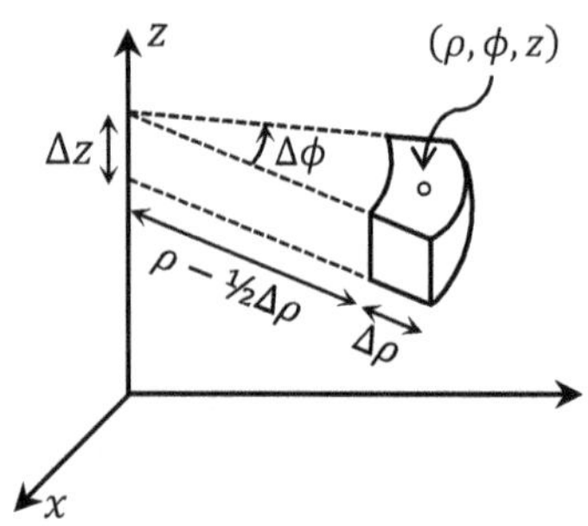

Figure 12.4. To find the divergence of a vector field $\boldsymbol{V}(\boldsymbol{r})$ at an arbitrary point $\boldsymbol{r} = (\rho,\phi,z)$, one must integrate $\boldsymbol{V}(\boldsymbol{r})$ over the surface of an infinitesimally small volume element cut from a hollow cylinder in the vicinity of the point $\boldsymbol{r}$. The surface integral is subsequently normalized by the volume $\rho\Delta\rho\Delta\phi\Delta z$ of the element. In the diagram, the area of the front facet is $(\rho + ½\Delta\rho)\Delta\phi\Delta z$, whereas that of the rear facet is $(\rho - ½\Delta\rho)\Delta\phi\Delta z$. The areas of the top and bottom facets are $\rho\Delta\phi\Delta\rho$, and those of the side facets are $\Delta\rho\Delta z$.

Divergence in spherical coordinates can be determined in similar fashion, albeit with the aid of an infinitesimal volume element located at the point $\boldsymbol{r} = (\rho,\theta,\phi)$, which is cut from a hollow sphere centered at the origin and having inner and outer radii $\rho \pm ½\Delta\rho$ (see problem 10 at the end of the chapter).

Example 2. Consider the vector function $\boldsymbol{A}(\boldsymbol{r},t) = \boldsymbol{A}_0 \exp[\mathrm{i}(\boldsymbol{k}\cdot\boldsymbol{r} - \omega t)]$, where $\omega = \omega' + \mathrm{i}\omega''$ is a complex constant, while $\boldsymbol{k} = \boldsymbol{k}' + \mathrm{i}\boldsymbol{k}''$ and $\boldsymbol{A}_0 = \boldsymbol{A}_0' + \mathrm{i}\boldsymbol{A}_0''$ are constant complex vectors. Thus, associated with each point $(\boldsymbol{r},t)$ in space-time is the complex vector $\boldsymbol{A}(\boldsymbol{r},t)$. The roles of $\boldsymbol{k}$ and ω in defining the properties of such plane waves were discussed in a previous example. Here, we examine the role of the complex amplitude $\boldsymbol{A}_0$ of the plane wave. Let us choose a fixed location, $\boldsymbol{r}_0$, in space and consider the time dependence of $\boldsymbol{A}(\boldsymbol{r}_0,t)$ at that point. Since it is generally the real part of a complex entity that is of interest, we write the real part of this complex plane wave as follows:

$$\begin{aligned}\mathrm{Real}[\boldsymbol{A}(\boldsymbol{r}_0,t)] &= \exp(-\boldsymbol{k}''\cdot\boldsymbol{r}_0)\exp(\omega''t)\,\mathrm{Real}\{(\boldsymbol{A}_0' + \mathrm{i}\boldsymbol{A}_0'')\exp[\mathrm{i}(\boldsymbol{k}'\cdot\boldsymbol{r}_0 - \omega't)]\}\\ &= \exp(-\boldsymbol{k}''\cdot\boldsymbol{r}_0)\exp(\omega''t)\,[\boldsymbol{A}_0'\cos(\omega't - \boldsymbol{k}'\cdot\boldsymbol{r}_0) + \boldsymbol{A}_0''\sin(\omega't - \boldsymbol{k}'\cdot\boldsymbol{r}_0)]. \quad (22)\end{aligned}$$

As time progresses, the tip of the real-valued vector $\boldsymbol{C}(t) = \boldsymbol{A}_0'\cos(\omega't - \boldsymbol{k}'\cdot\boldsymbol{r}_0) + \boldsymbol{A}_0''\sin(\omega't - \boldsymbol{k}'\cdot\boldsymbol{r}_0)$ describes a closed trajectory, as follows. At some point in time, say t_0, we will have $\sin(\omega't_0 - \boldsymbol{k}'\cdot\boldsymbol{r}_0) = 0$ and $\cos(\omega't_0 - \boldsymbol{k}'\cdot\boldsymbol{r}_0) = 1$, at which point $\boldsymbol{C}(t_0)$ coincides with $\boldsymbol{A}_0'$. A quarter of a period later, the sine will be unity and the cosine zero; then $\boldsymbol{C}(t)$ will coincide with $\boldsymbol{A}_0''$. Another quarter of a period later, $\boldsymbol{C}(t)$ will become $-\boldsymbol{A}_0'$, then $-\boldsymbol{A}_0''$, then again $\boldsymbol{A}_0'$, and so on.

The process thus repeats itself with frequency ω'. One can readily show that the tip of the vector $\boldsymbol{C}(t)$ repeatedly describes a perfect ellipse, with the repetition having a period $T = 2\pi/\omega'$. If $\boldsymbol{A}_0'$ and $\boldsymbol{A}_0''$ happen to be orthogonal to each other, they will be the major and minor semi-axes of the ellipse thus described (although not necessarily in that order). Orthogonal or not, one can always determine the axes of the ellipse and their orientation in 3D space from the knowledge of $\boldsymbol{A}_0'$ and $\boldsymbol{A}_0''$ alone.

The divergence of the vector field $\boldsymbol{A}(\boldsymbol{r},t) = \boldsymbol{A}_0 \exp[\mathrm{i}(\boldsymbol{k}\cdot\boldsymbol{r} - \omega t)]$ is found by a straightforward application of Eq.(20), namely,

$$\boldsymbol{\nabla}\cdot\{\boldsymbol{A}_0 \exp[\mathrm{i}(\boldsymbol{k}\cdot\boldsymbol{r} - \omega t)]\} = \mathrm{i}(A_{0x}k_x + A_{0y}k_y + A_{0z}k_z)\exp[\mathrm{i}(\boldsymbol{k}\cdot\boldsymbol{r} - \omega t)]$$
$$= \mathrm{i}\boldsymbol{k}\cdot\boldsymbol{A}_0 \exp[\mathrm{i}(\boldsymbol{k}\cdot\boldsymbol{r} - \omega t)]. \qquad (23)$$

The divergence operator $\boldsymbol{\nabla}\cdot$ acting on the vector plane wave $\boldsymbol{A}_0 \exp[\mathrm{i}(\boldsymbol{k}\cdot\boldsymbol{r} - \omega t)]$ thus yields a scalar field that is also a plane wave with the same frequency ω and the same wave vector (or k-vector) $\boldsymbol{k}$, but its amplitude is obtained by dot-multiplying $\boldsymbol{A}_0$ with the complex vector $\mathrm{i}\boldsymbol{k}$.

Another simple operation that can be performed on the vector plane wave $\boldsymbol{A}_0 \exp[\mathrm{i}(\boldsymbol{k}\cdot\boldsymbol{r} - \omega t)]$ is time differentiation, namely,

$$\partial\boldsymbol{A}(\boldsymbol{r},t)/\partial t = (\partial/\partial t)\boldsymbol{A}_0 \exp[\mathrm{i}(\boldsymbol{k}\cdot\boldsymbol{r} - \omega t)] = -\mathrm{i}\omega\boldsymbol{A}_0 \exp[\mathrm{i}(\boldsymbol{k}\cdot\boldsymbol{r} - \omega t)]. \qquad (24)$$

This yields yet another vector plane wave with the same $\boldsymbol{k}$ and ω, but with a different amplitude.

12.6. Theorem of Gauss. Consider a (generally complex-valued) vector field $\boldsymbol{V}(\boldsymbol{r},t)$ and an arbitrary closed surface S in 3D space. Gauss's theorem states that, at fixed time t, the integral of $\boldsymbol{\nabla}\cdot\boldsymbol{V}(\boldsymbol{r},t)$ over the volume enclosed by S is precisely equal to the surface integral of $\boldsymbol{V}(\boldsymbol{r},t)$ over S; that is,

$$\int_{\text{volume}} \boldsymbol{\nabla}\cdot\boldsymbol{V}(\boldsymbol{r},t)\mathrm{d}v = \oint_{\text{surface}} \boldsymbol{V}(\boldsymbol{r},t)\cdot\mathrm{d}\boldsymbol{s}. \qquad (25)$$

To verify this theorem, divide the volume enclosed by S into a large number of small (i.e., infinitesimal) volume elements, then use the definition of $\boldsymbol{\nabla}\cdot\boldsymbol{V}(\boldsymbol{r},t)$ in Eq.(19) to see that the volume integral in Eq.(25) is nothing but the sum of surface integrals over the infinitesimally small surfaces of the various volume elements contained within S. When two such surfaces happen to be in contact with each other, their contributions to the volume integral cancel out. (Recall that surface normals of a closed surface always point outward; hence adjacent surface elements will have their surface normals pointing in opposite directions.) The only surfaces that could contribute to the volume integral (i.e., not get canceled out by their adjacent neighbors) are those *without* adjacent neighbors; all such surfaces, of course, reside on the enclosing surface S itself. The equality of the volume integral on the left-hand side of Eq.(25) with the surface integral on its right-hand side is thus established.

12.7. Curl of a vector field. The curl operator acts on a vector field $\boldsymbol{V}(\boldsymbol{r},t)$ to produce another vector field $\boldsymbol{\nabla}\times\boldsymbol{V}(\boldsymbol{r},t)$. To find the curl of $\boldsymbol{V}(\boldsymbol{r},t)$ at a given point $(\boldsymbol{r}_0,t_0)$ in space-time, first freeze the time everywhere at $t = t_0$. Draw an infinitesimally small, flat loop around $\boldsymbol{r}_0$—with $\boldsymbol{r}_0$ itself somewhere in the middle, sitting on the flat surface of the loop. The orientation of the loop in 3D space and its shape are arbitrary, so long as every point on the loop is sufficiently close to $\boldsymbol{r}_0$. Choose a direction of travel around the loop and, in accordance with the right-hand rule, orient the loop's surface normal $\hat{\boldsymbol{n}}$ to correspond with the chosen direction of travel. The product of the surface area of the loop Δs and $\hat{\boldsymbol{n}}$ now defines the surface vector $\boldsymbol{\Delta s}$ of the infinitesimal loop. The projection of $\boldsymbol{\nabla} \times \boldsymbol{V}(\boldsymbol{r}_0, t_0)$ on the surface normal $\hat{\boldsymbol{n}}$ is thus given by the line integral of $\boldsymbol{V}(\boldsymbol{r}, t_0)$ around the loop, normalized by its surface area Δs, namely,

$$\hat{\boldsymbol{n}} \cdot [\boldsymbol{\nabla} \times \boldsymbol{V}(\boldsymbol{r}_0, t_0)] = lim_{\Delta s \to 0}(1/\Delta s) \oint_{\text{loop}} \boldsymbol{V}(\boldsymbol{r}, t_0) \cdot \mathrm{d}\boldsymbol{\ell} \cdot \tag{26}$$

Note that the above procedure produces only the *projection* of $\boldsymbol{\nabla} \times \boldsymbol{V}(\boldsymbol{r}_0, t_0)$ along $\hat{\boldsymbol{n}}$. To find the complete vector at $(\boldsymbol{r}_0, t_0)$, one must repeat the process three times, once for each of the three independent, orthogonal directions in 3D space. The normalization by Δs in Eq.(26) is absolutely necessary, otherwise the line integral vanishes as the loop shrinks around $\boldsymbol{r}_0$. There is nothing in this procedure that requires $\boldsymbol{V}(\boldsymbol{r}, t)$ to be real valued; the curl operator therefore may be generally applied to complex-valued vector fields. When $\boldsymbol{\nabla} \times$ is applied to a vector field at all points $(\boldsymbol{r}, t)$ in space-time, it generates a new and altogether different vector field.

The curl of a vector field at a given point $(\boldsymbol{r}_0, t_0)$ in space-time is a property of the field at and around $\boldsymbol{r}_0$ at the time t_0, which is completely independent of any coordinate system that might be used to describe the field. In a Cartesian coordinate system, one could choose the three small loops to be infinitesimal squares with surface normals parallel to the coordinate axes. Using the first two terms in the Taylor series expansion of the field around $\boldsymbol{r}_0$, one finds

$$\boldsymbol{\nabla} \times \boldsymbol{V}(\boldsymbol{r}, t) = (\partial V_z/\partial y - \partial V_y/\partial z)\hat{\boldsymbol{x}} + (\partial V_x/\partial z - \partial V_z/\partial x)\hat{\boldsymbol{y}} + (\partial V_y/\partial x - \partial V_x/\partial y)\hat{\boldsymbol{z}}. \tag{27}$$

Corresponding expressions in cylindrical and spherical coordinates may be found by a similar procedure, provided that the chosen loops in each case conform to the preferred geometry of the coordinate system. Next, we derive the curl operator in a spherical coordinate system. The case of cylindrical coordinates is left as an exercise for the reader.

12.7.1. Curl operator in spherical coordinates. Computing the curl of a vector field, such as $\boldsymbol{V}(\boldsymbol{r})$, requires that we compute the three components of $\boldsymbol{\nabla} \times \boldsymbol{V}(\boldsymbol{r})$ along the $\hat{\boldsymbol{r}}$, $\hat{\boldsymbol{\theta}}$, and $\hat{\boldsymbol{\phi}}$ directions. For each such component, we choose an infinitesimally small closed loop perpendicular to the desired direction, as shown in figure 12.5. The corresponding component of the curl is then obtained by evaluating the line integral of $\boldsymbol{V}(\boldsymbol{r})$ around the loop, followed by normalizing the result by the area of the loop, as follows:

$$\begin{aligned}[\boldsymbol{\nabla} \times \boldsymbol{V}(\boldsymbol{r})]_r &= lim_{\Delta\theta,\Delta\phi \to 0}\big[V_\phi(r, \theta + ½\Delta\theta, \phi) r \sin(\theta + ½\Delta\theta)\, \Delta\phi \\ &\qquad - V_\phi(r, \theta - ½\Delta\theta, \phi) r \sin(\theta - ½\Delta\theta)\, \Delta\phi \\ &\qquad - V_\theta(r, \theta, \phi + ½\Delta\phi) r\Delta\theta + V_\theta(r, \theta, \phi - ½\Delta\phi) r\Delta\theta\big]/(r^2 \sin\theta\, \Delta\theta\Delta\phi) \\ &= (r \sin\theta)^{-1}\big[\partial_\theta(\sin\theta\, V_\phi) - \partial_\phi V_\theta\big].\end{aligned} \tag{28a}$$

$$\begin{aligned}[\boldsymbol{\nabla} \times \boldsymbol{V}(\boldsymbol{r})]_\theta &= lim_{\Delta r,\Delta\phi \to 0}[V_r(r, \theta, \phi + ½\Delta\phi)\Delta r - V_r(r, \theta, \phi - ½\Delta\phi)\Delta r \\ &\qquad + V_\phi(r - ½\Delta r, \theta, \phi)(r - ½\Delta r) \sin\theta\, \Delta\phi \\ &\qquad - V_\phi(r + ½\Delta r, \theta, \phi)(r + ½\Delta r) \sin\theta\, \Delta\phi]/(r \sin\theta\, \Delta r\Delta\phi) \\ &= (r \sin\theta)^{-1}(\partial_\phi V_r) - r^{-1}\partial_r(rV_\phi).\end{aligned} \tag{28b}$$

$$[\boldsymbol{\nabla} \times \boldsymbol{V}(\boldsymbol{r})]_\phi = lim_{\Delta r, \Delta\theta \to 0}[V_r(r, \theta - ½\Delta\theta, \phi)\Delta r - V_r(r, \theta + ½\Delta\theta, \phi)\Delta r$$
$$+V_\theta(r + ½\Delta r, \theta, \phi)(r + ½\Delta r)\Delta\theta$$
$$-V_\theta(r - ½\Delta r, \theta, \phi)(r - ½\Delta r)\Delta\theta]/(r\Delta r\Delta\theta)$$
$$= r^{-1}[\partial_r(rV_\theta) - \partial_\theta V_r]. \tag{28c}$$

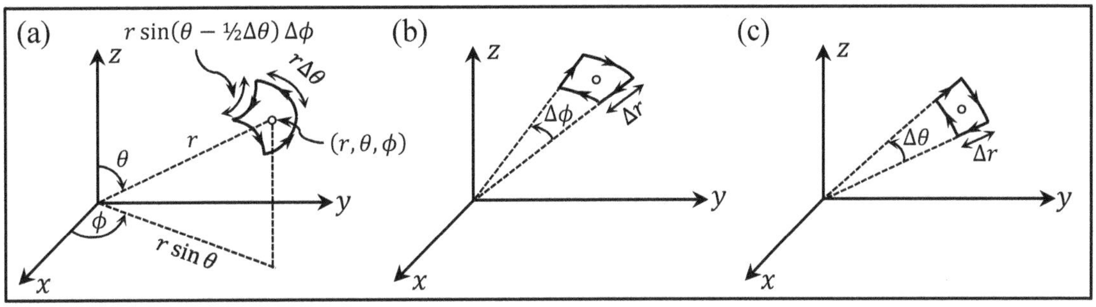

Figure 12.5. In a spherical coordinate system, the curl of the vector field $\boldsymbol{V}(\boldsymbol{r})$ at the arbitrary point $\boldsymbol{r} = (r, \theta, \phi)$ has three components along the $\hat{\boldsymbol{r}}$, $\hat{\boldsymbol{\theta}}$, and $\hat{\boldsymbol{\phi}}$ directions. To determine each component, the line integral of the field must be computed around an infinitesimal closed loop perpendicular to the direction of the desired curl component, then normalized by the area of the loop. In (a), the loop is perpendicular to $\hat{\boldsymbol{r}}$ and has an area equal to $r^2 \sin\theta\, \Delta\theta\Delta\phi$. The closed loop in (b) is perpendicular to $\hat{\boldsymbol{\theta}}$ and has an area equal to $r\sin\theta\,\Delta r\Delta\phi$. The loop in (c) is perpendicular to $\hat{\boldsymbol{\phi}}$ and has area $r\Delta r\Delta\theta$.

Example 3. The curl of $\boldsymbol{A}(\boldsymbol{r}, t) = \boldsymbol{A}_0 \exp[\mathrm{i}(\boldsymbol{k}\cdot\boldsymbol{r} - \omega t)]$, a plane wave vector field with $\omega = \omega' + \mathrm{i}\omega''$, $\boldsymbol{k} = \boldsymbol{k}' + \mathrm{i}\boldsymbol{k}''$, and $\boldsymbol{A}_0 = \boldsymbol{A}_0' + \mathrm{i}\boldsymbol{A}_0''$, is found by a straightforward application of Eq.(27), as follows:

$$\boldsymbol{\nabla} \times \{\boldsymbol{A}_0 \exp[\mathrm{i}(\boldsymbol{k}\cdot\boldsymbol{r} - \omega t)]\}$$
$$= \mathrm{i}\big[(A_{0z}k_y - A_{0y}k_z)\hat{\boldsymbol{x}} + (A_{0x}k_z - A_{0z}k_x)\hat{\boldsymbol{y}} + (A_{0y}k_x - A_{0x}k_y)\hat{\boldsymbol{z}}\big]\exp[\mathrm{i}(\boldsymbol{k}\cdot\boldsymbol{r} - \omega t)]$$
$$= \mathrm{i}\boldsymbol{k} \times \boldsymbol{A}_0 \exp[\mathrm{i}(\boldsymbol{k}\cdot\boldsymbol{r} - \omega t)]. \tag{29}$$

The curl operator $\boldsymbol{\nabla}\times$ acting on the vector plane wave $\boldsymbol{A}_0 \exp[\mathrm{i}(\boldsymbol{k}\cdot\boldsymbol{r} - \omega t)]$ thus yields another vector field that is also a plane wave having the same frequency ω and the same wave vector $\boldsymbol{k}$, but its amplitude is obtained by cross-multiplying $\boldsymbol{A}_0$ into the complex vector $\mathrm{i}\boldsymbol{k}$.

12.8. Theorem of Stokes. Consider a (generally complex-valued) vector field $\boldsymbol{V}(\boldsymbol{r}, t)$, and an arbitrary surface S (not necessarily flat and not necessarily closed) in 3D space. The surface S is generally bounded by a closed loop, which vanishes when S is a closed surface. The Stokes theorem states that, at fixed time t, the surface integral of $\boldsymbol{\nabla} \times \boldsymbol{V}(\boldsymbol{r}, t)$ over S is precisely equal to the line integral of $\boldsymbol{V}(\boldsymbol{r}, t)$ around the boundary of S; that is,

$$\int_{\text{surface}} [\boldsymbol{\nabla} \times \boldsymbol{V}(\boldsymbol{r}, t)] \cdot \mathrm{d}\boldsymbol{s} = \oint_{\text{boundary}} \boldsymbol{V}(\boldsymbol{r}, t) \cdot \mathrm{d}\boldsymbol{\ell}. \tag{30}$$

To verify this theorem, divide the surface S into a large number of small surface elements, then use the definition of $\boldsymbol{\nabla} \times \boldsymbol{V}(\boldsymbol{r}, t)$ in Eq.(26) to see that the surface integral in Eq.(30) is nothing but the sum of line integrals over the infinitesimally small boundaries of the various surface elements $\Delta\boldsymbol{s}$ that comprise S. When two such boundaries, belonging to adjacent surface elements, happen to be in contact with each other, their contributions to the surface integral cancel out. (Recall that the direction of travel around two adjacent loops whose surface normals point in nearly the same direction is such that the line elements on those legs of the loops that are in contact with each other point in opposite directions). The only segments of the infinitesimal loop boundaries that could contribute to the surface integral in Eq.(30) (i.e., not cancelled out by their adjacent neighbors) are those *without* adjacent neighbors. All such segments turn out to be segments of the boundary of S itself, their totality constituting the entire closed boundary of S. It is thus clear that the surface integral on the left-hand side of Eq.(30) must be equal to the line integral on its right-hand side.

In the special case of a closed surface S, the boundary vanishes and, with it, the surface integral of $\boldsymbol{\nabla} \times \boldsymbol{V}(\boldsymbol{r}, t)$ over S. The vector field $\boldsymbol{W}(\boldsymbol{r}, t) = \boldsymbol{\nabla} \times \boldsymbol{V}(\boldsymbol{r}, t)$ derived from the arbitrary vector field $\boldsymbol{V}(\boldsymbol{r}, t)$ thus has the unique property that its integral over *any* closed surface is precisely equal to zero. In particular, if the chosen closed surface is infinitesimally small, one finds that the divergence of $\boldsymbol{W}(\boldsymbol{r}, t)$ is zero everywhere. We conclude, therefore, that $\boldsymbol{\nabla} \cdot [\boldsymbol{\nabla} \times \boldsymbol{V}(\boldsymbol{r}, t)] = 0$, a conclusion that may also be verified analytically via Eqs.(20) and (27).

Any closed loop in 3D space may be associated with an infinite number of surfaces that are all bounded by the same loop. Think of a closed loop as a physical object made up of a wire frame. If this loop is dipped into a soap solution, a soap film will form over its wire frame, which film may be associated with one of the surfaces S that are bounded by this closed loop. Blowing into the soap film will produce all kinds of surfaces that share the property that they are all bounded by the same closed loop. Imagine now a vector field $\boldsymbol{V}(\boldsymbol{r}, t)$ permeating the region of space in which these soap films reside. At any given time $t = t_0$, the line integral of $\boldsymbol{V}(\boldsymbol{r}, t_0)$ around the closed loop is a well-defined entity (generally a complex number). The Stokes theorem then requires that the surface integral of $\boldsymbol{\nabla} \times \boldsymbol{V}(\boldsymbol{r}, t_0)$ over any and all surfaces bounded by this loop be exactly the same. In other words, no matter which surface one selects, the surface integral of $\boldsymbol{\nabla} \times \boldsymbol{V}(\boldsymbol{r}, t_0)$ always turns out to be the same.

The surface S and its closed-loop boundary depicted in figure 12.3 are fairly simple geometric structures to which one may readily apply the Stokes theorem for an arbitrary vector field $\boldsymbol{V}(\boldsymbol{r}, t)$. Not so is the double-loop depicted here on the right-hand side, which nevertheless is associated with an infinite number of continuous surfaces, all of which are bounded by the same double loop. Think of a fairly simple wire frame loop, say, in the shape of a single circle or a single rectangle. Imagine dipping it into a soap solution to form a continuous surface across its wire frame boundary. Now try bending and twisting the wire into complicated and arbitrary patterns, without causing the soap film to break. So long as you do not attempt to pierce the soap film, or to combine and coalesce various patches of it into a new patch, there will be a well-defined surface with a well-defined boundary to which the Stokes theorem may be applied.

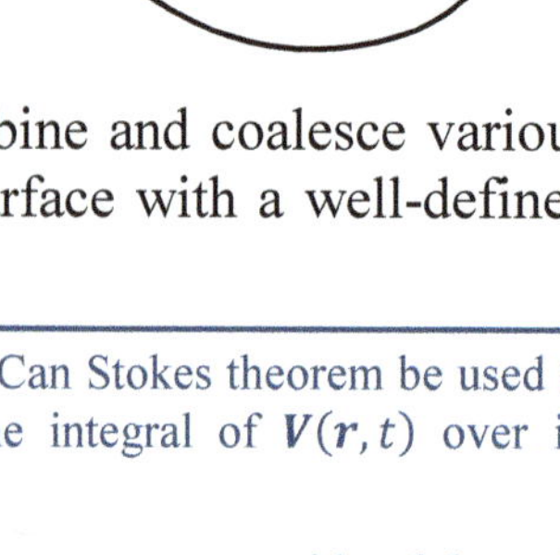

Digression. Imagine a Möbius strip having one or more twists in its fabric. Can Stokes theorem be used to convert the integral of $\boldsymbol{\nabla} \times \boldsymbol{V}(\boldsymbol{r}, t)$ over the surface of the strip into the integral of $\boldsymbol{V}(\boldsymbol{r}, t)$ over its boundary?

Hint: You may want to take a long rectangular strip of paper, draw small squares on one side of the strip only, in such a way as to cover the entire surface with these squares, then give the same clockwise or counter-clockwise direction to each and every little square on the paper strip. Twist the paper one or more times before joining its narrow ends together. Which way are the surface normals $\Delta\boldsymbol{s}$ associated with the little squares pointing? Are the directions of these surface normals varying continuously everywhere in space, or are there discontinuities? In which direction should the line integral over the boundary of the strip be taken? Is there an abrupt change in the direction of the line element $\Delta\boldsymbol{\ell}$ as one moves continuously along the boundary of the strip? Suppose now that after twisting the strip, its two narrow ends are brought close to each other but not glued together; in other words, let there be a small gap where the narrow ends of the strip normally join each other. How does this slight modification to the Möbius strip affect Stokes's theorem? Do the line integrals over the narrow edges of the strip cancel each other out?

12.9. Vector identities. The gradient, divergence, and curl operators of vector calculus, when applied in tandem, or when applied to combinations of various scalar and vector fields, yield identities such as $\boldsymbol{\nabla}\cdot(\boldsymbol{\nabla}\times\boldsymbol{A})=0$ and $\boldsymbol{\nabla}\times(\psi\boldsymbol{A})=(\boldsymbol{\nabla}\psi)\times\boldsymbol{A}+\psi\boldsymbol{\nabla}\times\boldsymbol{A}$. Table 12.2 lists several useful identities of vector calculus, where $\psi(\boldsymbol{r})$ or $\psi(\boldsymbol{r},t)$ is an arbitrary scalar field, while $\boldsymbol{A}(\boldsymbol{r})$ and $\boldsymbol{B}(\boldsymbol{r})$—or $\boldsymbol{A}(\boldsymbol{r},t)$ and $\boldsymbol{B}(\boldsymbol{r},t)$—are arbitrary vector fields. (For proofs of these identities, see the problems at the end of the chapter.) Note that the identities listed in table 12.2 apply equally to static as well as dynamic fields, simply because the derivatives appearing in the gradient, divergence, and curl operators are differentiations with respect to the *space* coordinates—and not the time coordinate t.

$$\boldsymbol{\nabla}\cdot(\boldsymbol{\nabla}\times\boldsymbol{A})=0.$$

$$\boldsymbol{\nabla}\times(\boldsymbol{\nabla}\psi)=0.$$

$$\boldsymbol{\nabla}\cdot(\psi\boldsymbol{A})=(\boldsymbol{\nabla}\psi)\cdot\boldsymbol{A}+\psi\boldsymbol{\nabla}\cdot\boldsymbol{A}.$$

$$\boldsymbol{\nabla}\times(\psi\boldsymbol{A})=(\boldsymbol{\nabla}\psi)\times\boldsymbol{A}+\psi\boldsymbol{\nabla}\times\boldsymbol{A}.$$

$$\boldsymbol{\nabla}\cdot(\boldsymbol{A}\times\boldsymbol{B})=\boldsymbol{B}\cdot(\boldsymbol{\nabla}\times\boldsymbol{A})-\boldsymbol{A}\cdot(\boldsymbol{\nabla}\times\boldsymbol{B}).$$

$$\boldsymbol{\nabla}\times(\boldsymbol{\nabla}\times\boldsymbol{A})=\boldsymbol{\nabla}(\boldsymbol{\nabla}\cdot\boldsymbol{A})-\boldsymbol{\nabla}^2\boldsymbol{A}.$$

$$\boldsymbol{\nabla}(\boldsymbol{A}\cdot\boldsymbol{B})=(\boldsymbol{A}\cdot\boldsymbol{\nabla})\boldsymbol{B}+(\boldsymbol{B}\cdot\boldsymbol{\nabla})\boldsymbol{A}+\boldsymbol{A}\times(\boldsymbol{\nabla}\times\boldsymbol{B})+\boldsymbol{B}\times(\boldsymbol{\nabla}\times\boldsymbol{A}).$$

$$\boldsymbol{\nabla}\times(\boldsymbol{A}\times\boldsymbol{B})=\boldsymbol{A}(\boldsymbol{\nabla}\cdot\boldsymbol{B})-\boldsymbol{B}(\boldsymbol{\nabla}\cdot\boldsymbol{A})+(\boldsymbol{B}\cdot\boldsymbol{\nabla})\boldsymbol{A}-(\boldsymbol{A}\cdot\boldsymbol{\nabla})\boldsymbol{B}.$$

If $\boldsymbol{r}$ is a point in three-dimensional Euclidean space, then

$$\boldsymbol{\nabla}(1/|\boldsymbol{r}|)=-\boldsymbol{r}/|\boldsymbol{r}|^3.$$

$$\boldsymbol{\nabla}\cdot(\boldsymbol{r}/|\boldsymbol{r}|^3)=4\pi\delta(\boldsymbol{r}).$$

Table 12.2. Important identities of vector calculus.

The expression $(\boldsymbol{A}\cdot\boldsymbol{\nabla})\boldsymbol{B}$ appearing in table 12.2 and involving the vector fields $\boldsymbol{A}(\boldsymbol{r})$ and $\boldsymbol{B}(\boldsymbol{r})$, is defined in Cartesian coordinates as $(\boldsymbol{A}\cdot\boldsymbol{\nabla})\boldsymbol{B}=A_x\partial_x\boldsymbol{B}+A_y\partial_y\boldsymbol{B}+A_z\partial_z\boldsymbol{B}$. This operation yields the derivative of the vector field $\boldsymbol{B}(\boldsymbol{r})$, taken at an arbitrary point $\boldsymbol{r}$ along the direction of the local vector $\boldsymbol{A}(\boldsymbol{r})$; the derivative is subsequently multiplied by the magnitude $|\boldsymbol{A}(\boldsymbol{r})|$ of $\boldsymbol{A}(\boldsymbol{r})$. To understand the reasoning behind this interpretation, let $\Delta\boldsymbol{r}=(\Delta x)\hat{\boldsymbol{x}}+(\Delta y)\hat{\boldsymbol{y}}+(\Delta z)\hat{\boldsymbol{z}}=\boldsymbol{A}(\boldsymbol{r})\Delta r/|\boldsymbol{A}(\boldsymbol{r})|$ be a short vector of length Δr aligned with $\boldsymbol{A}(\boldsymbol{r})$ at point $\boldsymbol{r}$. A Taylor series expansion of $\boldsymbol{B}(\boldsymbol{r})$ at the same point $\boldsymbol{r}$ now yields

$$\begin{aligned}\boldsymbol{B}(\boldsymbol{r}+\Delta\boldsymbol{r})-\boldsymbol{B}(\boldsymbol{r})&\cong(\partial_x\boldsymbol{B})\Delta x+(\partial_y\boldsymbol{B})\Delta y+(\partial_z\boldsymbol{B})\Delta z\\&=[(\partial_x\boldsymbol{B})A_x+(\partial_y\boldsymbol{B})A_y+(\partial_z\boldsymbol{B})A_z]\,\Delta r/|\boldsymbol{A}(\boldsymbol{r})|.\qquad(31)\end{aligned}$$

In the limit when $\Delta r\to 0$, the above equation shows that $(\boldsymbol{A}\cdot\boldsymbol{\nabla})\boldsymbol{B}$ is the product of $|\boldsymbol{A}(\boldsymbol{r})|$ and the limiting value of $[\boldsymbol{B}(\boldsymbol{r}+\Delta\boldsymbol{r})-\boldsymbol{B}(\boldsymbol{r})]/\Delta r$. (See problem 24 for a proper expression of $(\boldsymbol{A}\cdot\boldsymbol{\nabla})\boldsymbol{B}$ in cylindrical coordinates.)

The various operators of vector calculus expressed in Cartesian, cylindrical, and spherical coordinate systems are listed in table 12.3. Also shown in this table are formulas for the Laplacian operator $\nabla^2 = \boldsymbol{\nabla} \cdot \boldsymbol{\nabla}$ acting on a scalar field $\psi(\boldsymbol{r})$ or $\psi(\boldsymbol{r}, t)$. The Laplacian operator for vector fields is described in section 11.

Cartesian (x, y, z):

$$\boldsymbol{\nabla}\psi = \frac{\partial \psi}{\partial x}\hat{\boldsymbol{x}} + \frac{\partial \psi}{\partial y}\hat{\boldsymbol{y}} + \frac{\partial \psi}{\partial z}\hat{\boldsymbol{z}}$$

$$\boldsymbol{\nabla} \cdot \boldsymbol{A} = \frac{\partial A_x}{\partial x} + \frac{\partial A_y}{\partial y} + \frac{\partial A_z}{\partial z}$$

$$\boldsymbol{\nabla} \times \boldsymbol{A} = \left(\frac{\partial A_z}{\partial y} - \frac{\partial A_y}{\partial z}\right)\hat{\boldsymbol{x}} + \left(\frac{\partial A_x}{\partial z} - \frac{\partial A_z}{\partial x}\right)\hat{\boldsymbol{y}} + \left(\frac{\partial A_y}{\partial x} - \frac{\partial A_x}{\partial y}\right)\hat{\boldsymbol{z}}$$

$$\boldsymbol{\nabla}^2\psi = \frac{\partial^2 \psi}{\partial x^2} + \frac{\partial^2 \psi}{\partial y^2} + \frac{\partial^2 \psi}{\partial z^2}$$

Cylindrical (ρ, ϕ, z):

$$\boldsymbol{\nabla}\psi = \frac{\partial \psi}{\partial \rho}\hat{\boldsymbol{\rho}} + \frac{\partial \psi}{\rho \partial \phi}\hat{\boldsymbol{\phi}} + \frac{\partial \psi}{\partial z}\hat{\boldsymbol{z}}$$

$$\boldsymbol{\nabla} \cdot \boldsymbol{A} = \frac{\partial(\rho A_\rho)}{\rho \partial \rho} + \frac{\partial A_\phi}{\rho \partial \phi} + \frac{\partial A_z}{\partial z}$$

$$\boldsymbol{\nabla} \times \boldsymbol{A} = \left(\frac{\partial A_z}{\rho \partial \phi} - \frac{\partial A_\phi}{\partial z}\right)\hat{\boldsymbol{\rho}} + \left(\frac{\partial A_\rho}{\partial z} - \frac{\partial A_z}{\partial \rho}\right)\hat{\boldsymbol{\phi}} + \frac{1}{\rho}\left(\frac{\partial(\rho A_\phi)}{\partial \rho} - \frac{\partial A_\rho}{\partial \phi}\right)\hat{\boldsymbol{z}}$$

$$\boldsymbol{\nabla}^2\psi = \frac{\partial(\rho \partial\psi/\partial\rho)}{\rho \partial \rho} + \frac{\partial^2 \psi}{\rho^2 \partial \phi^2} + \frac{\partial^2 \psi}{\partial z^2}$$

Spherical (r, θ, ϕ):

$$\boldsymbol{\nabla}\psi = \frac{\partial \psi}{\partial r}\hat{\boldsymbol{r}} + \frac{\partial \psi}{r \partial \theta}\hat{\boldsymbol{\theta}} + \frac{1}{r\sin\theta}\frac{\partial \psi}{\partial \phi}\hat{\boldsymbol{\phi}}$$

$$\boldsymbol{\nabla} \cdot \boldsymbol{A} = \frac{\partial(r^2 A_r)}{r^2 \partial r} + \frac{1}{r\sin\theta}\frac{\partial(\sin\theta A_\theta)}{\partial \theta} + \frac{1}{r\sin\theta}\frac{\partial A_\phi}{\partial \phi}$$

$$\boldsymbol{\nabla} \times \boldsymbol{A} = \frac{1}{r\sin\theta}\left(\frac{\partial(\sin\theta A_\phi)}{\partial \theta} - \frac{\partial A_\theta}{\partial \phi}\right)\hat{\boldsymbol{r}} + \left(\frac{1}{r\sin\theta}\frac{\partial A_r}{\partial \phi} - \frac{\partial(rA_\phi)}{r \partial r}\right)\hat{\boldsymbol{\theta}} + \frac{1}{r}\left(\frac{\partial(rA_\theta)}{\partial r} - \frac{\partial A_r}{\partial \theta}\right)\hat{\boldsymbol{\phi}}$$

$$\boldsymbol{\nabla}^2\psi = \frac{\partial(r^2 \partial\psi/\partial r)}{r^2 \partial r} + \frac{1}{r^2\sin\theta}\frac{\partial(\sin\theta\, \partial\psi/\partial\theta)}{\partial \theta} + \frac{1}{r^2\sin^2\theta}\frac{\partial^2 \psi}{\partial \phi^2}$$

$$\left[\text{Note that } \frac{\partial(r^2 \partial\psi/\partial r)}{r^2 \partial r} = \frac{\partial^2(r\psi)}{r \partial r^2}\right]$$

Table 12.3. Vector operations in Cartesian, cylindrical, and spherical Coordinates.

12.10. Scalar and vector fields in the Fourier domain. A scalar field such as $\psi(\boldsymbol{r},t)$ can be Fourier transformed, as follows:

$$\psi(\boldsymbol{k},\omega)=\int_{-\infty}^{\infty}\psi(\boldsymbol{r},t)\exp[-\mathrm{i}(\boldsymbol{k}\cdot\boldsymbol{r}-\omega t)]\,\mathrm{d}x\mathrm{d}y\mathrm{d}z\mathrm{d}t. \tag{32}$$

The inverse Fourier transform of $\psi(\boldsymbol{k},\omega)$ then yields the original function $\psi(\boldsymbol{r},t)$, namely,

$$\psi(\boldsymbol{r},t)=(2\pi)^{-4}\int_{-\infty}^{\infty}\psi(\boldsymbol{k},\omega)\exp[\mathrm{i}(\boldsymbol{k}\cdot\boldsymbol{r}-\omega t)]\,\mathrm{d}k_x\mathrm{d}k_y\mathrm{d}k_z\mathrm{d}\omega. \tag{33}$$

Note in the above equations that the Fourier kernel is written in terms of the k-vector $k_x\hat{\boldsymbol{x}}+k_y\hat{\boldsymbol{y}}+k_z\hat{\boldsymbol{z}}$ and the angular frequency ω, rather than the usual spatial frequency $\boldsymbol{s}=s_x\hat{\boldsymbol{x}}+s_y\hat{\boldsymbol{y}}+s_z\hat{\boldsymbol{z}}$ and temporal frequency f. Considering that $\boldsymbol{k}=2\pi\boldsymbol{s}$ and $\omega=2\pi f$, it is easy to see that a simple change of variables explains the appearance of the pre-factor $(2\pi)^{-4}$ in Eq.(33). If we now differentiate both sides of Eq.(33) with respect to time, we find that the Fourier transform of $\partial_t\psi(\boldsymbol{r},t)$ is $-\mathrm{i}\omega\psi(\boldsymbol{k},\omega)$. In other words, differentiation of a function with respect to time results in the multiplication of its Fourier transform by $-\mathrm{i}\omega$. Similarly, if we operate on $\psi(\boldsymbol{r},t)$ with the gradient operator, we will find $\boldsymbol{\nabla}\psi(\boldsymbol{r},t)\leftrightarrow\mathrm{i}\boldsymbol{k}\psi(\boldsymbol{k},\omega)$. In words, the Fourier transform of the gradient of a scalar field is the product of the Fourier transform of the field and the complex vector $\mathrm{i}\boldsymbol{k}$.

In similar fashion, one may define the Fourier integrals for a vector field $\boldsymbol{V}(\boldsymbol{r},t)$ as follows:

$$\boldsymbol{V}(\boldsymbol{k},\omega)=\int_{-\infty}^{\infty}\boldsymbol{V}(\boldsymbol{r},t)\exp[-\mathrm{i}(\boldsymbol{k}\cdot\boldsymbol{r}-\omega t)]\,\mathrm{d}x\mathrm{d}y\mathrm{d}z\mathrm{d}t. \tag{34}$$

$$\boldsymbol{V}(\boldsymbol{r},t)=(2\pi)^{-4}\int_{-\infty}^{\infty}\boldsymbol{V}(\boldsymbol{k},\omega)\exp[\mathrm{i}(\boldsymbol{k}\cdot\boldsymbol{r}-\omega t)]\,\mathrm{d}k_x\mathrm{d}k_y\mathrm{d}k_z\mathrm{d}\omega. \tag{35}$$

Thus, the Fourier transform of the time derivative of $\boldsymbol{V}(\boldsymbol{r},t)$ is $-\mathrm{i}\omega\boldsymbol{V}(\boldsymbol{k},\omega)$. Likewise, the divergence and curl operators acting on $\boldsymbol{V}(\boldsymbol{r},t)$ yield the Fourier transform pairs $\boldsymbol{\nabla}\cdot\boldsymbol{V}(\boldsymbol{r},t)\leftrightarrow\mathrm{i}\boldsymbol{k}\cdot\boldsymbol{V}(\boldsymbol{k},\omega)$ and $\boldsymbol{\nabla}\times\boldsymbol{V}(\boldsymbol{r},t)\leftrightarrow\mathrm{i}\boldsymbol{k}\times\boldsymbol{V}(\boldsymbol{k},\omega)$. Some of the vector calculus identities are easier to derive in the Fourier domain than in the usual xyz space. For instance, it is easy to verify that

$$\boldsymbol{\nabla}\times[\boldsymbol{\nabla}\psi(\boldsymbol{r})]\leftrightarrow\mathrm{i}\boldsymbol{k}\times[\mathrm{i}\boldsymbol{k}\psi(\boldsymbol{k})]=-(\boldsymbol{k}\times\boldsymbol{k})\psi(\boldsymbol{k})=0. \tag{36}$$

$$\boldsymbol{\nabla}\cdot[\boldsymbol{\nabla}\times\boldsymbol{V}(\boldsymbol{r})]\leftrightarrow\mathrm{i}\boldsymbol{k}\cdot[\mathrm{i}\boldsymbol{k}\times\boldsymbol{V}(\boldsymbol{k})]=-(\boldsymbol{k}\times\boldsymbol{k})\cdot\boldsymbol{V}(\boldsymbol{k})=0. \tag{37}$$

Similarly, one may go about evaluating the divergence of $\boldsymbol{A}(\boldsymbol{r})\times\boldsymbol{B}(\boldsymbol{r})$ in the Fourier domain in the following way:

shorthand for $\mathrm{d}k_x\mathrm{d}k_y\mathrm{d}k_z$

$$\boldsymbol{\nabla}\cdot[\boldsymbol{A}(\boldsymbol{r})\times\boldsymbol{B}(\boldsymbol{r})]=\boldsymbol{\nabla}\cdot\left[(2\pi)^{-6}\int_{-\infty}^{\infty}\boldsymbol{A}(\boldsymbol{k})\exp(\mathrm{i}\boldsymbol{k}\cdot\boldsymbol{r})\,\mathrm{d}\boldsymbol{k}\times\int_{-\infty}^{\infty}\boldsymbol{B}(\boldsymbol{k}')\exp(\mathrm{i}\boldsymbol{k}'\cdot\boldsymbol{r})\,\mathrm{d}\boldsymbol{k}'\right]$$

$$=\boldsymbol{\nabla}\cdot\left\{(2\pi)^{-6}\int_{-\infty}^{\infty}\boldsymbol{A}(\boldsymbol{k})\times\boldsymbol{B}(\boldsymbol{k}')\exp[\mathrm{i}(\boldsymbol{k}+\boldsymbol{k}')\cdot\boldsymbol{r}]\,\mathrm{d}\boldsymbol{k}\mathrm{d}\boldsymbol{k}'\right\}$$

$$=\mathrm{i}(2\pi)^{-6}\int_{-\infty}^{\infty}(\boldsymbol{k}+\boldsymbol{k}')\cdot[\boldsymbol{A}(\boldsymbol{k})\times\boldsymbol{B}(\boldsymbol{k}')]\exp[\mathrm{i}(\boldsymbol{k}+\boldsymbol{k}')\cdot\boldsymbol{r}]\,\mathrm{d}\boldsymbol{k}\mathrm{d}\boldsymbol{k}'$$

$$=\mathrm{i}(2\pi)^{-6}\{\int_{-\infty}^{\infty}\boldsymbol{k}\cdot[\boldsymbol{A}(\boldsymbol{k})\times\boldsymbol{B}(\boldsymbol{k}')]\exp[\mathrm{i}(\boldsymbol{k}+\boldsymbol{k}')\cdot\boldsymbol{r}]\,\mathrm{d}\boldsymbol{k}\mathrm{d}\boldsymbol{k}'$$

$$+\int_{-\infty}^{\infty}\boldsymbol{k}'\cdot[\boldsymbol{A}(\boldsymbol{k})\times\boldsymbol{B}(\boldsymbol{k}')]\exp[\mathrm{i}(\boldsymbol{k}+\boldsymbol{k}')\cdot\boldsymbol{r}]\,\mathrm{d}\boldsymbol{k}\mathrm{d}\boldsymbol{k}'\}$$

$\boldsymbol{a}\cdot(\boldsymbol{b}\times\boldsymbol{c})=(\boldsymbol{a}\times\boldsymbol{b})\cdot\boldsymbol{c}$

$$=(2\pi)^{-6}\{\int_{-\infty}^{\infty}[\mathrm{i}\boldsymbol{k}\times\boldsymbol{A}(\boldsymbol{k})]\cdot\boldsymbol{B}(\boldsymbol{k}')\exp[\mathrm{i}(\boldsymbol{k}+\boldsymbol{k}')\cdot\boldsymbol{r}]\,\mathrm{d}\boldsymbol{k}\mathrm{d}\boldsymbol{k}'$$

$$- \int_{-\infty}^{\infty} [\mathrm{i}\boldsymbol{k}' \times \boldsymbol{B}(\boldsymbol{k}')] \cdot \boldsymbol{A}(\boldsymbol{k}) \exp[\mathrm{i}(\boldsymbol{k}+\boldsymbol{k}')\cdot \boldsymbol{r}]\, \mathrm{d}\boldsymbol{k}\mathrm{d}\boldsymbol{k}'\}$$

$$= (2\pi)^{-6}\{\int_{-\infty}^{\infty} \mathrm{i}\boldsymbol{k} \times \boldsymbol{A}(\boldsymbol{k}) \exp(\mathrm{i}\boldsymbol{k}\cdot\boldsymbol{r})\, \mathrm{d}\boldsymbol{k} \cdot \int_{-\infty}^{\infty} \boldsymbol{B}(\boldsymbol{k}') \exp(\mathrm{i}\boldsymbol{k}'\cdot\boldsymbol{r})\, \mathrm{d}\boldsymbol{k}'$$

$$- \int_{-\infty}^{\infty} \boldsymbol{A}(\boldsymbol{k}) \exp(\mathrm{i}\boldsymbol{k}\cdot\boldsymbol{r})\, \mathrm{d}\boldsymbol{k} \cdot \int_{-\infty}^{\infty} \mathrm{i}\boldsymbol{k}' \times \boldsymbol{B}(\boldsymbol{k}') \exp(\mathrm{i}\boldsymbol{k}'\cdot\boldsymbol{r})\, \mathrm{d}\boldsymbol{k}'\}$$

$$= \boldsymbol{B}(\boldsymbol{r}) \cdot [\boldsymbol{\nabla} \times \boldsymbol{A}(\boldsymbol{r})] - \boldsymbol{A}(\boldsymbol{r}) \cdot [\boldsymbol{\nabla} \times \boldsymbol{B}(\boldsymbol{r})]. \tag{38}$$

12.11. The Laplacian operator. For a scalar field $\psi(\boldsymbol{r})$, the Laplacian is defined as the divergence of the gradient of $\psi(\boldsymbol{r})$; that is,

$$\nabla^2\psi(\boldsymbol{r}) = \boldsymbol{\nabla} \cdot [\boldsymbol{\nabla}\psi(\boldsymbol{r})] = (\partial_x^2 + \partial_y^2 + \partial_z^2)\psi(\boldsymbol{r}). \tag{39}$$

We thus arrive at the Fourier transform pair $\nabla^2\psi(\boldsymbol{r}) \leftrightarrow \mathrm{i}\boldsymbol{k} \cdot [\mathrm{i}\boldsymbol{k}\psi(\boldsymbol{k})] = -k^2\psi(\boldsymbol{k})$. Table 12.3 contains general expressions for the Laplacian operator in Cartesian, cylindrical, and spherical coordinate systems.

For the vector field $\boldsymbol{V}(\boldsymbol{r})$, it is tempting to define the Laplacian as $\boldsymbol{\nabla}[\boldsymbol{\nabla} \cdot \boldsymbol{V}(\boldsymbol{r})]$, or, alternatively, as $\boldsymbol{\nabla} \times [\boldsymbol{\nabla} \times \boldsymbol{V}(\boldsymbol{r})]$. However, neither option yields the simple Fourier transform $-k^2\boldsymbol{V}(\boldsymbol{k})$, which, in analogy with the aforementioned result for scalar fields, one would hope to find for the Laplacian of $\boldsymbol{V}(\boldsymbol{r})$. Instead, we have

$$\boldsymbol{\nabla}[\boldsymbol{\nabla} \cdot \boldsymbol{V}(\boldsymbol{r})] \leftrightarrow -\boldsymbol{k}[\boldsymbol{k} \cdot \boldsymbol{V}(\boldsymbol{k})]. \tag{40}$$

$$\boldsymbol{\nabla} \times [\boldsymbol{\nabla} \times \boldsymbol{V}(\boldsymbol{r})] \leftrightarrow -\boldsymbol{k} \times [\boldsymbol{k} \times \boldsymbol{V}(\boldsymbol{k})] \overset{\boxed{\boldsymbol{a}\times(\boldsymbol{b}\times\boldsymbol{c}) = (\boldsymbol{a}\cdot\boldsymbol{c})\boldsymbol{b} - (\boldsymbol{a}\cdot\boldsymbol{b})\boldsymbol{c}}}{=} (\boldsymbol{k} \cdot \boldsymbol{k})\boldsymbol{V}(\boldsymbol{k}) - [\boldsymbol{k} \cdot \boldsymbol{V}(\boldsymbol{k})]\boldsymbol{k}. \tag{41}$$

A quick glance at the above equations reveals that their combination in the form of $\boldsymbol{\nabla}[\boldsymbol{\nabla} \cdot \boldsymbol{V}(\boldsymbol{r})] - \boldsymbol{\nabla} \times [\boldsymbol{\nabla} \times \boldsymbol{V}(\boldsymbol{r})]$ will have the desired Fourier transform, namely $-k^2\boldsymbol{V}(\boldsymbol{k})$. Indeed, this is how the Laplacian of vector fields is generally defined; that is,

$$\boldsymbol{\nabla}^2\boldsymbol{V}(\boldsymbol{r}) = \boldsymbol{\nabla}[\boldsymbol{\nabla} \cdot \boldsymbol{V}(\boldsymbol{r})] - \boldsymbol{\nabla} \times [\boldsymbol{\nabla} \times \boldsymbol{V}(\boldsymbol{r})] \leftrightarrow -k^2\boldsymbol{V}(\boldsymbol{k}). \tag{42}$$

In Cartesian coordinates, one will have $-k^2\boldsymbol{V}(\boldsymbol{k}) = -k^2[V_x(\boldsymbol{k})\hat{\boldsymbol{x}} + V_y(\boldsymbol{k})\hat{\boldsymbol{y}} + V_z(\boldsymbol{k})\hat{\boldsymbol{z}}]$, which, upon inverse Fourier transformation, yields

$$\boldsymbol{\nabla}^2\boldsymbol{V}(\boldsymbol{r}) = [\nabla^2 V_x(\boldsymbol{r})]\hat{\boldsymbol{x}} + [\nabla^2 V_y(\boldsymbol{r})]\hat{\boldsymbol{y}} + [\nabla^2 V_z(\boldsymbol{r})]\hat{\boldsymbol{z}}. \tag{43}$$

Thus, in Cartesian coordinates, the Laplacians of the x, y, z components of a vector field become the x, y, z components of the Laplacian of the vector field. In general, however, this simple relationship does *not* hold in other coordinate systems, and one must rely on Eq.(42) to arrive at expressions for the Laplacian of a vector field in non-Cartesian systems.

Suggestions for further reading

1. H. M. Schey, *Div, Grad, Curl, and All That: An Informal Text on Vector Calculus*, 3rd edition, W. W. Norton, San Francisco (1996).
2. J. E. Marsden and A. Tromba, *Vector Calculus* (6th edition), W. H. Freeman and Co., New York (2012).
3. J. H. Hubbard and B. B. Hubbard, *Vector Calculus, Linear Algebra, and Differential Forms: A Unified Approach*, 5th edition, Published by Matrix Editions, Ithaca, New York (2015).
4. P. C. Matthews, *Vector Calculus*, Springer Undergraduate Mathematics Series, Springer-Verlag, London, United Kingdom (1998).
5. G. B. Arfken and H. J. Weber, *Mathematical Methods for Physicists*, 6th edition, Academic Press, Cambridge, Massachusetts (2005).

Problems

*In the following problems, $\boldsymbol{A} = (A_x, A_y, A_z)$, $\boldsymbol{B} = (B_x, B_y, B_z)$ and $\boldsymbol{C} = (C_x, C_y, C_z)$ are ordinary vectors in 3D Cartesian space. Do *not* confuse them with complex vectors in the complex plane. Recall that the dot product is defined as $\boldsymbol{A} \cdot \boldsymbol{B} = A_x B_x + A_y B_y + A_z B_z$ and the cross product is $\boldsymbol{A} \times \boldsymbol{B} = [(A_y B_z - A_z B_y), (A_z B_x - A_x B_z), (A_x B_y - A_y B_x)]$.

Problem 1) Show that $\boldsymbol{A} \cdot \boldsymbol{B} = |\boldsymbol{A}||\boldsymbol{B}| \cos\theta$, where $|\boldsymbol{A}|$ is the length of $\boldsymbol{A}$, $|\boldsymbol{B}|$ is the length of $\boldsymbol{B}$, and θ is the angle between $\boldsymbol{A}$ and $\boldsymbol{B}$.

Problem 2) Show that $\boldsymbol{A} \times \boldsymbol{B}$ is a vector perpendicular to the plane of $\boldsymbol{A}$ and $\boldsymbol{B}$, with a length equal to the area of the parallelogram.

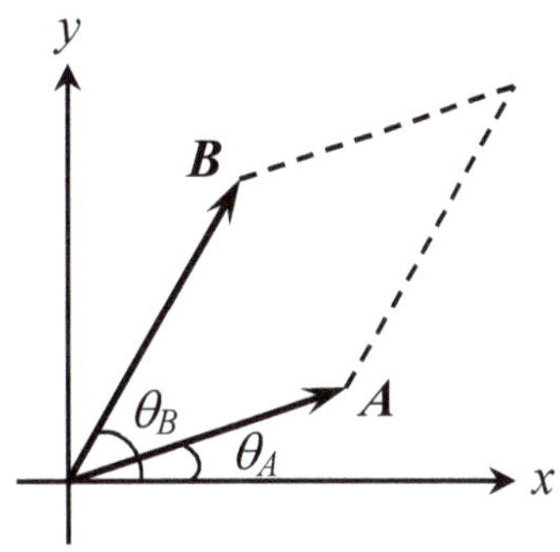

Problem 3)

(a) Show that $(\boldsymbol{A} \times \boldsymbol{B}) \cdot \boldsymbol{C}$ is equal to the volume of the parallelepiped shown.

(b) Without doing extensive calculations, show that $(\boldsymbol{A} \times \boldsymbol{B}) \cdot \boldsymbol{C} = \boldsymbol{A} \cdot (\boldsymbol{B} \times \boldsymbol{C}) = \boldsymbol{B} \cdot (\boldsymbol{C} \times \boldsymbol{A})$.

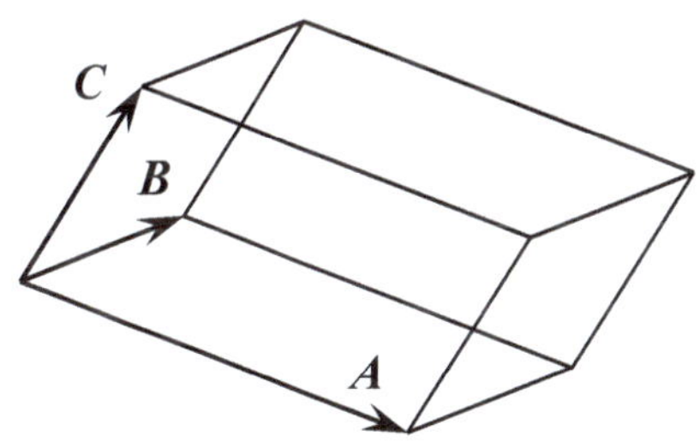

Problem 4) Prove the following vector identities. Whenever possible use geometric arguments instead of (or in combination with) algebraic arguments.

(a) $\boldsymbol{A} \cdot \boldsymbol{B} = \boldsymbol{B} \cdot \boldsymbol{A}$.

(b) $\boldsymbol{A} \cdot (\boldsymbol{B} + \boldsymbol{C}) = \boldsymbol{A} \cdot \boldsymbol{B} + \boldsymbol{A} \cdot \boldsymbol{C}$.

(c) $\boldsymbol{A} \times \boldsymbol{B} = -\boldsymbol{B} \times \boldsymbol{A}$.

(d) $\boldsymbol{A} \times (\boldsymbol{B} + \boldsymbol{C}) = \boldsymbol{A} \times \boldsymbol{B} + \boldsymbol{A} \times \boldsymbol{C}$.

(e) $(\boldsymbol{A} + \boldsymbol{B}) \times \boldsymbol{C} = \boldsymbol{A} \times \boldsymbol{C} + \boldsymbol{B} \times \boldsymbol{C}$.

Hint: To prove part (e), you may invoke the identities given in parts (c) and (d).

Problem 5) (a) Use a counter-example to show that $(\boldsymbol{A} \times \boldsymbol{B}) \times \boldsymbol{C}$ is *not* necessarily equal to $\boldsymbol{A} \times (\boldsymbol{B} \times \boldsymbol{C})$.

(b) Prove that $\boldsymbol{A} \times (\boldsymbol{B} \times \boldsymbol{C}) = (\boldsymbol{A} \cdot \boldsymbol{C})\boldsymbol{B} - (\boldsymbol{A} \cdot \boldsymbol{B})\boldsymbol{C}$. Can you give a geometric interpretation of this identity?

Problem 6) Construct the new vector $\boldsymbol{C} = \boldsymbol{B} - [(\boldsymbol{A} \cdot \boldsymbol{B})\boldsymbol{A}]/|\boldsymbol{A}|^2$ from two arbitrary vectors $\boldsymbol{A}$ and $\boldsymbol{B}$. Prove that $\boldsymbol{C}$ is orthogonal to $\boldsymbol{A}$.

Problem 7) Show that $(\boldsymbol{A} \times \boldsymbol{B}) \cdot (\boldsymbol{C} \times \boldsymbol{D}) = (\boldsymbol{A} \cdot \boldsymbol{C})(\boldsymbol{B} \cdot \boldsymbol{D}) - (\boldsymbol{A} \cdot \boldsymbol{D})(\boldsymbol{B} \cdot \boldsymbol{C})$.

Problem 8) Show that $[\boldsymbol{A} \times (\boldsymbol{B} \times \boldsymbol{A})] \cdot \boldsymbol{B} = (\boldsymbol{A} \times \boldsymbol{B}) \cdot (\boldsymbol{A} \times \boldsymbol{B})$.

Problem 9) Find expressions for $\boldsymbol{\nabla} f(\boldsymbol{r})$ in Cartesian, cylindrical, and spherical coordinate systems starting with the definition of the *gradient* operator, namely that which produces a vector field in 3D space that represents the direction and magnitude of the maximum "slope" of $f(\boldsymbol{r})$ at each location $\boldsymbol{r}$.

Problem 10) Find expressions for $\boldsymbol{\nabla} \cdot \boldsymbol{f}(\boldsymbol{r})$ in Cartesian, cylindrical, and spherical coordinate systems starting with the definition of the *divergence* operator, namely that which produces a scalar field in 3D space that represents the normalized surface integral of $\boldsymbol{f}(\boldsymbol{r})$ over a small volume element Δv centered at $\boldsymbol{r}$ (normalization is by Δv).

Problem 11) Find expressions for $\boldsymbol{\nabla} \times \boldsymbol{f}(\boldsymbol{r})$ in Cartesian, cylindrical, and spherical coordinate systems starting with the definition of the *curl* operator, namely that which produces a vector field in 3D space that represents the normalized line integral of $\boldsymbol{f}(\boldsymbol{r})$ around the boundary of a small surface element Δs centered at $\boldsymbol{r}$ (normalization is by Δs).

Problem 12) The figure shows three triangles in the yz-, xz-, and xy-planes of a Cartesian coordinate system. These triangles form the sides of a pyramid, with their surface vectors, denoted by $\Delta\boldsymbol{S}_x$, $\Delta\boldsymbol{S}_y$, $\Delta\boldsymbol{S}_z$, adding up to yield the surface vector $\Delta\boldsymbol{S}$ corresponding to the base of the pyramid, as follows:

$$\Delta\boldsymbol{S} = \Delta\boldsymbol{S}_x + \Delta\boldsymbol{S}_y + \Delta\boldsymbol{S}_z.$$

Consider a vector field $\boldsymbol{A}(\boldsymbol{r})$, whose curl at an interior point $\boldsymbol{r}_0$ of the pyramid is given by $\boldsymbol{C} = \boldsymbol{\nabla} \times \boldsymbol{A}(\boldsymbol{r}_0) = C_x\hat{\boldsymbol{x}} + C_y\hat{\boldsymbol{y}} + C_z\hat{\boldsymbol{z}}$. In the limit of a small pyramid, show that $\boldsymbol{C} \cdot \Delta\boldsymbol{S}$ approaches the integral of $\boldsymbol{A}(\boldsymbol{r})$ around the base of the pyramid.

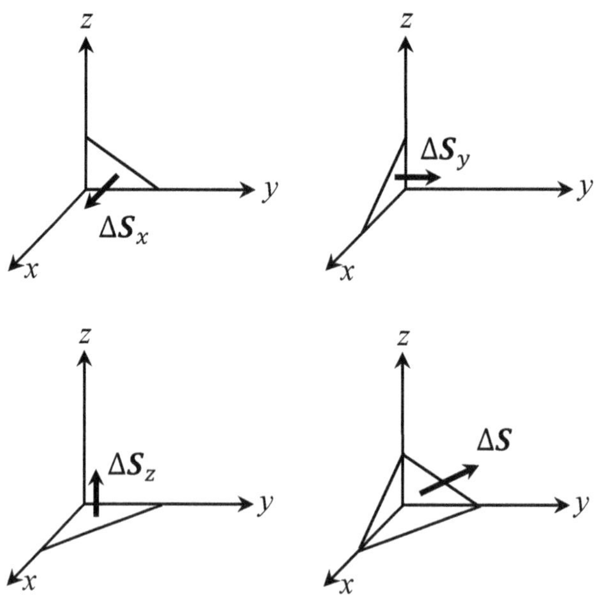

Problem 13) In the xyz Cartesian coordinate system, $\boldsymbol{r} = x\hat{\boldsymbol{x}} + y\hat{\boldsymbol{y}} + z\hat{\boldsymbol{z}}$ is the position vector, and $\boldsymbol{n} = \boldsymbol{r}/r$ (where $r = |\boldsymbol{r}|$) is a unit vector parallel to $\boldsymbol{r}$.

a) Determine $\boldsymbol{\nabla} \cdot \boldsymbol{r}$ and $\boldsymbol{\nabla} \cdot \boldsymbol{n}$.

b) Determine $\boldsymbol{\nabla} \times \boldsymbol{r}$ and $\boldsymbol{\nabla} \times \boldsymbol{n}$.

c) For an arbitrary vector field $\boldsymbol{A}(\boldsymbol{r})$, show that $(\boldsymbol{A} \cdot \boldsymbol{\nabla})\boldsymbol{n} = [\boldsymbol{A} - (\boldsymbol{A} \cdot \boldsymbol{n})\boldsymbol{n}]/r = \boldsymbol{A}_\perp/r$.

Note: $(\boldsymbol{A} \cdot \boldsymbol{\nabla})\boldsymbol{n} = A_x(\partial\boldsymbol{n}/\partial x) + A_y(\partial\boldsymbol{n}/\partial y) + A_z(\partial\boldsymbol{n}/\partial z)$.

Problem 14) Verify the following identities:

a) $\boldsymbol{\nabla} \cdot (\psi \boldsymbol{A}) = (\boldsymbol{\nabla}\psi) \cdot \boldsymbol{A} + \psi \boldsymbol{\nabla} \cdot \boldsymbol{A}$.

b) $\boldsymbol{\nabla} \cdot (\boldsymbol{A} \times \boldsymbol{B}) = \boldsymbol{B} \cdot (\boldsymbol{\nabla} \times \boldsymbol{A}) - \boldsymbol{A} \cdot (\boldsymbol{\nabla} \times \boldsymbol{B})$.

Problem 15) Verify the following identities:

a) $\boldsymbol{\nabla} \times (\boldsymbol{\nabla}\psi) = 0$.

b) $\boldsymbol{\nabla} \times (\psi \boldsymbol{A}) = (\boldsymbol{\nabla}\psi) \times \boldsymbol{A} + \psi \boldsymbol{\nabla} \times \boldsymbol{A}$

c) $\boldsymbol{\nabla} \times (\boldsymbol{\nabla} \times \boldsymbol{A}) = \boldsymbol{\nabla}(\boldsymbol{\nabla} \cdot \boldsymbol{A}) - \boldsymbol{\nabla}^2 \boldsymbol{A}$.

d) $\boldsymbol{\nabla} \times (\boldsymbol{A} \times \boldsymbol{B}) = \boldsymbol{A}(\boldsymbol{\nabla} \cdot \boldsymbol{B}) - \boldsymbol{B}(\boldsymbol{\nabla} \cdot \boldsymbol{A}) + (\boldsymbol{B} \cdot \boldsymbol{\nabla})\boldsymbol{A} - (\boldsymbol{A} \cdot \boldsymbol{\nabla})\boldsymbol{B}$.

Problem 16) The closed loop C defines the boundary of a surface S in 3D space. The surface S is not necessarily flat, and the loop C is not necessarily planar. For an arbitrary vector field $\boldsymbol{A}(\boldsymbol{r})$, let the vector field $\boldsymbol{B}(\boldsymbol{r})$ be defined as $\boldsymbol{B} = \boldsymbol{\nabla} \times \boldsymbol{A}$.

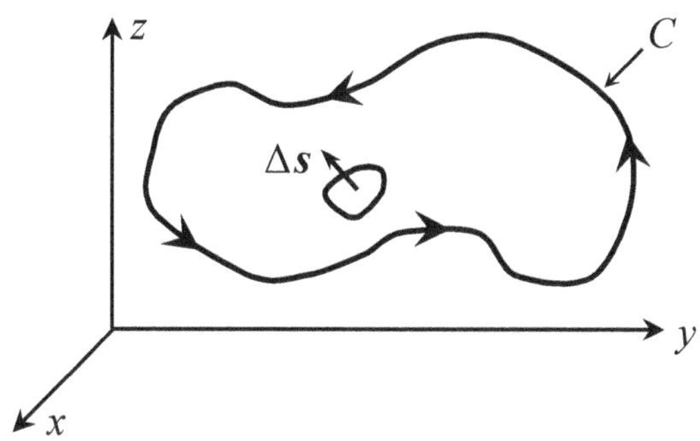

a) By dividing S into small segments $\Delta \boldsymbol{s}$ and using the conceptual definition of the curl operator, show that $\int_S \boldsymbol{B} \cdot \mathrm{d}\boldsymbol{s} = \oint_C \boldsymbol{A} \cdot \mathrm{d}\boldsymbol{\ell}$ (Stokes theorem).

b) If $\boldsymbol{S}$ happens to be a closed surface (i.e., the surface surrounding a finite volume), show that $\int_S \boldsymbol{B} \cdot \mathrm{d}\boldsymbol{s} = 0$.

c) From part (b), conclude that $\boldsymbol{\nabla} \cdot \boldsymbol{B} = 0$ is an immediate consequence of the definition of $\boldsymbol{B}$ as the curl of some other vector field.

Problem 17) Let $\psi(\boldsymbol{r})$ be an arbitrary scalar field and $\boldsymbol{C}$ a constant vector. Show that $\boldsymbol{\nabla} \cdot (\psi \boldsymbol{C}) = (\boldsymbol{\nabla}\psi) \cdot \boldsymbol{C}$. (This is a special case of the identity verified in problem 14(a), when the vector field $\boldsymbol{C}$ is constant.) Use Gauss's theorem to prove that

$$\int_{\text{volume}} \boldsymbol{\nabla}\psi(\boldsymbol{r})\mathrm{d}v = \oint_{\text{surface}} \psi(\boldsymbol{r})\mathrm{d}\boldsymbol{s}.$$

As usual, $\mathrm{d}v$ is the volume element and $\mathrm{d}\boldsymbol{s}$ is the surface element along the outward surface normal.

Problem 18) Let $\boldsymbol{A}(\boldsymbol{r})$ be an arbitrary vector field and $\boldsymbol{C}$ a constant vector. As a special case of the identity verified in problem 14(b), show that $\boldsymbol{\nabla} \cdot (\boldsymbol{A} \times \boldsymbol{C}) = (\boldsymbol{\nabla} \times \boldsymbol{A}) \cdot \boldsymbol{C}$. Use Gauss's theorem to prove that $\int_{\text{volume}} \boldsymbol{\nabla} \times \boldsymbol{A}(\boldsymbol{r})\mathrm{d}v = -\oint_{\text{surface}} \boldsymbol{A}(\boldsymbol{r}) \times \mathrm{d}\boldsymbol{s}$. As usual, $\mathrm{d}v$ is the volume element and $\mathrm{d}\boldsymbol{s}$ is the surface element along the outward surface normal.

Problem 19) Let $\psi(\boldsymbol{r})$ be an arbitrary scalar field and $\boldsymbol{C}$ a constant vector. Show that $\boldsymbol{\nabla} \times (\psi \boldsymbol{C}) = (\boldsymbol{\nabla}\psi) \times \boldsymbol{C}$. (This is a special case of the identity verified in problem 15(b), when the vector field $\boldsymbol{C}$ is constant.) Use the Stokes theorem to prove that $\int_{\text{surface}} \boldsymbol{\nabla}\psi(\boldsymbol{r}) \times \mathrm{d}\boldsymbol{s} = -\oint_{\text{boundary}} \psi(\boldsymbol{r})\mathrm{d}\boldsymbol{\ell}$. As usual, $\mathrm{d}\boldsymbol{s}$ is the surface element along the surface normal and $\mathrm{d}\boldsymbol{\ell}$ is the oriented line element. The right-hand rule determines the direction of integration around the boundary in relation to the direction chosen for the surface normals.

Problem 20) Let $\phi(\boldsymbol{r})$ and $\psi(\boldsymbol{r})$ be two arbitrary scalar fields. Applying Gauss's theorem to the vector field $\phi\boldsymbol{\nabla}\psi$, prove Green's first identity, namely,

$$\int_{\text{volume}}\{\phi(\boldsymbol{r})\boldsymbol{\nabla}^2\psi(\boldsymbol{r}) + [\boldsymbol{\nabla}\phi(\boldsymbol{r})]\cdot[\boldsymbol{\nabla}\psi(\boldsymbol{r})]\}\mathrm{d}v = \oint_{\text{surface}}\phi(\boldsymbol{r})[\partial\psi(\boldsymbol{r})/\partial\boldsymbol{n}]\mathrm{d}s.$$

In this equation $\boldsymbol{\nabla}^2\psi = \boldsymbol{\nabla}\cdot\boldsymbol{\nabla}\psi$, and $(\partial\psi/\partial\boldsymbol{n})\mathrm{d}s = (\boldsymbol{\nabla}\psi)\cdot\mathrm{d}\boldsymbol{s}$, where $\partial\psi/\partial\boldsymbol{n}$ is the derivative of $\psi(\boldsymbol{r})$ along the outward unit surface normal $\boldsymbol{n} = \mathrm{d}\boldsymbol{s}/\mathrm{d}s$.

Problem 21) Let $\phi(\boldsymbol{r})$ and $\psi(\boldsymbol{r})$ be two arbitrary scalar fields. Using Green's first identity derived in problem 20, prove Green's theorem, namely,

$$\int_{\text{volume}}[\phi(\boldsymbol{r})\boldsymbol{\nabla}^2\psi(\boldsymbol{r}) - \psi(\boldsymbol{r})\boldsymbol{\nabla}^2\phi(\boldsymbol{r})]\mathrm{d}v = \oint_{\text{surface}}\left[\phi(\boldsymbol{r})\frac{\partial\psi(\boldsymbol{r})}{\partial n} - \psi(\boldsymbol{r})\frac{\partial\phi(\boldsymbol{r})}{\partial n}\right]\mathrm{d}s.$$

Problem 22) Let $\boldsymbol{\alpha} = \boldsymbol{\alpha}' + \mathrm{i}\boldsymbol{\alpha}''$, $\boldsymbol{\beta} = \boldsymbol{\beta}' + \mathrm{i}\boldsymbol{\beta}''$, and $\boldsymbol{\gamma} = \boldsymbol{\gamma}' + \mathrm{i}\boldsymbol{\gamma}''$ be three arbitrary complex-valued vectors. Identify the real and imaginary components of the following vector products:

(a) $\boldsymbol{\alpha}\cdot\boldsymbol{\beta}$. (b) $\boldsymbol{\alpha}\times\boldsymbol{\beta}$. (c) $\boldsymbol{\gamma}\times\boldsymbol{\gamma}$. (d) $\boldsymbol{\alpha}\cdot(\boldsymbol{\beta}\times\boldsymbol{\gamma})$.

Problem 23) Let $\psi(\boldsymbol{r},t) = \psi_0\exp[\mathrm{i}(\boldsymbol{k}_1\cdot\boldsymbol{r} - \omega_1 t)]$ be a scalar plane wave, and let $\boldsymbol{A}(\boldsymbol{r},t) = \boldsymbol{A}_0\exp[\mathrm{i}(\boldsymbol{k}_2\cdot\boldsymbol{r} - \omega_2 t)]$ and $\boldsymbol{B}(\boldsymbol{r},t) = \boldsymbol{B}_0\exp[\mathrm{i}(\boldsymbol{k}_3\cdot\boldsymbol{r} - \omega_3 t)]$ be two vector plane waves. In general, ψ_0, ω_1, ω_2, and ω_3 are arbitrary complex constants, while $\boldsymbol{A}_0$, $\boldsymbol{B}_0$, $\boldsymbol{k}_1$, $\boldsymbol{k}_2$, and $\boldsymbol{k}_3$ are arbitrary complex vectors. Show that these plane waves satisfy the following vector identities:

a) $\boldsymbol{\nabla}\times(\boldsymbol{\nabla}\psi) = 0$.

b) $\boldsymbol{\nabla}\cdot(\psi\boldsymbol{A}) = (\boldsymbol{\nabla}\psi)\cdot\boldsymbol{A} + \psi(\boldsymbol{\nabla}\cdot\boldsymbol{A})$.

c) $\boldsymbol{\nabla}\times(\psi\boldsymbol{B}) = (\boldsymbol{\nabla}\psi)\times\boldsymbol{B} + \psi\boldsymbol{\nabla}\times\boldsymbol{B}$.

d) $\boldsymbol{\nabla}\cdot(\boldsymbol{A}\times\boldsymbol{B}) = \boldsymbol{B}\cdot(\boldsymbol{\nabla}\times\boldsymbol{A}) - \boldsymbol{A}\cdot(\boldsymbol{\nabla}\times\boldsymbol{B})$.

Hint: $\boldsymbol{U}\cdot(\boldsymbol{V}\times\boldsymbol{W}) = \boldsymbol{V}\cdot(\boldsymbol{W}\times\boldsymbol{U}) = \boldsymbol{W}\cdot(\boldsymbol{U}\times\boldsymbol{V})$.

Problem 24) The operation $(\boldsymbol{A}\cdot\boldsymbol{\nabla})\boldsymbol{B}$ involving the vector fields $\boldsymbol{A}(\boldsymbol{r})$ and $\boldsymbol{B}(\boldsymbol{r})$ is defined in Cartesian coordinates as follows:

$$(\boldsymbol{A}\cdot\boldsymbol{\nabla})\boldsymbol{B} = A_x\frac{\partial\boldsymbol{B}}{\partial x} + A_y\frac{\partial\boldsymbol{B}}{\partial y} + A_z\frac{\partial\boldsymbol{B}}{\partial z}.$$

Confirm that the same operation in a cylindrical coordinate system should be written as

$$(\boldsymbol{A}\cdot\boldsymbol{\nabla})\boldsymbol{B} = A_\rho\frac{\partial\boldsymbol{B}}{\partial\rho} + A_\phi\frac{\partial\boldsymbol{B}}{\rho\partial\phi} + A_z\frac{\partial\boldsymbol{B}}{\partial z},$$

where

$$\frac{\partial\boldsymbol{B}}{\partial\phi} = \left(\frac{\partial B_\rho}{\partial\phi}\hat{\boldsymbol{\rho}} + B_\rho\frac{\partial\hat{\boldsymbol{\rho}}}{\partial\phi}\right) + \left(\frac{\partial B_\phi}{\partial\phi}\hat{\boldsymbol{\phi}} + B_\phi\frac{\partial\hat{\boldsymbol{\phi}}}{\partial\phi}\right) + \frac{\partial B_z}{\partial\phi}\hat{\boldsymbol{z}}$$

$$= \left(\frac{\partial B_\rho}{\partial\phi} - B_\phi\right)\hat{\boldsymbol{\rho}} + \left(\frac{\partial B_\phi}{\partial\phi} + B_\rho\right)\hat{\boldsymbol{\phi}} + \frac{\partial B_z}{\partial\phi}\hat{\boldsymbol{z}}.$$

Problem 25) Prove the following vector identities using the Fourier domain techniques:

a) $\boldsymbol{\nabla}\times\boldsymbol{\nabla}\times\boldsymbol{A}(\boldsymbol{r}) = \boldsymbol{\nabla}[\boldsymbol{\nabla}\cdot\boldsymbol{A}(\boldsymbol{r})] - \boldsymbol{\nabla}^2\boldsymbol{A}(\boldsymbol{r})$.

b) $\boldsymbol{B}(\boldsymbol{r})\cdot[\boldsymbol{\nabla}\times\boldsymbol{A}(\boldsymbol{r})] - \boldsymbol{A}(\boldsymbol{r})\cdot[\boldsymbol{\nabla}\times\boldsymbol{B}(\boldsymbol{r})] = \boldsymbol{\nabla}\cdot[\boldsymbol{A}(\boldsymbol{r})\times\boldsymbol{B}(\boldsymbol{r})]$.

Problem 26) Find the Fourier transforms of the following functions defined in 3D space specified by the radial coordinate $\boldsymbol{r} = r\hat{\boldsymbol{r}} = x\hat{\boldsymbol{x}} + y\hat{\boldsymbol{y}} + z\hat{\boldsymbol{z}}$. Here, α is a complex constant whose real part is positive, but is otherwise arbitrary. Explore the limit when $\alpha \to 0$.

$$f_1(\boldsymbol{r}) = \exp(-\alpha r)/r; \qquad \boldsymbol{f}_2(\boldsymbol{r}) = \exp(-\alpha r)\,\hat{\boldsymbol{r}}/r;$$

$$f_3(\boldsymbol{r}) = 1/r^2; \qquad \boldsymbol{f}_4(\boldsymbol{r}) = \boldsymbol{\nabla}(1/r) = -\hat{\boldsymbol{r}}/r^2.$$

Problem 27) Let the 3D Fourier transforms of the scalar functions $f(\boldsymbol{r})$, $g(\boldsymbol{r})$, and $h(\boldsymbol{r})$ be defined as follows:

$$F(\boldsymbol{k}) = \int_{-\infty}^{\infty} f(\boldsymbol{r}) \exp(-\mathrm{i}\boldsymbol{k}\cdot\boldsymbol{r})\,\mathrm{d}\boldsymbol{r},$$

$$G(\boldsymbol{k}) = \int_{-\infty}^{\infty} g(\boldsymbol{r}) \exp(-\mathrm{i}\boldsymbol{k}\cdot\boldsymbol{r})\,\mathrm{d}\boldsymbol{r},$$

$$H(\boldsymbol{k}) = \int_{-\infty}^{\infty} h(\boldsymbol{r}) \exp(-\mathrm{i}\boldsymbol{k}\cdot\boldsymbol{r})\,\mathrm{d}\boldsymbol{r}.$$

a) Assuming that $f(\boldsymbol{r})$ is the 3D convolution of $g(\boldsymbol{r})$ and $h(\boldsymbol{r})$, namely,

$$f(\boldsymbol{r}) = g(\boldsymbol{r}) * h(\boldsymbol{r}) = \int_{-\infty}^{\infty} g(\boldsymbol{r}')h(\boldsymbol{r}-\boldsymbol{r}')\mathrm{d}\boldsymbol{r}',$$

show that $F(\boldsymbol{k}) = G(\boldsymbol{k})H(\boldsymbol{k})$.

b) Extend this convolution theorem to cases where either $g(\boldsymbol{r})$ or $h(\boldsymbol{r})$ or both are vector functions of the spatial coordinates $\boldsymbol{r}$.

Problem 28) Use simple geometric arguments to show the following:

a) $\boldsymbol{\nabla}(1/|\boldsymbol{r}|) = -\boldsymbol{r}/|\boldsymbol{r}|^3$.

b) $\boldsymbol{\nabla}\cdot(\boldsymbol{r}/|\boldsymbol{r}|^3) = 4\pi\delta(\boldsymbol{r})$.

Here, $\delta(\boldsymbol{r})$ is a Dirac delta-function located at the origin of the coordinate system.

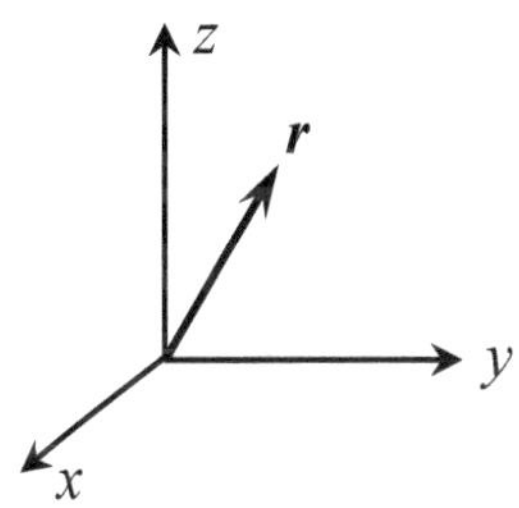

Chapter 13

Probability, Random Variables, and Stochastic Processes

"The test of science is not whether you are reasonable — there would not be much of physics if that was the case — the test is whether it works."

Sir Hermann Bondi (1919–2005)

"I would rather have questions that can't be answered than answers that can't be questioned."

Richard P. Feynman (1918–1988)

13.1. Introduction. The present chapter is devoted to a brief overview of some of the fundamental concepts in the theories of probability and stochastic processes. Noise, an ever-present feature of measurement systems (as well as other physical instruments), is a random process that submits only to probabilistic analysis with the aid of the theories of random functions and stochastic processes. Image processing and image analysis often employ noise suppression techniques to mitigate the deleterious effects of various random deformations and fluctuations on the quality of a reconstructed image. In photodetection, one is typically concerned with several types of noise, such as dark-current noise and thermal noise, originating in the body of the detector as well as in the electronic circuitry surrounding the detector; shot noise, which, roughly speaking, is associated with the random arrival of individual photons at the detector; and gain (or amplification) noise, which is related to the statistical nature of photo-multiplication processes in avalanche photodiodes and photomultiplier tubes.

Lasers and fiber amplifiers suffer from spontaneous emission noise, as well as gain noise, which is inevitably present in any kind of gain medium. The speckle effect deteriorates the quality of received signals and images in various coherent optical systems. A branch of the optical sciences, known as statistical optics, addresses the problems associated with coherence, partial coherence, and incoherence, and also partial polarization and depolarization, which are inherently random in nature. In quantum optics, there exist phase and amplitude fluctuations that are ultimately rooted in the Heisenberg uncertainty principle and can be investigated only with the aid of probabilistic methods. The capacity of an information transmission system (e.g., a fiber optical communication network) for fast and accurate transmission of information, and also the storage capacity of a data storage medium (e.g., magnetic hard disk, optical disk, magnetic tape) are intimately tied to the signal-to-noise ratio that is achievable in such systems. It is for these reasons that the theories of probability and random processes are important topics of study for the student of optics and photonics. The following sections address a few elementary concepts and methods that are commonly encountered in the analyses of noisy and/or fluctuating optical systems.

13.2. Discrete and continuous random variables. The simplest example of a discrete random variable is what one gets in a coin toss. If the coin is fair, the outcome will be heads 50 percent of the time and tails 50 percent of the time. Assign a specific number to an arbitrary variable such as x for each outcome, say, $x = 0$ when the outcome is head, and $x = 1$ when it is tail. We will then have the *random variable* x with probability $p_0 = ½$ assigned to $x = 0$ and probability $p_1 = ½$ assigned to $x = 1$. If the coin happens to be loaded, the probabilities will not be equal. For instance, if heads were twice as likely an outcome as tails, then $p_0 = ⅔$ and $p_1 = ⅓$. The numbers associated with the possible outcomes of a coin toss are thus $x = 0$ and $x = 1$ for the

outcomes themselves, and p_0 and p_1 for the respective probabilities of these outcomes. Clearly, one can imagine any pair of real numbers for p_0 and p_1, provided that each number is between 0 and 1, and that $p_0 + p_1 = 1$. We say that the random variable x represents the outcome of a coin-toss experiment, with probabilities p_0 and p_1 associated with the two possible outcomes $x = 0$ (i.e., heads) and $x = 1$ (i.e., tails).

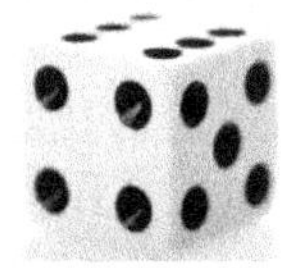

As a second example, suppose the experiment consists of throwing a fair dice. One may call the random variable in this case y, to which one assigns the value $y = 1$ when the outcome of the throw is 1, $y = 2$, when the outcome is 2, and so on, up to $y = 6$. For a fair dice, the various outcomes will be equally probable; that is, $p_1 = p_2 = p_3 = p_4 = p_5 = p_6 = 1/6$. If the dice happens to be loaded, the six probabilities will no longer be identical, but they will surely be nonnegative, and their sum will be 1; that is, $p_n \geq 0$ and $\sum_{n=1}^{6} p_n = 1$.

As an example of a continuous random variable, suppose a police officer, standing at a street corner, is monitoring the speed of cars as they approach her radar. If we use the random variable v to indicate the detected velocity of each car that passes by, the actual values of v on any given day will fall somewhere in the range of, say, zero to 150 miles per hour (mph). In principle, of course, v can be any positive real number between 0 and 150, but, in practice, it is highly unlikely that a specific velocity, say, 87.2376 mph, is ever detected. Instead, one typically looks at small intervals of v, say, $\Delta v = 2.0$ mph, and examines the number of cars that had velocities somewhere in the range between 50 and 52 mph, relative to those that had velocities between 52 and 54 mph, or those that travelled in the range of 78 to 80 mph, and so on. If we now divide the number of cars that had velocities in the range of $v_0 \pm \frac{1}{2}\Delta v$ by the total number of cars that were monitored during a day, we will find values for the fraction (or percentage) of cars that had velocities in the "vicinity" of specific values of v_0, where the "vicinity" in the present example means $v_0 \pm 1$ mph. Here, the random variable v is a continuous variable that can assume *any* value on the real axis within a given range, say, $[v_{\min}, v_{\max}]$. However, it does not make sense to speak of the probability of occurrence of any specific value of v inside the allowed range. Rather, one speaks of the probability that the observed v will be in the vicinity of any particular value of v, where the "vicinity" is understood to be a specific narrow interval Δv surrounding v.

13.3. The probability density function. For a single random variable x, the probability density (or probability distribution) function is a real function $p(x)$ with two properties: (i) $p(x) \geq 0$ for all x; (ii) the area under $p(x)$ equals 1.0; that is, $\int_{-\infty}^{\infty} p(x)\mathrm{d}x = 1.0$.[1-3†]

In the example of a coin-toss experiment brought up in the preceding section, the probability distribution function is $p(x) = p_0\delta(x) + p_1\delta(x-1)$. The function is zero everywhere except at $x = 0$ and $x = 1$, where the delta-functions are located. The area under $p(x)$ is the area p_0 of the first δ-function, which is located at $x = 0$, plus the area p_1 of the second δ-function, which is located at $x = 1$. Considering that both p_0 and p_1 are nonnegative, and that $p_0 + p_1 = 1$, we see that the above $p(x)$ satisfies both requirements for a probability density function. It is not difficult now to show that the probability density function for the dice-throwing experiment of section 2 is given by

$$p(x) = p_1\delta(x-1) + p_2\delta(x-2) + p_3\delta(x-3) + p_4\delta(x-4) + p_5\delta(x-5) + p_6\delta(x-6).$$

For a continuous random variable, such as the velocity v of automobiles monitored by a police radar on a given day, as described in the preceding section, one must first

† Superscript numbers throughout the chapter identify references listed at the end of the chapter.

determine the probability of occurrence of v in a narrow interval Δv surrounding v_0, then divide (or normalize) this probability by Δv to determine the probability density function $p(v)$ at $v = v_0$. The probability that the observed value of v will fall within a Δv neighborhood of v_0 will then be $p(v_0)\Delta v$. Needless to say, these statements become accurate in the limit when $\Delta v \to 0$. The requirement that the area under the function $p(v)$ must equal 1, that is, $\int_{-\infty}^{\infty} p(v)\mathrm{d}v = 1$, guarantees that, if one were to add up the number of cars that traveled within *all* the allowed (albeit non-overlapping) intervals Δv, then one would find the *total* number of cars that have been monitored on the given day.

With the above definition of the probability density function, we now see that there is hardly any difference between discrete and continuous random variables. The use of Dirac's δ-function enables one to transition smoothly between continuous and discrete random variables. For instance, let the (continuous) random variable x have the probability density $p(x) = p_1/(\alpha_2 - \alpha_1)$ in the interval $\alpha_1 \le x \le \alpha_2$ and $p(x) = p_2/(\beta_2 - \beta_1)$ in the interval $\beta_1 \le x \le \beta_2$, where $p_1 + p_2 = 1$, and the intervals $[\alpha_1, \alpha_2]$ and $[\beta_1, \beta_2]$ are non-overlapping. Outside these two intervals, the probability density function vanishes. Now, suppose α_2 becomes extremely close to α_1. Then, in the limit when $\alpha_2 \to \alpha_1$, the first part of the probability-density function approaches $p_1\delta(x - \alpha_1)$. The random variable x now has a discrete part, where the discrete value $x = \alpha_1$ has a probability of p_1, whereas the continuous part, which occurs when $\beta_1 \le x \le \beta_2$, has uniform probability density $p(x) = p_2/(\beta_2 - \beta_1)$. It is possible, of course, for β_2 to approach β_1 as well, in which case the second part of the probability density function approaches $p_2\delta(x - \beta_1)$. In the limit when $\alpha_2 \to \alpha_1$ and $\beta_2 \to \beta_1$, the random variable x becomes a purely discrete variable that can assume either the value α_1 with probability p_1, or the value β_1 with probability p_2.

13.4. Average, variance, standard deviation, and entropy of a random variable. By definition, the average (or mean, or expected value) of the random variable x, whose probability density function is given by $p(x)$, is evaluated as follows:

$$\langle x \rangle = \int_{-\infty}^{\infty} x p(x)\mathrm{d}x. \tag{1}$$

This expected value of x is also referred to as the first moment of the random variable x. In general, the kth moment of the random variable is defined as

$$\langle x^k \rangle = \int_{-\infty}^{\infty} x^k p(x)\mathrm{d}x. \tag{2}$$

The variance of a random variable is the expected value of the square of its deviation from the mean value of the variable; that is,

$$\sigma_x^2 = \langle (x - \langle x \rangle)^2 \rangle = \int_{-\infty}^{\infty} (x - \langle x \rangle)^2 p(x)\mathrm{d}x. \tag{3}$$

Straightforward algebraic manipulation of this expression yields

$$\sigma_x^2 = \int_{-\infty}^{\infty} (x^2 - 2x\langle x \rangle + \langle x \rangle^2)p(x)\mathrm{d}x = \langle x^2 \rangle - 2\langle x \rangle\langle x \rangle + \langle x \rangle^2 = \langle x^2 \rangle - \langle x \rangle^2. \tag{4}$$

Thus, the variance σ_x^2 of the random variable x is obtained by subtracting the square of its mean value from the expected value of x^2 (i.e., the second moment of x).

The standard deviation of the random variable x is defined as the square root of its variance, namely, σ_x. The standard deviation is generally used as a measure of the distance away from $\langle x \rangle$, within which the random variable x is expected to be found. Depending on the detailed shape of the probability density function, one might say that x

is reasonably (i.e., with high confidence) expected to be within $\pm 2\sigma_x$, or $\pm 3\sigma_x$, or $\pm 5\sigma_x$, etc., of the mean value $\langle x \rangle$.

The entropy $H(x)$ of the random variable x is a measure of the uncertainty in the outcome of the experiment associated with x.[4,5] For a discrete random variable x, whose values x_n occur with probability p_n, the entropy is defined as follows:[4-8]

$$H(x) = -\sum_n p_n \ln p_n. \tag{5}$$

The graphical depiction of the function $-x \ln x$ over the interval $0 \le x \le 1$ on the right shows that the entropy of a discrete random variable x is well defined, and that, in general, $H(x) \ge 0$. For a random variable x with a finite number of outcomes, that is, when x belongs to the set $\{x_1, x_2, \cdots, x_n, \cdots, x_N\}$, it is easy to show that the maximum entropy occurs when all the outcomes are equally likely (see chapter 5, problem 8). Under such circumstances, $p_n = 1/N$ and $H_{\max}(x) = \ln N$.

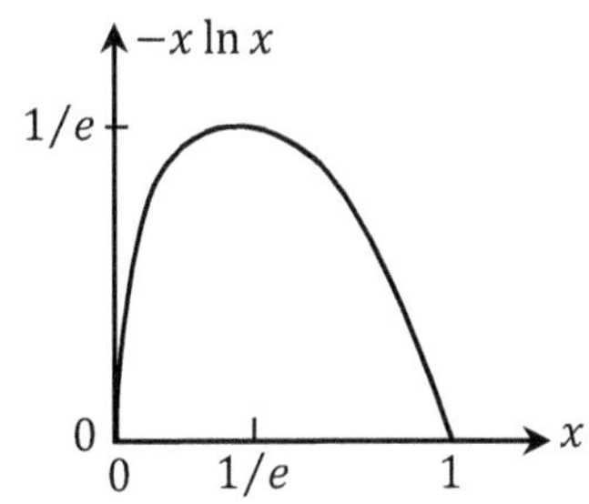

13.5. Jensen's inequality. The Danish mathematician Johan Jensen proved the famous inequality now named after him in 1906. Given a random variable x and a convex function $f(x)$, Jensen's inequality relates the expected value of $f(x)$ to the value of the function $f(\cdot)$ at the expected value of x. For a convex cup function, such as that shown in figure 13.1a, the Jensen inequality is $\langle f(x) \rangle \ge f(\langle x \rangle)$, whereas for a convex cap function, such as that in figure 13.1b, the direction of the inequality reverses, so that $\langle f(x) \rangle \le f(\langle x \rangle)$.[7]

The proof of Jensen's inequality for a random variable with only two possible outcomes, say, x_1 and x_2 having probabilities p_1 and p_2, is rather trivial. With reference to figure 13.1a, it is easy to see that $f(\langle x \rangle) = f(p_1 x_1 + p_2 x_2) = f[x_1 + p_2(x_2 - x_1)]$ is directly below the point $\langle f(x) \rangle = p_1 f(x_1) + p_2 f(x_2) = f(x_1) + p_2[f(x_2) - f(x_1)]$. In the general case when the random variable has three or more outcomes, one must resort to the method of proof by induction, outlined below, to prove Jensen's inequality.

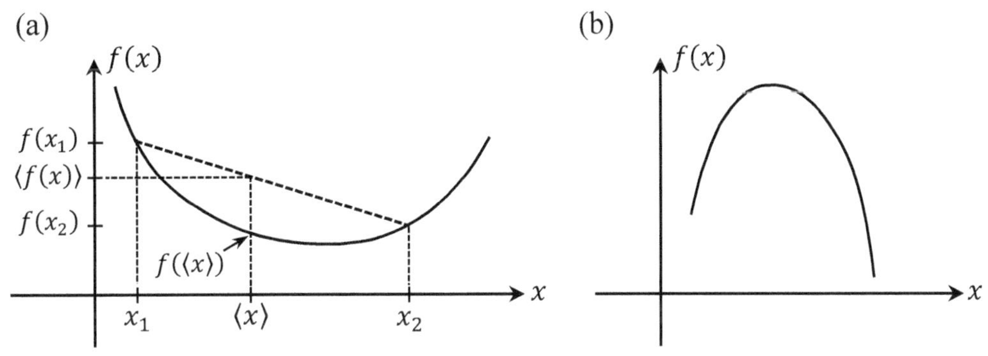

Figure 13.1. (a) A convex cup function transforms a random variable x to a new random variable $f(x)$, such that the expected value of $f(x)$ exceeds the value of $f(\cdot)$ at the expected value of x, namely, $\langle f(x) \rangle \ge f(\langle x \rangle)$. (b) A convex cap function has a similar property, except for a reversal of the direction of inequality; that is, $\langle f(x) \rangle \le f(\langle x \rangle)$.

Suppose Jensen's inequality is valid for any random variable that has N possible outcomes. If we now consider a random variable with $N+1$ possible outcomes, say, $x_1, x_2, \cdots, x_N, x_{N+1}$, with corresponding probabilities $p_1, p_2, \cdots, p_N, p_{N+1}$, there must exist at least one outcome x_m with a probability $p_m < 1$. We can then write

$$\langle x\rangle = p_m x_m + (1-p_m)\sum_{n\neq m}[p_n/(1-p_m)]x_n. \tag{6}$$

Consequently,

$$f(\langle x\rangle) \le p_m f(x_m) + (1-p_m) f\{\sum_{n\neq m}[p_n/(1-p_m)]x_n\}. \tag{7}$$

In this inequality, the expression inside the curly brackets is the average of the N values $x_1, \cdots, x_{m-1}, x_{m+1}, \cdots, x_{N+1}$, with corresponding probabilities $p_n/(1-p_m)$, where $n \neq m$ and $\sum_{n\neq m} p_n/(1-p_m) = 1$. Since, by assumption, Jensen's inequality is applicable to the second term on the right-hand side of inequality (7), we proceed to write

$$f(\langle x\rangle) \le p_m f(x_m) + (1-p_m)\sum_{n\neq m}[p_n/(1-p_m)]f(x_n) = \sum_{n=1}^{N+1} p_n f(x_n) = \langle f(x)\rangle. \tag{8}$$

This completes our proof-by-induction of Jensen's inequality for convex cup functions. The proof for convex cap functions is essentially identical, except for a change in the direction of the inequality.

As a trivial application of Jensen's inequality, consider the convex cup function $f(x) = x^2$. One sees immediately that $\langle x^2\rangle \ge \langle x\rangle^2$, and that, therefore, the variance σ_x^2 of any and all random variables must be nonnegative. Generalizing this result, we find that, for integer k, the even moment $\langle x^{2k}\rangle$ of any random variable x must be greater than or equal to the mean value $\langle x\rangle$ of the random variable raised to the power of $2k$. Another application of Jensen' inequality will be seen in section 10, where we demonstrate that the conditional entropy $H(x|y)$ of the random variable x given the outcome of another random variable y cannot exceed the entropy $H(x)$ of x in the absence of information about the random variable y.

13.6. Binomial, Gaussian, and Poisson distributions. A binomial probability distribution is obtained when an experiment with a binary outcome (i.e., $x = x_1$ with probability p, and $x = x_2$ with probability $1-p$), is repeated N times. The repetitions of the experiment must be independent of each other, so that, in each experiment, the outcomes x_1 and x_2 will occur with the same probabilities p and $1-p$. The probability that, after N repeats of the experiment, one will end up with exactly n occurences of x_1 (and, therefore, $N-n$ occurences of x_2) is given by the following binomial distribution:[1-3]

$$p_n = \binom{N}{n} p^n (1-p)^{N-n}; \qquad (n = 0, 1, 2, \cdots, N). \tag{9}$$

Considering the binomial expansion $(a+b)^N = \sum_{n=0}^{N}\binom{N}{n} a^n b^{N-n}$ (see chapter 1, section 1.6.1, Example 2), it is now easy to see that $\sum_{n=0}^{N} p_n = [p + (1-p)]^N = 1$, which is a required property of any probability distribution. The average and variance of the random variable n with binomial distribution are found straightforwardly, as follows:

$$\begin{aligned}
\langle n\rangle &= \sum_{n=0}^{N} n p_n = \sum_{n=1}^{N} n\binom{N}{n} p^n (1-p)^{N-n} \\
&= \sum_{n=1}^{N} \frac{n(N!)}{n!(N-n)!} p^n (1-p)^{N-n} \\
&= Np \sum_{n=1}^{N} \frac{(N-1)!}{(n-1)!(N-n)!} p^{n-1} (1-p)^{N-n} \\
&= Np \sum_{n'=0}^{N-1} \frac{(N-1)!}{n'!(N-1-n')!} p^{n'} (1-p)^{N-1-n'} \\
&= Np[p + (1-p)]^{N-1} = Np.
\end{aligned} \tag{10}$$

To determine the variance of the random variable n, it is somewhat easier to begin by evaluating the expected value of $n(n-1)$, as follows:

$$\begin{aligned}\langle n(n-1)\rangle &= \sum_{n=0}^{N} n(n-1)p_n = \sum_{n=2}^{N} n(n-1)\binom{N}{n} p^n(1-p)^{N-n} \\ &= \sum_{n=2}^{N} \frac{n(n-1)(N!)}{n!(N-n)!} p^n(1-p)^{N-n} \\ &= N(N-1)p^2 \sum_{n=2}^{N} \frac{(N-2)!}{(n-2)!(N-n)!} p^{n-2}(1-p)^{N-n} \\ &= N(N-1)p^2 \sum_{n'=0}^{N-2} \frac{(N-2)!}{n'!(N-2-n')!} p^{n'}(1-p)^{N-2-n'} \\ &= N(N-1)p^2[p+(1-p)]^{N-2} = N(N-1)p^2. \end{aligned} \tag{11}$$

We thus have $\langle n(n-1)\rangle = \langle n^2\rangle - \langle n\rangle = \langle n^2\rangle - Np = N(N-1)p^2$. Consequently, the variance of n is given by

$$\sigma_n^2 = \langle n^2\rangle - \langle n\rangle^2 = N^2p^2 - Np^2 + Np - N^2p^2 = Np(1-p). \tag{12}$$

Another way to arrive at the same results is to recognize that the random variable n is simply the sum of identical random variables x, when the corresponding experiment is repeated N times, assuming that the outcome of each experiment is defined as $x_1 = 1$ and $x_2 = 0$. For a single experiment, the mean value of x is $\langle x\rangle = p$ and the variance is $\sigma_x^2 = \langle x^2\rangle - \langle x\rangle^2 = p - p^2 = p(1-p)$. When the same experiment is repeated N (independent) times, the corresponding averages and also the corresponding variances will add up, resulting in $\langle n\rangle = N\langle x\rangle = Np$ and $\sigma_n^2 = N\sigma_x^2 = Np(1-p)$.

A continuous random variable x is said to be Gaussian if its probability density function is given by[1-3]

$$p(x) = \frac{1}{\sqrt{2\pi}\sigma_x}\exp[-(x-x_0)^2/(2\sigma_x^2)]. \tag{13}$$

The well-known integrals of $\exp(-x^2)$, $x\exp(-x^2)$, and $x^2\exp(-x^2)$ may be used to arrive at the following important properties of a Gaussian random variable:

$$\int_{-\infty}^{\infty}\exp(-x^2)\,dx \overset{\text{chapter 1, problem 4}}{=} \sqrt{\pi} \quad \rightarrow \quad \int_{-\infty}^{\infty} p(x)dx = 1. \tag{14}$$

$$\int_{-\infty}^{\infty} x\exp(-x^2)\,dx \overset{\text{odd function of } x}{=} 0 \quad \rightarrow \quad \langle x\rangle = \int_{-\infty}^{\infty} xp(x)dx = x_0. \tag{15}$$

$$\int_{-\infty}^{\infty} x^2\exp(-x^2)\,dx \overset{\text{integration by parts}}{=} \sqrt{\pi}/2 \quad \rightarrow \quad \mathrm{Var}(x) = \int_{-\infty}^{\infty}(x-x_0)^2p(x)dx = \sigma_x^2. \tag{16}$$

Thus, the area under $p(x)$ equals 1, as it should be, x_0 is the mean (or average) value of x, while σ_x^2 is its variance, and σ_x is its standard deviation. In the limit when $\sigma_x \rightarrow 0$, $p(x)$ approaches a δ-function, that is, $p(x) \rightarrow \delta(x-x_0)$, indicating that x is no longer a random variable, but rather a well-defined entity whose value is fixed at $x = x_0$ with a probability of 1.

A discrete random variable x that assumes nonnegative integer values $x = 0, 1, 2, \cdots, n, \cdots$ with probability $p_n = \exp(-\alpha)\,\alpha^n/n!$ is said to have a Poissanian distribution.[1-3] Here, $\alpha > 0$ is a constant parameter that turns out to be both the average and the variance of the random variable x. Given the Taylor series expansion $\exp(\alpha) = \sum_{n=0}^{\infty}(\alpha^n/n!)$, it is straightforward to demonstrate the following identities:

$$\sum_{n=0}^{\infty} p_n = \exp(-\alpha) \sum_{n=0}^{\infty} \alpha^n / n! = 1. \tag{17}$$

$$\langle x \rangle = \sum_{n=0}^{\infty} n p_n = \exp(-\alpha) \sum_{n=1}^{\infty} \alpha^n / (n-1)! = \alpha. \tag{18}$$

$$\begin{aligned} \langle x^2 \rangle &= \sum_{n=0}^{\infty} n^2 p_n = \exp(-\alpha) \sum_{n=1}^{\infty} n \alpha^n / (n-1)! \\ &= \alpha \exp(-\alpha) \frac{\mathrm{d}}{\mathrm{d}\alpha} \sum_{n=1}^{\infty} \alpha^n / (n-1)! = \alpha \exp(-\alpha) \frac{\mathrm{d}}{\mathrm{d}\alpha} [\alpha \exp(\alpha)] \\ &= \alpha \exp(-\alpha) [\alpha \exp(\alpha) + \exp(\alpha)] = \alpha^2 + \alpha. \end{aligned} \tag{19}$$

Thus, the variance of x is seen to be $\sigma_x^2 = \langle x^2 \rangle - \langle x \rangle^2 = \alpha$, which yields the standard deviation $\sigma_x = \sqrt{\alpha}$.

Example 1. Starting at $t = 0$, a pulse of light, having optical power $P(t)$ for $t \geq 0$, shines on a photodetector whose efficiency is specified as η. Thus, during a short time interval Δt, the probability of a single photoelectron release within the detector is $\eta P(t) \Delta t$. For sufficiently short Δt, the probability of more than one photoelectron release is negligible. Consequently, the probability that no photoelectrons are registered during this short interval is $1 - \eta P(t) \Delta t$. If the probability that n electrons are released during the time interval $(0, t)$ is denoted by $p_n(t)$, then the probability that n electrons are released during the interval $(0, t + \Delta t)$ will be given by

$$p_n(t + \Delta t) = p_{n-1}(t) \eta P(t) \Delta t + p_n(t) [1 - \eta P(t) \Delta t]. \tag{20}$$

Considering that for $n = 0$ we have $p_{n-1}(t) = 0$, the above equation yields

$$p_0(t + \Delta t) - p_0(t) = -\eta P(t) p_0(t) \Delta t \quad \rightarrow \quad p_0'(t) = -\eta P(t) p_0(t), \tag{21}$$

which has a solution in the form of

$$p_0(t) = c_0 \exp[-\eta \int_0^t P(t') \mathrm{d}t']. \tag{22}$$

Here, c_0 is the arbitrary constant of integration. Defining $\Lambda(t) = \eta \int_0^t P(t') \mathrm{d}t'$, we will have, for $n > 0$,

$$p_n'(t) + \eta P(t) p_n(t) = \eta P(t) p_{n-1}(t) \tag{23}$$

$$\rightarrow \quad \frac{\mathrm{d}}{\mathrm{d}t} [e^{\Lambda(t)} p_n(t)] = \eta P(t) e^{\Lambda(t)} p_{n-1}(t). \tag{24}$$

Thus, for $n = 1$, we will have $e^{\Lambda(t)} p_1(t) = c_1 + c_0 \eta \int_0^t P(t') \mathrm{d}t' = c_1 + c_0 \Lambda(t)$. Given that $p_1(t = 0) = 0$, it is immediately clear that the constant of integration c_1 must be zero. Consequently,

$$p_1(t) = c_0 \Lambda(t) \exp[-\Lambda(t)]. \tag{25}$$

Substitution into Eq.(24) now yields $p_2(t)$ and repeating the procedure gives the following expression for arbitrary values of n:

$$p_n(t) = c_0 \exp[-\Lambda(t)] \Lambda^n(t) / n!. \tag{26}$$

Given that $\sum_{n=0}^{\infty} p_n(t) = 1.0$, it is now easy to see that the integration constant c_0 must be equal to 1.0. The probability distribution function for the number of released photoelectrons at any instant of time t is thus seen to be Poissonian, with $\alpha = \Lambda(t)$.

Brain teaser. A game-show contestant is shown three closed doors to choose from. He is told that an expensive car awaits him behind one of the doors, whereas a goat lurks behind each of the remaining doors. The contestant will have to choose a door, and his prize will be whatever happens to be behind the door. Once the contestant announces his choice, the host opens one of the other two doors to reveal a goat behind it. She then asks the contestant if he would like to switch his choice to the remaining door. Now, the contestant's chances of winning at first were one out of three. However, after the host opened one of the doors and revealed a goat behind it, his chances of winning would double (i.e., become two out of three) if indeed he chose to switch. Explain the rationale behind this increased chance of winning.

Solution. Suppose he chose door 1 in the first place. He would be losing if the car happened to be behind door 2 or door 3. This is a 2/3 chance of losing. But now, if he did switch, one of three things could happen: (i) The car is behind door 1, in which case the contestant loses. (ii) The car is behind door 2. The host has opened door 3. The contestant switches to door 2 and wins. (iii) The car is behind door 3. The host has opened door 2. The contestant switches to door 3 and wins. There is a 2/3 chance of winning and 1/3 chance of losing if he chooses to switch. It is wiser, therefore, to switch!

13.7. Characteristic function. The characteristic function $\psi(s)$ of a random variable x is defined as the Fourier transform of its probability density function $p(x)$; that is,[1-3]

$$\psi(s) = \int_{-\infty}^{\infty} p(x) \exp(-\mathrm{i}2\pi s x)\, \mathrm{d}x. \tag{27}$$

Considering that the nth derivative of $\psi(s)$, when evaluated at $s = 0$, yields

$$\begin{aligned} \psi^{(n)}(s)\big|_{s=0} &= \int_{-\infty}^{\infty} (-\mathrm{i}2\pi x)^n p(x) \exp(-\mathrm{i}2\pi s x)\, \mathrm{d}x\Big|_{s=0} \\ &= (-\mathrm{i}2\pi)^n \int_{-\infty}^{\infty} x^n p(x) \mathrm{d}x = (-\mathrm{i}2\pi)^n \langle x^n \rangle, \end{aligned} \tag{28}$$

it is seen immediately that the coefficients of the Taylor series expansion of $\psi(s)$ around the point $s = 0$ are related to the various moments $\langle x^n \rangle$ of the random variable x in the following way:

$$\psi(s) = 1 - \mathrm{i}2\pi\langle x\rangle s - 2\pi^2\langle x^2\rangle s^2 + \mathrm{i}(4\pi^3/3)\langle x^3\rangle s^3 + (2\pi^4/3)\langle x^4\rangle s^4 + \cdots. \tag{29}$$

The characteristic function of the Gaussian distribution given in Eq.(13) is readily computed as follows:

$$\begin{aligned} \psi(s) &= \frac{1}{\sqrt{2\pi}\sigma_x} \int_{-\infty}^{\infty} \exp[-(x - x_0)^2/(2\sigma_x^2)] \exp(-\mathrm{i}2\pi s x)\, \mathrm{d}x \\ &= \exp(-\mathrm{i}2\pi x_0 s) \int_{-\infty}^{\infty} \exp(-\pi y^2) \exp[-\mathrm{i}2\pi(\sqrt{2\pi}\sigma_x s) y]\, \mathrm{d}y \quad \leftarrow \boxed{y = (x - x_0)/\sqrt{2\pi}\sigma_x} \\ &= \exp(-2\pi^2\sigma_x^2 s^2 - \mathrm{i}2\pi x_0 s). \quad \leftarrow \boxed{\text{Fourier transform of } e^{-\pi y^2} \text{ is } e^{-\pi s^2}} \end{aligned} \tag{30}$$

It is now easy to verify that $\psi(0) = 1$, $\psi'(0) = -\mathrm{i}2\pi x_0$, and $\psi''(0) = -4\pi^2(\sigma_x^2 + x_0^2)$, in agreement with Eq.(29). For a derivation of the characteristic function of the binomial distribution given in Eq.(9), see problem 6 at the end of the chapter. The characteristic function of the Poisson distribution is the subject of problem 8.

13.8. Addition of random variables. Let the independent random variables x and y assume values on the real axis and define the new random variable z as the result of adding the x and y values obtained in any given trial; that is, $z = x + y$. The probability density functions for x and y are denoted by $p_x(x)$ and $p_y(y)$, respectively. Suppose we finely divide the real axis into tiny intervals Δ, then allow the random variable x to acquire only the discrete values $n\Delta$ with probability $p_x(n\Delta)\Delta$, and the random variable y to acquire the discrete values $m\Delta$ with probability $p_y(m\Delta)\Delta$. Here, m and n are arbitrary integers ranging from $-\infty$ to ∞. It is easy to see that the random variable z may now acquire the discrete values $k\Delta$ with probability $\sum_{n=-\infty}^{\infty} p_x(n\Delta)p_y(k\Delta - n\Delta)\Delta^2$. In the limit when $\Delta \rightarrow 0$, we normalize the probability of z being $k\Delta$ by the width Δ of the tiny intervals to arrive at the following probability density function for the random variable z:

$$p_z(z) = \int_{-\infty}^{\infty} p_x(x)p_y(z - x)\mathrm{d}x. \tag{31}$$

This equation reveals the probability density function $p_z(z)$ of the sum of two independent random variables x and y as the convolution between their individual probability density functions.[1-3] Considering that $p_x(x) \geq 0$ and $p_y(y) \geq 0$, it is clear that the result of the convolution is a similarly nonnegative function; that is, $p_z(z) \geq 0$. Also, the area under $p_z(z)$ is the product of the areas under $p_x(x)$ and $p_y(y)$. Given that $\int_{-\infty}^{\infty} p_x(x)\mathrm{d}x = 1$ and also $\int_{-\infty}^{\infty} p_y(y)\mathrm{d}y = 1$, we conclude that $\int_{-\infty}^{\infty} p_z(z)\mathrm{d}z = 1$.

The characteristic function $\psi_z(s)$ of the sum z of two independent random variables x and y is obtained by Fourier transforming the convolution of the probability density functions $p_x(\cdot)$ and $p_y(\cdot)$. It is well known that the Fourier transform of the convolution is the product of the individual Fourier transforms. Consequently, $\psi_z(s) = \psi_x(s)\psi_y(s)$.

13.9. Joint probability and conditional probability. Suppose the outcome of an experiment is denoted by the random variable x, and the outcome of a second experiment, conducted in parallel, is denoted by the random variable y. Both x and y are real numbers, and the corresponding probability density functions are given by $p_x(x)$ and $p_y(y)$. If we now consider the points (x, y) in the xy-plane as the results of conducting both experiments jointly, we may assign a two-dimensional probability density function $p_{xy}(x, y)$ to the outcome of the joint experiment. The probability that the outcome x of the joint experiment is within a small neighborhood Δx of x, while that of the outcome y is within a small neighborhood Δy of y is given by $p_{xy}(x, y)\Delta x\Delta y$. In the special case where x and y are independent of each other, we will have $p_{xy}(x, y) = p_x(x)p_y(y)$. In general, however, the variables x and y could have some sort of connection (or correlation) with each other, in which case their joint probability density function $p_{xy}(x, y)$ cannot be so neatly split into the product of their individual density functions.

As a simple example, consider the experiment in which a dice and a coin are thrown together. The outcome of the dice throw will be a number x belonging to the set $\{1, 2, 3, 4, 5, 6\}$, while the coin toss will yield a number y, which is either 0 (for heads) or 1 (for tails). The outcome (x, y) of the joint experiment thus belong to the set

$$\{(1,0), (2,0), (3,0), (4,0), (5,0), (6,0), (1,1), (2,1), (3,1), (4,1), (5,1), (6,1)\}.$$

If the dice and the coin happen to be fair and uncoupled to each other, then each of these outcomes will occur with probability $1/12$ (i.e., the product of the probability $1/6$ for each outcome of a dice throw and $1/2$ for each outcome of a coin toss). If, however, the coin and the dice are somehow coupled to each other (say, via magnetic coupling), then each of the 12 possible outcomes will have its own probability $p_{xy}(x, y)$. All one can say in such cases is that $p_{xy}(x, y) \geq 0$ for all possible outcomes, and that the sum over all (x, y) of $p_{xy}(x, y)$ is equal to 1.0.

If one is interested in finding the probability of each outcome x for the dice throw, one must add the probabilities $p_{xy}(x, 0)$ and $p_{xy}(x, 1)$ together. Similarly, to find the probability of each outcome y of the coin toss, one must add up the individual probabilities $p_{xy}(1, y)$, $p_{xy}(2, y)$, $p_{xy}(3, y)$, $p_{xy}(4, y)$, $p_{xy}(5, y)$, and $p_{xy}(6, y)$.

In general, when the joint probability density function $p_{xy}(x, y)$ is specified, one determines the density functions of the individual random variables, x and y, as follows:

$$p_x(x) = \int_{-\infty}^{\infty} p_{xy}(x, y)\mathrm{d}y, \quad (32a)$$

$$p_y(y) = \int_{-\infty}^{\infty} p_{xy}(x, y)\mathrm{d}x. \quad (32b)$$

Needless to say, the joint probability density function must satisfy the universal properties $p_{xy}(x, y) \geq 0$ and $\iint_{-\infty}^{\infty} p_{xy}(x, y)\mathrm{d}x\mathrm{d}y = 1$. Equations (32a) and (32b) are thus assured to satisfy the requirements that $p_x(x) \geq 0$, $p_y(y) \geq 0$, and

$$\int_{-\infty}^{\infty} p_x(x)\mathrm{d}x = \int_{-\infty}^{\infty} p_y(y)\mathrm{d}y = \iint_{-\infty}^{\infty} p_{xy}(x, y)\mathrm{d}x\mathrm{d}y = 1. \quad (33)$$

In the context of a joint experiment, say, one in which the outcome is (x, y), conditional probability arises when the value of one of the random variables is revealed. For instance, if the outcome of y is revealed to be y_0, then the conditional probability for any outcome of x will be given by[1-3]

$$p_{x|y}(x|y = y_0) = p_{xy}(x, y = y_0) / \int_{-\infty}^{\infty} p_{xy}(x, y_0)\mathrm{d}x. \quad (34)$$

As an example, consider the experiment in which both a dice and a coin are thrown. If it is revealed that the coin toss has resulted in heads (i.e., $y = 0$), then the probability of any particular showing of the dice will be

$$p_{x|y}(x = n|y = 0) = p_{xy}(x = n, y = 0) / \sum_{m=1}^{6} p_{xy}(x = m, y = 0). \quad (35)$$

In similar fashion, one could define the conditional probability distribution $p_{y|x}(y|x)$ for situations in which the outcome of x is revealed and, subsequently, it is desired to know the probability distribution for the outcome of y.

13.10. Conditional entropy and mutual information. In the present section, we consider a pair (x, y) of discrete random variables having a joint probability distribution function $\sum_m \sum_n p_{xy}(x_n, y_m)\delta(x - x_n)\delta(y - y_m)$. The conditional entropy of the random variable x, given that the specific value of y has been revealed, is defined as follows:[4,5]

$$H(x|y) = -\sum_m p_y(y_m) \sum_n p_{x|y}(x_n|y_m) \ln[p_{x|y}(x_n|y_m)]. \quad (36)$$

Considering that $p_{xy}(x_n, y_m) = p_y(y_m)p_{x|y}(x_n|y_m)$, one may equivalently write this conditional entropy as $H(x|y) = -\sum_m \sum_n p_{xy}(x_n, y_m) \ln[p_{x|y}(x_n|y_m)]$. We now prove that the conditional entropy $H(x|y)$ is always less than (or at best equal to) the entropy $H(x)$ of the random variable x in the absence of additional information revealed by the disclosure of y. This feature of the entropy of random variables is crucial, of course, if entropy is intended as a measure of the uncertainty surrounding a random variable, or, equivalently, as a measure of the amount of information revealed about a random variable once the outcome of the corresponding experiment becomes known.[4-8]

The following proof of the identity $H(x) \geq H(x|y)$ draws on Jensen's inequality, which was the subject of section 5. (For an alternative proof, see problem 11 at the end of the chapter.) Starting with the definitions of $H(x)$ and $H(x|y)$ given in Eqs.(5) and (36), we write

$$H(x) - H(x|y) = -\sum_n p_x(x_n)\ln p_x(x_n) + \sum_m \sum_n p_{xy}(x_n, y_m)\ln[p_{x|y}(x_n|y_m)]$$

$$= -\sum_m \sum_n p_{xy}(x_n, y_m)\ln[p_x(x_n)/p_{x|y}(x_n|y_m)]$$

$$= -\sum_m \sum_n p_{xy}(x_n, y_m)\ln[p_x(x_n)p_y(y_m)/p_{xy}(x_n, y_m)]$$

[$-\ln x$ is convex cup] →

$$\geq -\ln\{\sum_m \sum_n p_{xy}(x_n, y_m) \times [p_x(x_n)p_y(y_m)/p_{xy}(x_n, y_m)]\}$$

$$= -\ln[\sum_m \sum_n p_x(x_n)p_y(y_m)]$$

$$= -\ln\{[\sum_n p_x(x_n)] \times [\sum_m p_y(y_m)]\} = -\ln(1) = 0. \quad (37)$$

Defining the joint entropy of the pair of random variables x and y by a straightforward extension of Eq.(5), namely $H(x, y) = -\sum_m \sum_n p_{xy}(x_n, y_m)\ln[p_{xy}(x_n, y_m)]$, and recalling that $p_{xy}(x_n, y_m) = p_y(y_m)p_{x|y}(x_n|y_m)$, it is now easy to verify that[4-8]

$$H(x|y) = H(x, y) - H(y). \quad (38)$$

The mutual information $I(x, y)$ between the pair of random variables x and y is defined as follows:[4,5]

$$I(x, y) = H(x) - H(x|y) = H(x) + H(y) - H(x, y) = H(y) - H(y|x). \quad (39)$$

It is seen that the mutual information $I(x, y)$ is a symmetric function that can be interpreted as the amount of information about x when the outcome of the random variable y is revealed, or, alternatively, as the amount of information about y when the outcome of the random variable x becomes known.

13.11. The central limit theorem.[1-3] The important central limit theorem states that the sum of a large number of independent random variables is a new random variable whose probability density function approaches that of a Gaussian distribution. We saw in section 8 that, if x_1 and x_2 are two independent random variables having probability densities $p_1(x)$ and $p_2(x)$, respectively, then their sum will be a random variable with the density function $p_1(x) * p_2(x)$. The characteristic function of the random variable $y = x_1 + x_2$ thus turns out to be $\psi_y(s) = \psi_1(s)\psi_1(s)$. These statements are readily extended to the sum of N independent random variables $x_1, x_2, \cdots, x_N$, so that the characteristic function of the random variable $y = \sum_{n=1}^{N} x_n$ is given by $\psi_y(s) = \prod_{n=1}^{N} \psi_n(s)$.

In general, $\psi_n(s)$ peaks at $s = 0$, where its value equals 1, simply because $p_n(x)$ is real, nonnegative and has an area equal to 1. Each (Hermitian) characteristic function may be approximated as $\psi_n(s) \cong 1 - \mathrm{i}2\pi\langle x_n\rangle s - 2\pi^2\langle x_n^2\rangle s^2$ in the vicinity of $s = 0$ (see Eq.(29)). If we now define a normalized random variable $\eta = y/N$, its probability density function will be $p_\eta(x) = Np_y(Nx)$ and, by the scaling theorem of Fourier transform theory (see chapter 7, section 6.2), its characteristic function will be $\psi_\eta(s) = \psi_y(s/N)$. We conclude that the normalized sum of N independent random variables, namely, $\eta = (x_1 + x_2 + \cdots + x_N)/N$ will have the characteristic function $\psi_\eta(s) = \prod_{n=1}^{N} \psi_n(s/N)$. Consequently,

$$\psi_\eta(s) \cong \prod_{n=1}^{N}[1 - \mathrm{i}2\pi\langle x_n\rangle(s/N) - 2\pi^2\langle x_n^2\rangle(s/N)^2]$$

[write $e^x \cong 1 + x + ½x^2$, then retain terms of order s and s^2.]

$$\cong \prod_{n=1}^{N} \exp[-\mathrm{i}2\pi\langle x_n\rangle(s/N) - 2\pi^2(\langle x_n^2\rangle - \langle x_n\rangle^2)(s/N)^2]$$

$$= \exp[-\mathrm{i}2\pi(N^{-1}\sum_{n=1}^{N}\langle x_n\rangle)s - 2\pi^2(N^{-2}\sum_{n=1}^{N}\sigma_n^2)s^2]. \quad (40)$$

A comparison with Eq.(30) now reveals that the random variable η is Gaussian, with a mean value of $\langle\eta\rangle = N^{-1}\sum_{n=1}^{N}\langle x_n\rangle$ and a variance of $\sigma_\eta^2 = N^{-2}\sum_{n=1}^{N}\sigma_n^2$.

13.12. Stochastic (or random) processes.[9-13] Suppose that a nondeterministic process generates a function $f(t)$ that has a different manifestation each time that the process is repeated. The outcome of the process is thus a member of a set (or ensemble) of functions, each of which could be produced during a given run of the experiment, but it is not known in advance which function $f(t)$ would be observed. If the statistical properties of $f(t)$ happen to be independent of the specific time t chosen to monitor the process, then the process is said to be stationary. For instance, for a stationary process, the average value $\langle f(t)\rangle$ of the function measured at any time t should be the same for all t. Similarly, the variance $\sigma_f^2 = \langle f^2(t)\rangle - \langle f(t)\rangle^2$ must also be time independent.

The autocorrelation function of a random process is defined as the expected value of the product $f(t)f(t')$, where t and t' are a pair of arbitrarily chosen instants of time. In general, the autocorrelation function $\varphi(t,t') = \langle f(t)f(t')\rangle$ depends on both t and t'. In the case of a stationary process, however, since the origin of time is irrelevant, the autocorrelation function will depend solely on the distance $\tau = t' - t$ between t and t'. As a matter of fact, the sign of τ is irrelevant as well, hence the autocorrelation function $\varphi(\tau)$ of a stationary process is always going to be an even function.

A stationary random process is said to be *ergodic* if one could determine the statistical averages of the process by integrating over any one of its representative sample functions.[9-13] Thus, for a stationary ergodic process, the following relations will hold:

$$\langle f^n(t)\rangle = lim_{T\to\infty} T^{-1}\int_{-T/2}^{T/2} f^n(t)\mathrm{d}t; \qquad (n = 1, 2, 3, \cdots). \tag{41}$$

$$\langle f(t)f(t+\tau)\rangle = lim_{T\to\infty} T^{-1}\int_{-½T+|\tau|}^{½T-|\tau|} f(t)f(t+\tau)\mathrm{d}t. \tag{42}$$

The general definition of stationarity requires that, for any set of chosen instants of time, say, $\{t_1, t_2, t_3, \cdots\}$, the joint probability distribution of $f(t_1), f(t_2), f(t_3), \cdots$ be independent of the choice of the origin of time. In practice, it is only necessary that the statistical properties of interest, such as average, variance, autocorrelation function, etc., be independent of the chosen origin of time. Similarly, a practical notion of ergodicity only demands that the statistical properties that are of interest in a particular application be derivable by time averaging of individual representatives of the random process.

13.13. The Wiener-Khinchin theorem.[3] Consider a stationary and ergodic random process exemplified by the function $f(t)$. The autocorrelation function of this process is denoted by $\varphi(\tau)$ and defined as $\varphi(\tau) = \langle f(t)f(t+\tau)\rangle$, where the angled brackets represent ensemble averaging. The ergodicity of the process then allows one to substitute the time average of a single representative function for the ensemble average. We begin by truncating $f(t)$ and examining its behavior in the interval $-½T \le t \le ½T$. Eventually, of course, we will let $T \to \infty$ to obtain the exact expression of the Wiener-Khinchin theorem. The squared absolute value of the Fourier transform of the truncated function, when ensemble averaged and normalized by the duration T of the function, yields the power spectral density $S_T(s)$ of the process, as follows:

$$\begin{aligned} S_T(s) &= T^{-1}\langle|\int_{-T/2}^{T/2} f(t)\exp(-\mathrm{i}2\pi st)\,\mathrm{d}t|^2\rangle \\ &= T^{-1}\int_{t=-T/2}^{T/2}\int_{t'=-T/2}^{T/2}\langle f(t)f(t')\rangle\exp[\mathrm{i}2\pi s(t'-t)]\,\mathrm{d}t\,\mathrm{d}t'. \end{aligned} \tag{43}$$

The integrand of this two-dimensional integral is a function only of a single variable, namely $\tau = t' - t$. Consequently, the integral is reducible to a one-dimensional integral, as shown in figure 13.2. The variable τ' depicted in figure 13.2 equals $(t' - t)/\sqrt{2} = \tau/\sqrt{2}$. In the direction perpendicular to the τ' axis, the integrand of Eq.(43) remains constant, which means that $\varphi(\tau') \exp(\mathrm{i}2\pi s\tau')$ needs to be multiplied by the area of the shaded band shown in figure 13.2, namely $(\sqrt{2}T - 2|\tau'|)\mathrm{d}\tau'$. All in all, we will have

$$
\begin{aligned}
S_T(s) &= T^{-1} \iint_{t,t'=-T/2}^{T/2} \langle f(t) f(t+\tau)\rangle \exp(\mathrm{i}2\pi s\tau)\, \mathrm{d}t\mathrm{d}t' \\
&= T^{-1} \int_{\tau'=-T/\sqrt{2}}^{T/\sqrt{2}} \varphi(\sqrt{2}\tau') \exp(\mathrm{i}2\pi s\sqrt{2}\tau')(\sqrt{2}T - 2|\tau'|)\mathrm{d}\tau' \\
&= T^{-1} \int_{\tau=-T}^{T} (T - |\tau|)\varphi(\tau) \exp(\mathrm{i}2\pi s\tau)\mathrm{d}\tau \\
&= \int_{-T}^{T} (1 - |\tau/T|)\varphi(\tau) \exp(\mathrm{i}2\pi s\tau)\mathrm{d}\tau. \qquad (44)
\end{aligned}
$$

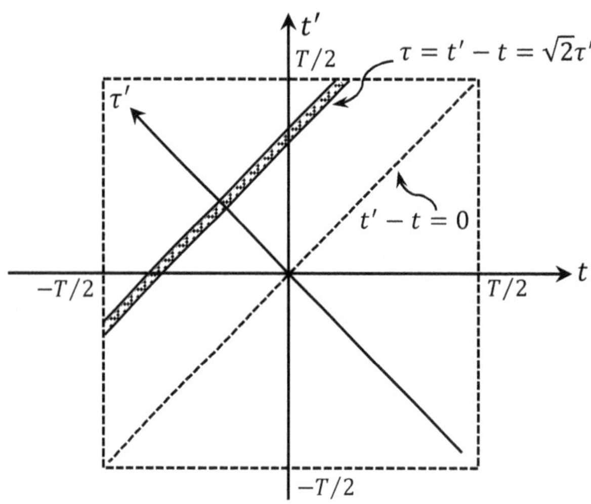

Figure 13.2. The integration range of Eq.(43) is a $T \times T$ squre within the tt'-plane. Since the integrand depends only on $\tau = t' - t$, we define a new variable $\tau' = \tau/\sqrt{2}$, then integrate over τ' from $-T/\sqrt{2}$ to $T/\sqrt{2}$. At each τ', the integrand must be multiplied by the area of the shaded band, namely $(\sqrt{2}T - 2|\tau'|)\mathrm{d}\tau'$, to preserve the area of integration.

In the limit when $T \to \infty$, the triangular function $1 - |\tau/T|$ approaches 1 and the integral in Eq.(44) reduces to the Fourier transform of the autocorrelation function $\varphi(\tau)$. The function $S_T(s)$ is then written simply as $S(s)$ and referred to as the power spectral density of the random process $f(t)$. Note that, since $\varphi(\tau)$ is a real and even function of τ, its Fourier transform $S(s)$ will also be real and even. Moreover, considering the definition of $S(s)$ as the expected value of a nonnegative entity, the power spectral density always turns out to be a nonnegative function of the frequency s.

Example 1. Figure 13.3 shows a random process $f(t)$ that jumps between the values of $+1$ and -1 at random times t_k. The timing of each jump does not depend on how long it has been since the preceding jump. During each narrow time window Δt, the probability of occurrence of a jump is $\alpha\Delta t$, where the parameter $\alpha > 0$ is a fixed but otherwise arbitrary constant. The autocorrelation function $\langle f(t)f(t+\tau)\rangle$ for this process is readily determined by noting that the product $f(t)f(t+\tau)$ either equals $+1$ or -1, depending on whether the number of jumps, n, during a time interval τ is even or odd. The jumps obey Poissanian statistics, with $p_n = \exp(-\alpha|\tau|)\,(\alpha|\tau|)^n/n!$. Consequently,

$$\langle f(t)f(t+\tau)\rangle = (\sum_{n=\text{even}} p_n) - (\sum_{n=\text{odd}} p_n)$$

$$= \exp(-\alpha|\tau|)\,[\cosh(\alpha|\tau|) - \sinh(\alpha|\tau|)] = \exp(-2\alpha|\tau|). \quad (45)$$

The power spectral density of the random process, which is the Fourier transform of its autocorrelation function, is now found as follows:

$$S(s) = \int_{-\infty}^{0} \exp(2\alpha\tau)\exp(-\mathrm{i}2\pi s\tau)\,\mathrm{d}\tau + \int_{0}^{\infty} \exp(-2\alpha\tau)\exp(-\mathrm{i}2\pi s\tau)\,\mathrm{d}\tau$$

$$= \frac{1}{2\alpha - \mathrm{i}2\pi s} + \frac{1}{2\alpha + \mathrm{i}2\pi s} = \frac{\alpha}{\alpha^2 + \pi^2 s^2}. \quad (46)$$

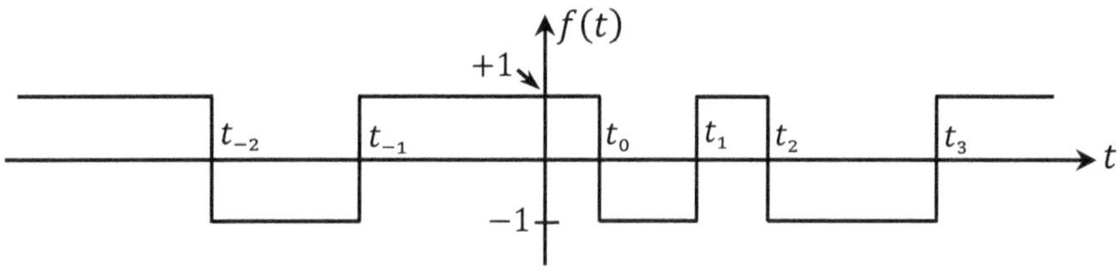

Figure 13.3. The stationary and ergodic random process $f(t)$ jumps between $+1$ and -1 values at random times t_k, where the various t_k are independent of one another. The timing of each jump does not depend on how long it has been since the preceding jump, occurring with probability $\alpha\Delta t$ during each narrow time window Δt, where $\alpha > 0$ is a fixed but otherwise arbitrary constant.

Example 2. The stationary ergodic process depicted in figure 13.4 approaches the so-called white Gaussian noise in the limit when $\Delta_0 \to 0$. Individual manifestations of this process have a random but fixed amplitude during each time interval Δ_0, then switch to a different fixed random value at the end of the allotted interval. The first jump after $t = 0$ occurs at t_0, which is a uniformly distributed random variable in the interval $(0, \Delta_0]$. The amplitude of the waveform $f(t)$ during each time interval Δ_0 is $1/\sqrt{\Delta_0}$ multiplied by a random variable chosen from a Gaussian distribution with zero mean and a variance of σ_0^2. The values of $f(t)$ in different intervals have the same average (zero) and the same variance (σ_0^2/Δ_0), but they are chosen independently of each other.

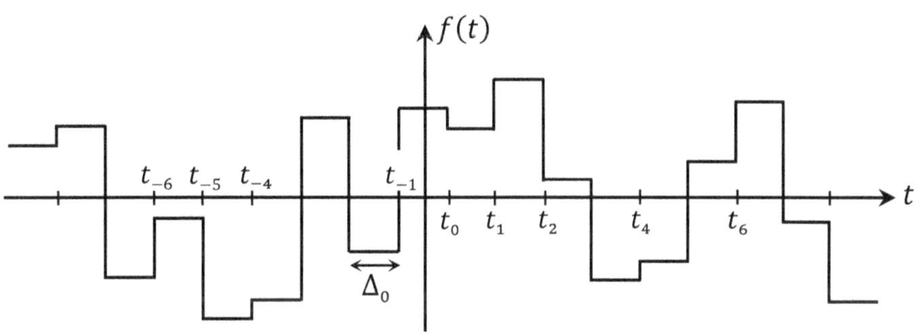

Figure 13.4. The stationary ergodic random process $f(t)$ assumes a fixed but random value during each time interval Δ_0, then jumps to a different random value and remains constant at its new value until the next jump. The value of $f(t)$ during each time interval Δ_0 is a Gaussian random variable whose average (or expected value) is $\langle f(t_n)\rangle = 0$, and whose variance is $\langle f^2(t_n)\rangle = \sigma_0^2/\Delta_0$. The first jump of $f(t)$ immediately after $t = 0$ occurs at t_0, which is a uniformly distributed random variable in the interval $(0, \Delta_0]$. In other words, each $f(t)$ has its own (fixed) t_0, but different manifestations of $f(t)$ could exhibit their first jump at any time $t = t_0$, where $0 < t_0 \le \Delta_0$. In the limit when $\Delta_0 \to 0$, the random process $f(t)$ approaches the so-called white Gaussian noise, having a mean value of zero and a power spectral density of σ_0^2.

Figure 13.5a shows the autocorrelation function $\varphi(\tau) = \langle f(t)f(t+\tau)\rangle$ of the random process $f(t)$. Clearly, the value of the autocorrelation at $\tau = 0$ must be σ_0^2/Δ_0, and for $\tau \geq \Delta_0$ it must drop to zero, simply because $f(t)$ and $f(t+\tau)$ are independently chosen random variables, each having a mean value of zero. In between $\tau = 0$ and $\tau = \Delta_0$, the expected value of $f(t)f(t+\tau)$ drops linearly, from its maximum of σ_0^2/Δ_0 to zero, because the probability of a jump grows linearly, from 0 to 1, as τ increases from 0 to Δ_0. We thus have $\varphi(\tau) = (\sigma_0^2/\Delta_0)\mathrm{Tri}(\tau/\Delta_0)$, whose Fourier transform is $S(s) = \sigma_0^2 \mathrm{sinc}^2(\Delta_0 s)$. In the limit when $\Delta_0 \to 0$, we will have $\mathrm{sinc}^2(\Delta_0 s) \to 1$. Consequently, $S(s)$ becomes a constant function of the frequency s, whose value is σ_0^2 at all frequencies. This is the spectrum of white noise with a power spectral density of σ_0^2. The fact that the instantaneous values of $f(t)$ were chosen in accordance with a Gaussian random variable makes this particular white noise a Gaussian white noise.

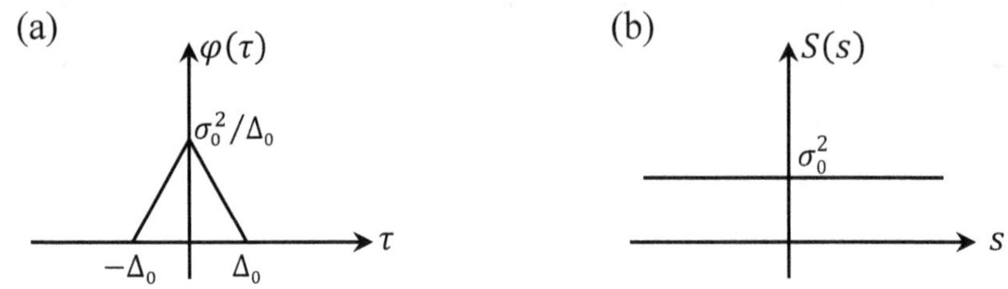

Figure 13.5. (a) The autocorrelation function $\varphi(\tau) = \langle f(t)f(t+\tau)\rangle$ for the random process $f(t)$ depicted in figure 13.4. In the limit when $\Delta_0 \to 0$, the function $\varphi(\tau)$ approaches $\delta(\tau)$. (b) Power spectral density $S(s)$ of the white noise obtained by Fourier transforming the autocorrelation function $\varphi(\tau)$ of the random process, in the limit when $\Delta_0 \to 0$.

Brain teaser. One hundred passengers, each with an assigned seat, board a plane that has exactly one hundred seats. The first passenger, ignoring his or her seat assignment, chooses one of the available seats at random. The remaining passengers, upon boarding the plane, sit on their assigned seats, unless the seat has already been occupied, in which case they choose, again randomly, one of the available seats. What is the probability that the last passenger, upon boarding, will find his or her assigned seat still available?

13.14. Conceptual basis of error-correction coding. In digital signal processing, various kinds of information are translated into binary sequences consisting of 1s and 0s. For example, each letter of the alphabet can be represented by an 8-bit sequence such as 01000001 for the letter A, or 01011010 for the letter Z. Similarly, an audio or video signal can be sampled, discretized, and converted to binary, so that an entire song, or a picture, or a movie, or a telephone conversation, etc., is translated into a string of 0s and 1s. Such binary sequences can then be manipulated, with the aid of modern electronic circuitry and digital devices, for subsequent display, or transmission to different locations, or storage. Because of the ever-present noise, and perhaps due to other uncertain processes, some of the binary digits may be flipped during signal processing (or transmission, or retrieval); that is, a 0 may be received as a 1, or a 1 may be received as a 0. The purpose of error-correction coding is to embed additional information into any given binary sequence—commonly referred to as "user data"—so that such errors can be identified and subsequently corrected.[14]

Perhaps the simplest error-correcting code is the Hamming (7,4) code, which was discovered (or invented) by Richard Hamming in 1947. Figure 13.6 describes the basic construction and logic of this code using a circle diagram. Suppose the four-bit binary sequence $b_1b_2b_3b_4$ (constituting a small fragment of the user data) is to be encoded in such a way that a *single* bit-flip error (i.e., 0 to 1, or 1 to 0) that might corrupt any one of its bits can be detected. Placing the user bits $b_1, b_2, b_3,$ and b_4 in the cross-sectional regions of the circle diagram, one proceeds to construct a three-bit sequence $b_5b_6b_7$ (the so-called "check bits"), such that the total number of 1s in each circle becomes an even number. Thus, for instance, if $b_1b_2b_3b_4 = 0100$, then setting $b_5b_6b_7 = 101$ ensures that the number of 1s in the blue circle is even (2 in the present example), as is the number of 1s in the red circle (also 2), as well as the total number of 1s in the green circle (zero).

Upon transmission (or processing, or retrieval) of the seven-bit-long sequence $b_1b_2b_3b_4b_5b_6b_7$, if it so happens that any single digit of the sequence is flipped (whether the erroneous digit is within the information-carrying part or in the check-bits part), then a simple inspection of the circle diagram will reveal the location of the flipped bit. This is possible because the even-parity constraint on each circle can be used to pinpoint the location of the erroneous bit. For instance, a flipped b_2 will cause the even-parity violation of the red and blue circles, whereas a flipped b_1 violates the even-parity constraint on all three circles. In contrast, the flipping of b_6 will violate the green circle only. In this way, the Hamming (7,4) code can fix all single-bit errors that might occur in a properly constructed seven-bit-long sequence.

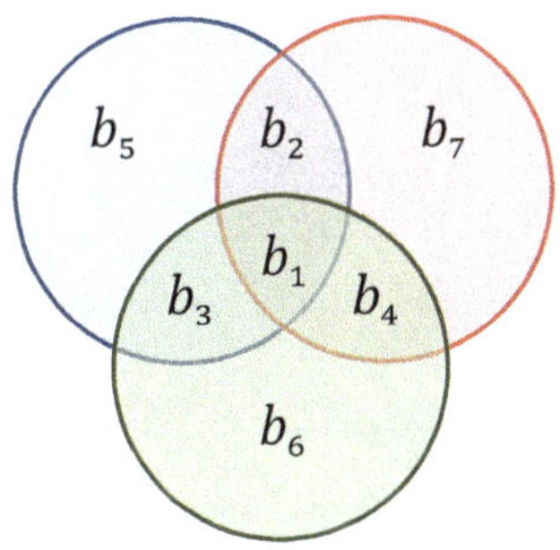

Figure 13.6. Diagram showing the bits of the seven-bit sequence $b_1b_2b_3b_4b_5b_6b_7$ placed in different regions of three intersecting circles. The first four bits, $b_1b_2b_3b_4$, constitute the information-carrying (or user data) portion of the seven-bit sequence, while the last three bits, $b_5b_6b_7$, are the check bits that are needed for error correction. The four user-data bits are placed in the regions of overlap between two or three circles. Each of these bits could be either 0 or 1, as all possible four-bit binary sequences are used to represent the user data. The check bits b_5, b_6, and b_7 are placed in the non-overlapping regions of each circle. These check bits are then chosen in such a way that the total number of 1s placed in each and every circle becomes an even number. Thus, for instance, if $b_1b_2b_3b_4 = 1001$, then $b_5b_6b_7 = 100$. During the reconstruction of a seven-bit block, one can determine if a single bit has flipped by examining all three circles and finding out if the constraint of an even number of 1s for each circle is satisfied. Depending on which circle(s) violates these constraints, one can determine the flipped bit and, subsequently, proceed to correct the error.

The price one pays for the ability to correct such single-bit errors is the addition of three check bits to every four-bit user-data sequence, or a 75 percent overhead on the actual information being processed. Starting with Hamming's pioneering work, many sophisticated as well as powerful error-correcting codes have been designed and implemented in the intervening decades. Modern codes typically operate on much longer sequences (or blocks) of user data, are capable of correcting a large number of errors that might occur in any given sequence, and add an overhead (in the form of additional check bits) that constitutes only a small fraction (say, 10 percent or 20 percent) of the size of the user-data block. For more information on the theory and practice of error-correcting codes, the reader is encouraged to consult the vast literature of the subject.[14]

References

1. G. R. Cooper and C. D. McGillem, *Probabilistic Methods of Signal and System Analysis*, Holt, Rinehart and Winston, New York (1971).
2. C. W. Helstrom, *Probability and Stochastic Processes for Engineers*, Macmillan, New York (1984).
3. Papoulis, *Probability, Random Variables, and Stochastic Processes*, 2nd edition, McGraw-Hill, New York (1984).
4. C. E. Shannon, "A Mathematical Theory of Communication," *Bell System Technical Journal* **27**(3), pp 379-423 (July 1948), and **27**(4), pp 623-666 (October 1948).
5. C. E. Shannon and W. Weaver, *The Mathematical Theory of Communication*, University of Illinois Press, Chicago (1963).
6. N. Abramson, *Information Theory and Coding*, McGraw-Hill, New York (1963).
7. R. G. Gallager, *Information Theory and Reliable Communication*, Wiley, Hoboken, New Jersey (1968).
8. M. Mansuripur, *Introduction to Information Theory*, Prentice-Hall, Upper Saddle River, New Jersey (1987).
9. S. Karlin and H. M. Taylor, *A First Course in Stochastic Processes*, 2nd edition, Academic Press, Cambridge, Massachusetts (1975).
10. H. M. Taylor and S. Karlin, *An Introduction to Stochastic Modeling*, Academic Press, Cambridge, Massachusetts (1984).
11. P. G. Hoel, S. C. Port and C. J. Stone, *Introduction to Stochastic Processes*, Houghton Mifflin, Boston, Massachusetts (1972).
12. E. Çinlar, *Introduction to Stochastic Processes*, Prentice-Hall, New Jersey (1975).
13. R. M. Gray and L. D. Davisson, *Random Processes: A Mathematical Approach for Engineers*, Prentice-Hall, Upper Saddle River, New Jersey (1986).
14. R. E. Blahut, *Theory and Practice of Error Control Codes*, Addison-Wesley, New York (1983).

Problems

Problem 1) The random variable x is assigned to an experiment whose outcomes are $x_1 \le x_2 \le x_3 \le \cdots \le x_N$, with probabilities $p_1, p_2, p_3, \cdots, p_N$. Show that the average $\langle x \rangle$ of the random variable x lies somewhere in the range between x_1 and x_N.

Problem 2) Let $h(x) = f(x) * g(x)$, where the asterisk stands for the convolution operation. Show that the area under $h(x)$ is the product of the areas under $f(x)$ and $g(x)$. A consequence of this result is that the probability-density function $p_z(\cdot)$ of the random variable $z = x + y$, where x and y are independent random variables, has an area equal to 1.0, since $p_z(\cdot)$ is the convolution of the probability-density functions $p_x(\cdot)$ and $p_y(\cdot)$.

Problem 3) Let x and y be two independent random variables with respective probability density functions $p_x(x)$ and $p_y(y)$. Find the average $\langle z \rangle$ and the variance σ_z^2 of the sum z of the two random variables; that is, $z = x + y$. Confirm your results by evaluating the first few terms in the Taylor series expansion of the characteristic function $\psi_z(s) = \psi_x(s)\psi_y(s)$.

Problem 4) Let the independent random variables x and y have mean values $\langle x \rangle$ and $\langle y \rangle$, and variances σ_x^2 and σ_y^2, respectively. The distance between x and y is denoted by a new random variable z; that is, $z = x - y$. Find the average $\langle z \rangle$ and the variance σ_z^2 of the random variable z.

Hint: You may evaluate $\langle z \rangle$ and σ_z^2 directly, or resort to the characteristic function $\psi_z(s)$. In the latter case, you must first demonstrate that $\psi_z(s) = \psi_x(s)\psi_y^*(s)$, where $\psi_y^*(s) = \psi_{-y}(s)$.

Problem 5) Prove that the necessary and sufficient conditions for a twice differentiable function $f(x)$ to be convex cup over a given domain is that $f''(x)$ be positive over the entire domain. (By the same token the necessary and sufficient conditions for a twice differentiable function $g(x)$ to be convex cap is that $g''(x)$ be negative over the entire domain of interest.)

Hint: $f(x)$ is convex cup if, for any arbitrary pair of points x_1, x_2 within the domain of the function, the inequality $\lambda f(x_1) + (1-\lambda)f(x_2) \ge f[\lambda x_1 + (1-\lambda)x_2]$, where $0 \le \lambda \le 1$ is an arbitrary parameter, is satisfied. The direction of inequality is reversed for a convex cap function.

Problem 6) a) Given the binomial distribution $p_n = \binom{N}{n} p^n (1-p)^{N-n}$, where $n = 0, 1, 2, \cdots, N$, and $0 < p < 1$, find the characteristic function $\psi(s)$.

b) Confirm that the first few terms in the Taylor series expansion of $\psi(s)$ yield the average and the variance of the binomial distribution.

c) Considering the discrete nature of the binomial distribution, show that its characteristic function must be periodic with a period of 1, then confirm that $\psi(s)$ found in part (a) is indeed a periodic function—with a period of 1.

d) For a fixed value of p, the binomial distribution function becomes broader with increasing N, which indicates that the Fourier transform of the distribution (i.e., the characteristic function $\psi(s)$) must become narrower as N increases. Since $\psi(s)$ is a periodic function, this narrowing of $\psi(s)$ occurs for individual repetitions of the function. Show that, for sufficiently large N, each period of the characteristic function approaches that of a Gaussian distribution, having the anticipated values of the average, Np, and the variance, $Np(1-p)$, for the corresponding random variable. This behavior, of course, is consistent with the central limit theorem of section 11.

Problem 7) Two independent random variables x_1 and x_2 have binomial probability distributions. The parameters of x_1 are p and N_1, whereas those of x_2 are p and N_2. Show that the sum of x_1 and x_2 is another random variable with a binomial distribution, having parameters p and $N_1 + N_2$.

Problem 8) a) Given the Poisson distribution $p_n = \exp(-\alpha)\,\alpha^n/n!$, where $\alpha > 0$ is a fixed real-valued constant, find the characteristic function $\psi(s)$.

b) Confirm that the first few terms in the Taylor series expansion of $\psi(s)$ yield the average and the variance of the Poisson distribution.

c) Considering the discrete nature of the Poisson distribution, show that its characteristic function must be periodic with a period of 1, then confirm that $\psi(s)$ found in part (a) is indeed a periodic function—with a period of 1.

d) For a fixed value of α, the Poisson distribution function becomes broader with increasing N, which indicates that the Fourier transform of the distribution (i.e., the characteristic function $\psi(s)$) must become narrower as N increases. Since $\psi(s)$ is a periodic function, this narrowing of $\psi(s)$ occurs for individual repetitions of the function. Show that, for sufficiently large α, each period of the characteristic function approaches that of a Gaussian distribution, having the anticipated values of the average α and the variance α for the corresponding random variable.

Problem 9) Let x_1 and x_2 be two independent random variables, having Poisson distributions with parameters $\alpha_1 > 0$ and $\alpha_2 > 0$, respectively. Show that the sum $x_1 + x_2$ of these random variables is another random variable with a Poisson distribution whose parameter is $\alpha_1 + \alpha_2$.

Problem 10) a) Prove that the function $f(x) = \exp(x)$ is a convex cup function and that $g(x) = \ln x$ is a convex cap function.

b) For an arbitrary random variable x having probability distribution function $p(x)$, show that the expected value $\langle e^x \rangle$ of $f(x)$ is greater than or equal to $e^{\langle x \rangle}$ and that the expected value $\langle \ln x \rangle$ of $g(x)$ is less than or equal to $\ln\langle x \rangle$.

Problem 11) a) Show that $\ln x \le x - 1$ for all values of x between 0 and ∞.

b) Use the inequality $\ln x \le x - 1$ to prove that the mutual information content $I(x, y)$ of a pair of discrete random variables, x, y, whose joint probability distribution is specified as $p_{xy}(x_m, y_n)$, is a positive entity.

Problem 12) Use Jensen's inequality to prove that the entropy $H(x)$ of the discrete random variable x whose outcomes $x_1, x_2, \cdots, x_N$ occur with probabilities $p_1, p_2, \cdots, p_N$ is less than or equal to $\ln N$, with $H(x) = \ln N$ only when all outcomes are equi-probable.

Problem 13) Find the probability that, among a group of n individuals, at least two will turn out to have the same birthday. (Ignore the possibility of a February 29 birthday.)

Index

Printed in the USA
CPSIA information can be obtained
at www.ICGtesting.com
LVHW080130051224
798241LV00008B/52
9781516535880